Geometric Algebra with Applications in Science and Engineering

D1202405

Geometric Algebra with Applications in Science and Engineering

Eduardo Bayro Corrochano
Garret Sobczyk

Editors

With 127 Figures

Birkhäuser
Boston • Basel • Berlin

QA
564
.G463
2001

Eduardo Bayro Corrochano
CINVESTAV
Centro de Investigacion y de Estudios
  Avanzados
Apartado Postal 31-438
Plaza la Luna, Guadalajara
Jalisco 44550
Mexico

Garret Sobczyk
Departamento de Fisica y Matematica
Universidad de las Americas-Puebla
72820 Cholula
Mexico

Library of Congress Cataloging-in-Publication Data
Bayro Corrochano, Eduardo.
    Geometric algebra with applications in science and engineering
Eduardo Bayro Corrochano, Garret Sobczyk.
      p. cm.
    Includes bibliographical references and index.
    ISBN 0-8176-4199-8 (alk. paper)
      1. Geometry, Algebraic.   I. Sobczyk, Garrett, 1943–   II. Title.
    QA564 .C686   2001
    516.3'5—dc21
                                                                00-046854
                                                                CIP

Printed on acid-free paper.
© 2001 Birkhäuser Boston   *Birkhäuser* ®

All rights reserved. This work may not be translated or copied in whole or in part without
the written permission of the publisher (Birkhäuser Boston, c/o Springer-Verlag New York,
Inc., 175 Fifth Avenue, New York, NY 10010, USA), except for brief excerpts in connection
with reviews or scholarly analysis. Use in connection with any form of information storage
and retrieval, electronic adaptation, computer software, or by similar or dissimilar methodology now known or hereafter developed is forbidden.
The use of general descriptive names, trade names, trademarks, etc., in this publication, even
if the former are not especially identified, is not to be taken as a sign that such names, as
understood by the Trade Marks and Merchandise Marks Act, may accordingly be used freely
by anyone.

ISBN 0-8176-4199-8
ISBN 3-7643-4199-8   SPIN 10771027

Production managed by Louise Farkas; manufacturing supervised by Erica Bresler.
Typeset by the editors in LaTex2e.
Printed and bound by Maple-Vail Book Manufacturing Group, York, PA.
Printed in the United States of America.

9 8 7 6 5 4 3 2 1

45052642

Contents

IV   Robotics   209

V   Quantum and Neural Computing, and Wavelets   279

VII Computational Methods in Clifford Algebras 459

22 Clifford Algebras as Projections of Group Algebras
Vladimir M. Chernov   **461**

23 Counterexamples for Validation and Discovering of New Theorems
Pertti Lounesto   **477**

Preface

The goal of this book is to present a unified mathematical treatment of diverse problems in mathematics, physics, computer science, and engineering using geometric algebra. Geometric algebra was invented by William Kingdon Clifford in 1878 as a unification and generalization of the works of Grassmann and Hamilton, which came more than a quarter of a century before. Whereas the algebras of Clifford and Grassmann are well known in advanced mathematics and physics, they have never made an impact in elementary textbooks where the vector algebra of Gibbs-Heaviside still predominates. The approach to Clifford algebra adopted in most of the articles here was pioneered in the 1960s by David Hestenes. Later, together with Garret Sobczyk, he developed it into a unified language for mathematics and physics. Sobczyk first learned about the power of geometric algebra in classes in electrodynamics and relativity taught by Hestenes at Arizona State University from 1966 to 1967. He still vividly remembers a feeling of disbelief that the fundamental geometric product of vectors could have been left out of his undergraduate mathematics education. Geometric algebra provides a rich, general mathematical framework for the development of multilinear algebra, projective and affine geometry, calculus on a manifold, the representation of Lie groups and Lie algebras, the use of the horosphere and many other areas.

This book is addressed to a broad audience of applied mathematicians, physicists, computer scientists, and engineers. Its purpose is to bring together under a single cover the most recent advances in the applications of geometric algebra to diverse areas of science and engineering. Most articles in this book were presented at the Special Parallel Session ACACSE'99 of the 5th International Conference on Clifford Algebras and their Applications in Mathematical Physics, held in Ixtapa-Zihuatanejo, Mexico, in July 1999. ACACSE'99 was organized by the editors of this book in the belief that the time is ripe for the general recognition of the powerful tools of geometric algebra by the much larger scientific and engineering communities. Since the First International Conference on Clifford Algebras, held in Canterbury, England, in 1985, major advances continue to be made in the application of geometric algebra to mathematics and theoretical physics and to what has become known as Clifford analysis. The most recent advances in these more established areas can be found in the Conference Proceedings (Birkhäuser, Progress in Physics Series **18, 19**, Boston 2000) Volume I: *Algebra and Physics*, edited by Rafał Abłamowicz and Bertfried Fauser, and Volume II: *Clifford Analysis*, edited by John Ryan and Wolfgang Sprössig. See also the Special Issue: Volume 39, Number 7, of the International Journal of Theoretical Physics, a collection of papers of the Ixtapa Conference edited by Zbigniew Oziewicz and David Finkelstein. Instead of editing a Volume III of the Proceedings, addressed to specialists in Clifford algebra, we decided that the time had come to introduce the powerful methods of geometric algebra to the much larger community of scientists and engineers who are seeking new mathematical tools to solve the ever more complicated problems of the 21th century. The book consists of 25 chapters organized into seven parts, each chapter written by experts in their field of speciality.

Part I Advances in geometric algebra presents a series of four chapters on the most up-to-date work that has been done on the horosphere, the conformal group, and related topics. The horosphere is a nonlinear model of Euclidean and pseudo-Euclidean geometry that captured the interest and imagination of many of the participants at the Ixtapa Conference. The horosphere offers a host of new computational tools in projective and hyperbolic geometries, with potential applications in many different areas.

Part II Theorem proving offers perhaps one of the most tantalizing new applications of geometric algebra and the horosphere. Some of the most difficult problems of mathematics have been successfully attacked with the help of the computer. The most striking and well-known success was in the proof of the 4-color problem. The two chapters in this part present new approaches to geometric reasoning and automatic theorem proving using geometric algebra, including solutions to problems formulated by Erdös and S.S. Chern. Each of the chapters also presents a wealth of bibliographic material. The day may come, sooner than most mathematicians realize, when computers will successfully attack the most intractable and outstanding problems and theorems in mathematics.

Part III Computer vision researchers still underestimate the important role played by geometry in vision. A large amount of accumulated evidence shows that animals have some kind of internal geometric representation of external reality. The first two chapters in this part formulate the principals of computer vision in geometric algebra and address the key problems of camera calibration and localization. The third chapter uses Bayesian inference, showing how estimation can be done using geometric algebra. Felix Klein in his Erlangen program stressed the role of invariant theory in characterizing projective geometry. The last chapter in this part uses invariant theory for the projective reconstruction of shape and motion.

Part IV Robotics the first two chapters discuss kinematics and trajectory interpolation in robot design in a rich geometric language of points, lines, and planes in dual and double quaternion algebras. The topic of robotics is an old theme; however it is only in the last decade that researchers have begun to consider multidimensional representations to solve old problems in the field. The last chapter shows how the representation of Lie algebras in terms of bivectors can be applied to problems in low-level image processing, using Lie filters in the affine n-plane. It also develops an algebra of incidence for application to problems in robotics.

Part V Quantum and neural computing, and wavelets is devoted to the new fields of quantum computers, neurogeometry, and Clifford wavelets, which go beyond Haar wavelets. The concept of a quantum computer was first introduced by Richard Feynman in the 1970s. The first chapter in this part explores the use of geometric algebra for analyzing the quantum states and quantum logic that is necessary to build a quantum computer based on nuclear magnetic resonance. The second chapter employs the geometric product in a generalization of neural networks that have been constructed using complex, hyperbolic, and dual numbers. The third chapter discusses wavelets constructed from multivectors and is a generalization of the concept of a quaternion wavelet.

Part VI Applications to engineering and physics is aimed at exploring some of the many applications of geometric algebra to the problems of engineering and physics. By looking at diverse problems from the perspective of a common-language, the problems are often found to be related at a deeper level. The first chapter explores some of the mathematical aspects of geometric wave propagation as applied to objects in collision. The second chapter explores the hidden symmetries of crystallography that are only revealed by a geometric analysis in higher dimensions. The third didactic chapter considers optimization problems that commonly arise in engineering using quaternions. The fourth chapter treats the Maxwell-Lorentz equations in problems of electrical engineering, and shows how a relativistic point of view can be of practical value. The last chapter of this part seeks to find the common ground that exists between the down-to-earth problems faced by engineers and the problems of the stars contemplated by otherworldly cosmologists.

Part VII Computational methods in Clifford algebra explores some of the new tools made possible by the rich structure of geometric algebra, and the state of the art software that exists today for doing calculations. The first chapter explores a generalization of fast transform methods that takes advantage of the richer algebraic structure of geometric algebra. The second, innovative chapter of this part reports the results of an experiment that tests the feasibility of using the Internet as a forum for settling disagreements between experts. The last three chapters of this part discuss the software available for doing computer-aided calculations in geometric algebra. It is hoped that the inclusion of these chapters will spur the further development of urgently needed software to do symbolic calculations in geometric algebra.

The editors believe that the contributions in this book will prove invaluable to anyone interested in Euclidean and non-Euclidean geometries and to scientists and engineers who are seeking more sophisticated mathematical tools for solving the ever more complex problems of the 21st century.

Eduardo Bayro Corrochano would like to thank the Center for Research in Mathematics (CIMAT, Guanajuato, Mexico) and the Consejo Nacional de Ciencia y Tecnología (REDII - CONACYT, Mexico) for their support of this project. We are very grateful to our student Natividad María Aguilera for her painstaking work on the Latex technical problems of putting this book together. Sandra Cancino helped us enormously in the cover design and in the drawing of many of the figures. We thank Lauren Lavery at Birkhäuser, Boston, for her friendly, professional assistance and Louise Farkas at Birkhäuser, New York, for the excellent proofreading. Garret Sobczyk thanks CIMAT for their kind hospitality during his sabbatical in the Fall Semester 1999. In addition, he is grateful to INIP of the Universidad de Las Americas, Puebla, for his sabbatical, which made work on this project possible.

Eduardo Bayro Corrochano, Guadalajara, Mexico
Garret Sobczyk, Puebla, Mexico   October 2, 2000

Contributors

Prof. Rafal Ablamowicz
Department of Mathematics,
Tennessee Technological University
Box 5054, Cookeville, TN 38505, USA
Tel. +1 (931) 372-6353
e–mail: rablamowicz@tntech.edu
http://math.tntech.edu/rafal/cliff4/

Dr. Shawn G. Ahlers
Robotics and Automation Laboratory
University of California, Irvine
Irvine, CA 92697, USA
e–mail: sahlers@uci.edu

G. Aragón
Programa de Desarrollo
Profesional en Automatización,
Universidad Autónoma Metropolitana,
Azcapotzalco San Pablo, 108
Colonia Reynosa-Tamaulipas
02200 D.F. México, México.
e–mail: gag@hp9000a1.uam.mx

J.L. Aragón
Instituto de Física, UNAM
Laboratorio de Juriquilla
Apartado Postal 1-1010
76000 Querétaro, Qro., México
e–mail: aragon@fenix.ifisicacu.unam.mx

Doctoral student *Vladimir Banarer*
Computer Science Institut
Christian Albrechts Universität zu Kiel
Preusserstrasse 1–9, 24105
Kiel, Germany
e–mail: vlb@ks.informatik.uni-kiel.de

Prof. William E. Baylis
Department of Physics
University of Windsor
Windsor, Ontario, Canada N9B 3P4
Phone (519) 253-4232 x2673
Fax (519) 973-7075
e–mail: baylis@uwindsor.ca
htpp://www.uwindsor.ca/physics

Dr. Eduardo Bayro-Corrochano
CINVESTAV
Centro de Investigación y de Estudios Avanzados
Apartado Postal 31-438
Plaza la Luna, Guadalajara,
Jalisco 44550, Mexico
Tel. 0052 3 6841580
Fax. 0052 3 6841708
e–mail: edb@gdl.cinvestav.mx
http://www.gdl.cinvestav.mx/~edb

Tim Bouma
Research Institute
for Computer Science
University of Amsterdam Kruislaan 403
1098 SJ Amsterdam, The Netherlands
e–mail: timbouma@wins.uva.nl
http://www.wins.uva.nl/~timbouma/

Prof. Vladimir M.Chernov
Image Processing Systems Institute
of RAS (IPSI RAS), P.B. 347
443001, Samara, Russia
e-mail: vche@smr.ru

Dr. David G. Cory
Department of Nuclear Engineering
Massachusetts Institute of Technology
150 Albany Street
Cambridge, MA 02139, USA
e-mail: dcory@mit.edu
http://mrix4.mit.edu/Cory/Cory.html

F. Dávila
CIMAT
Centro de Investigación en Matemáticas
Apartado Postal 402

36000 Guanajuato, Gto., México
e–mail: fabio@cimat.mx

Dr. Chris Doran
Astrophysics Group
Cavendish Laboratory
Madingley Road,
Cambridge CB3 0HE, United Kingdom
Fax: 44 1223 354599
e–mail: C.Doran@mrao.cam.ac.uk
http://www.mrao.cam.ac.uk/~clifford

Dr. Leo Dorst
Research Institute for Computer Science
University of Amsterdam
Kruislaan 403, NL-1012 VE Amsterdam
The Netherlands
fax +31-20-525 7490
e–mail: leo@wins.uva.nl
http://www.wins.uva.nl/~leo/

A. Gómez
Instituto de Física, UNAM
Apartado Postal 20-364
01000 México, Distrito Federal, México
e–mail: alfredo@fenix.ifisicacu.unam.mx

Dr. Timothy F. Havel
Biological Chemistry and
Molecular Pharmacology
Harvard Medical School
240 Longwood Ave., Boston, MA 02115, USA
e–mail: timothy_havel@hms.harvard.edu
http://mrix4.mit.edu/havel.html

Prof. David Hestenes
Dept. of Physics and Astronomy
Arizona State University
Phone: (602) 965-6277
Fax: (602) 965-7331
Tempe, Arizona 85287-1504, USA
e–mail: Hestenes@asu.edu
http://phy.asu.edu/directory

Dr. Anthony Lasenby
Astrophysics Group,
Cambridge University
Cavendish Laboratories
Madingley Road
Cambridge CB3 0HE, United Kingdom
e–mail: a.n.lasenby@mrao.cam.ac.uk
http://www.mrao.cam.ac.uk/~anthony

Dr. Joan Lasenby
Signal Processing Group,
Cambridge University
Engineering Department
Trumpington Street
Cambridge CB2 1PZ United Kingdom
e–mail: jl@eng.cam.ac.uk
http://www-sigproc.eng.cam.ac.uk/~jl

Prof. Dr. Hongbo Li
Institute of Systems Science
Academy of Mathematics
and Systems Science
Chinese Academy of Science
Beijing 100080
P. R. China
e–mail: hli@mmrc.iss.ac.cn
http://www.mmrc.iss.ac.cn/~hli

Dr. Stephen Mann
Computer Science Department
University of Waterloo
200 University Ave W.
Waterloo N2L 3G1 Canada
e–mail: smann@cgl.uwaterloo.ca
http://www.cgl.uwaterloo.ca/~smann/

Prof. John Michael McCarthy
Robotics and Automation Laboratory
University of California, Irvine
Irvine, CA 92697, USA
e–mail: jmmccart@uci.edu

Dr. Ljudmila Meister
Technische Universitat Darmstadt
FB Mathematik, AG 12
Schlossgartenstr. 7

D-64289 Darmstadt, Germany
meister@mathematik.tu-darmstadt.de
http://www.mathematik.tu-darmstadt.de

Prof. Pertti Lounesto
Helsinki Institute of Technology
FIN-00180 Helsinki, Finland
Tel. +358-9-31083366
e–mail: Pertti.Lounesto@hit.fi
http://www.hit.fi/~lounesto

Doctoral student *José María Pozo*
Departament de Física Fundamental
Universitat de Barcelona
Diagonal 647, E-08028 Barcelona, Spain
e–mail: jpozo@ffn.ub.es

M.A. Rodríguez
Departamento de Matemáticas, ESFM
Instituto Politécnico Nacional
Edificio 9, UPALM,
07300 México, D.F. , México
e–mail: marco@esfm.ipn.mx

Dr. J.M. Selig
School of Computing,
Information Systems and Mathematics
South Bank University
Borough Road London SE1 0AA, United Kingdom
e–mail: seligjm@sbu.ac.uk
http://www.sbu.ac.uk/~seligjm

Prof. Garret Sobczyk
Departamento de Física y Matemáticas
Universidad de las Ámericas - Puebla
72820 Cholula, México
e–mail: sobczyk@mail.pue.mx

Dr. Shyamal S. Somaroo
Biological Chemistry and
Molecular Pharmacology
Harvard Medical School
240 Longwood Ave.
Boston, MA 02115, USA
e–mail: Shyamal_Somaroo@mckinsey.com

Doctoral student *Adam Stevenson*
Signal Processing Group
Cambridge University
Engineering Department
Trumpington Street
Cambridge CB2 1PZ United Kingdom
e–mail: jl@eng.cam.ac.uk
http://www-sigproc.eng.cam.ac.uk/~jl
axss@eng.cam.ac.uk

Dr. Leonardo Traversoni
Departamento de Ingeniería
de Procesos e Hidráulica
División de Ciencias Básicas e Ingeniería
Universidad Autónoma Metropolitana
(Iztapalapa), Av. Michoacán y La Purísima
CP 09340 México D.F.
México
e-mail: ltd@xanum.uam.mx

Dr. Ching-Hua Tseng
Department of Nuclear Engineering
Massachusetts Institute of Technology
150 Albany Street
Cambridge, MA 02139, USA
e-mail: ctseng@mit.edu

Doctoral student *R. Vallejo*
CIMAT
Centro de Investigación en Matemáticas
Apartado Postal 402
36000 Guanajuato, Gto., México
e–mail: vallejo@cimat.mx

Dr. Dongming Wang
Laboratoire d'Informatique de Paris 6
Université Pierre et Marie Curie - CNRS
4, place Jussieu
F-75252 Paris Cedex 05, France
Fax: +33 (0)1 44 27 40 42
e–mail: Dongming.Wang@lip6.fr
http://www-calfor.lip6.fr/~wang

Part I

Advances in Geometric Algebra

Chapter 1

Old Wine in New Bottles: A New Algebraic Framework for Computational Geometry

David Hestenes

1.1 Introduction

My purpose in this chapter is to introduce you to a powerful new algebraic model for Euclidean space with all sorts of applications to computer-aided geometry, robotics, computer vision and the like. A detailed description and analysis of the model is soon to be published elsewhere [1], so I can concentrate on highlights here, although with a slightly different formulation that I find more convenient for applications. Also, I can assume that this audience is familiar with Geometric Algebra, so we can proceed rapidly without belaboring the basics.

1.2 Minkowski Algebra

Let $\mathcal{R}_{p,q} = \mathcal{G}(\mathcal{R}^{p,q})$ denote the Geometric Algebra generated by a vector space $\mathcal{R}^{p,q}$ with non-degenerate signature (p, q), where p is the dimension of its largest subspace of vectors with positive signature. The signature is said to be *Euclidean* if $q = 0$ and *Minkowski* if $q = 1$. We will be concerned with the *Minkowski algebra* $\mathcal{R}_{n+1,1}$ and its Euclidean subalgebra $\mathcal{R}_n$ determined by a *designated* unit bivector (or blade) E.

First we consider the Minkowski plane $\mathcal{R}^{1,1}$ and the Minkowski algebra $\mathcal{R}_{1,1} = \mathcal{G}(\mathcal{R}^{1,1})$ that it generates. It is most convenient to introduce a null basis $\{e, e_*\}$ for the plane that satisfies

$$e^2 = e_*^2 = 0, \quad e \cdot e_* = 1. \tag{1.1}$$

This generates a basis $\{1, e, e_*, E\}$ for $\mathcal{R}_{1,1}$, where

$$E = e \wedge e_* \quad \Longrightarrow \quad E^2 = 1 \tag{1.2}$$

defines a unit pseudoscalar for the plane. It is of some interest to remark that the "$*$" notation has been adopted to indicate that e_* is "dual" to

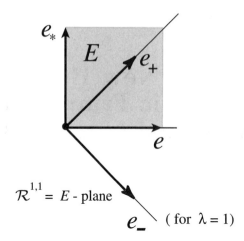

FIGURE 1.1. Minkowski Plane $\mathcal{R}^{1,1}$.

e in the sense of linear forms. That simplifies comparison of alternative mathematical representations.

Using the expansion $ee_* = e \cdot e_* + e \wedge e_*$, we can express the basis relations in the convenient form

$$ee_* = 1 + E, \tag{1.3}$$

whence

$$e_* e = 1 - E = 1 + E^\dagger = (ee_*)^\dagger. \tag{1.4}$$

We can also derive the "absorption property" for null vectors:

$$Ee = -eE = e, \quad e_* E = -Ee_* = e_*. \tag{1.5}$$

The above relations suffice for all our dealings with the E-plane. However, it is of some interest to compare the null basis with an orthonormal basis defined by

$$e_\pm \equiv \frac{1}{\sqrt{2}}(\lambda e \pm \lambda^{-1} e_*), \quad \lambda \neq 0. \tag{1.6}$$

We note that

$$e_\pm^2 = \pm 1, \quad e_+ \cdot e_- = 0, \quad E = e \wedge e_* = e_+ \wedge e_- = e_+ e_- \tag{1.7}$$

The two sets of base vectors are in Fig. 1.1 for $\lambda = 1$. As λ varies in eqn. (1.6), the directions of $e_\pm$ vary, but the orthonormality relations (1.7) remain fixed. On the other hand, a rescaling of the of the null vectors: $\{e, e_*\} \rightarrow \{\lambda e, \lambda^{-1} e_*\}$ does not affect any of the relations (1.1) to (1.5). Thus, we see that our definition of the null basis fixes directions but not scale, whereas the orthonormal basis has fixed scale but arbitrary direction. It is this difference that makes the null basis more suitable for our purposes.

1.3 Conformal Split

The *conformal split* was introduced in [2] to relate a Minkowski algebra to a maximal Euclidean subalgebra. The split can be defined in two different ways: additively and multiplicatively.

The *additive split* is defined as a direct sum:

$$\mathcal{R}^{n+1,1} = \mathcal{R}^n \oplus \mathcal{R}^{1,1}. \tag{1.8}$$

For those who would like to see this expressed in terms of a basis, we introduce an orthonormal basis $\{e_k \mid e_j \cdot e_k = \delta_{jk}; \, j, k = 1, 2, \ldots, n\}$ for the Euclidean vector space $\mathcal{R}^n$ and we note that the orthogonality conditions $e_* \cdot e_k = e \cdot e_k = 0$ are equivalent to the condition that the null vectors anticommute with the e_k, that is,

$$e_k e = -e e_k, \qquad e_k e_* = -e_* e_k. \tag{1.9}$$

Alternatively, the *multiplicative split* is defined as a direct product:

$$\mathcal{R}_{n+1,1} = \mathcal{R}_n \otimes \mathcal{R}_{1,1}. \tag{1.10}$$

Here $\mathcal{R}^n$ is actually a space of trivectors with a common bivector factor E. It is related to the vector space $\mathcal{R}^n$ by

$$\mathcal{R}^n = \mathcal{R}^n E \cong \mathcal{R}^n \tag{1.11}$$

and it generates the Euclidean algebra $\mathcal{R}_n = \mathcal{G}(\mathcal{R}^n)$. It has the basis

$$\{\mathbf{e}_k = e_k E = E e_k\} = \text{trivectors in } \mathcal{R}^3_{n+1,1}. \tag{1.12}$$

We still have the orthonormality conditions $\mathbf{e}_j \cdot \mathbf{e}_k = e_j \cdot e_k = \delta_{jk}$. However, in contrast to the e_k, the $\mathbf{e}_k$ commute with the null vectors, that is,

$$\mathbf{e}_k e = e \mathbf{e}_k, \qquad \mathbf{e}_k e_* = e_* \mathbf{e}_k. \tag{1.13}$$

This is one very good reason for preferring the multiplicative split over the additive split. The latter was employed in [1], but we will stick with the former.

The multiplicative split $\mathcal{R}_{4,1} = \mathcal{R}_3 \otimes \mathcal{R}_{1,1}$ has significant applications to computational geometry, robotics, computer vision, crytallography and molecular geometry. At a more sophisticated level, the split $\mathcal{R}_{4,2} = \mathcal{R}_{3,1} \otimes \mathcal{R}_{1,1}$ defines a conformal split of spacetime with potential applications to twistor theory and cosmological models in gauge gravity.

1.4 Models of Euclidean Space

We can model $\mathcal{E}^n$ as a set of points with algebraic properties. A standard way to do that is to identify each Euclidean point with a vector $\mathbf{x}$ in $\mathcal{R}^n$, as expressed by the isomorphism

$$\mathcal{E}^n \cong \mathcal{R}^n. \tag{1.14}$$

We call this the **inhomogeneous model** of $\mathcal{E}^n$, because the origin $\mathbf{0}$ is a distinguished point in $\mathcal{R}^n$, although all points in $\mathcal{E}^n$ are supposed to be identical.

To eliminate that drawback, we represent points in $\mathcal{E}^n$ by vectors $x \in \mathcal{R}^{n+1,1}$ and, to eliminate the extra degrees of freedom, we suppose that each point lies in the null cone

$$\mathcal{N}^{n+1} = \{x \,|\, x^2 = 0\} \tag{1.15}$$

and on the hyperplane

$$\mathcal{P}^{n+1}(e, e_*) = \{x \,|\, e \cdot x = 1\}, \tag{1.16}$$

where

$$e \cdot (x - e_*) = 0 \quad \Longleftrightarrow \quad e \cdot x = 1 \tag{1.17}$$

tells us that the plane passes through the point e_*. The intersection of these two surfaces is the *horosphere*:

$$\mathcal{N}_e^{n+1} = \mathcal{N}^{n+1} \cap \mathcal{P}^{n+1}(e, e_*). \tag{1.18}$$

Thus, we have the isomorphisms

$$\mathcal{R}^n \cong \mathcal{E}^n \cong \mathcal{N}_e^{n+1}. \tag{1.19}$$

We call the horosphere (Fig. 1.2) the **homogeneous model** of $\mathcal{E}^n$. It was first constructed by F. A. Wachter (1792–1817), but, as will become apparent, it is only by formulating it in terms of geometric algebra that it becomes a practical tool for computational geometry. To prove the isomorphism (1.19), we employ a conformal split to relate each homogeneous point $x \in \mathcal{N}_e^{n+1}$ to a unique inhomogeneous point $\mathbf{x} = x \wedge E \in \mathcal{R}^n$.

The conformal split proceeds as follows:

$$x = xE^2 = (x \wedge E + x \cdot E)E,$$

and the constraints on x imply

$$(x \cdot E)E = (x \cdot (e \wedge e_*))E = (\tfrac{1}{2}\mathbf{x}^2 e + e_*)E = -\tfrac{1}{2}\mathbf{x}^2 e + e_*.$$

Whence we obtain an explicit expression for the conformal split:

$$x = (\mathbf{x} + \tfrac{1}{2}\mathbf{x}^2 e + e_*)E = \mathbf{x}E - \tfrac{1}{2}\mathbf{x}^2 e + e_* . \tag{1.20}$$

From this we calculate

$$x \cdot y = \langle xEEy \rangle = \mathbf{x} \cdot \mathbf{y} - \tfrac{1}{2}(\mathbf{x}^2 + \mathbf{y}^2) e \cdot e_*.$$

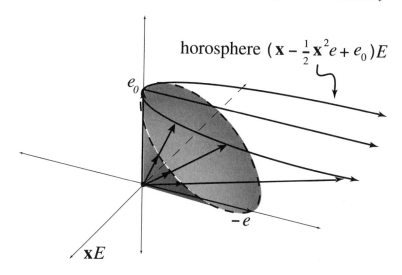

FIGURE 1.2. Horosphere.

Therefore, the inner product

$$x \cdot y = -\tfrac{1}{2}(\mathbf{x} - \mathbf{y})^2 = -\tfrac{1}{2}(\text{Euclidean distance})^2 \qquad (1.21)$$

specifies an intrinsic relation among points in $\mathcal{E}^n$.

Setting $\mathbf{x} = 0$ in (1.20), we see that e_* represents the origin in $\mathcal{R}^n$. From

$$\frac{x}{x \cdot e_*} = \frac{-2}{\mathbf{x}^2}\mathbf{x}^2\left(\frac{1}{\mathbf{x}} + \tfrac{1}{2}e + \frac{1}{\mathbf{x}^2}e_*\right)E \xrightarrow[\mathbf{x}^2 \to \infty]{} e,$$

we conclude that e represents a point at infinity.

To facilitate work with the homogenous model, we define I as the *unit pseudoscalar* for $\mathcal{R}^{n+1,1}$, and note the properties

$$|I|^2 = -II^\dagger = 1, \qquad I^\dagger = -I^{-1}, \qquad (1.22)$$

and $I^\dagger = I$ for $n = 2, 3$. The *dual* of a multivector A in $\mathcal{R}_{n+1,1}$ is defined by

$$\widetilde{A} \equiv AI^{-1} = -AI^\dagger \qquad (1.23)$$

Therefore, $\widetilde{E} = EI^{-1}$ is the *pseudoscalar* for $\mathcal{R}^n$.

1.5   Lines and Planes

Grassmann sought to identify the outer product $a \wedge b$ with the line determined by two points a and b [3]. However, he succeeded in doing that only in projective geometry [4]. The homogeneous model enables us to see why he failed to do it in Euclidean geometry: In the Euclidean case, it takes three points to determine a line, and one of them is the point e at infinity. Grassmann could not discover that because he did not have null vectors in his algebraic system.

With the conformal split, we can show that $a \wedge b \wedge e$ can be interpreted as a *line segment* in $\mathcal{E}^n$ or *line* through points a, b, e. The length of the segment is given by its square:

$$(a \wedge b \wedge e)^2 = (a - b)^2 = -2a \cdot b = (\text{length})^2. \tag{1.24}$$

Similarly, the outer product $a \wedge b \wedge c \wedge e$ represents a *plane segment* or *plane* in $\mathcal{E}^n$, and the area of the segment is given by its square:

$$(a \wedge b \wedge c \wedge e)^2 = \begin{vmatrix} 0 & 1 & 1 & 1 \\ 1 & 0 & a \cdot b & a \cdot c \\ 1 & b \cdot a & 0 & b \cdot c \\ 1 & c \cdot a & c \cdot b & 0 \end{vmatrix} = 4(\text{area})^2. \tag{1.25}$$

This is known, in a slightly different form, as the Cayley-Menger determinant. Cayley discovered it in 1841 and nearly a century later Menger [5] used it in a formulation of Euclidean geometry with interpoint distance as primitive. Dress and Havel [6] recognized its relation to Geometric Algebra.

Using eqn. (1.20), we can expand the geometric product of points a and b:

$$\begin{aligned} ab &= (aE)(Eb) = (\mathbf{a} + \tfrac{1}{2}\mathbf{a}^2 e + e_*)(\mathbf{b} - \tfrac{1}{2}\mathbf{b}^2 e - e_*) \\ &= -\tfrac{1}{2}(\mathbf{a} - \mathbf{b})^2 + \mathbf{a} \wedge \mathbf{b} + \tfrac{1}{2}(\mathbf{a}^2\mathbf{b} - \mathbf{b}^2\mathbf{a})e + (\mathbf{b} - \mathbf{a})e_* - \tfrac{1}{2}(\mathbf{a}^2 - \mathbf{b}^2)E. \end{aligned}$$

In expanding (1.26) we have used the relation $\mathbf{ab} = \mathbf{a} \cdot \mathbf{b} + \mathbf{a} \wedge \mathbf{b}$, which applies because $\mathbf{a}$ and $\mathbf{b}$ can be interpreted as vectors in $\mathcal{R}_n$, though they are trivectors in $\mathcal{R}_{n+1,1}$. In other words, we have *regraded* the elements of the subalgebra $\mathcal{R}_n$ to conform to our interpretation of $\mathcal{R}^n$ as an inhomogeneous model of $\mathcal{E}^n$. The use of boldface type should avoid confusion between the two different versions of outer product: $\mathbf{a} \wedge \mathbf{b}$ and $a \wedge b$.

The first term on the right side of (1.26) is recognized as the inner product $a \cdot b$, while the remaining terms make up $a \wedge b$. The profusion of terms in (1.26) is indicative of the extensive information inherent in the simple product ab. It is similar to the complexity of a spacetime split in physics [7].

From (1.26) we derive the projective split of a *line* (or *line segment*) through points a, b, e:

$$e \wedge a \wedge b = \mathbf{a} \wedge \mathbf{b} e + (\mathbf{b} - \mathbf{a}). \tag{1.27}$$

The coefficients on the right side of (1.27) will be recognized as the Plücker coordinates for a line with tangent $\mathbf{b} - \mathbf{a}$ and moment $\mathbf{a} \wedge (\mathbf{b} - \mathbf{a}) = \mathbf{a} \wedge \mathbf{b}$, as depicted in Fig. 1.3. Similarly, the projective split for a *plane* (or *plane segment*) is given by

$$e \wedge a \wedge b \wedge c = -\mathbf{a} \wedge \mathbf{b} \wedge \mathbf{c} e + (\mathbf{b} - \mathbf{a}) \wedge (\mathbf{c} - \mathbf{a}) E, \qquad (1.28)$$

where the coefficients are Plücker coordinates for a plane, as depicted in Fig. 1.3.

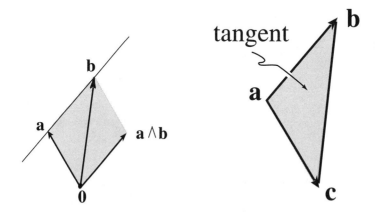

FIGURE 1.3. Line and plane.

1.6   Spheres and Hyperplanes

A *sphere* in $\mathcal{E}^n$ with radius ρ and center p is represented by a vector s in $\mathcal{R}^{n+1,1}$ with positive signature, where

$$\frac{s^2}{(s \cdot e)^2} = \rho^2, \qquad p = \frac{s}{s \cdot e} - \tfrac{1}{2}\rho^2 e. \qquad (1.29)$$

It is readily verified that $p^2 = 0$, so p is a homogeneous point. The constraint $s \cdot e = 1$ determines s uniquely and simplifies (1.29) to

$$s^2 = \rho^2 > 0, \quad p = s - \tfrac{1}{2}\rho^2 e. \qquad (1.30)$$

As depicted in Fig. 1.4, the equation for the sphere is

$$x \cdot s = 0. \qquad (1.31)$$

This is the equation for a hyperplane through the origin in $\mathcal{R}^{n+1,1}$, although only the vectors satisfying $x^2 = 0$ count as homogeneous points.

The conformal split gives us

$$s = \mathbf{p}E + \tfrac{1}{2}(\rho^2 - \mathbf{p}^2)e + e_*. \tag{1.32}$$

This helps us ascertain that the equation for a circle can be expressed in the alternative forms:

$$x \cdot p = -\tfrac{1}{2}\rho^2 \quad \Longrightarrow \quad \rho^2 = (\mathbf{x} - \mathbf{p})^2. \tag{1.33}$$

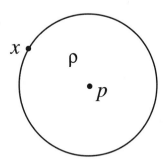

FIGURE 1.4. Sphere in $\mathcal{E}^n$ with radius ρ and center p.

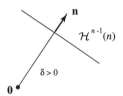

FIGURE 1.5. Hyperplane in $\mathcal{E}^n$.

Like a sphere, a *hyperplane* in $\mathcal{E}^n$ can be represented by a single vector n of positive signature. The vector can be normalized to unity, but it necessarily satisfies

$$n \cdot e = 0. \tag{1.34}$$

As depicted in Fig. 1.5, its conformal split has the form

$$n = \mathbf{n}E - \delta e, \tag{1.35}$$

where $n^2 = \mathbf{n}^2 = 1$.

The homogeneous model of $\mathcal{E}^n$ represents *all* spheres and hyperplanes in $\mathcal{R}^n$ as $(n+1)$-dim subspaces of $\mathcal{R}^{n+1,1}$ determined by their normal vectors, as expressed

$$\{x \,|\, x \cdot s = 0,\ s^2 > 0\,,\ s \cdot e \geq 0, x^2 = 0;\ x, s \in \mathcal{R}^{n+1,1}\}. \tag{1.36}$$

For a sphere the normal s satisfies $e \cdot s > 0$, but for a hyperplane it satisfies $e \cdot s = 0$. Thus, a hyperplane is a sphere through the point $e = \infty$.

A sphere determined by $n+1$ points $a_0, a_1, a_2, \ldots a_n$ is represented by the *tangent form*

$$\tilde{s} \equiv a_0 \wedge a_1 \wedge a_2 \wedge \cdots \wedge a_n \neq 0. \tag{1.37}$$

According to (1.30), its radius ρ and center p are easily obtained from its dual *normal form*

$$s = -(a_0 \wedge a_1 \wedge a_2 \wedge \cdots \wedge a_n)^{\tilde{}}. \tag{1.38}$$

It follows that the equation for a sphere can be given in the dual forms:

$$x \wedge \tilde{s} = 0 \quad \Longleftrightarrow \quad x \cdot s = 0. \tag{1.39}$$

These equations apply to a hyperplane as a sphere through ∞ by taking, say $a_0 = e$, to get the dual forms:

$$\tilde{n} = e \wedge a_1 \wedge a_2 \wedge \cdots \wedge a_n, \tag{1.40}$$

$$n = -(e \wedge a_1 \wedge a_2 \wedge \cdots \wedge a_n)^{\tilde{}}. \tag{1.41}$$

Example: Consider Simson's construction shown in Fig. 1.6. Given a triangle with vertices A, B, C and a point D in a Euclidean plane. Perpendiculars are dropped from D to the three sides of the triangle, intersecting them at points A_1, B_1, C_1.

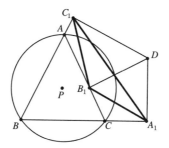

FIGURE 1.6. Simson's construction.

The circumcircle of triangle $e \wedge A \wedge B \wedge C$ is $\tilde{s} = A \wedge B \wedge C$, so we can obtain its radius from

$$\rho^2 = \left(\frac{s}{s \cdot e}\right)^2 = \frac{\tilde{s}^\dagger \tilde{s}}{(ts \wedge e)^2} = \frac{(C \wedge B \wedge A) \cdot (A \wedge B \wedge C)}{(e \wedge A \wedge B \wedge C)^2}. \tag{1.42}$$

The following identity can be derived:

$$e \wedge A_1 \wedge B_1 \wedge C_1 = \frac{A \wedge B \wedge C \wedge D}{2\rho^2}. \tag{1.43}$$

It follows that $A \wedge B \wedge C \wedge D = 0$ if and only if $e \wedge A_1 \wedge B_1 \wedge C_1 = 0$. In other words, D lies on the circumcircle if and only if A_1, B_1, C_1 are collinear. This is Simson's Theorem.

1.7   Conformal and Euclidean Groups

Orthogonal transformations on Minkowski space are called Lorentz transformations. Any Lorentz Transformation $\underline{G}$ can be expressed in the canonical form

$$\underline{G}(x) = \epsilon\, G\, x\, G^{-1} = \sigma x', \tag{1.44}$$

where G is a versor with parity $\epsilon = \pm 1$. G is the versor representation of $\underline{G}$, usually called the spin representation if $\epsilon = +1$.

Lorentz transformations leave the null cone $x^2 = x'^2 = 0$ invariant. However, the condition $e \cdot x' = e \cdot x = 1$ is not Lorentz invariant, so a point-dependent scale factor σ has been introduced into (1.44) to compensate for that.

The Lorentz group on $\mathcal{R}^{n+1,1}$ is isomorphic to the conformal group on $\mathcal{R}^n$, and the two groups are related by the conformal split

$$G\epsilon[\mathbf{x} + \tfrac{1}{2}\mathbf{x}^2 e + e_*]EG^{-1} = \sigma[\mathbf{x}' + \tfrac{1}{2}(\mathbf{x}')^2 e + e_*]E, \tag{1.45}$$

where

$$\mathbf{x}' = g(\mathbf{x}) \tag{1.46}$$

is a conformal transformation on $\mathcal{R}^n$.

The great advantage of the versor representation is that it reduces the composition of conformal transformations to versor multiplication, as expressed by the correspondence

$$g_3(\mathbf{x}) = g_2[g_1(\mathbf{x})] \qquad \Longleftrightarrow \qquad G_3 = G_2 G_1.$$

Every versor G can be expressed as the product of non-null vectors, as expressed by $G = s_k \ldots s_2 s_1$. A vector factor may represent either a hyperplane or a sphere in $\mathcal{R}^n$, as explained in the preceding section.

Reflection in the (hyper)plane specified by a vector n, has the simple form:

$$\underline{n}(x) = -nxn^{-1} = x', \tag{1.47}$$

with $\sigma = 1$. Rotations and translations can be generated multiplicatively from reflections. Thus, reflections in two planes m and n that intersect at a point c (Fig. 1.7), generate a **rotation** around the line of intersection, as specified by

$$
\begin{aligned}
G = mn \;&=\; (\mathbf{m}E - e\mathbf{m}\cdot\mathbf{c})(\mathbf{n}E - e\mathbf{n}\cdot\mathbf{c}) \\
&=\; \mathbf{mn} - e(\mathbf{m}\wedge\mathbf{n})\cdot\mathbf{c} \\
&=\; R - e(R\times\mathbf{c}).
\end{aligned}
$$

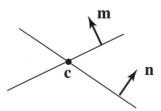

FIGURE 1.7. Generation of a rotation.

A **translation** by reflection in parallel planes m, n is specified by

$$
\begin{aligned}
G = mn \;&=\; (\mathbf{m}E - e\delta)(\mathbf{n}E + 0) \\
&=\; 1 + \tfrac{1}{2}\mathbf{a}e = T_{\mathbf{a}},
\end{aligned}
$$

where $\mathbf{a} = 2\mathbf{n}\delta$, as shown in Fig. 1.8.

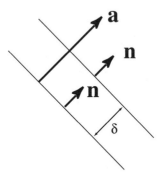

FIGURE 1.8. Generation of a translation.

The group of rigid displacements on $\mathcal{E}^3$ is called the (special) Euclidean group $SE(3)$. Each group element $\underline{D}$ can be expressed in the form

$$
\underline{D}(x) = DxD^{-1}, \tag{1.48}
$$

where the displacement versor $D = T_{\mathbf{a}}R$ specifies a rotation around an axis with direction $\mathbf{n} = R\mathbf{n}R^{\dagger}$ through the origin, followed by a translation $T_{\mathbf{a}} = 1 + \tfrac{1}{2}e\mathbf{a}$.

According to **Chasles' Theorem:** Any rigid displacement can be expressed as a *screw displacement*. This can be proved by finding a point **b** on the screw axis so that

$$D = T_{\mathbf{a}_\parallel} T_{\mathbf{a}_\perp} R = T_{\mathbf{a}_\parallel} R_{\mathbf{b}} \tag{1.49}$$

where $\mathbf{a}_\parallel = (\mathbf{a} \cdot \mathbf{n})\mathbf{n}$, and

$$R_{\mathbf{b}} = R + e\,\mathbf{b} \times R, \tag{1.50}$$

is a rotation that leaves **b** fixed. Equation (1.49) can be solved directly for

$$\mathbf{b} = \mathbf{a}_\perp (1 - R^{-2})^{-1} = \tfrac{1}{2}\mathbf{a}_\perp \frac{1 - R^2}{1 - \langle R^2 \rangle}. \tag{1.51}$$

This illustrates the computational power of geometric algebra.

The displacement versor can be put in the *screw* form

$$D = e^{\frac{1}{2}S}, \tag{1.52}$$

where

$$S = -i\mathbf{m} + e\mathbf{n}, \tag{1.53}$$

where $i = \tilde{E}$ is the pseudoscalar for E^3, and S is called a screw. The screws compose $se(3)$, the Lie algebra of $SE(3)$. It is a bivector algebra, closed under the commutator product:

$$S_1 \times S_2 = \tfrac{1}{2}(S_1 S_2 - S_2 S_1). \tag{1.54}$$

All the elements of screw theory are natural consequences of geometric algebra! Each screw has a unique decomposition into a null and a non-null part:

$$S_k = -i\mathbf{m}_k + e\mathbf{n}_k. \tag{1.55}$$

The geometric product of two screws has the decomposition

$$S_1 S_2 = S_1 \cdot S_2 + S_1 \times S_2 + S_1 \wedge S_2. \tag{1.56}$$

Under a rigid displacement $\underline{U}$, the transformation of a screw is given by

$$S_k' = \underline{U} S_k = U S_k U^{-1} = Ad_U S_k. \tag{1.57}$$

This is the "adjoint representation" of $SE(3)$, as indicated by the notation on the right. The transformation (1.57) preserves the geometric product:

$$S_1' S_2' = \underline{U}(S_1 S_2) = U(S_1 \cdot S_2 + S_1 \times S_2 + S_1 \wedge S_2)U^{-1}$$

The invariants

$$\underline{U}e = e, \qquad \underline{U}i = i \tag{1.58}$$

imply the invariants:

$$\underline{U}(S_1 \wedge S_2) = S_1 \wedge S_2 = -ie(\mathbf{m}_1 \cdot \mathbf{n}_2 + \mathbf{m}_2 \cdot \mathbf{n}_1), \qquad (1.59)$$

$$S_1' \cdot S_2' = S_1 \cdot S_2 = -\mathbf{m}_1 \cdot \mathbf{m}_2. \qquad (1.60)$$

The latter invariant may be recognized as the Killing Form for $se(3)$.

It is convenient to introduce the notion of a *coscrew* (Ball's reciprocal screw) defined by

$$S_k^* \equiv \langle\, S_k ie_* \,\rangle_2 = \tfrac{1}{2}(S_k ie_* + ie_* S_k) = i\mathbf{n}_k + \mathbf{m}_k e_*. \qquad (1.61)$$

Then the invariant (57) can be written in the scalar-valued form:

$$S_1^* \cdot S_2 = S_2^* \cdot S_1 = \langle\, (S_1 \wedge S_2) ie_* \,\rangle = \mathbf{m}_1 \cdot \mathbf{n}_2 + \mathbf{m}_2 \cdot \mathbf{n}_1. \qquad (1.62)$$

For a single screw we get the *pitch* invariant:

$$h = \tfrac{1}{2}\frac{S^* \cdot S}{S \cdot S} = \mathbf{n} \cdot \mathbf{m}^{-1} \qquad (1.63)$$

1.8   Screw Mechanics

Screw theory with geometric algebra enables us to combine the rotational and translational equations of motion for a rigid body into a single equation. The kinematics of a body point

$$x = Dx_0 D^{-1} \qquad (1.64)$$

is completely characterized by the displacement spinor $D = D(t)$, which obeys the kinematical equation

$$\dot{D} = \tfrac{1}{2} V D \qquad (1.65)$$

with

$$V = -i\boldsymbol{\omega} + \mathbf{v}e, \qquad (1.66)$$

where $\boldsymbol{\omega}$ is the angular velocity of the body and we can take $\mathbf{v}$ to be its center-of-mass velocity. It follows that $\dot{x} = V \cdot x$ and $\dot{\mathbf{x}} = \boldsymbol{\omega} \times \mathbf{x} + \mathbf{v}$.

A *comomentum* P is defined for the body by

$$P = \underline{M} V = i\underline{I}\boldsymbol{\omega} + m\mathbf{v}e_* = i\boldsymbol{\ell} + \mathbf{p}e_*. \qquad (1.67)$$

This defines a generalized "mass tensor" $\underline{M}$ in terms of the inertia tensor $\underline{I}$ and the body mass m. According to the transformation equations below, the comomentum is a coscrew.

The *coforce* or *wrench* W acting on a rigid body is defined in terms of the torque Γ and net force $\mathbf{f}$ by

$$W = i\Gamma + \mathbf{f}e_*. \tag{1.68}$$

The dynamical equation for combined rotational and translational motion then takes the compact form:

$$\dot{P} = W, \tag{1.69}$$

where the overdot indicates a time derivative. An immediate consequence is the conservation law

$$\dot{K} = V \cdot W = \boldsymbol{\omega} \cdot \Gamma + \mathbf{v} \cdot \mathbf{f} \tag{1.70}$$

for kinetic energy

$$K = \tfrac{1}{2} V \cdot P = \tfrac{1}{2}(\boldsymbol{\omega} \cdot \boldsymbol{\ell} + \mathbf{v} \cdot \mathbf{p}). \tag{1.71}$$

A change of reference frame, including a shift of base point, is expressed by

$$x \quad \longrightarrow \quad x' = \underline{U}x = UxU^{-1}. \tag{1.72}$$

We consider here only the case when the spinor U is constant. Then (1.72) induces the transformations

$$V' = \underline{U}V, \tag{1.73}$$

$$P = \bar{U}P'. \tag{1.74}$$

Thus, the transformation of V is *Covariant*, while the transformation of P is *Contravariant*. Their scalar product is the *Invariant*

$$P' \cdot V' = P \cdot V. \tag{1.75}$$

There is much more about all this in [8], [9] and [10], especially applications. For more screw theory, see [11] and [12].

1.9   Conclusions

We have seen that the homogeneous model for Euclidean space has at least three major advantages.

I. *Intrinsic properties of $\mathcal{E}^n$ are embedded in the algebraic properties of homogeneous points.* In other words we have

This was **Grassmann's great goal**, and he would surely be pleased to know that it has finally been achieved, although the path has not been straightforward.

$$\left\{ \begin{array}{c} \text{Synthetic} \\ \text{geometry} \end{array} \right\} \xleftrightarrow[\text{with}]{\text{integrated}} \left\{ \begin{array}{c} \text{Computational} \\ \text{geometry} \end{array} \right\}$$

II. **All** *spheres and hyperplanes in $\mathcal{E}^n$ are uniquely represented by vectors in $\mathcal{R}^{n+1,1}$.* This unifies and simplifies the treatment of spheres and hyperplanes, especially with respect to duality properties.

III.

$$\left\{ \begin{array}{c} \text{Conformal group} \\ \text{on } \mathcal{E}^n \text{ (or } \mathcal{R}^n) \end{array} \right\} \cong \left\{ \begin{array}{c} \text{Lorentz group} \\ \text{on } \mathcal{R}^{n+1,1} \end{array} \right\}$$

$$\cong_2 \left\{ \begin{array}{c} \text{Versor group} \\ \text{in } \mathcal{R}_{n+1,1} \end{array} \right\}$$

This isomorphism linearizes the conformal group and reduces composition of conformal transformations to versor multiplication.

Chapter 2

Universal Geometric Algebra

Garret Sobczyk

2.1 Introduction

Since Grassmann's original work "Ausdehnungslehre" in 1844, and William Kingdom Clifford's later discovery of "geometric algebra" in 1878, the mathematical community has been puzzled by exactly how these works fit into the main stream of mathematics. Certainly the importance of these works in the mathematics at the end of the 20th Century has been recognized, but there has been no general agreement about where and how the methods should be utilized. In this chapter, I wish to show how the works of Grassmann and Clifford can be integrated into the mainstream of mathematics in such a way as to require as little as possible changes to the main body of mathematics as we know it today. As has been often repeated by Hestenes and others, geometric algebra should be seen as a great unifier of the geometric ideas of mathematics.

Some of the opposition to the acceptance of geometric algebra into the mainstream is without doubt due to the more sophisticated algebraic skills and identities that must be mastered. Indeed, I have to admit my own frustration in not being able to do more than a line or two of computations without making a serious mistake. I believe that what is most needed in the area today is an efficient computer software package for carrying out symbolic calculations in geometric algebra. Lounesto's CLICAL has proven itself to be invaluable to the researcher for making the numeric calculations necessary to check theoretical work, and Ablamowicz's CLIFFOR has been successfully employed by a number of researchers. But a fully integrated symbolic computer software package to do geometric algebra is still waiting in the wings. A partial solution to this problem is to develop the basic ideas of geometric algebra in such a way that problems can be quickly reduced to their matrix equivalents for which computer software is readily available. Thus, one of the major objectives of this chapter is to develop the main ideas of geometric algebra in such a way that matrix methods can be employed almost immediately at any step of a calculation.

In section 2, we begin with an n-dimensional real vector space $\mathcal{N}$ which we call the *null cone*, since we are assuming that all vectors in $\mathcal{N}$ are *null vectors* (the square of each vector is zero). Taking all linear combinations of sums of products of vectors in $\mathcal{N}$ generates the 2^n-dimensional associative

Grassmann algebra $\mathcal{G}(\mathcal{N})$. This stucture is sufficiently rich to efficiently develop many of the basic notions of linear algebra, such as the matrix of a linear operator and the theory of determinants and their properties. We are careful to define and use notation which is fully compatible with traditional matrix theory, but more general in the sense that we are considering matrices over the Grassmann algebra $\mathcal{G}(\mathcal{N})$. Next, we introduce an n-dimensional null cone $\overline{\mathcal{N}}$ which is *reciprocal* or *dual* to $\mathcal{N}$, and its associated *reciprocal* or *dual* Grassmann algebra $\mathcal{G}(\overline{\mathcal{N}})$. By demanding that a few simple additional rules be satisfied, which relates the basis elements of the reciprocal null cones $\mathcal{N}$ and $\overline{\mathcal{N}}$, we obtain the 2^{2n}-dimensional *universal geometric algebra* $\mathcal{G}(\mathcal{N}, \overline{\mathcal{N}})$. It seems natural to allow a countable infinite number of basis elements, and to call the resulting general structure universal geometric algebra.

In section 3, we extend the familiar addition and multiplication of matrices over the real and complex numbers to more general matrices of elements over $\mathcal{G}_{n,n}$. This notation is useful for the compact formulation and proofs of theorems relating the matrix and geometric algebra formalisms. The *meet* and *join* in projective geometry are incorporated as new algebraic operations on simple k-blades in the geometric algebra.

In section 4, linear transformations are studied as mappings between null cones. The *generalized spectral decomposition* of a linear operator is presented and the principal correlation is discussed. The principal correlation is one step away from the more familiar *polar form* and singular value decompositions of a linear transformation. The principal correlation also makes possible a compact treatment of the generalized inverse of a linear transformation. The *bivector of a linear operator* is defined, making possible the *spinor representation* of a general linear transformation.

In section 5, projection operators are defined allowing us to move from the null cone to various subspaces where rotations and translations in can be carried out in pseudo-euclidean spaces and affine spaces of arbitrary signature. After these operations are completed, we can return to the null cone via an inverse projection operator.

In section 6, basic properties of affine and projective geometries are explored, including new proofs of Desargues' theorem and Simpson's theorem.

In section 7, an introduction is given to a *non-linear* model of euclidean space, called the *horosphere*, and its relationship to the affine plane. Various applications of the horosphere are discussed in chapters 4, 3 and 17 of the present volume.

2.2   The Universal Geometric Algebra

By an n-dimensional null cone $\mathcal{N}$, we mean a real linear n-dimensional vector space on which an associative multiplication is defined with the property that for each $x \in \mathcal{N}$, $x^2 = xx = 0$. It follows that

$$(x + y)^2 = x^2 + xy + yx + y^2 = xy + yx = 0$$

or $xy = -yx$ for all $x, y \in \mathcal{N}$. Decomposing the *geometric product* xy into symmetric and antisymmetric parts, we find that

$$xy = \frac{1}{2}(xy + xy) + \frac{1}{2}(xy - yx) = x \cdot y + x \wedge y \qquad (2.1)$$

where the *inner product*

$$x \cdot y = \frac{1}{2}(xy + yx) \equiv 0$$

vanishes, and the *outer product*

$$x \wedge y = \frac{1}{2}(xy - yx) = xy$$

reduces to the geometric product for all vectors $x, y \in \mathcal{N}$.

The 2^n-dimensional *Grassmann algebra* $\mathcal{G}_N$ of the null cone $\mathcal{N}$ is defined by taking the *associative algebra* of all geometric *sums* of *products* of the vectors in $\mathcal{N}$, subject to the one condition that $x^2 = 0$ for all $x \in \mathcal{N}$. We write

$$\mathcal{G}_N = gen\{\mathcal{N}\} = gen\{e_1, ..., e_n\}$$

The Grassmann algebra $\mathcal{G}_N$ of the null cone $\mathcal{N}$ is the linear space spanned by the 2^n-dimensional *basis of multivectors,*

$$\{1; \; e_1, \ldots, e_n; \; e_{12}, \ldots, e_{n-1n}; \ldots; \; e_{1\ldots k}, \ldots, e_{n-k+1\ldots n}; \ldots; \; e_{12\ldots n}\}$$

where $\{e_{1\ldots k}, \ldots, e_{n-k+1\ldots n}\}$ is the $\begin{pmatrix} n \\ k \end{pmatrix}$-dimensional basis of k-vectors

$$e_{j_1 j_2 \ldots j_k} \equiv e_{j_1} e_{j_2} \ldots e_{j_k}$$

for the $\begin{pmatrix} n \\ k \end{pmatrix}$ sets of indicies $1 \le j_1 < j_2 < \ldots < j_k \le n$. The Grassmann exterior product $x_1 x_2 \ldots x_k$ of k vectors is *antisymmetric* over the interchange of any two of its vectors;

$$x_1 \ldots x_i \ldots x_j \ldots x_k = -x_1 \ldots x_j \ldots x_i \ldots x_k$$

and has the geometric interpretation of a *directed k-vector* or k-element of volume. Note that the exterior product of null vectors in the null cone $\mathcal{N}$ is equivalent to the outer product of those vectors,

$$x_1 x_2 \ldots x_k \equiv x_1 \wedge x_2 \wedge \ldots \wedge x_k.$$

Given the null cone $\mathcal{N} = span\{e\}$ and its associated Grassmann algebra $\mathcal{G}_N$, we can always define a *reciprocal null cone* $\overline{\mathcal{N}} = span\{\overline{e}\}$ and its associated *reciprocal Grassmann algebra* $\mathcal{G}_{\overline{N}}$. The reciprocal null cone $\overline{\mathcal{N}}$ of the null cone $\mathcal{N}$ is defined in such a way as to encapsulate the familiar properties of the mathematical *dual space* of the vector space $\mathcal{N}$. Thus, in addition to the basic properties

$$e_i^2 = 0 = \overline{e}_i^2, \quad \text{and} \quad e_i e_j = -e_j e_i, \ \overline{e}_i \overline{e}_j = -\overline{e}_j \overline{e}_i,$$

we require that

$$\overline{e}_i \cdot e_j = \delta_{i,j} = e_j \cdot \overline{e}_i \tag{2.2}$$

for all $i, j = 1, 2, \ldots, n$. With this definition, the Grassmann algebra $\mathcal{G}_{\overline{N}}$ of the reciprocal cone $\overline{\mathcal{N}}$ becomes the natural dual of the Grassmann algebra $\mathcal{G}_N$.

The *neutral pseudoeuclidean space* $I\!\!R^{n,n}$ is the smallest linear space which contains *both* the null cones $\mathcal{N}$ and $\overline{\mathcal{N}}$. Thus,

$$I\!\!R^{n,n} = span\{\mathcal{N}, \overline{\mathcal{N}}\} = \{x + \overline{y} |\ x \in \mathcal{N}, \overline{y} \in \overline{\mathcal{N}}\}.$$

Likewise, the 2^{2n}–dimensional associative geometric algebra $\mathcal{G}_{n,n}$ is the smallest geometric algebra that contains *both* the Grassmann algebras $\mathcal{G}_N$ and $\mathcal{G}_{\overline{N}}$, see [6]. We write

$$G_{n,n} = \mathcal{G}_N \times \mathcal{G}_{\overline{N}} = gen\{e_1, \ldots, e_n, \overline{e}_1, \ldots, \overline{e}_n\}. \tag{2.3}$$

Whereas the geometric product (2.1) of the vectors x, y,

$$xy = x \cdot y + x \wedge y \tag{2.4}$$

reduces to the outer product when both x and y belong either to the null cone $\mathcal{N}$, or both belong to the reciprocal null cone $\overline{\mathcal{N}}$, because of the duality relationship (2.2), this is no longer true for arbitrary $x, y \in I\!\!R^{n,n}$. Indeed, in the context of the pseudoeuclidean space $I\!\!R^{n,n}$, the pseudo-inner product $x \cdot y$ becomes *non-degenerate* with *neutral signature*. A very useful geometric identity that we will need is the inner product of a vector with a bivector,

$$x \cdot (a \wedge b) = (x \cdot a)b - (x \cdot b)a = -(a \wedge b) \cdot x. \tag{2.5}$$

We call $\mathcal{G}_{n,n}$ the universal geometric algebra of order 2^{2n}. When n is countably infinite, we call $\mathcal{G} = \mathcal{G}_{\infty,\infty}$ the *universal geometric algebra*, [16]. The universal algebra $\mathcal{G}$ contains all of the algebras $\mathcal{G}_{n,n}$ as proper subalgebras. In [6], $\mathcal{G}_{n,n}$ is called the *mother algebra*.

2.2.1 The standard basis

We have used properties of the *Witte basis* of null vectors $\{e, \bar{e}\}$ to define the geometric algebra $\mathcal{G}_{n,n}$. From the Witte basis, we now construct the *standard orthonormal basis* $\{\sigma, \eta\}$ of $\mathcal{G}_{n,n}$,

$$\sigma_i = \frac{1}{\sqrt{2}}(e_i + \bar{e}_i), \quad \eta_i = \frac{1}{\sqrt{2}}(e_i - \bar{e}_i) \tag{2.6}$$

for $i = 1, 2, \ldots, n$. Using the defining relationships (2.2) of the reciprocal frames $\{e\}$ and $\{\bar{e}\}$, we find that these basis vectors satisfy

$$\sigma_i{}^2 = 1, \ \sigma_i \cdot \sigma_j = \delta_{i,j}, \ \sigma_i \sigma_j = -\sigma_j \sigma_i \text{ for } i \neq j,$$

$$\eta_i{}^2 = -1, \ \eta_i \cdot \eta_j = -\delta_{i,j}, \ \eta_i \eta_j = -\eta_j\, \eta_i \ \text{ for } \ i \neq j,$$

$$\eta_i \sigma_j = -\sigma_j \eta_i, \ \eta_i \cdot \sigma_j = 0. \tag{2.7}$$

The basis $\{\sigma\}$ spans a real *Euclidean vector space* $\mathbb{R}^n$ and generates the geometric subalgebra $\mathcal{G}_{n,0}$, whereas $\{\eta\}$ spans an antieuclidean space $\mathbb{R}^{0,n}$ and generates the geometric subalgebra $\mathcal{G}_{0,n}$. We can now express the geometric algebra $\mathcal{G}_{n,n}$ as the product of these geometric subalgebras

$$\mathcal{G}_{n,n} = \mathcal{G}_{n,0} \otimes \mathcal{G}_{0,n} = gen\{\sigma_1, ..., \sigma_n, \eta_1, ..., \eta_n\}. \tag{2.8}$$

2.3   Matrices of Geometric Numbers

In this section, we extend the familiar addition and multiplication of matrices over the real and complex numbers to more general matrices of elements over $\mathcal{G}_{n,n}$. This notation is useful for the compact formulation and proof of theorems, as well as for relating the matrix and geometric algebra formalisms.

We begin by writing the frame of basis vectors $\{e\}$ and the corresponding reciprocal frame of basis vectors $\{\bar{e}\}$ of $\mathcal{G}_{n,n}$ in *row form* and in *column form*, respectively,

$$\{e\} \equiv \begin{pmatrix} e_1 & e_2 & \cdot & \cdot & e_n \end{pmatrix}, \quad \{\bar{e}\} \equiv \begin{pmatrix} \bar{e}_1 \\ \bar{e}_2 \\ \cdot \\ \cdot \\ \bar{e}_n \end{pmatrix}.$$

We say that the basis $\{e\}$ *spans* $\mathcal{N}$ and write $\mathcal{N} = span\{e\}$. Similarly, $\overline{\mathcal{N}} = span\{\bar{e}\}$.

In terms of these bases, any *vector* or *point* $x \in \mathcal{N}$ can be written

$$x = \{e\}x_{\{e\}} = \{e_1, e_2, \ldots, e_n\} \begin{pmatrix} x_1 \\ x_2 \\ \cdot \\ \cdot \\ x_n \end{pmatrix} = \sum_{i=1}^{n} x_i e_i \qquad (2.9)$$

for $x_i \in \mathbb{R}$. The column vector

$$x_{\{e\}} = \begin{pmatrix} x_1 \\ x_2 \\ \cdot \\ \cdot \\ x_n \end{pmatrix} = \begin{pmatrix} \bar{e}_1 \cdot x \\ \bar{e}_2 \cdot x \\ \cdot \\ \cdot \\ \bar{e}_n \cdot x \end{pmatrix} = \{\bar{e}\} \cdot x$$

consist of the components of x with respect to the basis $\{e\}$. Note that in the basis representation of x, we are employing the natural matrix multiplication between a *row* and a *column*, and we will do this for matrices of even more general elements. Since vectors in $\mathcal{N}^n$ behave like column vectors, and vectors $\bar{y} \in \overline{\mathcal{N}}^n$ behave like row vectors, we define the operation of *transpose* of the vector x by

$$x^t = (\{e\}x_{\{e\}})^t = x^t_{\{e\}}\{\bar{e}\} = \begin{pmatrix} x_1 & x_2 & \ldots & x_n \end{pmatrix} \begin{pmatrix} \bar{e}_1 \\ \bar{e}_2 \\ \cdot \\ \cdot \\ \bar{e}_n \end{pmatrix} \qquad (2.10)$$

The transpose operation allows us to move between the reciprocal null cones. The closely related Hermitian transpose will be used later on to define an Hermitian inner product.

Taking advantage of the usual matrix multiplication between a *row* and a *column*, and the properties of the geometric product listed in (2.1), and (2.2), we get

$$\{\bar{e}\}\{e\} = \{\bar{e}\} \cdot \{e\} + \{\bar{e}\} \wedge \{e\} = id + \begin{pmatrix} \bar{e}_1 \wedge e_1 & \bar{e}_1 \wedge e_2 & \ldots & \bar{e}_1 \wedge e_n \\ \bar{e}_2 \wedge e_1 & \bar{e}_2 \wedge e_2 & \ldots & \bar{e}_2 \wedge e_n \\ \ldots & \ldots & \ldots & \ldots \\ \ldots & \ldots & \ldots & \ldots \\ \bar{e}_n \wedge e_1 & \bar{e}_n \wedge e_2 & \ldots & \bar{e}_n \wedge e_n \end{pmatrix}$$

where id is the $n \times n$ identity matrix, computed by taking all inner products $\bar{e}_i \cdot e_j$ between the basis vectors of $\{\bar{e}\}$ and $\{e\}$. Similarly,

$$\{e\}\{\bar{e}\} = \{e\} \cdot \{\bar{e}\} + \{e\} \wedge \{\bar{e}\} = \sum_{i=1}^{n} e_i \cdot \bar{e}_i + \sum_{i=1}^{n} e_i \wedge \bar{e}_i = n + \sum_{i=1}^{n} e_i \wedge \bar{e}_i,$$

giving the useful formulas

$$\{e\} \cdot \{\overline{e}\} = \sum_{i=1}^{n} e_i \cdot \overline{e}_i = n \tag{2.11}$$

and

$$\{e\} \wedge \{\overline{e}\} = \sum_{i=1}^{n} e_i \wedge \overline{e}_i \tag{2.12}$$

2.3.1 Reciprocal basis

The basis $\{e\} \in \mathcal{N}^n$ and $\{\overline{e}\} \in \overline{\mathcal{N}}^n$ are said to be *reciprocal*, or *dual*, because they satisfy the relationship $\{\overline{e}\} \cdot \{e\} = id$, where id is the $n \times n$ identity matrix. More generally, given a second basis $\{a\}$ of $\mathcal{N}$, the problem is to construct a dual basis $\{\overline{a}\}$ of $\overline{\mathcal{N}}$ such that $\{\overline{a}\} \cdot \{a\} = id$. For the construction below, we will need the the pseudoscalar $I = e_1 \wedge e_2 \ldots \wedge e_n$ of $\mathcal{N}$ and the corresponding reciprocal pseudoscalar element $\overline{I} = \overline{e}_n \wedge \ldots \wedge \overline{e}_2 \wedge \overline{e}_1$ of $\overline{\mathcal{N}}$ satisfying $I \cdot \overline{I} = 1$.

The new basis $\{a\}$ of $\mathcal{N}$ is related to the standard basis $\{e\}$ by the equation

$$\{a\} = \{e\}\mathcal{A} = \{e_1, e_2, \ldots, e_n\}\mathcal{A} \tag{2.13}$$

where $\mathcal{A}$ is called the *matrix of transition* from the basis $\{e\}$ to the basis $\{a\}$. Taking the outer product $\bigwedge_{i=1}^{n}\{a\}$ of the basis vectors $\{a\}$, we get

$$\bigwedge_{i=1}^{n} \{a\} \equiv a_1 \wedge a_2 \wedge \ldots \wedge a_n = \det(\mathcal{A})e_1 \wedge e_2 \wedge \ldots e_n, \tag{2.14}$$

where $\det \mathcal{A}$ is called the *determinant* of the matrix $\mathcal{A}$. We see from (2.14) that the determinant of the matrix of transition between two bases cannot be zero. Dotting both sides of (2.14) by $\overline{I}$, we get the explicit expression

$$\det(\mathcal{A}) = \bigwedge_{i=1}^{n} \{a\} \cdot \overline{I}$$

Because the *determinant function* is so important, we shall also use the alternative equivalent *bracket* notation

$$\det(\mathcal{A}) \equiv \det\{a\} \equiv [a_1 \; a_2 \; \ldots \; a_n] \tag{2.15}$$

The reciprocal basis $\{\overline{a}\}$ is now easily constructed:

$$\overline{a}_i = (-1)^{i+1} \frac{(a_1 \wedge \ldots \wedge i^* \wedge \ldots \wedge a_n) \cdot \overline{I}}{[a_1 \; a_2 \; \ldots \; a_n]} \tag{2.16}$$

Calculating

$$\{\overline{a}\} \cdot \{a\} = \begin{pmatrix} \overline{a}_1 \cdot a_1 & \overline{a}_1 \cdot a_2 & \ldots & \overline{a}_1 \cdot a_n \\ \overline{a}_2 \cdot a_1 & \overline{a}_2 \cdot a_2 & \ldots & \overline{a}_2 \cdot a_n \\ & \ldots & & \\ & \ldots & & \\ \overline{a}_n \cdot a_1 & \overline{a}_n \cdot a_2 & \ldots & \overline{a}_n \cdot a_n \end{pmatrix} = id$$

proves the construction. We have actually found the inverse of the matrix of transition $\mathcal{A}$! Writing $\{\overline{a}\} = \mathcal{B}\{\overline{e}\}$ for the transition matrix of the dual basis, we see that

$$\{\overline{a}\} \cdot \{a\} = \mathcal{B}\{\overline{e}\} \cdot \{e\}\mathcal{A} = \mathcal{B}\mathcal{A} = id$$

from which it follows that $\mathcal{B} = \mathcal{A}^{-1}$. From the expression $\{\overline{a}\} = \mathcal{A}^{-1}\{\overline{e}\}$, we easily find

$$\mathcal{A}^{-1} = \mathcal{A}^{-1}\{\overline{e}\} \cdot \{e\} = \{\overline{a}\} \cdot \{e\}$$

which is equivalent to the well known formula for the *inverse* of the matrix $\mathcal{A}$.

2.3.2 *Generalized inverse of a matrix*

Let $\mathcal{N}^n$ and $\overline{\mathcal{N}}^n$ be n-dimensional reciprocal null cones in $\mathbb{R}^{n,n}$ with the dual bases $\{e\}$ and $\{\overline{e}\}$. Let $\mathcal{N}^{n'} \subset \mathcal{N}^n$ and $\overline{\mathcal{N}}^{n'} \subset \overline{\mathcal{N}}^n$ be n'-dimensional reciprocal subspaces with the dual bases $\{e'\}$ and $\{\overline{e}'\}$. Suppose that an $n \times n'$ matrix $\mathcal{A}$ is given such that $n' = rank(\mathcal{A}) \leq n$ and $\{e'\} = \{e\}\mathcal{A}$. We can now define what is meant by a *generalized inverse* of the matrix $\mathcal{A}$.

Definition 1 *A matrix $\mathcal{A}^{inv}$ is called a generalized inverse of the matrix $\mathcal{A}$ if*

$$\{e'\} = \{e\}\mathcal{A} \quad and \quad \mathcal{A}^{inv}\{\overline{e}\} = \{\overline{e}'\}.$$

If $n' > n$ for the matrix $\mathcal{A}$, then a generalized inverse is defined to be the transpose of a generalized inverse of the matrix $\mathcal{A}^T$.

Whereas the inverse of a matrix exists only for square matrices with $\det \neq 0$, the generalized inverse as defined above will exist for non-square matrices $\mathcal{A}$ so long as $rank(\mathcal{A}) = min\{n, n'\}$. We will see in a later section that a generalized inverse, although not unique, can be defined for any nonzero matrix.

Given *any* $n \times n'$ real matrix $\mathcal{A}$ with $n' = rank(\mathcal{A}) \leq n$, the problem of finding a generalized inverse of $\mathcal{A}$ is equivalent to the problem of constructing a reciprocal dual basis $\{\overline{e}'\}$ (not unique!) for the basis $\{e'\} = \{e\}\mathcal{A}$ of the subspace $\mathcal{N}^{n'} = span(\{e'\})$ of $\mathcal{N}$. Since the construction is also important in the next section, we will present it here.

Since $I_\mathcal{A} = \bigwedge\{e'\} = \bigwedge\{e\}\mathcal{A} \neq 0$, we can find an n'-vector $\overline{I}_\mathcal{A}$ in the geometric algebra $\mathcal{G}(\overline{\mathcal{N}}^n)$ of the reciprocal cone $\overline{\mathcal{N}}^n$, with the property that

$\overline{I}_A \cdot I_A \neq 0$. We then use $\overline{I}_A$ to construct a *pseudo-determinant* function on the subspace $\mathcal{N}^{n'}$,

$$\det(\mathcal{A}) = \det\{e'\} = [e'_1 \; e'_2 \; \ldots \; e'_{n'}] = I_A \cdot \overline{I}_A \neq 0 \qquad (2.17)$$

The construction of the reciprocal basis $\{\overline{e}'\}$ for the subspace now proceeds exactly as in the previous section. We define

$$\overline{e}'_i = (-1)^{i+1} \frac{(e'_1 \wedge \ldots \wedge i^* \wedge \ldots \wedge e'_{n'}) \cdot \overline{I}_A}{[e'_1 \; e'_2 \; \ldots \; e'_{n'}]} \qquad (2.18)$$

With the reciprocal basis $\{\overline{e}'\}$ in hand, a generalized inverse $\mathcal{A}^{inv}$ of $\mathcal{A}$ can easily be found. It will be the matrix which satisfies

$$\{\overline{e}'\} = \mathcal{A}^{inv}\{\overline{e}\}$$

Dotting both sides of this equation on right by $\{e\}$ gives the generalized inverse of $\mathcal{A}$,

$$\mathcal{A}^{inv} = \mathcal{A}^{inv}\{\overline{e}\} \cdot \{e\} = \{\overline{e}'\} \cdot \{e\}. \qquad (2.19)$$

2.3.3 The meet and joint operations

Let $\mathcal{N}^{n+1}$ and $\overline{\mathcal{N}}^{n+1}$ be $(n+1)$-dimensional reciprocal null cones in $I\!R^{n+1,n+1}$. It is well known that the *directions* or *rays* of non-zero vectors in $\mathcal{N}^{n+1}$ can be identified with the points of the n-dimensional projective plane Π^n, [9]. To express this idea more precisely, we write

$$\Pi^n \equiv \mathcal{N}^{n+1}/I\!R^*$$

where $I\!R^* = I\!R - \{0\}$. We are thus led to identify *points, lines, planes, ...,* and higher dimensional *k-planes* in Π^n with 1, 2, 3, ..., $k+1$-dimensional subspaces $\mathcal{S}^r$ of $\mathcal{N}^{n+1}$, where $k \leq n$.

The meet and join operations of projective geometry are most easily characterized in terms of the *intersection* and *union* of the subspaces which name the objects in Π^n. On the other hand, each r-dimensional subspace $\mathcal{A}^r$ can be described by a non-zero r-blade $A_r \in \mathcal{G}_{\mathcal{N}^{n+1}}$. We say that an r-blade A_r represents, or is a *representant* of an r-subspace $\mathcal{A}^r$ of $\mathcal{N}^{n+1}$ if and only if

$$\mathcal{A}^r = \{x \in \mathcal{N}^{n+1}| \; x \wedge A_r = 0\}.$$

With this identification, the problem of finding the meet and join is reduced to a problem in geometric algebra of finding the corresponding *meet* and *join* of the $(r+1)$- and $(s+1)$-blades in the geometric algebra $\mathcal{G}(\mathcal{N}^{n+1})$ which represent these subspaces.

Let A_r, B_s and C_t be blades representing three subspaces $\mathcal{A}^r$, $\mathcal{B}^s$ and $\mathcal{C}^t$, respectively. We say that

Definition 2 *The t-blade $C_t = A_r \cap B_s$ is the meet of A_r and B_s if and only if C^t is the intersection of the subspaces A^r and B^s,*

$$C^t = A^r \cap B^s.$$

We say that

Definition 3 *The t-blade $C_t = A_r \cup B_s$ is the join of A_r and B_s if and only if C^t is the union of the supspaces A^r and B^s,*

$$C^{t+1} = A^{r+1} \cup B^{s+1}.$$

Suppose that an $(r-1)$-plane in Π^n is reprented by the r-blade

$$A_r = a_1 \wedge a_2 \wedge \ldots \wedge a_r$$

and an $(s-1)$-plane by

$$B_s = b_1 \wedge b_2 \wedge \ldots \wedge b_s.$$

Considering the a's and b's to be the basis elements spanning the respective subspaces A^r and B^s, they can be sorted in such a way that

$$A^r \cup B^s = span\{a_1, a_2, \ldots a_s, b_{\lambda_1}, \ldots, b_{\lambda_k}\},$$

where the λ's are chosen as small as possible and are ordered to satisfy $1 \le \lambda_1 < \lambda_2 < \ldots < \lambda_k \le s$, and

$$A^r \cap B^s = span\{b_{\alpha_1}, \ldots, b_{\alpha_{s-k}}\},$$

where

$$B_s = b_{\lambda_1} \wedge \ldots \wedge b_{\lambda_k} \wedge b_{\alpha_1} \wedge \ldots \wedge b_{\alpha_{s-k}}.$$

It follows that

$$A^r \cup B^s = A^r \wedge b_{\lambda_1} \wedge \ldots \wedge b_{\lambda_k},$$

and

$$A^r \cap B^s = b_{\alpha_1} \wedge \ldots \wedge b_{\alpha_{s-k}}.$$

The problem of "meet" and "join" has thus been solved by finding the union and intersection of linear subspaces and their equivalent $(s+k)$-blade and $(s-k)$-blade representants.

It is important to note that it is only in the special case when $A_r \cap B_s = 0$ that the join reduces to the outer product. That is

$$A_r \cap B_s = 0 \quad \Leftrightarrow \quad A_r \cup B_s = A_r \wedge B_s$$

However, after the join $I_{A_r \cup B_s} \equiv A_r \cup B_s$ has been found, it can be used to find the meet $A_r \cap B_s$. The idea is the same as in the previous section. If $\overline{I}_{A_r \cup B_s}$ is any $(r+s)$-blade in $\mathcal{G}(\overline{\mathcal{N}}^{n+1})$ for which $I_{A_r \cup B_s} \cdot \overline{I}_{A_r \cup B_s} \neq 0$, then

$$A_r \cap B_s = A_r \cdot [B_s \cdot \overline{I}_{A_r \cup B_s}] = [\overline{I}_{A_r \cup B_s} \cdot A_r] \cdot B_s$$

2.3.4   Hermitian inner product

We now come to the delicate subject of complexification. Up until now, we have only considered real geometric algebras and their corresponding real matrices. Any pseudoscalar of the geometric algebra $\mathcal{G}_{n,n}$ will always have a positive square, and will *anticommute* with the vectors in $I\!\!R^{n,n}$. If we insisted on dealing only with *real* geometric algebras, we might consider working in the geometric algebra $\mathcal{G}_{n,n+1}$ where the pseudoscalar element i can be chosen to have the desired $i^2 = -1$. In this algebra, i will *commute* with all the vectors in $I\!\!R^{n,n+1}$, and a *complex vector* $x + iy$ in $\mathcal{G}_{n,n+1}$ would consist of the real vector part x and a pseudovector or $(2n)$-blade part iy.

Instead, we choose to choose to directly *complexify* the geometric algebra $\mathcal{G}_{n,n}$ to get the *complex geometric algebra* $\mathcal{G}_{2n}(\boldsymbol{C})$. Whereas this algebra is isomorphic to $\mathcal{G}_{n,n+1}$, it is somewhat easier to work with than the former. A *complex* vector $z \in \boldsymbol{C}^{2n}$ will have the form $z = x + iy$ where $x, y \in I\!\!R^{2n}$. The imaginary unit i, where $i^2 = -1$, is defined to *commute* with all elements in the geometric algebra $\mathcal{G}_{2n}(\boldsymbol{C})$.

The complexified null cone $\mathcal{N}^n(\boldsymbol{C})$, and reciprocal cone $\overline{\mathcal{N}}^n(\boldsymbol{C})$, are spanned by the complex null vectors

$$e_j = \frac{1}{\sqrt{2}}(\sigma_j + i\sigma_{n+j}) \text{ and } \bar{e}_j = \frac{1}{\sqrt{2}}(\sigma_j - i\sigma_{n+j})$$

for $j = 1, 2, \ldots, n$. A complex null vector $x \in \mathcal{N}^n(\boldsymbol{C})$ has the form $x = \{e\}x_{\{e\}}$, where in this case $x_i \in \boldsymbol{C}$. For what follows, we shall adopt the convention that $x^* \equiv x^*_{\{e\}}\{\bar{e}\}$, so *Hermitian conjugation* is an operation which takes us from the complex null cone $\mathcal{N}^n(\boldsymbol{C})$ to the dual null cone $\overline{\mathcal{N}}^n(\boldsymbol{C})$. Notice that applied to the components $x_{\{e\}}$, $x^*_{\{e\}}$ is the usual Hermitian transpose of the column vector $x_{\{e\}}$,

$$x^*_{\{e\}} = \overline{\left(\begin{array}{cccc} x_1 & x_2 & \cdots & x_n \end{array} \right)} = \left(\begin{array}{c} \bar{x}_1 \\ \bar{x}_2 \\ . \\ . \\ \bar{x}_n \end{array} \right)^T \tag{2.20}$$

We are now ready to define the *Hermitian inner product* $\langle x, y \rangle$ on $\mathcal{N}^n(\boldsymbol{C})$. For all $x, y \in \mathcal{N}^n(\boldsymbol{C})$,

$$\langle x, y \rangle \equiv x^* \cdot y = x^*_{\{e\}} y_{\{e\}}$$

If the components of x and y are all real, the Hermitian conjuation operation reduces to the tranpose operation, defined in (2.10). The Hermitian inner product has all of the usual properties.

2.4   Linear Transformations

Let $\mathcal{N}^n \cup \mathcal{N}^{n'}$ and $\overline{\mathcal{N}}^n \cup \overline{\mathcal{N}}^{n'}$ be $(n+n')$-dimensional reciprocal null cones in $\mathbb{R}^{n+n',n+n'}$ with the dual bases $\{e\} \cup \{e'\}$ and $\{\overline{e}\} \cup \{\overline{e}'\}$. Let $f : \mathcal{N} \to \mathcal{N}'$ be a linear transformation from the null cone $\mathcal{N}$ into the null cone $\mathcal{N}'$. In light of the previous section, we can consider the null cones $\mathcal{N}$ and $\mathcal{N}'$ to be over the real or complex numbers. Let

$$Hom(\mathcal{N}, \mathcal{N}') = \{f : \mathcal{N} \to \mathcal{N}'| \ f \text{ is a linear transformation}\}$$

denote the linear space of all homomorphisms from $\mathcal{N}$ to $\mathcal{N}'$, with the usual operation of addition of transformations. Of course, only when $\mathcal{N} = \mathcal{N}'$ is the operation of multiplication (composition) defined.

Given an operator $f \in Hom(\mathcal{N}, \mathcal{N}')$, $y' = f(x)$, the *matrix* $\mathcal{F}$ of f with respect to the bases $\{e\}$ and $\{e'\}$ is defined by

$$f\{e\} = \{fe\} = \{fe_1, fe_2, \ldots, fe_n\} = \{e'_1, e'_2, \ldots, e'_{n'}\}\mathcal{F} = \{e'\}\mathcal{F} \quad (2.21)$$

Of course, the matrix $\mathcal{F} = (f_{ij})$ is defined by its $n \times n'$ components $f_{ij} = \overline{e}_i \cdot f(e_j) \in \mathcal{C}$ for $i = 1, 2, \ldots, n'$ and $j = 1, 2, \ldots, n$. It follows that $f(e_j) \equiv fe_j = \sum_{i=1}^{n'} e_i f_{ij}$. By dotting both sides of the above equation on the left by $\{\overline{e}'\}$, we find the explicit expression $\mathcal{F} = \{\overline{e}'\} \cdot \{e'\}\mathcal{F} = \{\overline{e}'\} \cdot f\{e\}$.

We wish to pursue this material only far enough to show that traditional linear algebra fits very nicely into the far richer geometric algebra framework, which brings to bear both new geometric insight and new computational tools.

The transpose (or Hermitian transpose (2.20)) $f^*(y'^*)$ of the mapping $y' = f(x)$ is defined by the requirement that for all $x \in \mathcal{N}$ and $y'^* \in \overline{\mathcal{N}}'$,

$$f(x) \cdot y'^* = x \cdot f^*(y'^*) \quad \Leftrightarrow \quad \{\overline{e}'\} \cdot f\{e\} = f^*(\{\overline{e}'\}) \cdot \{e\}.$$

2.4.1   Spectral decomposition of a linear operator

We shall now briefly consider linear operators from the null cone $\mathcal{N}$ into itself. Let

$$End(\mathcal{N}) = \{f : \mathcal{N} \to \mathcal{N}| \ f \text{ is a linear operator}\}$$

denote the *algebra* of all endomorphisms from $\mathcal{N}$ to itself. In this case, the operations of addition and composition of linear operators are well defined. Let $\{e\}$ be a basis of $\mathcal{N}$, and $\{\overline{e}\}$ be the corresponding reciprocal basis of $\overline{\mathcal{N}}$. Then

$$f\{e\} = \{e\}\mathcal{F} \quad \Leftrightarrow \quad \mathcal{F} = \{\overline{e}\} \cdot \{e\}\mathcal{F} = \{\overline{e}\} \cdot f\{e\}$$

gives the *matrix* $\mathcal{F}$ of f with respect to the basis $\{e\}$.

Recall that the *characteristic polynomial* of f is defined by

$$\varphi_f(x) = \det(x - f) = \{\bigwedge_{i=1}^{n}(x - f)\{e\}\} \cdot \overline{I}.$$

By the well-known *Caley-Hamilton theorem*, we know that $\varphi_f(f) = 0$, so that every linear operator satisfies its *characteristic equation*. The *minimal polynomial* $\psi_f(x)$ of f is the polynomial of least degree that has this property. Taken over the complex numbers C, we can express φ_f and ψ_f in the *factored form*

$$\varphi_f(x) = \prod_{i=1}^{r}(x - x_i)^{n_i} \quad \text{and} \quad \psi_f(x) = \prod_{i=1}^{r}(x - x_i)^{m_i}$$

where $1 \le m_i \le n_i \le n$ for $i = 1, 2, \ldots, r$.

The minimal polynomial uniquely determines, up to an ordering of the idempotents, the following *spectral decomposition theorem* of the linear operator f.

Theorem 1 *If f has the minimal polynomial $\psi(x)$, then a set of commuting mutually annihilating idempotents and corresponding nilpotents $\{(p_i, q_i) | i = 1, \ldots, r\}$ can be found such that*

$$f = \sum_{i=1}^{r}(x_i + q_i)p_i,$$

where $rank(p_i) = n_i$, and the index of nilpotency $index(q_i) = m_i$, for $i = 1, 2, \ldots, r$. Furthermore, when $m_i = 1$, $q_i = 0$.

Various forms of the spectral decomposition theorem are known, but they are certainly under-utilized, perhaps because of the clumsey form in which in which they are often presented. In [17, 18], the spectral decomposion theorem is used to derive the Jordan canonical form, and other basic canonical forms of a linear operator.

The spectral decomposition theorem has many different uses. Any function g defined on the spectrum $\{x_1, x_2, \ldots, x_r\}$ of the operator f, can be defined on the operator f,

$$g(f) \equiv \sum_{i=1}^{r} g(x_i + q_i)p_i$$

where

$$g(x_i + q_i) = \sum_{j=0}^{m_i - 1} \frac{f^{(k)}(x_i)}{k!}q_i^j]p_i$$

We say that an operator f is *diagonalizable* if and only if it has the spectral form

$$f = \sum_{i=1}^{r} x_i p_i$$

If f is diagonalizable, then a generalized inverse of f is defined by

$$f^{inv} = \sum_{x_i \neq 0} \frac{1}{x_i} p_i$$

2.4.2 Principal correlation

The idea of a *principal correlation* d of a linear transformation $f : \mathcal{N} \rightarrow \mathcal{N}'$, where $n' = dim(\mathcal{N}') \geq dim(\mathcal{N}) = n$, is basic to the construction of both the generalized *polar decomposition* and *singular value decomposition* of the transformation f. The transformation $d : \mathcal{N} \rightarrow \mathcal{N}'$ is a generalization of a *unitary operator* and becomes a unitary operator when $\mathcal{N} = \mathcal{N}'$ and $rank(f) = n$.

Definition 4 *A transformation d is said to be a principal correlation of a transformation f if $rank(d) = rank(f)$ and*

$$fd^* = df^* \quad and \quad d^*f = f^*d,$$

where $(dd^*)^2 = dd^*$, and $(d^*d)^2 = d^*d$.

The problem remains to show that given $f : \mathcal{N} \rightarrow \mathcal{N}'$, where $n \leq n'$, that a principal correlation will always exist. Assuming that it does exists with the specified properties, we will *solve* for d based upon the existence of the generalized inverse (2.19).

If it does exists, then

$$ff^* = df^*dd^*fd^* = df^*fd^* = (df^*)^2$$

and

$$f^*f = d^*fd^*df^*d = d^*ff^*d = (d^*f)^2$$

It follows that $df^* = \sqrt{ff^*}$ or $df^*f = \sqrt{ff^*}f$. Multiplying both sides of this last equality on right by the generalized inverse $(f^*f)^{inv}$ of f^*f gives the formula for a principal correlation d of f,

$$d = \sqrt{ff^*}f(f^*f)^{inv} \quad \Leftrightarrow \quad d^* = (f^*f)^{inv}f^*\sqrt{ff^*}$$

The fact that d and d^* have the desired properties as given in the definition, is a simple excercise in linear algebra.

In terms of the principal correlation d of f, a generalized inverse may be defined by $f^{inv} = (d^*f)^{inv}d^*$.

2.4.3   The bivector of a linear operator

With the help of the geometric algebra identity (2.5), and (2.12), we can express the vector x in the form

$$x = \{e\}x_{\{e\}} = (\{e\} \wedge \{\overline{e}\}) \cdot x,$$

where $x_{\{e\}}$ are the *column vector* components of x introduced in (2.9),

$$x_{\{e\}} = \begin{pmatrix} x_1 \\ x_2 \\ \cdot \\ \cdot \\ x_n \end{pmatrix} = \begin{pmatrix} \overline{e}_1 \cdot x \\ \overline{e}_2 \cdot x \\ \cdot \\ \cdot \\ \overline{e}_n \cdot x \end{pmatrix} = \{\overline{e}\} \cdot x. \tag{2.22}$$

This is the key idea to the bivector representation of a linear operator.

Let $f \in End(\mathcal{N})$. Then, we have

$$f(x) = f(\{e\}x_{\{e\}}) = \{e\}\mathcal{F}x_{\{e\}} = [(\{e\}\mathcal{F}) \wedge \{\overline{e}\}] \cdot x = F \cdot x \tag{2.23}$$

where the bivector $F \in \mathcal{G}_{n,n}$ is defined by

$$F = (\{e\}\mathcal{F}) \wedge \{\overline{e}\} = \sum_{i=1}^{n} \sum_{j=1}^{n} f_{ij} e_i \wedge \overline{e}_j = \{e\} \wedge (\mathcal{F}\{\overline{e}\})$$

Thus, every linear operator $f \in End(\mathcal{N})$ can be represented in the bivector form $f(x) = F \cdot x$, where $F = (\{e\}\mathcal{F}) \wedge \{\overline{e}\}$ is a bivector in the universal geometric algebra $\mathcal{G}_{n,n}$. The components of $f(x) = F \cdot x$ can be directly recovered from the bivector F,

$$f_{ij} = \overline{e}_i \cdot f(e_j) = F \cdot (e_j \wedge \overline{e}_i).$$

Consider now $f, g \in End(\mathbb{N})$, $f(x) = F \cdot x$ and $g(x) = G \cdot x$. The *commutator* $[f, g]$ of the linear operators f and g is defined by

$$[f, g](x) = (fg - gx)(x) = f(g(x)) - g(f(x)). \tag{2.24}$$

The linear operators in $End(\mathbb{N})$, taken together with the commutator product, make up the *general Lie algebra* $gl(\mathbb{N})$ of the vector space $\mathbb{N}$.

Using the bivector representation of f and g, we find that

$$[f, g](x) = F \cdot (G \cdot x) - G \cdot (F \cdot x) = (F \times G) \cdot x, \tag{2.25}$$

where the *commutator product of bivectors* $F \times G$ is defined by $F \times G = \frac{1}{2}[FG - GF]$, [7, p.14]. Thus the Lie bracket of the linear operators f and g becomes the commutator product of their respective bivectors F and G. The bivectors of all of the linear operators in $End(\mathbb{N})$, taken together with the commutator product, make up the Lie algebra $spin(\mathbb{N})$.

The significance of the Lie algebra $spin(N)$ will be discussed in the next section. We can also express the commutator $[f, g]$ directly in terms of the matrices of f and g. Writing $f\{e\} = \{e\}\mathcal{F}$, and $g\{e\} = \{e\}\mathcal{G}$, we get

$$[f, g]\{e\} = fg\{e\} - gf\{e\} = \{e\}(\mathcal{F}\mathcal{G} - \mathcal{G}\mathcal{F}) = \{e\}[\mathcal{F}, \mathcal{G}], \qquad (2.26)$$

where $[\mathcal{F}, \mathcal{G}]$ is the commutator product of the matrices $\mathcal{F}$ and $\mathcal{G}$.

2.4.4 Spinor representation

The set of all linear operators $f \in End(N)$, such that $\det f \neq 0$, make up the *general linear group* $GL(N)$ with the group operation being the usual composition of linear operators. Expressed in terms of the bivector representations (2.23) of f and g,

$$f \circ g(x) = F \cdot (G \cdot x) = (F : G) \cdot x$$

where the group operation of composition of bivectors $F : G$, the equivalent of multiplying matrices, is defined by

$$(F : G) \equiv \{F \cdot [G \cdot \{e\}]\} \wedge \{\bar{e}\}.$$

In terms of the bivector of a linear operator,

$$\det f \neq 0 \quad \Leftrightarrow \quad \wedge_{i=1}^n F = F \wedge F \wedge \cdots \wedge F \neq 0,$$

but we will not prove this fact.

Each bivector in $F \in spin(N)$ defines a corresponding *one parameter group* element in $gl(N)$. Recall that for $f \in End(\mathcal{N})$, a one parameter group is defined by the exponential mapping $g_t(x) = e^{tf}x$, where the group operation is the composition of linear operator [5, p.115]. We have

$$g_s(g_t x) = (e^{sf} e^{tf})x = e^{(s+t)f}x = g_{s+t}x$$

for all $s, t \in \mathbb{R}$. Using the bivector representation (2.23), $f(x) = F \cdot x$, we have the following important

Theorem 2 *The one parameter group $g_t x = e^{tf}x$ of the skew–symmetric transformation $f(x) = F \cdot x$, can be expressed in the spinor form*

$$g_t x = e^{tf}x \equiv e^{\frac{t}{2}F} x e^{-\frac{t}{2}F} \qquad (2.27)$$

Proof:
    We will prove the theorem by showing that the terms of the Taylor series expansion of both sides of the equation (2.27) are identical at $t = 0$.
    We begin with

$$e^{tf}x \doteq e^{\frac{t}{2}F} x e^{-\frac{t}{2}F} \qquad (2.28)$$

Clearly, for $t = 0$ we have

$$e^{0f}x = e^{0F}xe^{0F} = x$$

Next, taking the first derivative of both sides of (2.28), we get

$$
\begin{aligned}
e^{tf}fx &\doteq \frac{1}{2}Fe^{\frac{t}{2}F}xe^{-\frac{t}{2}F} - \frac{1}{2}Fe^{\frac{t}{2}F}xe^{-\frac{t}{2}F} \\
&= e^{\frac{t}{2}F}(F \cdot x)e^{-\frac{t}{2}F}.
\end{aligned}
\tag{2.29}
$$

Setting $t = 0$ gives the identity $f(x) = F \cdot x$.

Taking the derivative of both sides of (2.29), gives

$$
e^{tf}f^2x \doteq \frac{1}{2}Fe^{\frac{t}{2}F}(F \cdot x)e^{-\frac{t}{2}F} - \frac{1}{2}Fe^{\frac{t}{2}F}(F \cdot x)e^{-\frac{t}{2}F}
$$

$$
= e^{\frac{t}{2}F}[F \cdot (F \cdot x)]e^{-\frac{t}{2}F},
$$

and setting $t = 0$ gives the identity $f^2(x) = F \cdot (F \cdot x)$. Continuing to take successive derivatives of (2.28), gives

$$e^{tf}f^k(x) \doteq e^{\frac{t}{2}F}(F^k : x)e^{-\frac{t}{2}F} \tag{2.30}$$

where $F^k : x$ is defined recursively by $F^1 : x = F \cdot x$ and

$$F^k : x = F \cdot (F^{k-1} : x). \tag{2.31}$$

Finally, setting $t = 0$ in (2.30) gives the identity

$$f^k(x) = F^k : x$$

which completes the proof.

Q.E.D.

The expression (2.31) is interesting because it expresses the powers of a linear operator in terms of "powers" of its defining bivector. It is clear that each bivector defines a unique skew-symmetric linear operator, and conversely, each skew-symmetric linear operator defines a unique bivector, (2.23). Thus, the study of the structure of a bivector is determined by and uniquely determines the corresponding structure of the corresponding linear operator. The the proof of the above theorem is due to Marcel Riesz [15].

2.5   Pseudo-Euclidean Geometries

By construction, the real null cone $\mathcal{N} \subset \mathbb{R}^{n,n}$. Slightly more generally, we define the *complex null cone* $\mathcal{N}_i$,

$$\mathcal{N}_i = \{x + iy|\ x, y \in \mathcal{N}\} \subset \mathcal{G}_{n,n+1},$$

where the pseudoscalar $i \in \mathcal{G}_{n,n+1}$ commutes with all the elements of $\mathcal{G}_{n,n+1}$ and $i^2 = -1$. It is natural to work in the geometric algebra $\mathcal{G}_{n,n+1}$ when working on the null cone over the complex numbers. We have seen in the previous section that the bivectors in $\mathcal{G}_{n,n}^2$ make up the real Lie algebra $spin(n,n)$ when taken with the Lie bracket operation. A *complex bivector* $C \in \mathcal{G}_{n,n+1}$ has the form $C = A + iB$ for $A, B \in \mathcal{G}_{n,n}^2$. The complex bivectors make up the *complex Lie algebra $spin_i(n)$* under the Lie bracket operation.

The null cone $\mathcal{N}$ should be thought of as a *home base*. If we wish to do a rotation in a pseudoeuclidean space $\mathbb{R}^{p,q}$ where $p + q = n$, then we project the null cone $\mathcal{N}$ onto $\mathbb{R}^{p,q}$, perform the rotation using the bivectors of spinor algebra $spin(p,q)$, and then project *back* to the null cone. Suppose that $x = \{e\}x_{\{e\}} = \sum_{i=1}^{n} x_i e_i \in \mathcal{N}$. The projection $P_{p,q} : \mathcal{N} \to \mathbb{R}^{p,q}$ is defined by

$$x' = P_{p,q}(x) = I_{p,q} \cdot (\overline{I_{p,q}} \cdot x) = \sum_{i=1}^{p} x_i \sigma_i + \sum_{j=p+1}^{n} x_j \eta_j \in \mathbb{R}^{p,q}$$

where the reciprocal elements are specified by

$$I_{p,q} = \sigma_1 \ldots \sigma_p \eta_{p+1} \ldots \eta_n \quad \text{and}$$

$$\overline{I}_{p,q} = (2 - \sqrt{2})^n (\bar{e}_n - \eta_n) \ldots (\bar{e}_{p+1} - \eta_{p+1})(\bar{e}_p + \sigma_p) \ldots (\bar{e}_1 + \sigma_1).$$

The inverse projection $P'_{p,q} : \mathbb{R}^{p,q} \to \mathcal{N}$ is defined by

$$x = P'_{p,q}(x') = I'_{p,q} \cdot (\overline{I_{p,q}} \cdot x),$$

where $I'_{p,q} = e_1 e_2 \ldots e_n$ and $\overline{I}_{p,q}$ is defined as before, as can be verified by using CLICAL or by hand.

Note that an (p,q)-orthogonal transformation can be performed directly on $x \in \mathcal{N}$ using either its matrix representation, or its corresponding bivector representation $B \cdot x$.

2.6   Affine and Projective Geometries

Let $\mathcal{N} \in \mathbb{R}^{n,n}$ be the null cone in $\mathbb{R}^{n,n}$. It is often very desirable to extend the theory of linear transformations to include translations. This is most easily accomplished by introducing the concept of the *neutral affine n-plane* $\mathcal{A}_e(\mathcal{N})$ as a subset of the $(n+1)$-dimensional null cone $\mathcal{N}^{n+1} \subset \mathbb{R}^{n+1,n+1}$. We define

$$\mathcal{A}_e(\mathcal{N}) = \{x + e| \ x \in \mathcal{N} \ \} \subset \mathcal{N}^{n+1},$$

where $0 \neq e = e_{n+1} \in \mathcal{N}^{n+1}$. A slightly different, but equivalent definition is

$$\mathcal{A}_e(\mathcal{N}) = \{y| \ y \in \mathcal{N}^{n+1} \text{ and } y \cdot \bar{e} = 1 \ \} \subset \mathcal{N}^{n+1},$$

where $\bar{e} = \bar{e}_{n+1} \in \overline{\mathcal{N}^{n+1}}$ so that $e \cdot \bar{e} = 1$. This second definition is interesting because it brings us closer to the definition of the n-dimensional *projective plane* .

It is well known that the projective n-plane Π^n can be considered to be the set of all *points* determined by the *directions* or *rays* of nonzero vectors $y \in \mathcal{N}^{n+1}$. The projective n-plane Π^n can also be defined to be the set of all points of the affine plane $\mathcal{A}_e(\mathcal{N})$, taken together with idealized points at infinity. Each point $y \in \mathcal{A}_e(\mathcal{N})$ is called a *homogeneous representant* of the corresponding point in Π^n. To bring these different viewpoints closer together, points in the affine plane $\mathcal{A}_e(\mathcal{N})$ can be represented by rays in the space

$$\mathcal{A}_e^{rays}(\mathcal{N}) = \{y| \ y \in \mathcal{N}^{n+1} \text{ and } y \cdot \bar{e}_{n+1} \neq 0 \} \subset \mathcal{N}^{n+1} \qquad (2.32)$$

The set of rays $\mathcal{A}_e^{rays}(\mathcal{N})$ is yet another definition of the neutral affine plane, because each ray $y \in \mathcal{A}_e^{rays}(\mathcal{N})$ determines the unique point

$$\frac{y}{y \cdot \bar{e}_{n+1}} \in \mathcal{A}_e(\mathcal{N}),$$

and conversely, each point $y \in \mathcal{A}_e(\mathcal{N})$ determines a unique ray in $\mathcal{A}_e^{rays}(\mathcal{N})$. Thus, the affine plane of points $\mathcal{A}_e(\mathcal{N})$ is equivalent to the affine plane of rays $\mathcal{A}_e^{rays}(\mathcal{N})$. Some of these issues will be further discussed in the next chapter.

2.6.1 Pseudo-affine geometries

Just as we can move from the null cone $\mathcal{N}$ (embedded in neutral pseudoeuclidean space $\mathbb{R}^{n,n}$) to any of the pseudo-euclidean spaces $\mathbb{R}^{p,q}$ by a projection, by using essentially the same reciprocal projection elements, we can move from the neutral affine geometry $\mathcal{A}_e(\mathcal{N})$ on the null cone $\mathcal{N}^{n+1}$ to the pseudo-affine (p,q)-plane $\mathcal{A}_e(\mathbb{R}^{p,q}) \subset \mathbb{R}^{p+1,q+1}$. The underlying idea is, once again, the same. Certain kinds of transformations can be accomplished more easily in the different pseudo-affine planes. Once the transformation is performed, we can return to "home base" by using the inverse projection.

The projection from the affine plane $\mathcal{A}_e(\mathcal{N}^n)$ to the affine plane $\mathcal{A}_e(\mathbb{R}^{p,q})$ is specified $P_{p,q}(x) = I_{p,q} \cdot (\overline{I_{p,q}} \cdot x)$ where the reciprocal elements are defined by

$$I_{p,q} = \sigma_1 \ldots \sigma_p \eta_{p+1} \ldots \eta_n e_{n+1} \quad \text{and}$$

$$\overline{I}_{p,q} = (2 - \sqrt{2})^n \bar{e}_{n+1} (\bar{e}_n - \eta_n) \ldots (\bar{e}_{p+1} - \eta_{p+1})(\bar{e}_p + \sigma_p) \ldots (\bar{e}_1 + \sigma_1).$$

We can *project* the affine plane $\mathcal{A}_e(\mathbb{R}^{p,q})$ *back* to the affine plane $\mathcal{A}_e(\mathcal{N}^n)$ by replacing the reciprocal element $I_{p,q}$ by

$$I'_{p,q} = e_1 e_2 \ldots e_{n+1}$$

and using the same element $\overline{I}_{p,q}$ defined above. Using these concepts, we will explore the affine geometry of motion in detail in a later chapter.

2.6.2 Desargues' theorem

We will present a simple proof of the classical *Desargues' configuration* , a basic result of Projective Geometry. We do so to emphasize the point that even though geometric algebra is endowed with a metric, there is no reason why we cannot not use the tools of this structure to give a proof of this metric independent result. Indeed, as has been emphasized by Hestenes and others [1], all the results of linear algebra can be supplied with a projective interpretation. Another reason for presenting our proof, is to contrast it with the much more complicated proof that Hestenes has given in [9]. Whereas, it can be argued that Hestenes' proof is more profound, because it expresses the relationship as a duality relationship, it also assumes a much greater algebraic sophistication in its derivation. Also, Hestenes' proof is only valid when both triangles lie in the projective plane Π^2, our proof is equally valid when the triangles lie in Π^3.

Recall that points $a \in \Pi^3$ can be identified with nonzero *rays* $a \in I\!\!R^4$, and two rays a and b in $I\!\!R^4$ represent the same point if and only if $a \wedge b = 0$. Two distinct points $a, b \in \Pi^3$ define the line $a \wedge b \neq 0$, $c \in \Pi^3$ lies on this line (is colinear) if and only if $a \wedge b \wedge c = 0$. Suppose that $a, b, c, d \in \Pi^3$ such that no 3 of them are colinear, but that they are *coplanar* , $a \wedge b \wedge c \wedge d = 0$, then the *meet* of the projectives lines $a \wedge b$ and $c \wedge d$ is the unique point $d \in \Pi^3$ defined by

$$d = (a \wedge b) \cap (c \wedge d) = (a \wedge b) \cdot [(a \wedge b \wedge c) \cdot (c \wedge d)].$$

Refering to Figure 2.1, we are now ready to state and prove

Theorem 3 *(Desargues' Configuration:) Let a_1, a_2, a_3 and b_1, b_2, b_3 be the verticies of two triangles in Π^3, and suppose that*

$$(a_1 \wedge a_2) \cap (b_1 \wedge b_2) = c_3, \ (a_2 \wedge a_3) \cap (b_2 \wedge b_3) = c_1, \ (a_3 \wedge a_1) \cap (b_3 \wedge b_1) = c_2.$$

Then $c_1 \wedge c_2 \wedge c_3 = 0$ if and only if there is a point p such that

$$a_1 \wedge b_1 \wedge p = 0 = a_2 \wedge b_2 \wedge p = a_3 \wedge b_3 \wedge p.$$

Proof:

$$\left. \begin{array}{l} a_1 \wedge b_1 \wedge p = 0 \\ a_2 \wedge b_2 \wedge p = 0 \\ a_3 \wedge b_3 \wedge p = 0 \end{array} \right\} \ \leftrightarrow \ \left\{ \begin{array}{l} p = \alpha_1 a_1 + \beta_1 b_1 \\ p = \alpha_2 a_2 + \beta_2 b_2 \\ p = \alpha_3 a_3 + \beta_3 b_3 \end{array} \right.$$

but this in turn implies that

$$\begin{array}{l} \alpha_1 a_1 - \alpha_2 a_2 = -(\beta_1 b_1 - \beta_2 b_2) = c_3 \\ \alpha_2 a_2 - \alpha_3 a_3 = -(\beta_2 b_2 - \beta_3 b_3) = c_1 \\ \alpha_3 a_3 - \alpha_1 a_1 = -(\beta_3 b_3 - \beta_1 b_1) = c_2 \end{array}$$

Taking the sum of the last three equalities gives $c_1 + c_2 + c_3 = 0$, which implies that $c_1 \wedge c_2 \wedge c_3 = 0$. The other half of the proof follows by duality.

Q.E.D.

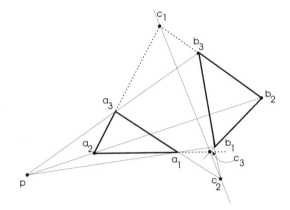

FIGURE 2.1. Desargue's Configuration.

2.6.3   Simpson's theorem for the circle

We have seen in the last section how geometric algebra can be used to prove theorems of Projective Geometry which do not depend on a metric. In this section, we will prove Simpson's theorem for the circle which *depends* upon the metric. We will prove this theorem in the affine plane of rays $\mathcal{A}_e^{rays}(\mathbb{R}^2)$, defined in (2.32). The operations of *meet* and *join* are defined in the affine plane of rays in *almost* the same way that they are in the the projective plane, with a slight modification to take into account that $e^2 = 0$.

For example, suppose that we are give two non-collinear points $a, b \in \mathcal{A}_e^{rays}(\mathbb{R}^2)$, then the line $\mathcal{L}_{ab}$ passing through the points $a, b \in \mathcal{A}_e^{rays}(\mathbb{R}^2)$ is uniquely defined by the 2-direction of the bivector $a \wedge b$. Suppose that we are given a third point $d \in \mathcal{A}_e^{rays}(\mathbb{R}^2)$, as in Figure 2.2, and we are asked to find the point p on the line $\mathcal{L}_{ab}$ such that $\mathcal{L}_{ab}$ is perpendicular to $\mathcal{L}_{pd}$. The point p we are looking for is of the form $p = d + si(a - b)$ for some $s \in \mathbb{R}$ and lies on the line $\mathcal{L}_{pd}$ which is uniquely defined by the bivector

$$p \wedge d = [d + si(a - b)] \wedge d = s[i(a - b)] \wedge d.$$

But the scalar $s \neq 0$ is unimportant since the line is uniquely defined by the *2-direction* of the bivector $p \wedge d$ and not by its *magnitude*. The point $p \in \mathcal{A}_e^{rays}(\mathbb{R}^2)$ is therefore uniquely specified by

$$p = (a \wedge b) \cap \{[i(a - b)] \wedge d\}, \tag{2.33}$$

where $i = \sigma_1 \sigma_2$ is the bivector tangent to $\mathcal{A}_e^{rays}(\mathbb{R}^2)$.

Evaluating (2.33) for the point $p \in \mathcal{A}_e^{rays}(\mathbb{R}^2)$, we find

$$p = (a \wedge b) \cap \{[i(a - b)] \wedge d\} = \{[\bar{e} \wedge (a - b)] \cdot d\} \cdot (a \wedge b)$$

$$= [\bar{e} \wedge (a - b)] \cdot (d \wedge a)\, b - [\bar{e} \wedge (a - b)] \cdot (d \wedge b)\, a$$

$$= (a - b) \cdot (b - d)\, a - (a - b) \cdot (a - d)\, b$$

The *normalized point*

$$p_h = \frac{p}{p \cdot \bar{e}} = -\frac{1}{(a-b)^2}[(a-b) \cdot (b-d)\, a - (a-b) \cdot (a-d)\, b] \quad (2.34)$$

will be in the affine plane $\mathcal{A}_e(\mathbb{R}^2)$.

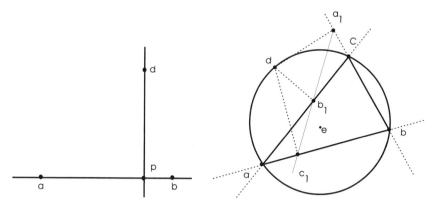

FIGURE 2.2. a) Perpendicular Point on the Line b) Simpson's Theorem for the Circle.

Refering to the Figure 2.2, we are now ready to state and prove

Theorem 4 *(Simpson's theorem for the circle.) Three non-conlinear points $a, b, c \in \mathcal{A}_e^2$ defines a unique circle. A fourth point $d \in \mathcal{A}_e^2$ will lie on this circle if and only if $a_1 \wedge b_1 \wedge c_1 = 0$, where*

$$a_1 = (b \wedge c) \cap \{[i(b-c)] \wedge d\}$$
$$b_1 = (c \wedge a) \cap \{[i(c-a)] \wedge d\}$$
$$c_1 = (a \wedge b) \cap \{[i(a-b)] \wedge d\}$$

Proof. Using the above formula above for evaluating the meets, we find that

$$a_1 = (b-c) \cdot (d-c)\, b - (b-c) \cdot (d-b)\, c$$

$$b_1 = (c-a) \cdot (d-a)\, c - (c-a) \cdot (d-c)\, a$$

and

$$c_1 = (a-b) \cdot (d-b)\, a - (a-b) \cdot (d-a)\, b.$$

These points will be collinear if and only if $a_1 \wedge b_1 \wedge c_1 = 0$, but

$$a_1 \wedge b_1 \wedge c_1 = \{[(b-c) \cdot (d-c)][(c-a) \cdot (d-a)][(a-b) \cdot (d-b)]$$

$$-[(c-a) \cdot (d-c)][(a-b) \cdot (d-a)][(b-c) \cdot (d-b)]\}a \wedge b \wedge c \quad (2.35)$$

Note that the right-hand side of the last equation only involves *differences* of the points a, b, c, d, and that these differences lies in the tangent plane of $\mathcal{A}_e^2$ which is $\mathbb{R}^2$. Without loss of generality, we can assume that the center of the circle is the point e and that the circle has radius ρ. Using normalized points (2.34), it is not difficult to show that

$$a_1^h \wedge b_1^h \wedge c_1^h = \frac{\rho^2 - d^2}{4\rho^2} (a^h \wedge b^h \wedge c^h) \qquad (2.36)$$

$$\Leftrightarrow (b_1 - a_1) \wedge (c_1 - b_1) = \frac{\rho^2 - d^2}{4\rho^2} (b - a) \wedge (c - a) \qquad (2.37)$$

Since the points a^h, b^h, c^h are not colinear, $a^h \wedge b^h \wedge c^h \neq 0$; it follows that $a_1^h \wedge b_1^h \wedge c_1^h = 0$ if and only if $(d^h)^2 = \rho^2$.

Q.E.D.

Whereas the identity (2.37) in the affine plane $\mathcal{A}_e^2$ is not trivial, it is much easier to prove than the corresponding identity used by (1) in his proof of Simpson's theorem in the non-linear horosphere $\mathcal{H}^2$. The issue of distance geometry has also been addressed by [4]. It appears correct to say that it is always easier to carry out calculations in the affine plane, except when conformal transformations are involved.

2.7   Conformal Transformations

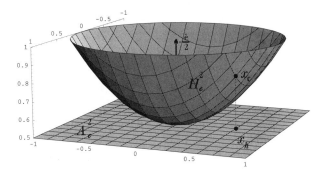

FIGURE 2.3. The affine plane $\mathcal{A}_e^2$ and horosphere $\mathcal{H}_e^2$.

We have already noted that translations can be easily effected in any of the affine planes. There is still another transformation of basic interest, the conformal transformation which preserves angles between tangent vectors [12, 14]. The most familiar example of a conformal transformation is that defined by any analytic function in the complex number plane. There are also conformal transformations which preserve angles of mappings in the pseudoeuclidean space $\mathbb{R}^{p,q}$. In the same way that we *linearized* a translation in $\mathbb{R}^{p,q}$ by moving up to the affine plane $\mathcal{A}_e(\mathcal{N})$, we can linearize a conformal transformation in $I\!\!R^{p,q}$ by moving up still further from the affine plane $\mathcal{A}_e(\mathcal{N})$ to the (p,q)-*horosphere* $\mathcal{H}_e^{p,q}$ in $I\!\!R^{p+1,q+1}$, (see, Chapter 3). The (p,q)-horosphere is most easily defined by

$$\mathcal{H}_e^{p,q} = \{\frac{1}{2}x_h \bar{e} x_h | \ x_h \in \mathcal{A}_e(I\!\!R^{p,q})\} \subset I\!\!R^{p+1,q+1},$$

where $e = e_{n+1}$ and $\bar{e} = \bar{e}_{n+1}$.

Let us call the point $x_c \in \mathcal{H}_e^{p,q}$,

$$x_c = \frac{1}{2}x_h \bar{e} x_h = \frac{1}{2}[(x_h \cdot \bar{e})x_h + (x_h \wedge \bar{e})x_h] = x_h - \frac{1}{2}x_h^2 \bar{e}$$

$$= \exp(\frac{1}{2}x\bar{e})e\exp(-\frac{1}{2}x\bar{e}) \tag{2.38}$$

the *conformal representant* of both the point $x_h = x + e \in \mathcal{A}_e(I\!\!R^{p,q})$ and $x \in I\!\!R^{p,q}$. Note that given x_c, it is easy to get back x_h by the simple projection,

$$x_h = (x_c \wedge \bar{e}) \cdot e$$

or to $x \in I\!\!R^{p,q}$,

$$x = (x_c \wedge \bar{e} \wedge e) \cdot (\bar{e} \wedge e).$$

The expression of the conformal representant x_c in the form (2.38) is interesting because it shows that all points on $\mathcal{H}_e^{p,q}$ can be obtained by a simple rotation of e in the plane of $x\bar{e}$. The affine plane $\mathcal{A}_e^2$ and horosphere $\mathcal{H}_e^2$ are pictured in Figure 2.3. The horosphere will be discussed in more detail in later chapters.

Acknowledgments

Garret Sobczyk gratefully acknowledges the support of INIP of the Universidad de Las Americas - Puebla, and CIMAT during his Sabbatical, Fall 1999.

Chapter 3

Realizations of the Conformal Group

Jose Maria Pozo and Garret Sobczyk

3.1 Introduction

Perhaps one of the first to consider the problems of projective geometry was Leonardo da Vinci (1452-1519). However, projective geometry as a self-contained discipline was not developed until the work "Traité des propriés projectives des figure" of the French mathematician Poncelet (1788-1867), published in 1822. The extrordinary generality and simplicity of projective geometry led the English mathematician Cayley to exclaim: "Projective Geometry is all of geometry" [16]. D. Hestenes in [8] showed how the methods of projective geometry, formulated in geometric algebra, can be effectively used to study basic properties of the conformal group. The purpose of this article is to further explore the deep relationships that exist beween projective geometry and the conformal group.

In section 2, we review some of the basic ideas of projective geometry that are needed in this work, and relate the projective plane to the affine plane by defining what we mean by a *point observer* in a higher dimensional space. In section 3, we define a *line observer* and show how it leads to the concept of a *conformal representant*, which is closely related to Hestenes' idea of a *conformal split* [8]. We also give the relationship between the conformal representant and the more familiar concept of *stereographic projection*. The non-linear model of a pseudoeuclidean space, called the *horosphere* is briefly discussed. The horosphere has recently attracted the attention of many workers, see for example, [4, 15, 6].

In section 4, we give a simple proof, using only basic concepts from differential geometry developed in [7], of the deep result relating conformal transformations in a pseudoeuclidean space to isometries in a pseudoeuclidean space of two higher dimensions. The original proof of this striking relationship was given by [5]. In section 5, we show that for any dimension greater than two, that any isometry on the *null cone* can be extended to all of the pseudoeuclidean space. In section 6, we discuss the interesting issue of the *conformal compactification* of the horosphere for the various signatures of the underlying pseudoeuclidean space $\mathbb{R}^{p,q}$, [15].

In the final section of the paper, we show the beautiful relationship that exists between Mobius transformations (linear fractional transformations) and their 2×2 matrix representation over a suitable geometric algebra.

3.2 Projective Geometry

Let $I\!\!R^{p,q}$ be a real n-dimensional pseudoeuclidean vector space over the real numbers $I\!\!R$, and suppose that the sums of geometric products of the vectors in $I\!\!R^{p,q}$ generate the geometric algebra $\mathcal{G}_{p,q} \equiv gen\{I\!\!R^{p,q}\}$, where $n = p + q$. A point $x \in I\!\!R^{p,q}$ is named by the vector x from the origin $0 \in I\!\!R^{p,q}$. Now let $A \in \mathcal{G}_{p,q}^{k}$ be any nonzero simple k-vector, or k-blade. By the k-direction of A in $I\!\!R^{p,q}$, we mean the k-dimensional subspace of $I\!\!R^{p,q}$ defined by

$$\mathcal{S}_A \equiv \{x \in I\!\!R^{p,q} \mid x {\wedge} A = 0\}.$$

When $k = 1$, we also say that $\mathcal{S}_a$ is a *line* or *ray* through the origin with the direction of $a \in I\!\!R^{p,q}$. The notation and geometric algebra identities used in this paper are from [1, p.37-43] and [7].

It is well-known that *points* in the *projective space* Π^{n-1} can be identified with rays in $I\!\!R^{p,q}$; thus we can write $\mathcal{S}_a \in \Pi^{n-1}$. Likewise, the 2-blade $A = a {\wedge} b \in \mathcal{G}_{p,q}^{2}$ determines the *projective line* $\mathcal{S}_A \subset \Pi^{n-1}$ which passes through the projective points $\mathcal{S}_a, \mathcal{S}_b \in \Pi^{n-1}$. In general, each k-direction in $\mathcal{G}_{p,q}^{k}$ determines a unique $(k-1)$-plane in Π^{n-1}. Note that two nonzero k-blades $A, B \in \mathcal{G}_{p,q}^{k}$ determine the same k-direction in $I\!\!R^{p,q}$,

$$\mathcal{S}_A = \mathcal{S}_B \quad \text{iff} \quad A = sB \text{ for some } s \in I\!\!R^{*},$$

where $I\!\!R^{*} \equiv I\!\!R - \{0\}$. We say that A and B are *representants* of the same k-direction in $\mathcal{G}_{p,q}^{k}$. This is the strict meaning of the word *representant* which will be used throughout this article. Hestenes and Ziegler in [9] have shown how the basic definitions and theorems of projective geometry can be efficiently formulated and proved in geometric algebra by the reinterpretation of the elements of geometric algebra given above. More details of the projective interpretation of geometric algebra can be found in that article, but they will not be needed here.

Let us now extend the pseudoeuclidean space $I\!\!R^{p,q}$ to the higher dimensional pseudoeuclidean space $I\!\!R^{p+1,q} = span\{I\!\!R^{p,q}, \sigma\}$ by introducing an orthonormal unit vector σ with the properties that $\sigma^2 = 1$ and σ is orthogonal to (*anticommutes with*) all the vectors $x \in I\!\!R^{p,q}$, i.e., $x\sigma = -\sigma x$. It follows that $\mathcal{G}_{p,q}$ is a subalgebra of $\mathcal{G}_{p+1,q} = gen(I\!\!R^{p+1,q})$.

By a *point observer* in $I\!\!R^{p+1,q}$, we mean any nonzero *fixed* point $a \in I\!\!R^{p+1,q}$ from which each point in $x \in I\!\!R^{p,q}$ is observed by a ray $\mathcal{S}_{x-a}$ passing through the points a and x. Choosing $a = -\sigma$ as our fixed point observer, the point x is observed by the ray $\mathcal{S}_{x-a} = \mathcal{S}_{x+\sigma}$. The vector $x_h = x + \sigma$ will be called the *homogeneous representant* of the ray $\mathcal{S}_{x_h}$. The homogeneous representant of a ray $\mathcal{S}_z$ is the unique representant $y \in \mathcal{S}_z$ which satisfies the condition that $y \cdot \sigma = 1$. Each vector $y \in I\!\!R^{p+1,q}$ is a representant of some point $x_h = x + \sigma \in I\!\!R^{p+1,q}$ if and only if $y \cdot \sigma \neq 0$. Conversely, given any representant $y \in \mathcal{S}_{x_h}$, the homogeneous representant is obtained from the equation $x_h = \frac{y}{y \cdot \sigma}$. The set of all homogeneous representants

$$\mathcal{A}_\sigma^{p,q} = \{x_h = x + \sigma | \ x \in I\!\!R^{p,q}\}$$

is the $p + q$-dimensional affine hyperplane in $I\!\!R^{p+1,q}$ which is orthogonal to σ and passes through the point σ.

In the Figure 3.1, it is *visually* evident that each point $x \in I\!\!R^{p,q}$ determines and is determined by a unique ray $\mathcal{S}_{x_h}$ in $I\!\!R^{p+1,q}$. Similarly, each 2-direction $\mathcal{S}_{x_h \wedge y_h}$ determines and is determined by a unique line passing through the points x and y in $I\!\!R^{p,q}$. In the same way, the plane defined by three points x, y and z in $I\!\!R^{p,q}$ determines and is determined by the unique 3-direction $\mathcal{S}_{x_h \wedge y_h \wedge z_h}$ in $I\!\!R^{p+1,q}$, and so on.

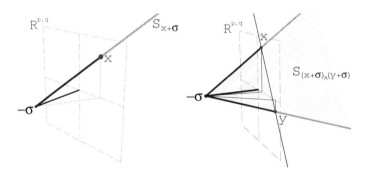

FIGURE 3.1. Points and lines in projective geometry.

3.3   The Conformal Representant and Stereographic Projection

In the previous section, we have seen how the methods of projective and affine geometry can be used in the study of $I\!\!R^{p,q}$ by choosing a point observer $a = -\sigma$ in a larger space $I\!\!R^{p+1,q}$ which contains $I\!\!R^{p,q}$ as a proper subspace. We now introduce the concept of *double projective geometry*, by choosing a second point observer $b = -\nu$ in a still larger space $I\!\!R^{p+1,q+1}$. The signature of the new orthonormal vector ν is chosen according to the requirements of the application. The appropriate signature for the study of conformal transformations is $\nu^2 = -1$. Thus, two point observers $a = -\sigma$

and $b = -\nu$ in $\mathbb{R}^{p+1,q+1}$ are chosen to satisfy

$$\sigma^2 = 1, \ \nu^2 = -1, \ \sigma \cdot \nu = \sigma \cdot x = \nu \cdot x = 0,$$

for all $x \in \mathbb{R}^{p,q}$. It follows that $\mathbb{R}^{p+1,q+1} = span\{\mathbb{R}^{p,q}, \sigma, \nu\}$.

The first observer a sees the second observer b as the unique line determined by the null direction $\bar{e} = b - a = \sigma - \nu$. Recall that for a *simple projection*, the point observer a sees each point $x \in \mathbb{R}^{p,q}$ as the ray $\mathcal{S}_{x-a}$. The *double projective line observer* $\bar{e} = b - a$ "sees" each point $x \in \mathbb{R}^{p,q}$ as the 2-direction

$$\mathcal{S}_{(b-a)\wedge(x-a)} = \mathcal{S}_{\bar{e}\wedge(x+\sigma)}.$$

Thus the line observer makes each point $x \in \mathbb{R}^{p,q}$ correspond to the unique 2-direction containing the line observer and intersecting $\mathbb{R}^{p,q}$ at the point x. See Figure 3.2.

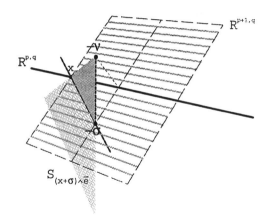

FIGURE 3.2. The double projective 2-direction of a point $x \in \mathbb{R}^{p,q}$.

Let $\mathcal{G}_{p+1,q+1} = gen(\mathbb{R}^{p+1,q+1})$ be the geometric algebra of $\mathbb{R}^{p+1,q+1}$. We will also need the *hyperbolic unit bivector* $u = \sigma\nu$, $u^2 = 1$, and the *null vectors* $e = \frac{1}{2}(\sigma + \nu)$ and $\bar{e} = \sigma - \nu$. We have the following simple but important relationships

$$e^2 = \bar{e}^2 = 0, \ e \cdot \bar{e} = 1, \ u = \bar{e}\wedge e = \sigma\wedge\nu, \ u^2 = 1 \tag{3.1}$$

The bivector $K_x = \bar{e}\wedge(x + \sigma)$ of the 2-direction $\mathcal{S}_{\bar{e}\wedge(x+\sigma)}$ determined by x is a unit hyperbolic bivector,

$$K_x^2 = (\bar{e}\wedge(x + \sigma))^2 = (\bar{e}\wedge x + \bar{e}\wedge\sigma)^2 = u^2 = 1,$$

so it uniquely determines two null directions, one of which is $\mathcal{S}_{\bar{e}}$. The second null vector x_c of the null direction $\mathcal{S}_{x_c}$ can be found by factoring the bivector K_x into $K_x = \bar{e} \wedge x_c$. This second null vector is easily found to be

$$x_c = x - \frac{1}{2}x^2 \bar{e} + e \in \mathbb{R}^{p+1,q+1}$$

and is called the *conformal representant* of the point $x \in \mathbb{R}^{p,q}$.

The conformal representant of the null ray $\mathcal{S}_z$ is characterized as the representant $y \in \mathcal{S}_z$ which satisfies $y \cdot \bar{e} = 1$. The set of all null vectors $0 \neq y \in \mathbb{R}^{p+1,q+1}$ make up the *null cone* $\mathbb{N} = \{y \in \mathbb{R}^{p+1,q+1} |\ y^2 = 0\}$. The subset of $\mathbb{N}$ containing all the representants $y \in \mathcal{S}_{x_c}$ for any $x \in \mathbb{R}^{p,q}$ is defined to be the set

$$\mathbb{N}_0 = \{y \in \mathbb{N} \mid y \cdot \bar{e} \neq 0\} = \cup_{x \in \mathbb{R}^{p,q}} \mathcal{S}_{x_c}$$

Let us summarize what we have accomplished. Each point $x \in \mathbb{R}^{p,q}$, when viewed from the point observer $-\sigma$, uniquely determines the ray $\mathcal{S}_{x_h}$ of the homogeneous representant $x_h = x + \sigma$ in $\mathbb{R}^{p+1,q}$. When a second point observer $-\nu$ is chosen in the still larger space $\mathbb{R}^{p+1,q+1}$, then each point $x \in \mathbb{R}^{p,q}$ determines a unique hyperbolic plane with the direction of the hyperbolic unit bivector $K_x = \bar{e} \wedge x_h$. Factoring K_x into the outer product of two null vectors, $K_x = \bar{e} \wedge x_c$ gives the conformal representant $x_c \in \mathbb{N}_0$ of the unique null ray $\mathcal{S}_{x_c} \in \mathbb{N}_0/\mathbb{R}^*$ corresponding to the point $x \in \mathbb{R}^{p,q}$.

The conformal representant $x_c = x - \frac{1}{2}x^2 \bar{e} + e \in \mathbb{R}^{p+1,q+1}$ has many nice properties. First, it is easily obtained from any other representant; for any $y \in \mathcal{S}_{x_c}$,

$$x_c = \frac{y}{y \cdot \bar{e}}.$$

Second, we can easily recover x from x_c by projecting x_c into $\mathbb{R}^{p,q}$,

$$x = (x_c \wedge u)u.$$

Third, the mapping $x_c :\ \mathbb{R}^{p,q} \hookrightarrow \mathbb{N}_0 \subset \mathbb{R}^{p+1,q+1}$ is continuous and infinitely differentiable (indeed, it's third differential vanishes), and it is also an isometric embedding.

$$dx_c = dx - x \cdot dx\bar{e} \quad \Rightarrow \quad (dx_c)^2 = (dx)^2 \qquad (3.2)$$

Finally, the points $x \in \mathbb{R}^{p,q}$ are represented by null rays $\mathcal{S}_{x_c}$ in $\mathbb{N}_0 \subset \mathbb{R}^{p+1,q+1}$, that is by the points of $\mathbb{N}_0/\mathbb{R}^*$.

The set of all conformal representants $\mathcal{H}^{p,q} = c(\mathbb{R}^{p,q})$ make up a non-linear model of the pseudoeuclidean space $\mathbb{R}^{p,q}$ called the *pseudo-horosphere*. The pseudoscalar I_{x_c} of the *tangent space* to $\mathcal{H}^{p,q}$ at the point x_c, determined by (3.2), is given by

$$I_{x_c} = IK_x = I\ \bar{e} \wedge x_c \qquad (3.3)$$

where $I = \sigma_1 \ldots \sigma_{p+1}\nu_1 \ldots \nu_{q+1}$ is the unit pseudoscalar of $\mathbb{R}^{p+1,q+1}$. The horosphere $\mathcal{H}^n$ for the Euclidean space $\mathbb{R}^n$ was first introduced by F.A. Wachter, a student of Gauss, [4], [6] and in chapter 1. The pseudohorosphere $\mathcal{H}^{p,q}$ is pictured in the Figure 3.3 crossing the circles and is a subset of the null cone N_0.

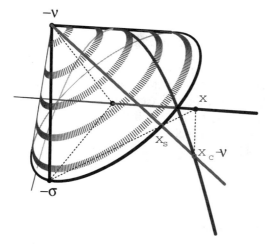

FIGURE 3.3. Conformal representant and stereographic projection.

The conformal representant x_c can be written in the form

$$x_c = x - \frac{1}{2}x^2 \bar{e} + e = \frac{1}{2}(x + \sigma)\bar{e}(x + \sigma) = \frac{1}{2}x_h \bar{e} x_h, \qquad (3.4)$$

which shows explicitly that x_c is null because it is obtained by a reflection and a dilation of the null vector $\bar{e}$. Since a dilation does not change the direction $\mathcal{S}_{x_c}$, another representant x_r can be defined by the reflexion,

$$x_r = (x + \sigma)^{-1}\bar{e}(x + \sigma) = (x + \sigma)^{-1}\sigma(x + \sigma) + \nu , \qquad \mathcal{S}_{x_c} = \mathcal{S}_{x_r}.$$

The first term $x_s = (x + \sigma)^{-1}\sigma(x + \sigma)$ of the last expression is the usual stereographic projection and is defined for the space $\mathbb{R}^{p,q}$ of any signature. The representant x_r is the image of the stereographic projection x_s as observed by the second point observer $-\nu$, $x_r = x_s - (-\nu)$, so clearly

$$\mathcal{S}_{x_r} = \mathcal{S}_{x_s + \nu}.$$

Evidently, both this representant and the stereographic projection, are not defined and are discontinuous when $(x + \sigma)$ does not have an inverse, in the hypersurface defined by $x^2 = -1$. Therefore, they have discontinuities only for non-euclidian spaces.

3.4   Conformal Transformations and Isometries

In this section we will show that every conformal transformation in $I\!\!R^{p,q}$ corresponds to two isometries on the null cone $I\!\!N_0$ in $I\!\!R^{p+1,q+1}$.

Definition 1 *A conformal transformation in $I\!\!R^{p,q}$ is any twicely differentiable mapping between two connected open subsets U and V,*

$$f : U \longrightarrow V, \quad x \longmapsto x' = f(x)$$

such that the metric changes by only a conformal factor

$$(df(x))^2 = \lambda(x)(dx)^2, \qquad \lambda(x) \neq 0.$$

If $p \neq q$ then $\lambda(x) > 0$. In the case $p = q$, there exists the posibility that $\lambda(x) < 0$, when the conformal transformations belong to two disjoint subsets. We will only consider the case when $\lambda(x) > 0$.

Recall that $I\!\!N_0 \equiv \{y \in I\!\!N \mid y \cdot \bar{e} \neq 0\}$ is the subset of $I\!\!N$ that contains all of the *nonzero* points on each of the rays $\mathcal{S}_{x_c}$ of the conformal representants x_c. Thus, each point $y \in I\!\!N_0$ can be coordinized by the parameter $t \in I\!\!R^*$ and $x \in I\!\!R^{p,q}$, by writing

$$y = y(x,t) = tx_c = t(x - \frac{1}{2}x^2\bar{e} + e).$$

Taking differentials, we immediately find

$$dy = dtx_c + tdx_c = dt\frac{y}{t} + t(dx - x \cdot dx \ \bar{e}).$$

Note that $x_c^2 = 0$ implies that $dx_c \cdot x_c = 0$, from which it easily follows with the help of (3.2) that

$$(dy)^2 = t^2(dx_c)^2 = t^2(dx)^2. \qquad (3.5)$$

Definition 1.1 *An isometry F on $I\!\!N_0$ is any twicely differentiable mapping between two connected open subsets U_0 and V_0 in the relative topology of $I\!\!N_0$,*

$$F : U_0 \longrightarrow V_0, \quad y \longmapsto y' = F(y)$$

which satisfies $(dF(y))^2 = (dy)^2$.

Using the "coordinates" $y(x,t) = tx_c$, any mapping in $I\!\!N_0$ can be expressed in the form

$$y' = F(y) = t'x_c' = \phi(x,t)f(x,t)_c$$

where $t' = \phi(x,t)$ and $x_c' = f(x,t)_c$ are defined implicitly by F. Using (3.5), we obtain the result that $y' = F(y)$ is an isometry if and only if

$$(dy')^2 = (dy)^2 \ \leftrightarrow \ t'^2(dx')^2 = t^2(dx)^2 \ \text{ or } \ (df(x,t))^2 = \frac{t^2}{\phi(x,t)^2}(dx)^2.$$

Since $f(x,t), x \in \mathbb{R}^{p,q}$ (non degenerate metric), and the right hand side of this equation does not contain dt, it follows that $f(x,t) = f(x)$ is independent of t. It then follows that $\phi(x) = \frac{\phi(x,t)}{t}$ is also indendent of t. Thus, we can express any isometry $y' = F(y)$ in the form $y' = t\phi(x)f(x)_c$, where $f(x)_c \in \mathbb{N}_0$ is the conformal representant of $f(x) \in \mathbb{R}^{p,q}$. This implies that $y' = F(y)$ is an isometry iff

$$y' = t\phi(x)f(x)_c \quad \text{and} \quad (df(x))^2 = (\phi(x))^{-2}(dx)^2.$$

Therefore, $f(x)$ is a conformal transformation with

$$\lambda(x) = \phi(x)^{-2} > 0 \;\leftrightarrow\; \phi(x) = \pm\frac{1}{\sqrt{\lambda(x)}}.$$

Since the functions $\phi(x)$ and $f(x)$ are independent of t and $F(y)$ is linear in t, we can always extend the open subsets U_0 and V_0 of $\mathbb{N}_0$ to t independent open subsets

$$U_0' = \cup_{y \in U_0} S_y, \quad \text{and} \quad V_0' = \cup_{y \in V_0} S_y.$$

The domain U is defined by U_0' as the preimage of the mapping S_{x_c}.

$$U = \{\; x \in \mathbb{R}^{p,q} \mid S_{x_c} \subset U_0'\} = \{\frac{y \wedge u}{y \cdot \overline{e}}u | y \in U_0\}.$$

To sum up, we have obtained the following results:

- Any isometry in $\mathbb{N}_0$ defines a unique conformal transformation in $\mathbb{R}^{p,q}$.

- Any conformal transformation $x' = f(x)$ in $\mathbb{R}^{p,q}$ defines two unique isometries in $\mathbb{N}_0$ given by

$$F(y) = \pm\frac{t}{\sqrt{\lambda(x)}}f(x)_c$$

3.5   Isometries in $\mathbb{N}_0$

In this section we will show that, for dimension greater than 2, any isometry in $\mathbb{N}_0$ is the restriction of an isometry in $\mathbb{R}^{p+1,q+1}$. The inverse of the statement is obvious. From the definition of an isometry, $(dF(x))^2 = (dy)^2$. Since $dF(y)$ and dy are vectors in $\mathbb{R}^{p+1,q+1}$, $dF(y)$ can be obtained as the result of applying a field of orthogonal transformations to dy,

$$dF(y) = R(y)dyR(y)^{*^{-1}}$$

where $R(y) \in Pin_{p+1,q+1} = \{X = a_1a_2 \cdots a_n \in \mathcal{G}_{p+1,q+1} \mid a_i^2 = \pm 1\}.$

Since $y^2 = 0$ for $y \in N_0$, it follows that

$$0 = y \cdot dy = tx_c \cdot (dtx_c + tdx_c) = t^2 x_c.dx_c.$$

This implies that $x_c \cdot v = 0$ for all vectors v in the tangent space of N_0 at the point y. The $(n+1)$-pseudoscalar I_y of the tangent space to N_0 at the point y can be defined by $I_y = Ix_c$ where I is the pseudoscalar of $\mathbb{R}^{p+1,q+1}$, given after equation (3.3). We have

$$x_c \cdot v = 0 \quad \Leftrightarrow \quad 0 = I(x_c \cdot v) = (Ix_c) \wedge v = I_y \wedge v \qquad (3.6)$$

The fact that the tangent space has dimension $n+1$ and a metrically degenerate null direction x_c is sufficient to guarantee that the image of $dF(y)$ defines a unique orthogonal transformation in $\mathbb{R}^{p+1,q+1}$, which determines (up to a sign) the versor $R(y)$. Note that $R(y)^{*-1} = \pm R(y)^\dagger$, where R^* and $R^\dagger$ denote the *main involution* and the *reversion* respectively.

To better manipulate the vectors in the tangent space of the null cone N_0, let us define the vector e_0, with $e_0^2 = 1$, as the direction of $\mathbb{R}$ orthogonal to $\mathbb{R}^{p,q}$ in the chart $\mathbb{R}^{p,q} \oplus \mathbb{R}$. In doing so, we have transformed $\mathbb{R}^{p,q} \oplus \mathbb{R}$ into $\mathbb{R}^{p+1,q}$ so that $dy(\mathbb{R}^{p+1,q})$ becomes the *tangent space* of N_0. We will also use the notation

$$\underline{a} = a + \alpha e_0 \quad \text{with} \quad a \in \mathbb{R}^{p,q} \quad \text{and} \quad \underline{a} \in \mathbb{R}^{p+1,q}.$$

Also, let

$$\partial \equiv \partial_x + e_0 \partial_t \quad \text{so that} \quad \underline{a} \cdot \partial = a \cdot \partial_x + \alpha \partial_t$$

With this notation,

$$dy(\underline{a}) = \underline{a} \cdot \partial \, tx_c = (\alpha \partial_t + a \cdot \partial_x) \, tx_c = \alpha x_c + t(a - a \cdot x \, \bar{e})$$

and the expression $dF = R \, dy \, R^{*-1}$ takes the form

$$\underline{a} \cdot \partial F \equiv dF(\underline{a}) = Rdy(\underline{a})R^{*-1} \equiv R\underline{a} \cdot \partial y R^{*-1}$$

Previously, we found that any isometry $F(y) = t\phi(x)f(x)_c$ in N_0 is linear in the scalar coordinate t. Taking the exterior derivative, we get

$$dF(y) \quad = \quad \tfrac{dt}{t} F(y) \quad + \quad td\left(\phi(x)f(x)_c\right)$$

or

$$R(y)dyR(y)^{*-1} \quad = \quad \tfrac{dt}{t} R(y)yR(y)^{*-1} \quad + \quad tR(y)dx_c R(y)^{*-1} \, ,$$

from which it follows that $F(y) = R(y)yR(y)^{*-1}$ and $R(y)$ is independent of t. Our aim now is to show that $R(y) = R(x) = R$ is also independent of x so that $F(y) = RyR^{*-1}$ will be a global orthogonal transformation in $N_0 \subset \mathbb{R}^{p+1,q+1}$.

Let us first impose the integrability condition that the second exterior differential ddF must vanish.

$$0 = ddF(\underline{a}\wedge\underline{b}) = \frac{1}{2}\left(\underline{a}\cdot\partial(Rdy(\underline{b})R^{*-1}) - \underline{b}\cdot\partial(Rdy(\underline{a})R^{*-1})\right)$$

$$= \frac{1}{2}\left(dR(\underline{a})dy(\underline{b})R^{*-1} + Rdy(\underline{b})dR^{*-1}(\underline{a})\right.$$

$$\left. -dR(\underline{b})dy(\underline{a})R^{*-1} - Rdy(\underline{a})dR^{*-1}(\underline{b})\right)$$

$$= \frac{1}{2}R\left(R^{-1}dR(\underline{a})dy(\underline{b}) - dy(\underline{b})R^{-1}dR(\underline{a})\right.$$

$$\left. - R^{-1}dR(\underline{b})dy(\underline{a}) + dy(\underline{a})R^{-1}dR(\underline{b})\right)R^{*-1}$$

We define $\Omega(\underline{a}) \equiv 2R^{-1}dR(\underline{a})$, which is a linear function of just $a \in \mathbb{R}^{p,q}$, since $R = R(x)$ is independent of t. It follows that

$$\Omega(e_0) = 0 , \quad \text{and} \quad \Omega(\underline{a}) = \Omega(a).$$

Thus, we get

$$\frac{1}{2}R\left(\Omega(\underline{a}) \times dy(\underline{b}) - \Omega(\underline{b}) \times dy(\underline{a})\right)R^{*-1} =$$

$$= \frac{1}{2}R\left(\Omega(\underline{a})\cdot dy(\underline{b}) - \Omega(\underline{b})\cdot dy(\underline{a})\right)R^{*-1}$$

$$= R\Omega\cdot dyR^{*-1}(\underline{a}\wedge\underline{b}) = 0 \quad \Rightarrow \Omega\cdot dy(\underline{a}\wedge\underline{b}) = 0. \ (3.7)$$

Equation (3.7) can then be separated into two parts,

$$\Omega\cdot dy(\underline{a}\wedge\underline{b}) = 0 \Rightarrow \begin{cases} 2\Omega\cdot dy(a\wedge b) = \Omega(a)\cdot dy(b) - \Omega(b)\cdot dy(a) = 0 \\ 2\Omega\cdot dy(a\wedge e_0) = \Omega(a)\cdot dy(e_0) = \Omega(a)\cdot x_c = 0 \end{cases} \ (3.8)$$

Secondly, we take take the *exterior derivative* of $\Omega = 2R(x)^{-1}dR(x)$, to find the integrability condition,

$$d\Omega = 2dR^{-1}dR = 2dR^{-1}RR^{-1}dR = -\frac{1}{2}\left(2R^{-1}dR\right)\left(2R^{-1}dR\right)$$

or

$$d\Omega(a\wedge b) = -\frac{1}{2}\Omega^2(a\wedge b) = -\frac{1}{2}\Omega(a) \times \Omega(b)$$

$$\Rightarrow d\Omega(x) + \frac{1}{2}\Omega(a) \times \Omega(b) = 0. \quad (3.9)$$

Since $\Omega(x, a) \cdot x_c = 0$, it follows by (3.6) that it is a bivector in the tangent space of $\mathbb{N}_0$ at the point $y = tx_c$. Thus, it can be written in the form

$$\Omega(x,a) = v(x,a) \wedge x_c + B(x,a)$$

where $v(x,a)$ is a vector in the tangent space of the horosphere $c(I\!\!R^{p,q})$, $v(x,a) \in dx_c(I\!\!R^{p,q})$, and $B(x,a)$ is a bivector over the same tangent space $B(x,a) \in dx_c(\mathcal{G}_{p,q}^2) = dx_c(I\!\!R^{p,q}) \wedge dx_c(I\!\!R^{p,q})$. From now on, we will not write the dependence on the position x : $\Omega(a) \equiv \Omega(x,a)$. We will also write $h(a) \equiv dx_c(a)$.

Imposing the first equation (3.8) we get

$$\Omega(a) \cdot dx_c(b) - \Omega(b) \cdot dx_c(a) = 0 \Rightarrow \begin{cases} v(a) \cdot h(b) - v(b) \cdot h(a) = 0 \\ B(a) \cdot h(b) - B(b) \cdot h(a) = 0 \Rightarrow \end{cases}$$

$$\Rightarrow B(a) \cdot \Big(h(b) \wedge h(c) \Big) = B(b) \cdot \Big(h(a) \wedge h(c) \Big) \Rightarrow B(a) = 0 \quad \forall a \in I\!\!R^{p,q}$$

$$\Rightarrow \Omega(a) = v(a) \wedge x_c$$

Imposing the second equation (3.9) we get

$$0 = d\Omega + \frac{1}{2}\Omega\Omega = d\Omega \Rightarrow \begin{cases} dv(a \wedge b) \wedge x_c = 0 \\ v(a) \wedge h(b) = v(b) \wedge h(a) \end{cases}$$

This last equation differentiates between the cases when dimension $p + q \geq 3$ or when $p + q < 3$. Wedging both sides of this equation with $h(a)$ gives

$$\left. \begin{array}{c} \Rightarrow v(a) \wedge h(a) \wedge h(b) = 0 \ \forall a, b \in I\!\!R^{p,q} \\ p + q \geq 3 \end{array} \right\} \Rightarrow v(a) \wedge h(a) = 0 \Rightarrow v(a) = \rho \, h(a)$$

It follows that

$$\rho \, h(a) \wedge h(b) = \rho \, h(b) \wedge h(a) \Rightarrow \rho = 0 \Rightarrow v(a) = 0 \quad \forall a \in I\!\!R^{p,q}$$

Therefore, R(y) is constant

$$\Omega(y) = 0 \Rightarrow dR(y) = 0 \Rightarrow R(y) = R = \text{constant}$$

Thus, $F(y)$ is a global orthogonal transformation in $I\!\!R^{p+1,q+1}$.

$$F(y) = R y R^{*-1} \qquad \qquad R \in Pin_{p+1,q+1}$$

Since group of isometries in $I\!\!N_0$ is a double covering of the group of Conformal transformations $Con_{p,q}$ in $I\!\!R^{p,q}$, and the group $Pin_{p+1,q+1}$ is a double covering of the group of orthogonal transformations $O(p+1, q+1)$, it follows that $Pin_{p+1,q+1}$ is a four-fold covering of $Con_{p,q}$.

For dimension $p + q = 2$, the equations can also be solved. The resulting expressions for $F(y)$ and $\Omega(y)$ can be found explicitly in terms of analytic and antianalytic functions in the case of complex numbers (with $i^2 = -1$ for $I\!\!R^{2,0}$ and $I\!\!R^{0,2}$, or $i^2 = 1$ for $I\!\!R^{1,1}$). The expression of $\Omega(y)$ involves the Schwarzian derivative [3, p.47],[10], and its geometric interpretation is currently under investigation.

3.6   Compactification

We have seen how the space $\mathbb{R}^{p,q}$ can be isometrically embedded as the hypersurface $c(\mathbb{R}^{p,q})$ in the higher dimensional space $\mathbb{R}^{p+1,q+1}$ by taking the conformal representant $c(x) = x_c$ of each of its points $x \in \mathbb{R}^{p,q}$. However, projective geometry is involved since we are identifying points $x \in \mathbb{R}^{p,q}$ with the corresponding *rays* $\mathcal{S}_{x_c}$ (or points) of the projective space $\mathbb{N}_0/\mathbb{R}^* \subset \Pi^{p+q+1}$. Whereas the limit

$$\lim_{x \to \infty} x_c = \lim_{x \to \infty} (x - \frac{1}{2}x^2\bar{e} + e)$$

always diverges in $\mathbb{R}^{p+1,q+1}$, as a limit of *directions* $\lim_{x \to \infty} \mathcal{S}_{x_c}$ may well exist in Π^{p+q}. The issue of *conformal compactification* has been discussed in [15].

Given $0 \neq v \in \mathbb{R}^{p,q}$, a reasonable condition for a sequence $\{x_n\} \in \mathbb{R}^{p,q}$ to have a *defined direction* v_∞ at infinity is that

$$\lim_{n \to \infty} x_n = v_\infty \quad \text{iff} \quad \left\{ \begin{array}{l} \lim_{n \to \infty} |x_n \cdot b| = \infty \text{ for some } b \in \mathbb{R}^{p,q} \\ \lim_{n \to \infty} \mathcal{S}_{x_n} = \mathcal{S}_v \end{array} \right\}$$

For any function $f : \mathbb{R}^{p,q} \to \mathcal{A}$, where $\mathcal{A}$ is a Haussdorf space, we write

$$\lim_{x \to v_\infty} f(x) = w \in \mathcal{A}$$

if $\lim_{n \to \infty} f(x_n) = w$ for any sequence $\{x_n\}$ in $\mathbb{R}^{p,q}$ such that $\lim_{n \to \infty} x_n = v_\infty$.

We will now study $\lim_{x \to v_\infty} \mathcal{S}_{x_c}$ where $x_c = c(x) \in \mathbb{R}_{p+1,q+1}$ is the conformal representant. Note that

$$x \mapsto \mathcal{S}_{c(x)} \equiv \mathcal{S}(x_c) \quad \text{and} \quad \mathcal{S}_c : \mathbb{R}^{p,q} \to \mathbb{N}_0/\mathbb{R}^* \subset \mathbb{N}/\mathbb{R}^* \subset \Pi^{p+q+1},$$

Our objective is to *compactify* $c(\mathbb{R}^{p,q})$ in $\mathbb{N}$, getting

$$\overline{\mathcal{S}(c(\mathbb{R}^{p,q}))} = \overline{\mathbb{N}_0/\mathbb{R}^*} = \mathbb{N}/\mathbb{R}^*$$

by adding all null directions in $\mathbb{N}$ which are not in $\mathbb{N}_0$. There are two cases to consider, when $v^2 \neq 0$, and when $v^2 = 0$.

Let $v, x \in \mathbb{R}^{p,q}$, and $v^2 \neq 0$. We have the two identities

$$x = \frac{x \cdot v}{v^2}v + \frac{x \wedge v}{v^2}v = \frac{x \cdot v}{v^2}(1 + \frac{x \wedge v}{x \cdot v})v,$$

and

$$x^2 = \frac{1}{v^2}(x \cdot v + x \wedge v)(x \cdot v - x \wedge v) = \frac{(x \cdot v)^2}{v^2}[1 - (\frac{x \wedge v}{x \cdot v})^2].$$

It follows from the first identity that $\lim_{x \to v_\infty} x = v_\infty$ if and only if

$$x \cdot v \to \infty, \quad \text{and} \quad \frac{x \wedge v}{x \cdot v} \to 0 \ .$$

These two conditions imply, with the help of the second identity, that

$$\lim_{x \to v_\infty} \frac{x^2}{(x \cdot v)^2} = \frac{1}{v^2}.$$

We now easily find that

$$\lim_{x \to v_\infty} x_c = \lim_{x \to v_\infty} (x - \frac{1}{2}x^2\bar{e} + e)$$

$$= \lim_{x \to v_\infty} \frac{(x \cdot v)^2}{v^2} \left[\frac{x \cdot v + x \wedge v}{(x \cdot v)^2} v - \frac{1}{2} \frac{x^2 v^2}{(x \cdot v)^2}\bar{e} + \frac{v^2}{(x \cdot v)^2}e \right] = \bar{e}_\infty,$$

so $\mathcal{S}_{x_c} \to \mathcal{S}_{\bar{e}}$ whenever $x \to v_\infty$ and $v^2 \neq 0$.

The analysis of the case $x \to v_\infty$ for nonzero $v \in I\!\!R^{p,q}$ with $v^2 = 0$ is more difficult. For such a v, since the pseudoeuclidean space $I\!\!R^{p,q}$ is nondegenerate, we can always find an $\bar{v} \in I\!\!R^{p,q}$ with the property that $\bar{v}^2 = 0$ and $\bar{v} \cdot v = 1$. Then, for any $x \in I\!\!R^{p,q}$, we have

$$x = (x \cdot \bar{v})v + (x \wedge v) \cdot \bar{v} = (x \cdot \bar{v})\left[v + \frac{(x \wedge v) \cdot \bar{v}}{(x \cdot \bar{v})}\right],$$

and

$$x^2 = (x \cdot \bar{v})^2 \left[\frac{2(x \wedge v) \cdot (\bar{v} \wedge v)}{x \cdot \bar{v}} + \frac{[(x \wedge v) \cdot \bar{v}]^2}{(x \cdot \bar{v})^2} \right].$$

From the first of these identities, it follows that $\lim_{x \to v_\infty} x = v_\infty$ if and only if

$$x \cdot \bar{v} \to \infty, \quad \text{and} \quad \frac{(x \wedge v) \cdot \bar{v}}{x \cdot \bar{v}} \to 0 .$$

These two conditions imply, with the help of the second identity, that

$$\lim_{x \to v_\infty} \frac{x^2}{(x \cdot \bar{v})^2} = 0.$$

However,

$$\lim_{x \to v_\infty} \frac{x^2}{x \cdot \bar{v}}$$

is indeterminant, as follows by considering the sequence $\{x_n\}$ for $x_n = nv + \frac{1}{2}\beta\bar{v}$, where $\beta \in I\!\!R$. For this sequence, we find that

$$\lim_{n \to \infty} \frac{x_n^2}{x_n \cdot \bar{v}} = \beta.$$

Indeed, the three possibilities for this limit are

$$\lim_{x \to v_\infty} \frac{x^2}{x \cdot \bar{v}} = \left\{ \begin{array}{l} \beta \in I\!\!R, \\ \pm\infty, \\ \text{doesn't exist} \end{array} \right\}.$$

We will now evaluate $\lim_{x \to v_\infty} \mathcal{S}_{x_c}$ for each of the three possibilities above. We find

$$\lim_{x \to v_\infty} \mathcal{S}_{x_c} = \lim_{x \to v_\infty} \mathcal{S}(x - \frac{1}{2}x^2\bar{e} + e)$$

$$= \lim_{x \to v_\infty} \mathcal{S}\left(x \cdot \bar{v}\left[v + \frac{(x \wedge v) \cdot \bar{v}}{x \cdot \bar{v}} - \frac{1}{2}\frac{x^2}{x \cdot \bar{v}}\bar{e} + \frac{e}{x \cdot \bar{v}}\right]\right)$$

$$= \lim_{x \to v_\infty} \mathcal{S}\left(v - \frac{1}{2}\frac{x^2}{x \cdot \bar{v}}\bar{e}\right) = \left\{ \begin{array}{c} \mathcal{S}_{v - \beta\bar{e}} \\ \mathcal{S}_{\bar{e}} \\ \text{doesn't exit} \end{array} \right\} \quad \text{respectively.}$$

Since each point $y \in \mathbb{N}$, such that $y \notin \mathbb{N}_0$, is of the form $y = v + \beta\bar{e}$ we have succeeded in showing that each point in $\mathbb{N}$ which is not in $\mathbb{N}_0$ is the limit point of a sequence $\{c(x_n)\}$ in $\mathbb{N}_0$, so

$$\overline{\mathcal{S}(c(\mathbb{R}^{p,q}))} = \overline{\mathbb{N}_0/\mathbb{R}^*} = \mathbb{N}/\mathbb{R}^*$$

3.7   Mobius Transformations

The *conformal split* of $y \in \mathbb{R}^{p+1,q+1}$ is made with respect to the bivector $u = \sigma\nu$,

$$y = (yu)u = (y \cdot u + y \wedge u)u = (y \cdot u + \mathbf{y})u,$$

where $\mathbf{y} \equiv y \wedge u$. It was introduced by Hestenes in [8] in his study of conformal transformations and has the nice property that the *relative components* of y with respect to u commute, that is

$$(y \cdot u)(y \wedge u) = (y \wedge u)(y \cdot u),$$

for all $y \in \mathbb{R}^{p+1,q+1}$ as is easily verified.

The conformal split has the disadvantage that in dealing with the relative geometric algebra $\mathbf{G_{p,q}}$ of the relative pseudoeuclidean space

$$\mathbf{R^{p,q}} = \{\mathbf{y} = y \wedge u | \, y \in \mathbb{R}^{p+1,q+1}\},$$

new inner and outer products must be introduced in $\mathbf{G_{p,q}}$ that differ from the inner and outer products in $G_{p+1,q+1}$. For this reason, we choose instead to deal directly with the subalgebra $\mathbb{R}^{p,q}$ of $\mathbb{R}^{p+1,q+1}$ as discussed in Section 2. Recall from Section 2 that

$$\mathbb{R}^{p+1,q+1} = span\{\mathbb{R}^{p,q}, \sigma, \nu\} = \mathbb{R}^{p,q} \oplus \mathbb{R}^{1,1}$$

where $\mathbb{R}^{1,1} = span\{\sigma, \nu\}$. The geometric algebras of $\mathbb{R}^{p,q}$ and $\mathbb{R}^{1,1}$ are defined by

$$\mathcal{G}_{p,q} = gen\{\mathbb{R}^{p,q}\} \quad \text{and} \quad \mathcal{G}_{1,1} = gen\{\mathbb{R}^{1,1}\}.$$

We now show how any element $y \in \mathbb{R}^{p+1,q+1}$ can be represented by a 2×2 matrix over the module $\mathcal{G}_{p,q}$. To do so, we define the *idempotents* $u_\pm = \frac{1}{2}(1 \pm u)$ satisfying $u_\pm^2 = u_\pm$, and note the additional defining algebraic relationships

$$u_+ + u_- = 1, \ u_+ - u_- = u, \ u_+ u_- = 0 = u_- u_+, \ \sigma u_+ = u_- \sigma,$$

and

$$u\bar{e} = \bar{e} = -\bar{e}u, \ eu = e = -ue, \ \sigma u_+ = e, \ 2\sigma u_- = \bar{e}.$$

Since any $y \in \mathbb{R}^{p+1,q+1}$ can be expressed in the form $y = x + \alpha e + \beta \bar{e}$ for $\alpha, \beta \in \mathbb{R}$, it follows by using these relationships that

$$y = xu_+ + \alpha e + \beta \bar{e} + xu_- = xu_+ + 2\beta u_+ \sigma + \alpha u_- \sigma + xu_-. \quad (3.10)$$

The 2×2 matrix form of y follows directly from (3.10). We have

$$y = \begin{pmatrix} 1 & \sigma \end{pmatrix} u_+ \begin{pmatrix} x & 2\beta \\ \alpha & -x \end{pmatrix} \begin{pmatrix} 1 \\ \sigma \end{pmatrix}, \quad (3.11)$$

as can be easily verified by employing ordinary matrix multiplication of non-commutative elements, and the algebraic relationships given above. We define the matrix $[y]$ of y to be the matrix

$$[y] = \begin{pmatrix} x & 2\beta \\ \alpha & -x \end{pmatrix}.$$

By the *isomorphism* proved below, it follows that any element $G \in \mathcal{G}_{p+1,q+1}$, being a product and linear combination of vectors in $\mathbb{R}^{p+1,q+1}$, can be written in the form

$$G = Au_+ + Bu_+ \sigma + C^* u_- \sigma + D^* u_-,$$

for $A, B, C, D \in \mathcal{G}_{p,q}$, and where $Z^* \equiv \sigma Z \sigma$ is the main involution in the algebra $\mathcal{G}_{p,q}$.

The matrix form of G is specified by

$$G = Au_+ + Bu_+ \sigma + C^* u_- \sigma + D^* u_- = \begin{pmatrix} 1 & \sigma \end{pmatrix} u_+ \begin{pmatrix} A & B \\ C & D \end{pmatrix} \begin{pmatrix} 1 \\ \sigma \end{pmatrix}$$

For

$$G_i = \begin{pmatrix} 1 & \sigma \end{pmatrix} u_+ \begin{pmatrix} A_i & B_i \\ C_i & D_i \end{pmatrix} \begin{pmatrix} 1 \\ \sigma \end{pmatrix},$$

we calculate the product $G_1 G_2$ as follows

$$G_1 G_2 = \begin{pmatrix} 1 & \sigma \end{pmatrix} u_+ \begin{pmatrix} A_1 & B_1 \\ C_1 & D_1 \end{pmatrix} \begin{pmatrix} 1 \\ \sigma \end{pmatrix} \begin{pmatrix} 1 & \sigma \end{pmatrix} u_+ \begin{pmatrix} A_2 & B_2 \\ C_2 & D_2 \end{pmatrix} \begin{pmatrix} 1 \\ \sigma \end{pmatrix},$$

$$= (1 \quad \sigma) \begin{pmatrix} A_1 & B_1 \\ C_1 & D_1 \end{pmatrix} u_+ \begin{pmatrix} 1 & \sigma \\ \sigma & 1 \end{pmatrix} u_+ \begin{pmatrix} A_2 & B_2 \\ C_2 & D_2 \end{pmatrix} \begin{pmatrix} 1 \\ \sigma \end{pmatrix},$$

$$= (1 \quad \sigma) u_+ \begin{pmatrix} A_1 & B_1 \\ C_1 & D_1 \end{pmatrix} \begin{pmatrix} A_2 & B_2 \\ C_2 & D_2 \end{pmatrix} \begin{pmatrix} 1 \\ \sigma \end{pmatrix},$$

which proves the isomorphism $[G_1 G_2] = [G_1][G_2]$. Note in the steps above that the idempotent u_+ commutes with the elements in the algebra $\mathcal{G}_{p,q}$. This shows that the geometric algebra $G_{p+1,q+1} = G_{p,q} \otimes G_{1,1}$ can be represented by 2×2 matrices over the geometric algebra $\mathcal{G}_{p,q}$, even through the elements in the algebras $\mathcal{G}_{1,1}$ and $\mathcal{G}_{p,q}$ don't commute. Whereas all the rules of matrix multiplication remain valid over the relative algebra $\mathcal{G}_{p,q}$, it must be rememebered that multiplication in $\mathcal{G}_{p,q}$ is generally non-commutative.

We will show how the matrix form, defined above, can be used to define the most general conformal transformation in $I\!\!R^{p,q}$ as a linear fractional transformation of elements of the geometric algebra $G_{p,q}$. This beautiful result is based on the isomorphism between $Pin_{p+1,q+1}$ and $Conf(p,q)$, established in the Section 3. Using the matrix form (3.11) of the conformal representant x_c given in (3.4), we get

$$x_c = x - \frac{1}{2} x^2 \bar{e} + e = (1 \quad \sigma) u_+ \begin{pmatrix} x & -x^2 \\ 1 & -x \end{pmatrix} \begin{pmatrix} 1 \\ \sigma \end{pmatrix}$$

from which it follows that

$$[x_c] = \begin{pmatrix} x & -x^2 \\ 1 & -x \end{pmatrix} = \begin{pmatrix} x \\ 1 \end{pmatrix} (1 \quad -x).$$

The last equality on the right gives the factored *spinor form* of x_c and is very interesting because we can write any other spinor factorization in the form

$$[x_c] = \begin{pmatrix} x \\ 1 \end{pmatrix} (1 \quad -x) = \begin{pmatrix} x \\ 1 \end{pmatrix} H^{-1} H (1 \quad -x)$$

$$= \begin{pmatrix} x H^{-1} \\ H^{-1} \end{pmatrix} (H \quad -Hx),$$

where H is any invertible element in $\mathcal{G}_{p,q}$.

Each element R in the versor group $Pin(p+1,q+1)$ defines an orthogonal transformation $y' = Ry(R^*)^{-1}$ for $y \in I\!\!R^{p+1,q+1}$, where $R = v_1 v_2 \ldots v_r$ is a product of invertible vectors in $I\!\!R^{p+1,q+1}$, and $R^* \equiv (-1)^r R$ is the main involution of the element R. We will also need to define $R^\dagger$ by *reversing* the order of the products of vectors in R. Thus, $R^\dagger = v_r v_{r-1} \ldots v_2 v_1$. When the orthogonal transformation, defined by R, is applied to the conformal representant x_c of the point $x \in I\!\!R^{p,q}$, it induces a conformal transformation in $I\!\!R^{p,q}$. Since $\beta_R \equiv RR^\dagger = \pm 1$ is a scalar, we can write this induced transformation in the form $x'_c = \alpha_R R x_c R^\dagger$, where the scalar $\alpha_R \equiv \alpha_R(x) \neq$

0 is chosen so that $x'_c \in \mathbb{R}^{p+1,q+1}$ is the conformal representant of a corresponding point $x' \in \mathbb{R}^{p,q}$, i.e., $x'_c \cdot \bar{e} = 1$.

Taking the matrix representation, we can express this transformation in the relative space $\mathbb{R}^{p,q}$ as $x' = g(x)$ in

$$[x'_c] = \begin{pmatrix} x' & -x'^2 \\ 1 & -x' \end{pmatrix} = \alpha_R[R]x_c[R^\dagger] = \alpha_R \begin{pmatrix} g(x) & -g(x)^2 \\ 1 & -g(x) \end{pmatrix}.$$

Now suppose for $R \in Pin\{p+1, q+1\}$, that $[R] = \begin{pmatrix} A & B \\ C & D \end{pmatrix}$, so that

$$R = \begin{pmatrix} 1 & \sigma \end{pmatrix} u_+ \begin{pmatrix} A & B \\ C & D \end{pmatrix} \begin{pmatrix} 1 \\ \sigma \end{pmatrix} = Au_+ + Bu_+\sigma + C^*u_-\sigma + D^*u_-,$$

from which it follows that

$$\begin{aligned} R^\dagger &= u_+ D^{*\dagger} + u_+\sigma B^\dagger + u_-\sigma C^{*\dagger} + u_- A^\dagger \\ &= D^{*\dagger}u_+ + B^{*\dagger}u_+\sigma + (C^{*\dagger})^*u_-\sigma + (A^{*\dagger})^*u_- \\ &= \begin{pmatrix} 1 & \sigma \end{pmatrix} u_+ \begin{pmatrix} D^{*\dagger} & B^{*\dagger} \\ C^{*\dagger} & A^{*\dagger} \end{pmatrix} \begin{pmatrix} 1 \\ \sigma \end{pmatrix}. \end{aligned}$$

This leads us to define the *transpose-like* operation

$$\begin{pmatrix} A & B \\ C & D \end{pmatrix}^\dagger = \begin{pmatrix} D^{*\dagger} & B^{*\dagger} \\ C^{*\dagger} & A^{*\dagger} \end{pmatrix}.$$

Note that $\dagger$ is the operation of *reversal* in $\mathcal{G}^{p+1,q+1}$, as well as in the subalgebra $\mathcal{G}^{p,q}$.

Taking advantage of the factored spinor form of x_c, we now calculate

$$\begin{aligned} [x'_c] &= \begin{pmatrix} x' \\ 1 \end{pmatrix} \begin{pmatrix} 1 & -x' \end{pmatrix} \\ &= \alpha_R \begin{pmatrix} A & B \\ C & D \end{pmatrix} \begin{pmatrix} x \\ 1 \end{pmatrix} \begin{pmatrix} 1 & -x \end{pmatrix} \begin{pmatrix} D^{*\dagger} & B^{*\dagger} \\ C^{*\dagger} & A^{*\dagger} \end{pmatrix} \\ &= \begin{pmatrix} Ax + B \\ Cx + D \end{pmatrix} H^{-1} H \begin{pmatrix} \alpha_R(D^{*\dagger} - xC^{*\dagger}) & \alpha_R(B^{*\dagger} - xA^{*\dagger}) \end{pmatrix} \end{aligned}$$

We can now choose $H = Cx + D$ to give the relationships

$$\frac{1}{\alpha_R} = (Cx + D)(D^{*\dagger} - xC^{*\dagger}), \quad \text{or} \quad D^{*\dagger} - xC^{*\dagger} = \frac{1}{\alpha_R(Cx + D)}.$$

and the desired linear fractional form

$$x' = g(x) = (Ax + B)(Cx + D)^{-1},$$

of the conformal transformation $x' = g(x)$ in $\mathcal{G}_{p,q}$. The linear fraction form, or *Mobius transformation* has been studied by many authors, [13, 11, 12], [8], [14],[2], and [15], to name only a few.

From the factored spinor form $[x_c] = \begin{pmatrix} x \\ 1 \end{pmatrix} \begin{pmatrix} 1 & -x \end{pmatrix}$, we easily calculate

$$[dx_c] = \begin{pmatrix} dx \\ 0 \end{pmatrix} \begin{pmatrix} 1 & -x \end{pmatrix} + \begin{pmatrix} x \\ 1 \end{pmatrix} \begin{pmatrix} 0 & -dx \end{pmatrix},$$

and

$$
\begin{aligned}
[x_c dx_c] &= \begin{pmatrix} x \\ 1 \end{pmatrix} \begin{pmatrix} 1 & -x \end{pmatrix} \begin{pmatrix} dx \\ 0 \end{pmatrix} \begin{pmatrix} 1 & -x \end{pmatrix} \\
&+ \begin{pmatrix} x \\ 1 \end{pmatrix} \begin{pmatrix} 1 & -x \end{pmatrix} \begin{pmatrix} x \\ 1 \end{pmatrix} \begin{pmatrix} 0 & -dx \end{pmatrix} \\
&= \begin{pmatrix} x \\ 1 \end{pmatrix} dx \begin{pmatrix} 1 & -x \end{pmatrix}.
\end{aligned}
\tag{3.12}
$$

We will use this last relationship to verify that $x' = g(x)$ is conformal.

Since $[x'_c] = \alpha_R[R][x_c][R^\dagger]$, we find that

$$[dx'_c] = d\alpha_g[R][x_c][R^\dagger] + \alpha_R[R][dx_c][R^\dagger],$$

and

$$[x'_c dx'_c] = \alpha_R^2 \beta_R[R][x_c dx_c][R^\dagger].$$

We can now easily calculate

$$[x'_c dx'_c] = \alpha_R^2 \beta_R[R] \begin{pmatrix} x \\ 1 \end{pmatrix} dx \begin{pmatrix} 1 & -x \end{pmatrix} [R^\dagger]$$

$$= \alpha_R^2 \beta_R \begin{pmatrix} Ax + B \\ Cx + D \end{pmatrix} dx \begin{pmatrix} D^\dagger - xC^\dagger & B^\dagger - xA^\dagger \end{pmatrix}$$

$$= \alpha_R^2 \beta_R \begin{pmatrix} x' \\ 1 \end{pmatrix} (Cx + D)dx \begin{pmatrix} D^\dagger - xC^\dagger & 1 - x^\dagger \end{pmatrix}$$

A little more work, using (3.12), gives the desired relationship

$$dx' = \alpha_R^2 \beta_R(Cx + D)dx(D^\dagger - xC^\dagger) = \beta_R(D^\dagger - xC^\dagger)^{-1}dx(Cx + D)^{-1}.$$

Squaring this last identity gives $(dx')^2 = \alpha_R^2(dx)^2$, which shows that $x' = g(x)$ is conformal.

Acknowledgments

José Pozo acknowledges the support of the Spanish Ministry of Education (MEC), grant AP96-52209390, the project PB96-0384, and the Catalan Physics Society (IEC).

Garret Sobczyk gratefully acknowledges the support of INIP of the Universidad de Las Americas-Puebla, and CIMAT-Guanajuato during his Sabbatical, Fall 1999.

Chapter 4

Hyperbolic Geometry

Hongbo Li

4.1 Introduction

Hyperbolic geometry is an important branch of mathematics and physics. For hyperbolic n-space, there are five important analytic models: the Poincaré ball model, the Poinca
ré half-space model, the Klein ball model, the hemisphere model and the hyperboloid model. The hyperboloid model is defined to be one branch $\mathcal{H}^n$ of the set

$$\{x \in \mathcal{R}^{n,1} | x \cdot x = -1\}.$$

Every model has its advantages and disadvantages. In hyperbolic geometry, some typical geometric entities are points, tangent directions, straight lines (geodesics), planes, circles, spheres, the distance between two points, and the angle between two intersecting lines. Lorentz transformations are typical geometric transformations. Compared with other models, the hyperboloid model has the following features in representing these geometric entities and transformations:

- The model is isotropic in that at every point of $\mathcal{H}^n$ the metric of the tangent space is the same.

- A straight line AB is the intersection of $\mathcal{H}^n$ with the plane determined by vectors A, B and the origin of $\mathcal{R}^{n,1}$. When viewed from the origin, it can be identified with a projective line in $\mathcal{P}^n$.

  Similarly, an r-plane in $\mathcal{H}^n$ can be identified with a projective r-plane in $\mathcal{P}^n$, where $0 \le r \le n-1$.

  These identifications enable us to study r-planes in the framework of linear subspaces of $\mathcal{R}^{n,1}$.

Academy of Mathematics and Systems Science, Chinese Academy of Sciences, Beijing 100080, P. R. China. This paper is written during the author's visit to the Institut für Informatik und Praktische Mathematik, Christian-Albrechts-Universität, D-24105 Kiel, Germany. It is supported partially by the DFG Foundations and the AvH Foundations of Germany, the Grant NKBRSF of China, the Hundred People Program of the Chinese Academy of Sciences, and the Qiu Shi Science and Technology Foundations of Hong Kong.

- The tangent direction of a line l at a point A is a vector orthogonal to the vector A in the plane determined by l and the origin of $\mathcal{R}^{n,1}$.

  The angle between two intersecting lines is the Euclidean angle between their tangent directions at the intersection. This is the conformal property of the model.

- Let A, B be two points, and let $d(A, B)$ be their hyperbolic distance. Then $A \cdot B = -\cosh d(A, B)$.

  This reduces a geometric problem of distances to an algebraic problem involving the inner product.

- A generalized circle is either a hyperbolic circle, or a horocycle, or a hypercycle (equidistant curve). A generalized circle is the intersection of $\mathcal{H}^n$ with an affine plane in $\mathcal{R}^{n,1}$.

  Similarly, a generalized r-sphere is the intersection of $\mathcal{H}^n$ with an affine $(r + 1)$-plane.

  This enables us to study generalized r-spheres in the framework of affine $(r + 1)$-planes in $\mathcal{R}^{n,1}$.

- Hyperbolic isometries are orthogonal transformations in $\mathcal{R}^{n,1}$ which leave $\mathcal{H}^n$ invariant. In particular, they are all linear transformations.

- The model is closely related to the model of an n-sphere in $\mathcal{R}^{n+1}$.

These features make it natural to apply Clifford algebra in hyperbolic geometry, just as Clifford algebra was applied to projective geometry (Hestenes and Ziegler, 1991) and spherical geometry (Hestenes, 1987). Some applications of Clifford algebra in hyperbolic 3-space can be found in (Iversen, 1992).

In this chapter, we present some of the results of our research on hyperbolic geometry with Clifford algebra. In the first section, we discuss our work on hyperbolic plane geometry with Clifford algebra (Li, 1997). In the second section we deal with hyperbolic conformal geometry with Clifford algebra (Li, Hestenes and Rockwood, 1999c). In the third section we discuss a universal model for conformal geometries of Euclidean, spherical and double-hyperbolic spaces (Li, Hestenes and Rockwood, 1999a, b, c). We show that with Clifford algebra we can not only reformulate old results with improvements and generalizations, but also discover new theorems.

4.2 Hyperbolic Plane Geometry with Clifford Algebra

We are concerned here with generalized triangles and convex polygons. The concept of a generalized triangle is a natural extension of the concept of a hyperbolic triangle. It naturally includes right-angled pentagons and right-angled hexagons (Fenchel, 1989). This extension is possible because algebraically these geometric objects have the same representation. Convex polygons correspond to polygons in Euclidean geometry. Using the spinor representation, we are able to extend the classical result on representing the area of a triangle in terms of the lengths of its three sides (Greenberg, 1980), to a nice formula which represents the area of a convex n-polygon in terms of the lengths of its sides.

4.2.1 Generalized triangles

Definition 4.1. A generalized point is either a point, or a point at infinity (end), or an imaginary point (tangent direction). A point at infinity is a one-dimensional null subspace of $\mathcal{R}^{n,1}$; an imaginary point is a one-dimensional Euclidean subspace of $\mathcal{R}^{n,1}$.

Algebraically, a point at infinity can be represented by a null vector (vector of zero square); an imaginary point can be represented by a unit vector (vector of square one), see Figure 4.1.

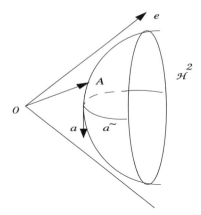

FIGURE 4.1. Generalized points: A is a point, e is a point at infinity and a is an imaginary point. $a^{\sim}$ represents the straight line of $\mathcal{H}^2$ normal to a.

Definition 4.2. A generalized triangle is composed of three non-collinear generalized points and the three lines connecting them, assuming that the lines exist.

There are all together 16 different kinds of generalized triangles, as shown in Figure 4.2 and Figure 4.3.

Below we assume that the dimension of the hyperbolic space is 2.

Corollary 4.1. Let A, B, C be three generalized points. Then they form a generalized triangle if and only if $A \wedge B \wedge C \neq 0$, $(A \cdot B)(B \cdot C)(C \cdot A) \neq 0$ and the three blades $A \wedge B, B \wedge C, A \wedge C$ are all Minkowski.

We can easily recognize that $(A \wedge B \wedge C)^{\sim}$ is the magnitude (Greenberg, 1980) of triangle ABC, when A, B, C are points. What is the geometric meaning of $(A \cdot B)(B \cdot C)(C \cdot A)$? We shall see that its sign characterizes the convexity of generalized triangle ABC.

Definition 4.3. A generalized triangle is said to be convex if any two of its three sides are on the same side of the third side.

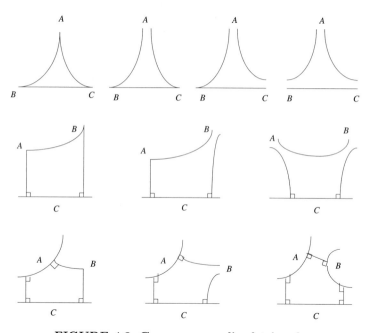

FIGURE 4.2. Convex generalized triangles.

Theorem 4.1. Let ABC be a generalized triangle. Then it is convex if and only if $(A \cdot B)(B \cdot C)(C \cdot A) < 0$.

In Euclidean plane geometry, we have right-angled triangles. In hyperbolic plane geometry we have a similar concept.

Definition 4.4. A generalized triangle is said to be right-angled if at least one of its vertices is a point and the inner angle at a vertex which is a point is $90°$, see Figure 4.4.

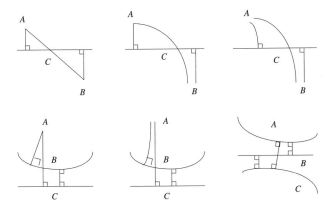

FIGURE 4.3. Non-convex generalized triangles.

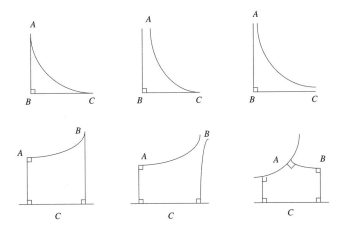

FIGURE 4.4. Right-angled generalized triangles.

Proposition 4.1. Let ABC be a generalized triangle. Then it is right-angled if and only if $((A \wedge B) \cdot (B \wedge C))\,((B \wedge C) \cdot (C \wedge A))\,((C \wedge A) \cdot (A \wedge B)) = 0$.

The sign of $((A \wedge B) \cdot (B \wedge C))\,((B \wedge C) \cdot (C \wedge A))\,((C \wedge A) \cdot (A \wedge B))$ characterizes another geometric invariant described below.

Definition 4.5. A generalized triangle is said to be acute-angled if it is convex and its inner angle at every vertex which is a point is acute.

Theorem 4.2. Let ABC be a generalized triangle. Then it is acute-angled if and only if $((A \wedge B) \cdot (B \wedge C))\,((B \wedge C) \cdot (C \wedge A))\,((C \wedge A) \cdot (A \wedge B)) < 0$.

4.2.2 The area and perimeter of a convex n-polygon

For a convex n-polygon of vertices $A_1, \ldots, A_n$, let $K_{A_1 \cdots A_n}$ be its area and $L_{A_1 \cdots A_n}$ be its perimeter. When $n = 3$ a convex 3-polygon is just a triangle.

A classical result on representing the area of a triangle in terms of the lengths of its three sides is the following (Greenberg, 1980)

Proposition 4.2. Let ABC be a triangle. Then

$$
\begin{cases}
\cos \dfrac{K_{ABC}}{2} = 2 \dfrac{1 - A \cdot B - B \cdot C - C \cdot A}{|A + B||B + C||C + A|}, \\[2ex]
\sin \dfrac{K_{ABC}}{2} = 2 \dfrac{|A \wedge B \wedge C|}{|A + B||B + C||C + A|}.
\end{cases}
$$

The dual of this result represents the perimeter of a triangle in terms of its three inner angles (Fenchel, 1989)

Proposition 4.3. For a triangle ABC, let

$$
a_1 = \frac{(A \wedge B)^\sim}{|A \wedge B|}, \quad a_2 = \frac{(B \wedge C)^\sim}{|B \wedge C|}, \quad a_3 = \frac{(C \wedge A)^\sim}{|C \wedge A|}.
$$

Then

$$
\begin{cases}
\cosh \dfrac{L_{ABC}}{2} = 2 \dfrac{1 + a_1 \cdot a_2 + a_2 \cdot a_3 + a_3 \cdot a_1}{|a_1 + a_2||a_2 + a_3||a_3 + a_1|}, \\[2ex]
\sinh \dfrac{L_{ABC}}{2} = 2 \dfrac{|a_1 \wedge a_2 \wedge a_3|}{|a_1 + a_2||a_2 + a_3||a_3 + a_1|}.
\end{cases}
$$

We explain (2.1) and (2.2) in terms of hyperbolic trigonometry. From

$$
\begin{aligned}
|A + B|^2 &= -(A + B) \cdot (A + B) = 2(1 - A \cdot B) \\
&= 2(1 + \cosh d(A, B)) = 4 \cosh^2 \frac{d(A, B)}{2},
\end{aligned}
$$

we get

$$
|A + B| = 2 \cosh \frac{d(A, B)}{2}.
$$

On the other hand, $|A \wedge B \wedge C|$ is the non-negative square root of

$$
|A \wedge B \wedge C|^2 = 1 - (A \cdot B)^2 - (B \cdot C)^2 - (C \cdot A)^2 - 2(A \cdot B)(B \cdot C)(C \cdot A).
$$

The vectors a_1, a_2, a_3 are unit vectors normal to oriented lines $A \wedge B$, $B \wedge C$, $C \wedge A$ respectively. We have

$$
a_1 \cdot a_2 = - \cos B,
$$

where B denotes the inner angle of the triangle at vertex B. So

$$
|a_1 + a_2| = 2 \sin \frac{B}{2},
$$

and $|a_1 \wedge a_2 \wedge a_3|$ is the non-negative square root of

$$|a_1 \wedge a_2 \wedge a_3|^2 = -1 + (a_1 \cdot a_2)^2 + (a_2 \cdot a_3)^2 + (a_3 \cdot a_1)^2$$
$$- 2(a_1 \cdot a_2)(a_2 \cdot a_3)(a_3 \cdot a_1).$$

(2.1) and (2.2) can be generalized to the case of convex n-polygons in $\mathcal{H}^2$, by means of spinor representations of Lorentz transformations (Li, 1997c). For example for $n = 4$,

$$\cos \frac{K_{A_1 A_2 A_3 A_4}}{2}$$
$$= 2 \frac{1 - \sum_{i<j} A_i \cdot A_j + (A_1 \cdot A_2)(A_3 \cdot A_4) - (A_1 \cdot A_3)(A_2 \cdot A_4) + (A_1 \cdot A_4)(A_2 \cdot A_3)}{|A_1 + A_2||A_2 + A_3||A_3 + A_4||A_4 + A_1|},$$

$$\sin \frac{K_{A_1 A_2 A_3 A_4}}{2}$$
$$= 2 \frac{|(A_1 \wedge A_2 \wedge A_3)^{\sim} + (A_1 \wedge A_2 \wedge A_4)^{\sim} + (A_1 \wedge A_3 \wedge A_4)^{\sim} + (A_2 \wedge A_3 \wedge A_4)^{\sim}|}{|A_1 + A_2||A_2 + A_3||A_3 + A_4||A_4 + A_1|},$$

$$\cosh \frac{L_{A_1 A_2 A_3 A_4}}{2}$$
$$= 2 \frac{1 + \sum_{i<j} a_i \cdot a_j + (a_1 \cdot a_2)(a_3 \cdot a_4) - (a_1 \cdot a_3)(a_2 \cdot a_4) + (a_1 \cdot a_4)(a_2 \cdot a_3)}{|a_1 + a_2||a_2 + a_3||a_3 + a_4||a_4 + a_1|},$$

$$\sinh \frac{L_{A_1 A_2 A_3 A_4}}{2}$$
$$= 2 \frac{|(a_1 \wedge a_2 \wedge a_3)^{\sim} + (a_1 \wedge a_2 \wedge a_4)^{\sim} + (a_1 \wedge a_3 \wedge a_4)^{\sim} + (a_2 \wedge a_3 \wedge a_4)^{\sim}|}{|a_1 + a_2||a_2 + a_3||a_3 + a_4||a_4 + a_1|}.$$

4.3   Hyperbolic Conformal Geometry with Clifford Algebra

For hyperbolic conformal geometry, we need the double-hyperbolic space, which is a double covering of the hyperbolic space. In the Minkowski space $\mathcal{R}^{n,1}$, the set

$$\mathcal{D}^n = \{x \in \mathcal{R}^{n,1} | x \cdot x = -1\}$$

is called an n-dimensional double-hyperbolic space. It has two connected components, $\mathcal{H}^n$ and $-\mathcal{H}^n$.

In Euclidean conformal geometry, spheres and planes are conformally invariant geometric objects. Similarly, in hyperbolic conformal geometry, generalized spheres, and planes and spheres at infinity are conformally invariant objects which are called total spheres. A Clifford algebraic model for the double-hyperbolic n-space, called the homogeneous model, is introduced to simplify the algebraic representations and manipulations of total spheres and hyperbolic conformal transformations. The model in coordinate form can be found in Cecil (1992).

Bunches of total spheres, which extend and generalize the concept of pencils of spheres and hyperplanes in Euclidean geometry, are classified and studied within the homogeneous model. The model also makes possible the spinor representation of hyperbolic conformal transformations. A typical conformal transformation, called a tidal transformation, is given as an example of the spinor approach.

4.3.1   Double-hyperbolic space

The following concepts will be needed: oriented generalized point, plane, sphere at infinity, generalized sphere, total sphere and double-sphere.

Definition 4.6.   An oriented generalized point in $\mathcal{D}^n$ is either a point, or an oriented point at infinity, or an oriented imaginary point. A point is an element in $\mathcal{D}^n$. An oriented point at infinity is a one-dimensional null half-space of $\mathcal{R}^{n,1}$. An oriented imaginary point is a one-dimensional Euclidean half-space of $\mathcal{R}^{n,1}$.

Definition 4.7.   An r-plane of $\mathcal{D}^n$ is the intersection of $\mathcal{D}^n$ with an $(r+1)$-space of $\mathcal{R}^{n,1}$.

In $\mathcal{G}_{n,1}$, an r-plane is represented by an $(r+1)$-blade corresponding to the $(r+1)$-space of $\mathcal{R}^{n,1}$. When $r = 0$, a 0-plane is a pair of antipodal points; when $r = n - 1$, an $(n-1)$-plane is called a hyperplane.

Definition 4.8.   The sphere at infinity of $\mathcal{D}^n$ is a set of points at infinity. An r-sphere at infinity in $\mathcal{D}^n$ is the intersection of the sphere at infinity with an $(r+1)$-plane of $\mathcal{D}^n$.

Any r-sphere at infinity is the sphere at infinity on an $(r+1)$-plane in $\mathcal{D}^n$. When $r = 0$, a 0-sphere at infinity is a pair of points at infinity.

Definition 4.9.   A generalized sphere is either a sphere, or a horosphere, or a hypersphere. It is determined by a pair (c, ρ), where c is a vector in $\mathcal{R}^{n,1}$ representing an oriented generalized point, called the center of the generalized sphere, and $\rho > 0$ is called the generalized radius.

1. When c is a point, the set $\{p \in \mathcal{D}^n | p \cdot c = -(1 + \rho)\}$ is the sphere with center c and generalized radius ρ.

2. When c is an oriented point at infinity, the set $\{p \in \mathcal{D}^n | p \cdot c = -\rho\}$ is the horosphere with center c and generalized radius ρ.

3. When c is an oriented imaginary point, the set $\{p \in \mathcal{D}^n | p \cdot c = -\rho\}$ is the hypersphere with center c and generalized radius ρ. The hyperplane of $\mathcal{D}^n$ represented by $c^{\sim}$ is called the axis of the hypersphere.

Definition 4.10. A generalized r-sphere is a generalized sphere in an $(r+2)$-plane, by considering the $(r+2)$-plane to be an $(r+1)$-dimensional double-hyperbolic space.

When $r = 0$, a 0-sphere is a pair of points on the same branch, a 0-horosphere is a point and a point at infinity, and a 0-hypersphere is a pair of non-antipodal points on different branches.

Definition 4.11. A total sphere in $\mathcal{D}^n$ refers to a generalized sphere, or a hyperplane, or the sphere at infinity. A total r-sphere is an r-dimensional generalized sphere, plane, or sphere at infinity.

Definition 4.12. A double-sphere of $\mathcal{D}^n$ is a hypersphere together with its reflection with respect to the axis. An r-double-sphere is an r-hypersphere together with its reflection with respect to the axis, see Figure 4.5.

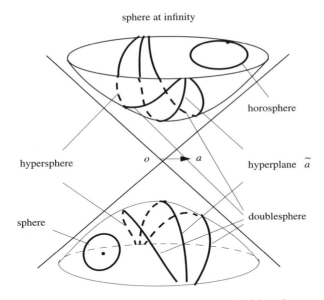

FIGURE 4.5. Total spheres and a doublesphere.

4.3.2 The homogeneous model of a double-hyperbolic space

The set $\mathcal{D}^n$ is in the Minkowski space $\mathcal{R}^{n,1}$. We now embed $\mathcal{R}^{n,1}$ into $\mathcal{R}^{n+1,1}$, and embed $\mathcal{D}^n$ into the null cone of $\mathcal{R}^{n+1,1}$ to obtain the homogeneous model of the double-hyperbolic space. This model makes possible a useful algebraic representation of total spheres and conformal transformations.

Let a_0 be a fixed unit vector in $\mathcal{R}^{n+1,1}$. The space represented by $a_0^{\sim}$ is a Minkowski $(n+1)$-space which we denote by $\mathcal{R}^{n,1}$. The mapping

$$x \mapsto x - a_0, \text{ for } x \in \mathcal{D}^n, \tag{3.3}$$

maps the set $\mathcal{D}^n$ in a one-to-one manner onto the set $\mathcal{N}_{a_0}^n = \{x \in \mathcal{R}^{n+1,1} | x \cdot x = 0, x \cdot a_0 = -1\}$. Conversely, from the orthogonal decomposition

$$x = P_{a_0}(x) + P_{a_0^{\sim}}(x) \tag{3.4}$$

of a vector $x \in \mathcal{N}_{a_0}^n$, we get a unique point $P_{a_0^{\sim}}(x) \in \mathcal{D}^n$. (3.4) is called the projective split of x with respect to a_0. The sphere at infinity of $\mathcal{D}^n$ is the set

$$\{x \in \mathcal{R}^{n+1,1} | x \cdot x = 0, x \cdot a_0 = 0\}. \tag{3.5}$$

Definition 4.13.   The set $\mathcal{N}_{a_0}^n$, together with the decomposition (3.4), defines the homogeneous model of the double-hyperbolic space $\mathcal{D}^n$.

Proposition 4.4.   Let p, q be two points (null vectors) on the same branch of $\mathcal{D}^n$ in the homogeneous model. Let $d(p, q)$ be the hyperbolic distance between the two points. Then

$$p \cdot q = 1 - \cosh d(p, q).$$

Corollary 4.2.   A point p is on the sphere with center c and generalized radius ρ, if and only if $p \cdot c = -\rho$.

In the homogeneous model, the oriented points at infinity of $\mathcal{D}^n$ are represented by the null vectors of $a_0^{\sim}$; the oriented imaginary points of $\mathcal{D}^n$ are represented by the vectors of $a_0^{\sim}$ with positive signature.

Corollary 4.3.   A point p is on the horosphere (or hypersphere) with center c and generalized radius ρ, when p is understood to be the null vector representing the point, if and only if $p \cdot c = -\rho$.

Comparing the above two corollaries with the definition (4.9), we can see clearly the advantage of the unified algebraic representation of thegeneralized spheres in the homogeneous model.

We have the following fundamental theorem for the homogeneous model.

Theorem 4.3.   Let $B_{r-1,1}$ be a Minkowski r-blade in $\mathcal{G}_{n+1,1}$, $2 \leq r \leq n+1$. Then $B_{r-1,1}$ represents a total $(r-2)$-sphere. We have the following cases.

   1. If $a_0 \cdot B_{r-1,1} = 0$, then $B_{r-1,1}$ represents an $(r-2)$-sphere at infinity.

2. If $a_0 \cdot B_{r-1,1}$ is Euclidean, then $B_{r-1,1}$ represents an $(r-2)$-sphere.

3. If $a_0 \cdot B_{r-1,1}$ is degenerate, then $B_{r-1,1}$ represents an $(r-2)$-horosphere.

4. If $a_0 \cdot B_{r-1,1}$ is Minkowski, but $a_0 \wedge B_{r-1,1} \neq 0$, then $B_{r-1,1}$ represents an $(r-2)$-hypersphere.

5. If $a_0 \wedge B_{r-1,1} = 0$, then $B_{r-1,1}$ represents an $(r-2)$-plane.

The dual form of the above theorem for $r = n + 1$ is

Theorem 4.4.   Let s be a vector of positive square in $\mathcal{R}^{n+1,1}$. Then $s^{\sim}$ represents a total sphere. We have the following cases.

1. If $a_0 \wedge s = 0$, then $s^{\sim}$ represents the sphere at infinity. The sphere at infinity is represented by $a_0^{\sim}$.

2. If $a_0 \wedge s$ is Minkowski, then $s^{\sim}$ represents a sphere. The sphere with center c and generalized radius ρ is represented by $(c - \rho a_0)^{\sim}$, where c is the null vector representing the center.

3. If $a_0 \wedge s$ is degenerate, then $s^{\sim}$ represents a horosphere. The horosphere with center c and generalized radius ρ is represented by $(c - \rho a_0)^{\sim}$.

4. If $a_0 \wedge s$ is Euclidean, but $a_0 \cdot s \neq 0$, then $s^{\sim}$ represents a hypersphere. The hypersphere with center c and generalized radius ρ is represented by $(c - \rho a_0)^{\sim}$.

5. If $a_0 \cdot s = 0$, then $s^{\sim}$ represents a hyperplane. A hyperplane with normal direction c is represented by $c^{\sim}$.

4.3.3 Bunches of total spheres

Various collections of total spheres are important geometric objects in hyperbolic conformal geometry.

Definition 4.14.   A bunch of total spheres is determined by $B_1, \ldots, B_r$ is the set of total spheres given by $\lambda_1 B_1 + \ldots + \lambda_r B_r$, where the λ's are scalars. When the meet $B_1 \vee \cdots \vee B_r \neq 0$, the integer $r - 1$ is called the dimension of the bunch. A pencil is a one-dimensional bunch.

The dimension of a bunch in $\mathcal{D}^n$ is between 1 and $n - 1$. When $B_1 \vee \cdots \vee B_r \neq 0$, we can use $B_1 \vee \cdots \vee B_r$ to represent the bunch.

The concept and classification of bunches are fundamental in the study of hyperbolic conformal geometry, because total spheres are invariants of conformal transformations.

Theorem 4.5. [Classification of bunches] Let $B_1, \ldots, B_r$ be total spheres. Let $A_{n-r+2} = B_1 \vee \cdots \vee B_r \neq 0$.

1. When $a_0 \cdot A_{n-r+2} = 0$, the bunch is called a concentric bunch. It is composed of the sphere at infinity and the generalized spheres whose centers are in the subspace $(a_0 \wedge A_{n-r+2})^\sim$ of $\mathcal{R}^{n,1}$.

   For example when $r = 2$, if A_n is Euclidean, it represents the pencil of spheres centered at $\pm(a_0 \wedge A_n)^\sim / |A_n|$; if A_n is null, it represents the pencil of horospheres centered at $\pm(a_0 \wedge A_n)^\sim$; if A_n is Minkowski, it represents the pencil of hyperspheres centered at $\pm(a_0 \wedge A_n)^\sim$, see Figure 4.6.

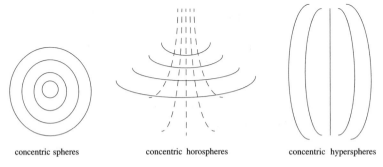

concentric spheres   concentric horospheres   concentric hyperspheres

FIGURE 4.6. Concentric pencil on one branch of $\mathcal{D}^n$ ($r = 2$).

2. When $a_0 \wedge A_{n-r+2} = 0$, the bunch is called a hyperplane bunch, since it is composed of hyperplanes only. There are three cases (see Figure 4.7):

   - When A_{n-r+2} is Euclidean, the bunch is composed of hyperplanes perpendicular to the $(r-1)$-plane $a_0 \wedge A_{n-r+2}^\sim$.

     For example, when $r = 2$, hyperplanes in the bunch are ultra-parallel to each other.

   - When A_{n-r+2} is degenerate, the bunch is composed of hyperplanes whose representations in the homogeneous model pass through the subspace A_{n-r+2} of $\mathcal{R}^{n+1,1}$.

     For example when $r = 2$, hyperplanes in the bunch are parallel to each other.

   - When A_{n-r+2} is Minkowski, the bunch is composed of hyperplanes passing through the $(n-r)$-plane A_{n-r+2}.

     For example, when $r = 2$, hyperplanes in the bunch have a common $(n-2)$-plane.

3. When A_{n-r+2} is Minkowski, the bunch is called a concurrent bunch, since every total sphere in the bunch includes the generalized $(n-r)$-sphere A_{n-r+2}. There are three cases (see Figure 4.8):

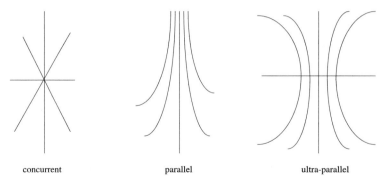

concurrent   parallel   ultra-parallel

FIGURE 4.7. Hyperplane pencil ($r = 2$).

- When $a_0 \cdot A_{n-r+2}$ is Euclidean, A_{n-r+2} represents an $(n-r)$-sphere.

- When $a_0 \cdot A_{n-r+2}$ is degenerate, A_{n-r+2} represents an $(n-r)$-horosphere.

- When $a_0 \cdot A_{n-r+2}$ is Minkowski, A_{n-r+2} represents an $(n-r)$-hypersphere, and the bunch is composed only of hyperspheres. In this case, $a_0 \wedge (a_0 \cdot A_{n-r+2})$ represents an $(n-r)$-plane, which is the axis of the $(n-r)$-hypersphere A_{n-r+2} and is the intersection of all axes of the hyperspheres in the bunch.

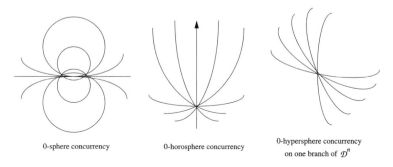

0-sphere concurrency   0-horosphere concurrency   0-hypersphere concurrency
                                                            on one branch of $\mathcal{D}^n$

FIGURE 4.8. Concurrent pencil ($r = 2$).

4. When A_{n-r+2} is degenerate, the bunch is called a tangent bunch. Any two non-intersecting total spheres in the bunch are tangent to each other. The tangency occurs at a point or a point at infinity, which corresponds to the unique one-dimensional null subspace in the space A_{n-r+2}. There are two cases (see Figure 4.9):

- When $a_0 \wedge A_{n-r+2}$ is degenerate, the tangency occurs at a point at infinity.

- When $a_0 \wedge A_{n-r+2}$ is Minkowski, the tangency occurs at a point.

tangency at a point   tangency at a point at infinity

FIGURE 4.9. Tangent pencil $(r = 2)$.

5. When A_{n-r+2} is Euclidean, the bunch is called a Poncelet bunch. $A_{n-r+2}^{\sim}$ represents a generalized $(r - 2)$-sphere, called a Poncelet sphere. There are three cases (see Figure 4.10):

- When $a_0 \wedge A_{n-r+2}$ is Minkowski, $A_{n-r+2}^{\sim}$ is an $(r - 2)$-sphere.
- When $a_0 \wedge A_{n-r+2}$ is degenerate, $A_{n-r+2}^{\sim}$ is an $(r - 2)$-horosphere.
- When $a_0 \wedge A_{n-r+2}$ is Euclidean, $A_{n-r+2}^{\sim}$ is an $(r - 2)$-hypersphere.

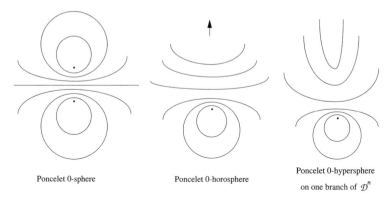

Poncelet 0-sphere   Poncelet 0-horosphere   Poncelet 0-hypersphere
on one branch of $\mathcal{D}^n$

FIGURE 4.10. Poncelet pencil $(r = 2)$.

4.3.4 Conformal transformations

The following theorem is fundamental in the study of conformal transformations in the homogeneous model.

Theorem 4.6. Any conformal transformation in $\mathcal{D}^n$ can be realized in the homogeneous model of $\mathcal{D}^n$ through the conjugation of a versor in $\mathcal{G}_{n+1,1}$, and vice versa. Two versors realize the same conformal transformation if and only they are the same up to a nonzero scalar or pseudoscalar factor.

We now use the versor representation to study a conformal transformation which is similar to dilation in Euclidean space. The tidal transformation is defined by the versor $1 + \lambda a_0 c$, where $\lambda \in \mathcal{R}$, $c \in \mathcal{R}^{n+1,1}$ and $c \cdot a_0 = 0$.

This transformation leaves the concentric pencil $(a_0 \wedge c)^\sim$ invariant. When c is a point or an oriented point at infinity, the set $\{c, -c\}$ is invariant; when c is an oriented imaginary point, the hyperplane $c^\sim$ is not invariant, but its sphere at infinity is.

Assume that p is a fixed point in $\mathcal{D}^n$, and is transformed to a point or point at infinity q. It can be proved that the parameter λ is a function of q on line $c \wedge p$. Below we give some of the properties of this function.

1. When c is a point (see Figure 4.11),

   (a) and q is any point or point at infinity q on the line $c \wedge p$, then for $C_c(q) = -c^{-1}qc$, $\lambda(-C_c(q)) = \dfrac{1}{\lambda(q)}$.

   (b) and q is any point on line $c \wedge p$, then $\lambda(q) = \dfrac{(q-p)^2}{(q-c)^2 - (p-c)^2}$.

   (c) then $\lambda(p - e^{\pm d(p,c)}c) = e^{\pm d(p,c)}$.

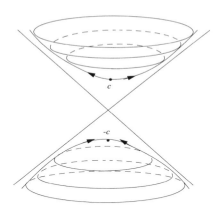

FIGURE 4.11. Tidal transformation when c is a point.

2. When c is an oriented point at infinity (see Figures 4.12 and 4.13),

   (a) then for any point q on the line $c \wedge p$, $\lambda(q) = \dfrac{1}{2}(\dfrac{1}{q \cdot c} - \dfrac{1}{p \cdot c})$.

   (b) and if $q \neq c$ is a point at infinity on line $c \wedge p$, then $\lambda(q) = -\dfrac{1}{2p \cdot c}$.

3. When c is an oriented imaginary point (see Figures 4.14, 4.15, and Figure 4.16),

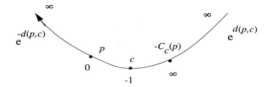

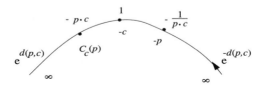

FIGURE 4.12. $\lambda = \lambda(q)$ **for Figure 4.11. The arrows indicate the increasing direction.**

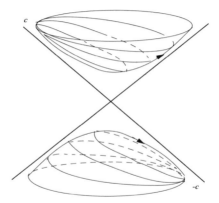

FIGURE 4.13. Tidal transformation when c is an oriented point at infinity.

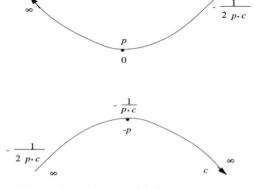

FIGURE 4.14. $\lambda = \lambda(q)$ **for Figure 4.13.**

(a) and q any point or point at infinity on the line $c \wedge p$, then
$$\lambda(-C_c(q)) = -\frac{1}{\lambda(q)}.$$

(b) and q any point on the line $c \wedge p$, then $\lambda(q) = \dfrac{(q-p)^2}{(q-c)^2 - (p-c)^2}.$

(c) and $p \cdot c < 0$, then for $d(p,c)$ the hyperbolic distance from p to the intersection t of the line $c \wedge p$ with the hyperplane $c^{\sim}$ on the branch of $\mathcal{D}^n$ containing p, we have

$$\lambda(C_c(p)) = -\sinh d(p,c), \quad \lambda(t) = -\tanh \frac{d(p,c)}{2},$$
$$\lambda(p + e^{d(p,c)}c) = -e^{d(p,c)}, \quad \lambda(p + e^{-d(p,c)}c) = e^{-d(p,c)}.$$

(d) and $p \cdot c = 0$ and q is on the branch of $\mathcal{D}^n$ containing p, then
$\lambda(q) = -\epsilon \tanh \dfrac{d(p,q)}{2}$, where ϵ is the sign of $q \cdot c$.

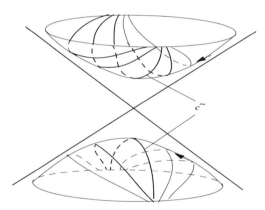

FIGURE 4.15. Tidal transformation when c is an oriented imaginary point.

4.4 A Universal Model for the Conformal Geometries of the Euclidean, Spherical, and Double-Hyperbolic Spaces

Here we introduce the homogeneous models for Euclidean and spherical spaces, and talk about the connections among these three homogeneous models. Hyperbolic, Euclidean and spherical geometries can be unified in such a way that we need only one Minkowski space, where null vectors represent points or points at infinity in any of the three geometric spaces,

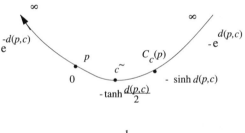

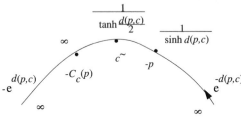

FIGURE 4.16. $\lambda = \lambda(q)$ for Figure 4.15. The arrows indicate its increase direction.

and where Minkowski subspaces represent spheres and planes. Furthermore, any theorem in one of the three geometries, when represented in the homogeneous model, is also a theorem in each of the other two geometries.

4.4.1   The homogeneous model of the Euclidean space

Let p_0 be a fixed point in $\mathcal{D}^n$. Then in $\mathcal{R}^{n+1,1}$, the blade $(p_0 \wedge a_0)^\sim$ represents a Euclidean n-space which we denote by $\mathcal{R}^n$. Let

$$e = p_0 + a_0, \ e_0 = \frac{p_0 - a_0}{2}. \tag{4.6}$$

Both e and e_0 are null vectors and $e \cdot e_0 = -1$.

The mapping

$$x \mapsto x + e_0 + \frac{x \cdot x}{2}e, \text{ for } x \in \mathcal{R}^n, \tag{4.7}$$

maps $\mathcal{R}^n$ in a one-to-one manner onto the set
$\mathcal{N}_e^n = \{x \in \mathcal{R}^{n+1,1} | x \cdot x = 0, x \cdot e = -1\}$. Conversely, from the orthogonal decomposition

$$x = P_{p_0 \wedge a_0}(x) + P_{(p_0 \wedge a_0)^\sim}(x) \tag{4.8}$$

of the vector $x \in \mathcal{N}_e^n$, we get the unique point $P_{(p_0 \wedge a_0)^\sim}(x) \in \mathcal{R}^n$. Equation (4.8) is the conformal split of x with respect to $p_0 \wedge a_0 = e_0 \wedge e$.

Definition 4.15. The set $\mathcal{N}_e^n$, together with the decomposition (4.8), defines the homogeneous model of the Euclidean space $\mathcal{R}^n$.

In the homogeneous model, e represents a point at infinity which is the one-point compactification of the Euclidean space. The point e_0 represents the zero vector in $\mathcal{R}^n$, and is called the origin.

Proposition 4.5. Let u, v be two points in $\mathcal{R}^n$ represented by null vectors in the homogeneous model. Let $d(u, v)$ be the Euclidean distance between the two points. Then

$$u \cdot v = -\frac{d^2(u, v)}{2}.$$

Theorem 4.7. Let $B_{r-1,1}$ be a Minkowski r-blade in $\mathcal{G}_{n+1,1}$, $2 \leq r \leq n+1$. Then $B_{r-1,1}$ represents an $(r-2)$-dimensional sphere or plane. If $e \wedge B_{r-1,1} = 0$, $B_{r-1,1}$ represents an $(r-2)$-plane, otherwise it represents an $(r-2)$-sphere.

When $r = n+1$, the dual form of the above theorem is

Theorem 4.8. Let s be a vector of positive square in $\mathcal{R}^{n+1,1}$. Then $s^\sim$ represents a sphere or a hyperplane.

1. If $e \cdot s = 0$, then $s^\sim$ represents a hyperplane. The hyperplane normal to unit vector n and has the signed distance δ from the origin in the direction of n, and is represented by $(n + \delta e)^\sim$.

2. If $e \cdot s \neq 0$, then $s^\sim$ represents a sphere. The sphere with center c and radius ρ is represented by $(c - e\rho^2/2)^\sim$.

The stereographic projection P_{DR} of $\mathcal{D}^n$, with pole at $-p_0$, to $\mathcal{R}^n$ maps $\mathcal{D}^n$ together with its sphere at infinity to $\mathcal{R}^n$ together with its point at infinity. It changes the hyperboloid model $\mathcal{H}^n$ into the Poincaré ball model. In the homogeneous models of $\mathcal{D}^n$ and $\mathcal{R}^n$, P_{DR} is just a rescaling of null vectors, taking $-x/(x \cdot a_0)$ into $-x/(x \cdot e)$.

Note that we could have chosen e and e_0 such that $e_0 \cdot e = \lambda$ for any fixed real number, and define $\mathcal{N}^n_{p_0}$ by the condition that $x \cdot e = \lambda$. It is only a matter of convention that we choose $\lambda = -1$.

4.4.2 The homogeneous model of the spherical space

Let p_0 be a fixed point in $\mathcal{D}^n$. Then $p_0^\sim$ represents a Euclidean $(n+1)$-space which we denote by $\mathcal{R}^{n+1}$. The unit sphere of the space $\mathcal{R}^{n+1}$ is the spherical n-space $\mathcal{S}^n$.

The mapping

$$x \mapsto x + p_0, \text{ for } x \in \mathcal{S}^n, \tag{4.9}$$

maps the set $\mathcal{S}^n$ in a one-to-one manner onto the set $\mathcal{N}^n_{p_0} = \{x \in \mathcal{R}^{n+1,1} | x \cdot x = 0, x \cdot p_0 = -1\}$. Conversely, from the orthogonal decomposition

$$x = P_{p_0}(x) + P_{p_0^\sim}(x) \tag{4.10}$$

of the vector $x \in \mathcal{N}_{p_0}^n$, we get the unique point $P_{p_0^{\sim}}(x) \in \mathcal{S}^n$. Equations (4.10) gives the projective split of x with respect to p_0.

Definition 4.16.   The set $\mathcal{N}_{p_0}^n$, together with the decomposition (4.10), defines the homogeneous model of the spherical space $\mathcal{S}^n$.

Proposition 4.6.   Let a, b be two points in $\mathcal{S}^n$ represented by null vectors in the homogeneous model. Let $d(a, b)$ be the spherical distance between the two points. Then
$$a \cdot b = \cos d(a, b) - 1.$$

Theorem 4.9.   Let $B_{r-1,1}$ be a Minkowski r-blade in $\mathcal{G}_{n+1,1}$, $2 \le r \le n + 1$. Then $B_{r-1,1}$ represents an $(r-2)$-dimensional sphere or plane. If $p_0 \wedge B_{r-1,1} = 0$, $B_{r-1,1}$ represents an $(r-2)$-plane, otherwise it represents an $(r-2)$-sphere.

When $r = n + 1$, the dual form of the above theorem is

Theorem 4.10.   Let s be a vector of positive square in $\mathcal{R}^{n+1,1}$. Then $s^{\sim}$ represents a sphere or hyperplane.

1. If $p_0 \cdot s = 0$, then $s^{\sim}$ represents a hyperplane. The hyperplane normal to the vector c is represented by $c^{\sim}$.

2. If $p_0 \cdot s \neq 0$, then $s^{\sim}$ represents a sphere. The sphere with center c and radius ρ is represented by $(c + p_0 \cos \rho)^{\sim}$.

The stereographic projection P_{SR} of $\mathcal{S}^n$, with the pole a_0, to $\mathcal{R}^n$ maps $\mathcal{S}^n$ to $\mathcal{R}^n$ together with its point at infinity. Let
$$e = p_0 + a_0, \quad e_0 = \frac{p_0 - a_0}{2}. \tag{4.11}$$

In the homogeneous models of $\mathcal{S}^n$ and $\mathcal{R}^n$, stereographic projection is just a rescaling of null vectors taking $-x/(x \cdot p_0)$ to $-x/(x \cdot e)$.

The composition of the inverse of the mapping P_{SR} with the mapping P_{DR} is denoted by P_{DS}. It changes the hyperboloid model of $\mathcal{H}^n$ into the hemisphere model. In the homogeneous model this is just a rescaling of null vectors.

Let P'_{SR} denote the stereographic projection of $\mathcal{S}^n$, with the pole a_0, to $\mathcal{R}^n$ where b_0 is a point in $\mathcal{S}^n$ normal to a_0. This projection changes the hemisphere model of $\mathcal{H}^n$ into the Poincaré half-space model. In the homogeneous model, this is just another rescaling of null vectors.

From the above discussion, we see that the hyperboloid, Poincaré ball, Poincaré half-space and hemisphere models are all unified in the homogeneous model. Changing from one model to another is just a rescaling of null vectors.

The derivation of the Poincaré ball model of $\mathcal{H}^n$ from the hyperboloid model, is realized by the diagram

$$\mathcal{H}^n \xdashrightarrow{\overset{\text{inverse of } P_{a_{\tilde{0}}}}{\hspace{3cm}}} \mathcal{N}^n_{a_0} \xdashrightarrow{\overset{\text{rescaling}}{\hspace{2cm}}} \mathcal{N}^n_e \xrightarrow{\overset{P_{(p_0 \wedge a_0)^\sim}}{\hspace{1.5cm}}} \text{Poincaré ball model.}$$

If we go from $\mathcal{N}^n_{a_0}$ to $\mathcal{R}^n$ by $P_{(p_0 \wedge a_0)^\sim}$ directly, we get the Klein ball model:

$$\mathcal{H}^n \xdashrightarrow{\overset{\text{inverse of } P_{a_{\tilde{0}}}}{\hspace{3cm}}} \mathcal{N}^n_{a_0} \xrightarrow{\overset{P_{(p_0 \wedge a_0)^\sim}}{\hspace{1.5cm}}} \text{Klein ball model.}$$

Since $\mathcal{N}^n_{a_0}$ is not a homogeneous model of $\mathcal{R}^n$, the Klein ball model fails to be conformal.

4.4.3 A universal model for three geometries

As was mentioned before, there are five important analytic models for hyperbolic n-space. The relations between these models (with the exception of the Klein ball model), together with Euclidean and spherical n-spaces, are realized by stereographic projections. Since the three geometric spaces correspond to the same null cone of $\mathcal{R}^{n+1,1}$ and the stereographic projections are just rescalings of null vectors, the three geometric spaces, together with four of the five models of hyperbolic geometry, can be unified in one Minkowski space, where null vectors represent points or points at infinity, and where Minkowski subspaces represent spheres and planes in any of the three geometries.

If a theorem in one of the three geometries is represented in the homogeneous model, it will be just as valid in all the three other geometries, because the three geometries are just different geometric interpretations of the same null vectors and the same Minkowski subspaces. Thus, a single theorem in one geometry generates many "new" theorems in the other geometries. We will see below that the homogeneous model also gives many new interpretations of a given theorem in the *same* geometry.

We illustrate this with Simson's Theorem in plane geometry (see Figure 4.17).

Theorem 4.11. [Simson's Theorem] Let ABC be a triangle, D be a point on the circumscribed circle of the triangle. Draw perpendicular lines from D to the three sides AB, BC, CA of triangle ABC. Let C_1, A_1, B_1 be intersections of the perpendicular lines with the corresponding sides. Then A_1, B_1, C_1 are collinear.

When $A, B, C, D, A_1, B_1, C_1$ are understood to be null vectors represen-

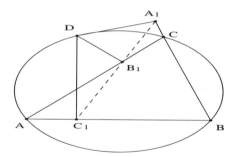

FIGURE 4.17. Simson's Theorem.

ting points in the plane, the hypothesis can be expressed as

$A \wedge B \wedge C \wedge D = 0$ A, B, C, D are on the same circle

$e \wedge A \wedge B \wedge C \neq 0$ ABC is a triangle

$e \wedge A_1 \wedge B \wedge C = 0$ A_1 is on line BC

$(e \wedge D \wedge A_1) \cdot (e \wedge B \wedge C) = 0$ Lines DA_1 and BC are perpendicular

$e \wedge A \wedge B_1 \wedge C = 0$ B_1 is on line CA

$(e \wedge D \wedge B_1) \cdot (e \wedge C \wedge A) = 0$ Lines DB_1 and CA are perpendicular

$e \wedge A \wedge B \wedge C_1 = 0$ C_1 is on line AB

$(e \wedge D \wedge C_1) \cdot (e \wedge A \wedge B) = 0$ Lines DC_1 and AB are perpendicular

The conclusion can be expressed as

$$e \wedge A_1 \wedge B_1 \wedge C_1 = 0.$$

Both the hypothesis and the conclusion are invariant under the rescaling of null vectors, so this theorem is valid for all three geometries, and is free of the requirement that $A, B, C, D, A_1, B_1, C_1$ represent points and e represents the point at infinity of $\mathcal{R}^n$. Various "new" theorems can be produced simply by interpreting the algebraic equalities and inequalities in the hypothesis and conclusion of the theorem differently.

For example, let us interchange the roles played by D and e in Euclidean geometry. The new constraints become

$e \wedge A \wedge B \wedge C = 0$ A, B, C are collinear

$A \wedge B \wedge C \wedge D \neq 0$ A, B, C, D are neither collinear

                         nor on the same circle

$A_1 \wedge B \wedge C \wedge D = 0$ A_1, B, C, D are on the same circle

$(e \wedge D \wedge A_1) \cdot (D \wedge B \wedge C) = 0$ line DA_1, circle DBC are perpendicular

$A \wedge B_1 \wedge C \wedge D = 0$ A, B_1, C, D are on the same circle

$(e \wedge D \wedge B_1) \cdot (D \wedge C \wedge A) = 0$ line DB_1, circle DCA are perpendicular

$A \wedge B \wedge C_1 \wedge D = 0$ A, B, C_1, D are on the same circle

$(e \wedge D \wedge C_1) \cdot (D \wedge A \wedge B) = 0$ line DC_1, circle DAB are perpendicular

The conclusion becomes

$$A_1 \wedge B_1 \wedge C_1 \wedge D = 0.$$

Using the facts that a line is "perpendicular" to a circle if and only if it passes through the center of the circle, and that any circular angle on a diameter is a right angle, we can restate the above "new" theorem as follows:

Theorem 4.12. Let DAB be a triangle, C be a point on line AB (see Figure 4.18). Let A_1B be perpendicular to DB, A_1C be perpendicular to CD, and AB_1 be perpendicular to AD. Let C_1 be the intersection of lines AB_1, and A_1B. Then D, A_1, B_1, C_1 are on the same circle.

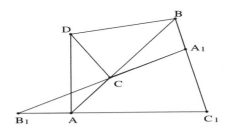

FIGURE 4.18. Theorem 4.12.

We can get another theorem by interchanging the roles of A, e. The new constraints become

$e \wedge B \wedge C \wedge D = 0$	B, C, D are collinear
$e \wedge A \wedge B \wedge C \neq 0$	A, B, C is a triangle
$A \wedge A_1 \wedge B \wedge C = 0$	A_1, A, B, C are collinear
$(A \wedge D \wedge A_1) \cdot (A \wedge B \wedge C) = 0$	circles ADA_1, ABC are perpendicular
$e \wedge A \wedge B_1 \wedge C = 0$	A, B_1, C are collinear
$(A \wedge D \wedge B_1) \cdot (e \wedge C \wedge A) = 0$	line CA, circle ADB_1 are perpendicular
$e \wedge A \wedge B \wedge C_1 = 0$	A, B, C_1 are collinear
$(A \wedge D \wedge C_1) \cdot (e \wedge A \wedge B) = 0$	line AB, circle ADC_1 are perpendicular

For these constraints the conclusion becomes

$$A \wedge A_1 \wedge B_1 \wedge C_1 = 0.$$

Using the fact that two circles are "perpendicular" if and only if the tangent lines to either one of circles at the points of intersection meet at the center of the other circle, we can restate the above "new" theorem as follows:

Theorem 4.13. Let ABC be a triangle, D be a point on line BC (see Figure 4.19). Let EF be the perpendicular bisector of line segment AD, which intersects lines AB and AC at points E and F respectively. Let C_1, B_1 be the symmetric points of A with respect to the points E, F respectively. Let AG be the tangent line of the circle ABC at A, which intersects EF at G. Let A_1G be the other tangent line of circle ABC passing through G, and A_1 the point of tangency. Then A, A_1, B_1, C_1 are on the same circle.

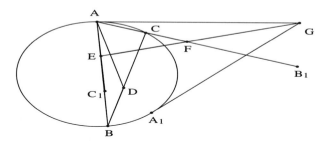

FIGURE 4.19. Theorem 4.13.

There are many new corresponding theorems in spherical geometry as well. We consider only one of them here. Let $e = -D$. The "new" theorem is the following:

Theorem 4.14. Let A, B, C, D be points on the same circle in the sphere (see Figure 4.20). Let A_1, B_1, C_1 be the symmetric points of the point $-D$ with respect to the centers of circles $(-D)BC$, $(-D)CA$, $(-D)AB$ respectively. Then $-D, A_1, B_1, C_1$ are on the same circle.

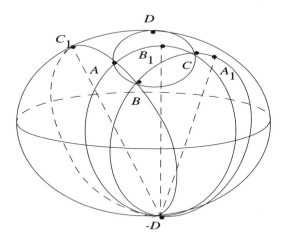

FIGURE 4.20. Theorem 4.14.

There are also various theorems in hyperbolic geometry that are equivalent to Simson's theorem. We present one of them here. Let A, B, C, D be points on the same branch of $\mathcal{D}^2$, $e = -D$. In the hyperbolic plane, a generalized sphere is usually called a generalized circle, and a hypersphere is usually called a hypercycle.

Theorem 4.15. Let A, B, C, D be points in the hyperbolic plane $\mathcal{H}^2$, and let them be on the same generalized circle. Let L_A, L_B, L_C be the axes of the hypercycles $(-D)BC$, $(-D)CA$, $(-D)AB$ respectively. Let A_1, B_1, C_1 be the symmetric points of D with respect to L_A, L_B, L_C respectively. Then the points $-D, A_1, B_1, C_1$ are on the same hypercycle.

4.5   Conclusion

Clifford algebra appears to play an important role in the study of hyperbolic geometry. It not only simplifies represention and computation, but also contributes to discovering and proving new theorems.

Part II

Theorem Proving

Chapter 5

Geometric Reasoning With Geometric Algebra

Dongming Wang

5.1 Introduction

Geometric (Clifford) algebra was motivated by geometric considerations and provides a comprehensive algebraic formalism for the expression of geometric ideas [11]. Recent research has shown that this formalism may be effectively used in algebraic approaches for automated geometric reasoning [7, 12, 14, 20, 24]. Starting with an introduction to Clifford algebra for n-dimensional Euclidean geometry, this chapter is mainly concerned with the automatic proving of theorems in geometry and identities in Clifford algebra. We explain how to express geometric concepts and relations and how to formulate geometric problems in the language of Clifford algebra. Several examples are given to illustrate a simple mechanism for deriving Clifford algebraic representations of constructed points, or other geometric objects, and how the representations may be used for proving theorems automatically. With explicit representations of geometric objects and simple substitutions, proving a theorem is reduced finally to verifying whether a Clifford algebraic expression is equal to 0. The latter is accomplished in our case by the techniques of term-rewriting for any fixed n.

We provide a short review of several coordinate-free approaches for automated theorem proving in geometry, including the work of White, Richter-Gebert, and Crapo at the level of bracket algebra [22, 17, 4], Mourrain's technique of deriving rewrite rules in Grassmann-Cayley algebra [15, 16], Li and Cheng's general approach of combining Clifford algebraic triangularization and reduction with Wu's coordinate-based method [14, 12], our study on proving theorems of constructive type using Clifford algebra and term-rewriting [20, 7, 1], and Havel's work [10] with Gibbs' vector algebra.

The above-mentioned approaches have been successful in proving a number of non-trivial geometric theorems. Clifford algebra approaches have the advantage of performing symbolic computations directly with geometric objects, so geometric meanings of the involved algebraic expressions may be easily interpreted. Geometric interpretation of bracket polynomials is possible by means of the Cayley factorization [22], but it is much more difficult to find the geometric interpretations of polynomials using coordinates. On the other hand, coordinate-based methods such as Wu's [23] are computationally more powerful and complete. To evaluate expressions in Clifford algebra, one cannot avoid the confluence problem and thus coordinates or pseudo-coordinates [1] may still have to be used. Therefore, coordinate-free techniques, though meriting attention for their geometric aspect, are unlikely to replace coordinate-based approaches for geometric problem solving.

The rewrite system that we have developed is independent of geometry and can be used or extended to prove all identities involving a *definite* number of (multi)vectors in any Clifford algebra of fixed dimension. We would also like to prove identities which involve an *arbitrary* number of (multi)vectors in a Clifford algebra of *arbitrary* dimension. By *arbitrary* we mean, for example, the outer product of r vectors and the Clifford algebra over an n-dimensional vector space, where r and n are arbitrary. We shall call such identities *indefinite identities*. This chapter initiates our investigation on proving indefinite identities in Clifford algebra automatically. We report an experiment in Maple V, which demonstrates the feasibility and effectiveness of proving indefinite identities on a computer. The program the author has written is based on the induction principle with heuristic simplification and was able to produce machine proofs for several nontrivial examples. The second part of this chapter is devoted to describing this work.

5.2   Clifford Algebra for Euclidean Geometry

Consider an n-dimensional Euclidean geometry $\mathbb{E}^n$ over the field $\mathcal{R}$ of real numbers. By taking a point O in $\mathbb{E}^n$ as the zero vector $\mathbf{0}$, any other point P in $\mathbb{E}^n$ may be represented by the vector $\mathbf{P}$ from O to P. We shall also call the vector $\mathbf{P}$ a point and make no distinction between P and $\mathbf{P}$ unless necessary.

For any two points P and Q, the line PQ connecting P and Q may be represented by the vector $\mathbf{Q} - \mathbf{P}$. In this way, PQ and $Q - P$ represent the same oriented line.

Let $\mathbf{a}$ and $\mathbf{b}$ be any two vectors. The geometric meaning of $\mathbf{a} + \mathbf{b}$ and $\mathbf{a} - \mathbf{b}$ is clear as in vector algebra. The *outer product* $\mathbf{a} \wedge \mathbf{b}$ is a *simple bivector*, which represents the oriented parallelogram spanned by $\mathbf{a}$ and $\mathbf{b}$. The outer product is *anti-commutative*: $\mathbf{a} \wedge \mathbf{b} = -\mathbf{b} \wedge \mathbf{a}$. Similarly, the *outer product*

$$\bigwedge_{i=1}^{r} \mathbf{a}_i = \mathbf{a}_1 \wedge \cdots \wedge \mathbf{a}_r$$

of r vectors is a *simple r-vector* representing the r-dimensional oriented simplex spanned by $\mathbf{a}_1, \ldots, \mathbf{a}_r$ for $3 \leq r \leq n$. For any scalar λ and simple

r-vector A,

$$\lambda \wedge A = A \wedge \lambda = \lambda A.$$

An r-*vector* is a linear combination of simple r-vectors (with coefficients in $\mathcal{R}$) for $0 \le r \le n$. The outer product of an r-vector A and an s-vector B

$$A \wedge B = (-1)^{rs} B \wedge A$$

is an $(r + s)$-vector if $r + s \le n$, and 0 otherwise.

A *multivector* is a linear combination of finitely many vectors, bivectors, ..., n-vectors, with coefficients in $\mathcal{R}$. Thus any multivector v may be written in the form

$$v = \sum_{i=0}^{n} \langle v \rangle_i,$$

where $\langle v \rangle_i$ is an i-vector, called the i-*vector part* of v. The multivector v is said to be *homogeneous* of *grade* r if it has only the r-vector part for some $0 \le r \le n$. The *magnitude* of v is defined by

$$|v| := |\langle v \rangle_0| + \sum_{i=1}^{n} \sqrt{|\langle v \rangle_i \cdot \langle v \rangle_i|}.$$

The *inner product* of two vectors **a** and **b** is a scalar:

$$\mathbf{a} \cdot \mathbf{b} = |\mathbf{a}|\,|\mathbf{b}| \cos \theta,$$

where θ denotes the angle formed by **a** and **b**. The inner product can be extended to any simple r-vector A and s-vector B as follows:

$$A \cdot B := \begin{cases} 0 & \text{if } r = 0, \\ (A \cdot \bar{B}) \wedge \mathbf{b} - (-1)^s (A \cdot \mathbf{b}) \bar{B} & \text{if } 1 = r < s, \\ \bar{A} \cdot (\mathbf{a} \cdot B) & \text{if } 1 < r \le s, \\ (-1)^{(r-1)s} B \cdot A & \text{if } r > s, \end{cases} \tag{2.1}$$

where **a** and **b** are vectors, $\bar{A}$ is a simple $(r - 1)$-vector such that $\bar{A} \wedge \mathbf{a} = A$ for $r > 1$, and $\bar{B}$ is a simple $(s - 1)$-vector such that $\bar{B} \wedge \mathbf{b} = B$ for $s > 1$. This recursive definition will be applied extensively in the inductive proof of indefinite identities (see Sections 5.4.1 and 5.4.2).

The *geometric product* Av of a simple r-vector A and a multivector v is defined as

$$Av := \begin{cases} A \wedge v + A \cdot v & \text{if } r \le 1, \\ \bar{A} \left\{ \left[\mathbf{a} - \dfrac{(\mathbf{a} \cdot \bar{A})\bar{A}}{\bar{A} \cdot \bar{A}} \right] v \right\} & \text{if } r = 2, \\ \bar{A} \left\{ \left[\mathbf{a} - \dfrac{(\mathbf{a} \cdot \bar{A}) \cdot \bar{A}}{\bar{A} \cdot \bar{A}} \right] v \right\} & \text{if } r > 2, \end{cases} \tag{2.2}$$

where $\mathbf{a}$ is a vector and $\bar{\mathsf{A}}$ a simple $(r-1)$-vector such that $\bar{\mathsf{A}} \wedge \mathbf{a} = \mathsf{A}$ for $r > 1$.

Moreover, for any scalars α, β and multivectors $\mathsf{u}, \mathsf{v}, \mathsf{w}, \mathsf{x}$, one has the following associativity, distributivity and linearity rules:

$$\mathsf{u} \wedge (\mathsf{v} \wedge \mathsf{w}) = (\mathsf{u} \wedge \mathsf{v}) \wedge \mathsf{w}, \qquad\qquad \mathsf{u}(\mathsf{vw}) = (\mathsf{uv})\mathsf{w};$$
$$\mathsf{w} \wedge (\alpha \mathsf{u} + \beta \mathsf{v}) = \alpha \mathsf{w} \wedge \mathsf{u} + \beta \mathsf{w} \wedge \mathsf{v}, \quad \mathsf{w}(\alpha \mathsf{u} + \beta \mathsf{v})\mathsf{x} = \alpha \mathsf{wux} + \beta \mathsf{wvx},$$
$$\mathsf{w} \cdot (\alpha \mathsf{u} + \beta \mathsf{v}) = \alpha \mathsf{w} \cdot \mathsf{u} + \beta \mathsf{w} \cdot \mathsf{v}.$$

Hence, the definitions of the outer, inner, and geometric products can be extended by using these rules to any multivectors.

The inner product is not associative, but for any r-vector A, s-vector B, and t-vector C, we have

$$\mathsf{A} \cdot (\mathsf{B} \cdot \mathsf{C}) = \begin{cases} (\mathsf{A} \cdot \mathsf{B}) \cdot \mathsf{C} & \text{if } r + t \leq s, \\ (\mathsf{A} \wedge \mathsf{B}) \cdot \mathsf{C} & \text{if } r, s \neq 0 \text{ and } r + s \leq t. \end{cases} \tag{2.3}$$

The above rules are not independent, and from them other rules may be derived.

All the multivectors under the addition and geometric multiplication form an associative algebra of dimension 2^n. It is called a *Clifford algebra* or *geometric algebra* (of positive-definite signature) [11] and may be used to model n-dimensional Euclidean geometry. There are other Clifford algebra models for Euclidean and other geometries, but this paper is concerned only with Euclidean geometry and this Clifford algebra model.

Two (multi)vectors are said to be *orthogonal* if their inner product is 0. Let $\mathit{I}\!\mathit{I}$ denote the outer product of n pairwise orthogonal unit vectors in $\mathbb{E}^n$; $\mathit{I}\!\mathit{I}$ is called a *pseudoscalar*.

For any r-vector V, we define $\mathsf{V}^\sim$, the *dual* of V, as follows:

$$\mathsf{V}^\sim := \begin{cases} (-1)^{\frac{n(n-1)}{2}} \mathsf{V} \mathit{I}\!\mathit{I} & \text{if } r = 0, \\ (-1)^{\frac{n(n-1)}{2}} \mathsf{V} \cdot \mathit{I}\!\mathit{I} & \text{otherwise.} \end{cases}$$

This definition extends naturally to an arbitrary multivector v:

$$\mathsf{v}^\sim := \sum_{i=0}^{n} \langle \mathsf{v} \rangle_i^\sim.$$

We shall prove that $\mathit{I}\!\mathit{I}^\sim = 1$.

Let us complete our introduction to Clifford algebra by making a few remarks on the soundness of the recursive definitions (2.1) and (2.2). In (2.1), the case $1 = r < s$ may be derived from the identity (1.41a) in [11] and the relation

$$\mathbf{b}\bar{\mathsf{B}} = \mathbf{b} \wedge \bar{\mathsf{B}} + \mathbf{b} \cdot \bar{\mathsf{B}},$$

and the case $1 < r \leq s$ follows from the second rule in (2.3). The other two cases in (2.1) and the cases for $r \leq 2$ in (2.2) may be easily verified. For the non-trivial case $r > 2$ in (2.2), we note that

$$[(\mathbf{a} \cdot \bar{\mathsf{A}}) \cdot \bar{\mathsf{A}}] \wedge \bar{\mathsf{A}} = 0 \quad \text{and} \quad (\mathsf{A} \cdot \bar{\mathsf{A}}) \cdot \bar{\mathsf{A}} = 0,$$

because A is a simple r-vector, $\mathbf{a} \cdot \bar{\mathsf{A}}$ is an $(r - 2)$-vector, and both $(\mathbf{a} \cdot \bar{\mathsf{A}}) \cdot \bar{\mathsf{A}}$ and $\mathsf{A} \cdot \bar{\mathsf{A}}$ are vectors (see [11, p. 20]). Moreover, it follows from (1.43) in [11] that

$$\bar{\mathsf{A}} \cdot \mathsf{A} = (\bar{\mathsf{A}} \cdot \bar{\mathsf{A}})\mathbf{a} - (\mathbf{a} \cdot \bar{\mathsf{A}}) \cdot \bar{\mathsf{A}} \quad \text{for} \quad r > 2.$$

The equality in (2.2) for $r > 2$ is thus established by (1.41c) and (1.41d) from [11]. In fact, our definition of the geometric product in this case provides a recursive treatment of the orthogonalization process.

We write *Clifford operators, identities, expressions, ...* for operators, identities, expressions, etc. in Clifford algebra. The reader is referred to the fundamental text [11] by Hestenes and Sobczyk for other important Clifford operators and a variety of Clifford identities relating these operators.

The Clifford algebra introduced above, together with its operators, provides a rich language for expressing concepts and relations in Euclidean geometry, and for automated theorem proving. For example:

- The distance between two points A and B is equal to $|A - B|$; the area of a triangle ABC is $|(A - B) \wedge (A - C)|/2$.

- The midpoint of A and B is $(A + B)/2$; the centroid of a triangle ABC is $(A + B + C)/3$.

- Two lines AB and CD are parallel iff $(A - B) \wedge (C - D) = 0$; they are perpendicular iff $(A - B) \cdot (C - D) = 0$.

- Three points A, B, C are collinear iff $(A - B) \wedge (A - C) = 0$, iff there exists a scalar λ such that $C = \lambda A + (1 - \lambda)B$.

- Point P lies on a circle centered at O with radius r iff

$$(P - O) \cdot (P - O) = r^2.$$

5.3 Geometric Theorem Proving

5.3.1 Deriving representations of geometric objects

One contribution of Li and Cheng [14, 12] to Clifford-algebra-based geometric reasoning is a set of solution formulas for systems of multi(vector) equations. These formulas are independent of any geometric knowledge and may be used to triangularize Clifford expressions and to establish certain representations of geometric objects. Here we present a different principle to derive formulas representing geometric objects. This simple principle makes use of geometric knowledge, and may be applied whenever a representation needs to be derived.

The idea is to take some basic Clifford representations of geometric relations in which parameters may be used. For any geometric object (typically a point) that may be constrained by several such relations, we proceed to eliminate the parameters in order to obtain an explicit representation for the object. We illustrate this idea by the following examples.

Example 5.1.

Let A, B, C, D be any four points in the plane. Represent the intersection point X of AB and CD in terms of A, B, C, D.

Since X is the intersection of AB and CD, X lies on AB. Thus there exists a scalar λ such that

$$X = \lambda A + (1 - \lambda)B. \tag{3.4}$$

On the other hand, X also lies on line CD, so

$$(D - C) \wedge [\lambda A + (1 - \lambda)B - C] = (D - C) \wedge (X - C) = 0.$$

The expression on the left-hand side may be simplified to $g\lambda + f$ with

$$g = D \wedge A - D \wedge B - C \wedge A + C \wedge B = (D - C) \wedge (A - B),$$
$$f = (D - C) \wedge B - D \wedge C.$$

Therefore, λ may be formally solved, $\lambda = -f/g$. However, the meaning of the formal expression f/g is not clear because one does not know what a fraction of two bivectors means. For this reason, we take the dual of

$$g\lambda + f : \quad g^{\sim}\lambda + f^{\sim} = 0.$$

It follows that $\lambda = -f^{\sim}/g^{\sim}$. By substituting this solution into (3.4), with simplification and arrangement, we find

$$X = \frac{A(C - B) \cdot D^{\sim} + A B \cdot C^{\sim} + B(A - C) \cdot D^{\sim} - B A \cdot C^{\sim}}{g^{\sim}},$$

which is equivalent to $\mathrm{int}(A, B, C, D)$ given in [20, 21].

Example 5.2.

Let C be the center of a circle, A a point on the circle and B any other point in the plane. Find the intersection point X of AB and the circle in terms of A, B, C.

Since X lies on AB, there exists a scalar λ such that

$$X = \lambda A + (1 - \lambda)B. \tag{3.5}$$

Since X also lies on the circle centered at C and passing through A, we have

$$(A - C) \cdot (A - C) = (X - C) \cdot (X - C). \tag{3.6}$$

It follows that

$$(\lambda - 1)[(A - B) \cdot (A - B)\lambda + A \cdot A - B \cdot B - 2C \cdot (A - B)] = 0.$$

The solution $\lambda = 1$ corresponds to the trivial case $X = A$. The other intersection point X is obtained by substituting the non-trivial solution of λ into (3.5):

$$
\begin{aligned}
X &= \text{int\_cir}(C, A, B) \tag{3.7}\\
&= \frac{[2C \cdot (A - B) - A \cdot A + B \cdot B] A + 2(A - B) \cdot (A - C) B}{(A - B) \cdot (A - B)},
\end{aligned}
$$

provided that A and B do not coincide.

Example 5.3.

Find the intersection point X of a circle centered at C and the line passing through a point A on the circle and perpendicular to a given vector $\mathbf{a}$ in the plane.

Since the line $X - A$ is perpendicular to $\mathbf{a}$, there exists a scalar λ such that $X = A + \lambda \mathbf{a}^\sim$. Substituting this expression into (3.6) and solving for λ, gives the two solutions $\lambda = 0$ and

$$\lambda = \frac{2(C - A) \cdot \mathbf{a}^\sim}{\mathbf{a} \cdot \mathbf{a}}.$$

The case $\lambda = 0$ corresponds to $X = A$. The other intersection point of interest is

$$X = \text{per\_int\_cir}(C, A, \mathbf{a}) = A + 2\frac{(C - A) \cdot \mathbf{a}^\sim}{\mathbf{a} \cdot \mathbf{a}} B, \tag{3.8}$$

provided that the vector $\mathbf{a}$ is non-zero.

Using the simple methods given in the above examples, expressions for many constructed geometric objects can be found. See [20, 7, 21] for more examples of such geometric constructions.

5.3.2 Examples of theorem proving

The following examples serve to illustrate how geometric theorems may be proved in the Clifford algebra formalism and how the above representations are used. They also demonstrate how to linearize geometric statements involving circles.

Example 5.4.

Let AB be the diameter of an arbitrary circle and C be any point on the circle. Then CA is perpendicular to CB.

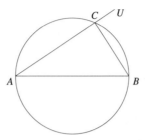

To simplify calculations, let the circle be centered at the origin. Then we have $B = -A$. Take a free point U on the plane and let the line AU intersect the circle at point C. According to (3.7), we have

$$C = \mathsf{int\_cir}(0, A, U) = \frac{(-A \cdot A + U \cdot U) A + 2 (A \cdot A - A \cdot U) U}{A \cdot A - 2 A \cdot U + U \cdot U}.$$

The conclusion of the theorem is

$$g = (C - A) \cdot (C - B) = C \cdot C - A \cdot A = 0.$$

This is easily proved by substituting the expression of C into g with simplification. The non-degeneracy condition for the theorem is $A \neq U$.

The proof of the above theorem is so simple that little algebraic computation and simplification need be performed. In most theorems in Euclidean geometry involving circles, the resulting expression cannot be easily reduced to 0 without using systematic means. Let us consider a couple of more examples.

From any point D on the circumcircle of an arbitrary triangle Δ, one may draw three perpendiculars to the three sides of Δ. Simson's theorem asserts that the intersections of the three perpendiculars are collinear (see, e.g., [20, 21]). Let the line determined by the intersections be called the *Simson line* of D for Δ. We have the following theorem.

Example 5.5.

The Simson line of any point D for a triangle ABC passes through the midpoint of D and the orthocenter of $\triangle ABC$.

Let $\mathsf{cir\_ctr}(A, B, C)$ and $\mathsf{ort\_ctr}(A, B, C)$ denote the circumcenter and orthocenter of $\triangle ABC$, respectively, and $\mathsf{per\_ft}(A, B, D)$ the intersection of the perpendicular from point D to the line AB. Their explicit expressions in terms of A, B, C, D may be found in [20, 21].

To simplify calculations, let A be located at the origin (i.e., $A = 0$). Then the hypothesis of the theorem can be stated constructively as follows:

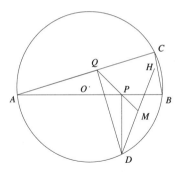

$$O = \mathsf{cir\_ctr}(A, B, C)$$
$$= \frac{B \cdot B\, B \cdot C^\sim B + [B \cdot B\, B \cdot C - (B \cdot C)^2 - (B \cdot C^\sim)^2] B^\sim}{2\, B \cdot B\, B \cdot C^\sim},$$

$$D = \mathsf{int\_cir}(O, A, U) = \frac{2\, O \cdot U}{U \cdot U} U,$$

$$H = \mathsf{ort\_ctr}(A, B, C) = \frac{B \cdot C\, [B \cdot C^\sim B + B \cdot (C - B) B^\sim]}{B \cdot B\, B \cdot C^\sim},$$

$$M = \mathsf{midp}(D, H) = \frac{D + H}{2},$$

$$P = \mathsf{per\_ft}(A, B, D) = \frac{B \cdot D}{B \cdot B} B,$$

$$Q = \mathsf{per\_ft}(C, A, D) = \frac{C \cdot D}{C \cdot C} C,$$

where U is a free point. The conclusion of the theorem to be proved is

$$g = (P - Q) \wedge (P - M) = 0.$$

Substituting the expressions of Q, P, M, H, D, O successively into g, one obtains an expression in B, C, U only. The numerator of this expression is

$$
\begin{aligned}
h = \{&[B \cdot B\, B \cdot C - (B \cdot C)^2 - (B \cdot C^\sim)^2]\, B \cdot U^\sim - B \cdot B\, B \cdot C^\sim B \cdot U\}^2 \\
&\{\ [B \cdot C\, (B \cdot U^\sim)^3 - B \cdot C^\sim B \cdot U\, (B \cdot U^\sim)^2 + B \cdot C\, (B \cdot U)^2\, B \cdot U^\sim \\
&\quad - B \cdot C^\sim (B \cdot U)^3]\, B \wedge U \\
&+ [B \cdot C^\sim (B \cdot U^\sim)^3 + B \cdot C\, B \cdot U\, (B \cdot U^\sim)^2 + B \cdot C^\sim (B \cdot U)^2\, B \cdot U^\sim \\
&\quad + B \cdot C\, (B \cdot U)^3]\, B \wedge U^\sim \\
&+ [B \cdot C\, (B \cdot U^\sim)^2 - 2\, B \cdot C^\sim B \cdot U\, B \cdot U^\sim \\
&\quad - B \cdot C\, (B \cdot U)^2]\, U \cdot U\, B \wedge B^\sim\}.
\end{aligned}
$$

The expression h does not automatically evaluate to 0, so it remains to show that h is identically equal to 0. For proving identities of this kind, we have developed a term-rewriting system by taking some of the laws for Clifford operators as rewrite rules. Using this system, the second factor of the expression h may be easily rewritten to 0. Therefore, the theorem is proved to be true provided that $B \cdot B\,C \cdot C\,B \cdot C^\sim U \cdot U \neq 0$, i.e., the triangle ABC does not degenerate to a line and $U \neq A$.

Example 5.6.

From a point D on the circumcircle of an arbitrary triangle ABC, draw three perpendiculars to the three sides BC, CA, AB of $\triangle ABC$ to meet the circle at points A_1, B_1, C_1 respectively. Then the lines AA_1, BB_1, CC_1 are parallel.

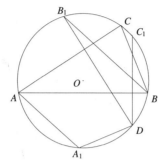

The constructions for the hypothesis of the theorem are

$$O = \mathsf{cir\_ctr}(A, B, C), \quad D = \mathsf{int\_cir}(O, A, U),$$
$$B_1 = \mathsf{per\_int\_cir}(O, D, C - A), \quad C_1 = \mathsf{per\_int\_cir}(O, D, A - B),$$

where U is again a free point. One of the conclusions to be proved is $BB_1 \parallel CC_1$, which may be expressed as

$$g = (B - B_1) \wedge (C - C_1) = 0.$$

Without loss of generality, let $A = 0$. Substitution of the expressions for C_1, B_1, D, O into g results in an expression involving only B, C and U. The numerator of this expression may then be rewritten to 0 by our system. Similarly, one may prove that $AA_1 \parallel BB_1$.

The above examples explicate a simple and efficient approach for proving geometric theorems automatically using Clifford algebra. The approach consists roughly of three steps:

(1) Formulate the hypothesis of the theorem in question constructively so that every introduced dependent point may be expressed in terms of the previously introduced points.

(2) Substitute the expressions for the dependent points into the Clifford equation to obtain an equality involving only the free points.

(3) Prove that the equality is an identity by term-rewriting.

A number of other well-known geometric theorems have been proved by using this approach (see [20, 7, 21]).

5.3.3   *Approaches to geometric reasoning*

Now we briefly summarize some of the available methods and techniques developed for automated geometric reasoning based on Clifford algebra. The reader may consult [13] for a more comprehensive review.

The use of geometric invariants in Clifford algebra may be seen in the early work of White [22] and Richter-Gebert [17]. The former gave a Cayley factorization algorithm for multilinear bracket polynomials that permits one to prove geometric theorems at the level of bracket algebra, and the latter developed a method that has proved a number of incidence theorems in projective geometry. The method of Richter-Gebert, based on bracket algebra, works by representing incidence relations using biquadratic equations. This method was extended later in [4] to prove theorems about circles in Euclidean geometry.

Under the operations of addition and the outer product, the multivectors form a Grassmann algebra of dimension 2^n. This Grassmann algebra, equipped with the *meet* operator, is what we call the *Grassmann-Cayley algebra*. Mourrain and Stolfi [15, 16] introduced a method that can generate a set of substitution rules from a set of constructions for points in projective geometry with the Grassmann-Cayley algebra formalism.

The Clifford algebra formalism has long been used to solve geometric problems in mathematics, physics, chemistry, and other areas. In the context of automated reasoning, Li and Cheng [14, 12] first proposed a general principle for proving geometric theorems and deriving geometric relations. The principle follows the paradigm of triangularization and reduction developed in Wu's coordinate-based method [23]. In order to make triangularization possible for Clifford expressions, Li and Cheng worked out a set of solution formulas for Clifford equations.

The author in [20] considered a class of constructive geometric theorems and devised a simple and effective method for their proof. This method avoids some redundant reductions that may occur in the method of Li and Cheng. The proof of a concrete theorem is simply to substitute the explicit expressions of the dependent geometric objects into the conclusion expression of the theorem and to evaluate the resulting expression h to 0 as shown by the examples in Section 5.3.2. Nevertheless, in Clifford algebra an expression that is identically equal to 0 does not necessarily evaluate to 0 automatically; it does when coordinates are used. An open question is how to prove identities in Clifford algebra automatically without using coordinates.

An approach suggested in [20] and developed in [7, 1] is based on term-rewriting. The basic idea is to use various fundamental relations and identities among different Clifford operators to rewrite Clifford expressions to normal forms systematically. The current version of the rewrite system for 2D and 3D as described in [1] consists of four major steps: basic simplification, reduction to normal forms, further simplification, and pseudo-coordinate expansion. The investigated issues include selection and grouping of rewrite rules, grading, normalization, termination, confluence, and completion, as well as strategic system design and combination to achieve efficiency.

The application of Clifford algebra to prove geometric theorems involving lines, circles, ellipses, parabolas, and hyperbolas has been studied by Yang *et al.* [24]. They also considered the normalization problem and presented several groups of rules to normalize and simplify Clifford expressions.

A Maple package called *Gibbs* was developed by Havel [10] and a coworker for the elementary expansion and simplification of expressions in Gibbs' vector algebra. The package was implemented during Havel's study of the local deformation problem in chemistry and is applicable to other reasoning problems in Euclidean geometry. The reader is referred to [10] as well as [22, 17].

An experiment with Maple V and Objective Caml on combining Clifford term-rewriting and algebraic computing for geometric theorem proving has been reported in [8]. Other related work on geometric reasoning using Clifford or vector algebra include the use of distances as coordinates [9], the application of non-commutative Gröbner bases in Grassmann algebra [20], the study of area in Grassmann geometry [6], the modeling of behavior of geometric objects using Euclidean ring [5], and the methods based on vector calculus described in [3, 19].

5.4   Proving Identities in Clifford Algebra

5.4.1   Introduction by examples

The rewrite system [1, 7, 8] we have developed is capable of proving identities in Clifford algebra of fixed dimension, for example, with $n = 2$ or 3. It is independent of geometry and can be used for automated reasoning in Clifford algebra. However, the system cannot be used to prove any Clifford identity for an indefinite n, nor any identity involving an indefinite number of vectors. There is a large number of such identities connecting different Clifford operators (see, for example, [11]). Our question is how to prove them automatically or semi-automatically on a computer. This is not an easy task. We propose to use the induction principle in combination with algebraic simplification, rewriting, heuristics, and other techniques. In this section, we present several examples to illustrate how the approach works. The machine proofs for the identities in these examples are produced automatically by a computer program that the author has written in Maple V.

Let us start with the following simple example. The proof is readable without need of an explanation.

Example 5.7.

Let $\mathbf{e}_1, \ldots, \mathbf{e}_n$ be n pairwise orthogonal unit vectors, i.e.,

$$\mathbf{e}_i \cdot \mathbf{e}_i = 1, \ \mathbf{e}_i \cdot \mathbf{e}_j = 0, \ i \neq j, \ 1 \leq i, j \leq n.$$

Show that

$$\mathbf{e}_m \cdot \bigwedge_{i=1}^{r} \mathbf{e}_i = 0 \ \text{ for } \ 1 \leq r < m \leq n.$$

Proof by induction on r.
Base case r = 1 :

$$0 = 0$$

reduces to True.
Assume that 1 ¡ r and the following induction hypothesis holds :

$$\mathbf{e}_m \cdot \left(\bigwedge_{k=1}^{r-1} \mathbf{e}_k \right) = 0$$

Proof of the induction case r :

$$\mathbf{e}_m \cdot \left(\bigwedge_{k=1}^{r} \mathbf{e}_k \right) = 0 \qquad\qquad (*)$$

The left-hand side of (*) reduces (by definition of . and/or simplification) to

$$(-1)^r \, \mathbf{e}_r \wedge \left(\mathbf{e}_m \cdot \left(\bigwedge_{j=1}^{r-1} \mathbf{e}_j \right) \right)$$

By the induction hypothesis, one gets

$$(-1)^r \, \mathbf{e}_r \wedge (0)$$

This reduces to

$$0$$

The above expression is equal to the right-hand side of (*) :

$$0$$

Q.E.D.
  One sees that in the above example and in what follows, that the recursive definition (2.1) of the inner product makes induction possible.

Example 5.8.

Let **a** and $\mathbf{b}_j$ all be vectors. Prove that

$$\mathbf{a} \cdot \bigwedge_{i=1}^{r} \mathbf{b}_i = \sum_{i=1}^{r}(-1)^{i+1}\mathbf{a} \cdot \mathbf{b}_i \left(\bigwedge_{j=1}^{i-1} \mathbf{b}_j\right) \wedge \left(\bigwedge_{j=i+1}^{r} \mathbf{b}_j\right) \text{ for } 1 < r \leq n.$$

This identity is numbered (1.38) in [11]. Our program proves it as follows.
  Proof by induction on r.
Base case r = 1 :

$$b_1 \cdot a = b_1 \cdot a$$

reduces to True.
Assume that 1 ¡ r and the following induction hypothesis holds :

$$a \cdot \left(\bigwedge_{i=1}^{r-1} b_i\right) = -\sum_{i=1}^{r-1}(-1)^i b_i \cdot a\left(\bigwedge_{j=1}^{i-1} b_j\right) \wedge \left(\bigwedge_{j=i+1}^{r-1} b_j\right)$$

Proof of the induction case r :

$$a \cdot \left(\bigwedge_{i=1}^{r} b_i\right) = -\sum_{i=1}^{r}(-1)^i b_i \cdot a\left(\bigwedge_{j=1}^{i-1} b_j\right) \wedge \left(\bigwedge_{j=i+1}^{r} b_j\right) \qquad (*)$$

The left-hand side of (*) reduces (by definition of . and/or simplification) to

$$\left(a \cdot \left(\bigwedge_{j=1}^{r-1} b_j\right)\right) \wedge b_r - (-1)^r a \cdot b_r \bigwedge_{j=1}^{r-1} b_j$$

By the induction hypothesis, one gets

$$\left(-\sum_{i=1}^{r-1}(-1)^i b_i \cdot a\left(\bigwedge_{j=1}^{i-1} b_j\right) \wedge \left(\bigwedge_{j=i+1}^{r-1} b_j\right)\right) \wedge b_r - (-1)^r a \cdot b_r \bigwedge_{j=1}^{r-1} b_j$$

This reduces to

$$\sum_{i=1}^{r-1} -(-1)^r b_i \cdot a\,(-1)^{r+i}\left(\bigwedge_{j=1}^{i-1} b_j\right) \wedge \left(\bigwedge_{j=i+1}^{r} b_j\right) - (-1)^r a \cdot b_r \bigwedge_{j=1}^{r-1} b_j$$

which simplifies to

$$-\sum_{i=1}^{r}(-1)^i b_i \cdot a\left(\bigwedge_{j=1}^{i-1} b_j\right) \wedge \left(\bigwedge_{j=i+1}^{r} b_j\right)$$

The above expression is equal to the right-hand side of (*) :

$$-\sum_{i=1}^{r}(-1)^i b_i \cdot a\left(\bigwedge_{j=1}^{i-1} b_j\right) \wedge \left(\bigwedge_{j=i+1}^{r} b_j\right)$$

Q.E.D.

Example 5.9.

Let **a** be a vector, B an r-vector, and C an s-vector. Prove that

$$\mathbf{a} \wedge (B \cdot C) = (\mathbf{a} \cdot B) \cdot C + (-1)^r B \cdot (\mathbf{a} \wedge C) \quad \text{for} \quad 1 < r \le s \le n. \quad (4.9)$$

This is the identity (1.43) given in [11] and mentioned before. Our machine proof proceeds by letting

$$B = \bigwedge_{i=1}^{r} \mathbf{b}_i$$

be a simple r-vector and making induction on r, where the $\mathbf{b}_i$ are vectors. The case in which B is an arbitrary r-vector follows from multilinearity.

Proof by induction on r.

Base case r = 2 :

$$a \wedge ((b_1 \wedge b_2) \cdot C) = a \cdot b_1 b_2 \cdot C - a \cdot b_2 b_1 \cdot C + (b_1 \wedge b_2) \cdot (a \wedge C)$$

The difference of the two sides reduces (by definition of . and/or simplification) to

$$a \wedge (b_1 \cdot (b_2 \cdot C)) - a \cdot b_1 b_2 \cdot C + a \cdot b_2 b_1 \cdot C - b_1 \cdot (b_2 \cdot (a \wedge C))$$

reduces (again by definition of . and/or simplification) to

$$a \wedge (b_1 \cdot (b_2 \cdot C)) - a \cdot b_1 b_2 \cdot C + b_1 \cdot (a \wedge (b_2 \cdot C))$$

reduces (again by definition of . and/or simplification) to

$$a \wedge (b_1 \cdot (b_2 \cdot C)) + (-1)^{s+|1-s|} a \wedge (b_1 \cdot (b_2 \cdot C))$$

reduces (again by definition of . and/or simplification) to

$$0$$

The base case is proved. Assume that $2 < r$ and the following induction hypothesis holds :

$$a \wedge ((\bigwedge_{i=1}^{r-1} b_i) \cdot C) = (a \cdot (\bigwedge_{i=1}^{r-1} b_i)) \cdot C + (-1)^{r-1} (\bigwedge_{i=1}^{r-1} b_i) \cdot (a \wedge C)$$

Proof of the induction case r :

$$a \wedge ((\bigwedge_{i=1}^{r} b_i) \cdot C) = (a \cdot (\bigwedge_{i=1}^{r} b_i)) \cdot C + (-1)^{r} (\bigwedge_{i=1}^{r} b_i) \cdot (a \wedge C) \quad (*)$$

The left-hand side of (*) reduces (by definition of . and/or simplification) to

$$a \wedge ((\bigwedge_{i=1}^{r-1} b_i) \cdot (b_r \cdot C))$$

By the induction hypothesis, one gets

$$(a \cdot (\bigwedge_{i=1}^{r-1} b_i)) \cdot (b_r \cdot C) + (-1)^{r-1} (\bigwedge_{i=1}^{r-1} b_i) \cdot (a \wedge (b_r \cdot C))$$

This reduces to

$$(a \cdot (\bigwedge_{i=1}^{r-1} b_i)) \cdot (b_r \cdot C) - (-1)^{r} (\bigwedge_{i=1}^{r-1} b_i) \cdot (a \wedge (b_r \cdot C))$$

The difference of the above expression and the right-hand side of (*) reduces (by definition of .) to

$$(a \cdot (\bigwedge_{i=1}^{r-1} b_i)) \cdot (b_r \cdot C) - (-1)^{r} (\bigwedge_{i=1}^{r-1} b_i) \cdot (a \wedge (b_r \cdot C)) - ((a \cdot (\bigwedge_{j=1}^{r-1} b_j)) \wedge b_r) \cdot C$$

$$+ (-1)^{r} a \cdot b_r (\bigwedge_{j=1}^{r-1} b_j) \cdot C - (-1)^{r} (\bigwedge_{i=1}^{r-1} b_i) \cdot (b_r \cdot (a \wedge C))$$

The above expression is reduced (by definition of . and/or simplification) to

$$0$$

Q.E.D.

Example 5.10.

Let $e_1, \ldots, e_n$ be as in Example 5.7. Prove that

$$\bigwedge_{i=1}^{r} e_i \cdot \bigwedge_{i=1}^{r} e_i = (-1)^{\frac{r(r-1)}{2}} \quad \text{for} \ 1 \le r \le n. \tag{4.10}$$

The following proof of (4.10) makes use of the identity shown in Example 5.7 as a lemma.

Proof by induction on r.

Base case r = 1 :

$$1 = 1$$

reduces to True.

Assume that $1 < r$ and the following induction hypothesis holds :

$$(\bigwedge_{i=1}^{r-1} e_i) \cdot (\bigwedge_{i=1}^{r-1} e_i) = (-1)^{\frac{(r-1)(r-2)}{2}}$$

Proof of the induction case r :

$$(\bigwedge_{i=1}^{r} e_i) \cdot (\bigwedge_{i=1}^{r} e_i) = (-1)^{\frac{r(r-1)}{2}} \qquad (*)$$

The left-hand side of (*) reduces (by definition of . and/or simplification) to

$$(\bigwedge_{i=1}^{r-1} e_i) \cdot (e_r \cdot (\bigwedge_{i=1}^{r} e_i))$$

It reduces (again by definition of .) to

$$(-1)^r (\bigwedge_{i=1}^{r-1} e_i) \cdot (e_r \wedge (e_r \cdot (\bigwedge_{j=1}^{r-1} e_j))) - (-1)^r (\bigwedge_{i=1}^{r-1} e_i) \cdot (\bigwedge_{j=1}^{r-1} e_j)$$

By the following lemma

$$e_m \cdot (\bigwedge_{k=1}^{r} e_k) = 0, \quad r < m$$

the expression is redued to

$$-(-1)^r (\bigwedge_{i=1}^{r-1} e_i) \cdot (\bigwedge_{j=1}^{r-1} e_j)$$

By the induction hypothesis, one gets

$$-(-1)^r (-1)^{\frac{(r-1)(r-2)}{2}}$$

This reduces to

$$(-1)^{\frac{r(r-1)}{2}}$$

The above expression is equal to the right-hand side of (*) :

$$(-1)^{\frac{r(r-1)}{2}}$$

Q.E.D.

With $r = n$ in (4.10), we have $I\!I \cdot I\!I = (-1)^{\frac{n(n-1)}{2}}$. It follows, from the definition of the dual operator, that $I\!I^\sim = 1$, which we have promised to prove.

Taking $I\!I$ for C in (4.9), we get

$$\mathbf{a} \wedge (\mathbf{B} \cdot I\!I) = (\mathbf{a} \cdot \mathbf{B}) \cdot I\!I,$$

that is,

$$\mathbf{a} \wedge \mathbf{B}^{\sim} = (\mathbf{a} \cdot \mathbf{B})^{\sim}.$$

This is one of the duality rules; the other duality rule is

$$\mathbf{a} \cdot \mathbf{B}^{\sim} = (\mathbf{a} \wedge \mathbf{B})^{\sim}.$$

The machine proofs in the above examples are formatted directly from files automatically generated by our program. The following window dump shows part of a proof session in Maple V.3.

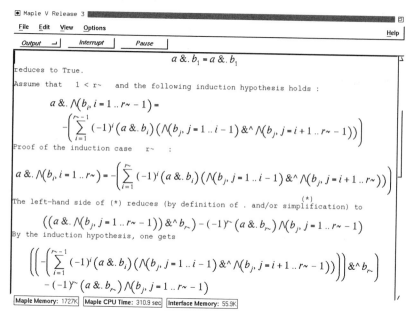

5.4.2 Principles and techniques

The problem is to build an effective prover that can produce machine proofs for sufficiently many Clifford algebra identities. We propose using mathematical induction because of the nature of identities in Clifford algebra, and the general applicability of the induction technique. Although induction is a standard in mathematical reasoning, its computer implementation for proving non-trivial theorems effectively and automatically is not a trivial task. In order to make inductive proofs possible, recursive definitions have to be introduced for Clifford operators and heuristics have to be implemented and used. Moreover, powerful routines of algebraic simplification and computation need to be incorporated. In what follows, we discuss some of these issues on the basis of our preliminary experiments.

The induction principle

Let $F(r) = 0$ be an identity that we want to prove. The induction scheme has the following form:

> If $F(r_0) = 0$, $r > r_0$, and $F(r-1) = 0$ implies that $F(r) = 0$ for all r, then $F(r) = 0$ holds for all $r \geq r_0$.

The base case $r = r_0$ is determined for the smallest value r_0 of the induction variable r satisfying the given conditions, and is proved by expanding definitions, simplification, and the application of lemmas. Note that the rewrite system described in [7, 1] may be used in this case when the rewrite rules are dimension-independent.

For the induction case r, definitions are expanded and simplified or rewritten in order to search for possibilities to use the induction hypothesis. The program then looks for expressions on the two sides of the identity that match.

It is possible to use well-developed, general induction theorem provers, such as the Boyer-Moore prover Nqthm [2], for our special purposes. We decided to experiment with Maple because of its capability for advanced algebraic computation and simplification and because of our previous experience with it.

Expanding definitions

One of the key points to the success of the proofs in Examples 5.7–5.10 is the recursive definition (2.1) of the inner product, which has been introduced to facilitate the application of induction. Expanding or opening up the definition of a function/operator is to replace the function call by its definition. Heuristics must be implemented to determine whether, when, and where a definition should be opened up. For example, in proving the induction case of (4.9), $a \cdot \bigwedge_{i=1}^{r} b_i$ is expanded to

$$(a \cdot \bigwedge_{i=1}^{r-1} b_i) \wedge b_r - (-1)^r (a \cdot b_r) \bigwedge_{i=1}^{r-1} b_i$$

and $b_r \cdot (a \wedge C)$ to

$$(-1)^s (b_r \cdot C) \wedge a + b_r \cdot a\, C,$$

but $a \cdot \bigwedge_{i=1}^{r-1} b_i$ is not expanded to

$$(a \cdot \bigwedge_{i=1}^{r-2} b_i) \wedge b_{r-1} - (-1)^{r-1} (a \cdot b_{r-1}) \bigwedge_{i=1}^{r-2} b_i.$$

An inappropriate expansion of a definition may lead to a failure in searching for a proof.

Simplification

This is one of the most complex and crucial processes in the proof of identities and other kinds of symbolic reasoning. The purpose of simplification is to replace terms and expressions by other equivalent and simpler ones. In our experiment, we take advantage of the powerful simplification mechanisms already available in Maple V. Additional rules are introduced to deal with some of the special properties of Clifford algebra. Pattern matching is needed for applying rewrite rules, induction hypotheses, and lemmas.

Manipulating indefinite objects

Although current computer algebra systems are powerful in dealing with definite symbolic objects, their ability to manipulate indefinite objects is quite limited. It is thus necessary to amend the computer algebra system in use for effective and correct manipulation of indefinite objects. Our implementation of new routines have benefited from our previous work [18] on the manipulation of indefinite sums and products. However, the enhanced capabilities of manipulating indefinite objects are still built on the top of the existing computer algebra systems.

Using lemmas

In order to maintain readability and structure, the proof of an identity may be shortened by applying lemmas. There is the general question of how many lemmas should be kept in the database. Using a large database of lemmas would not only increase the search time and space but also make proofs less interesting. Our program tests whether the lemmas can be used in the inductive proof.

The examples given in the preceding section do not involve the geometric product for multivectors of non-zero grade, nor the dual or other operators. However, the principles we have explained apply equally to proving identities involving other operators. Of course, an implementation covering more general cases will be very sophisticated involving recursive definitions. The recursive definition of the geometric product introduced in Section 5.2 is quite complicated. It may help to start with the definition of the geometric product for a special case, such as

$$\mathsf{B}\,\mathbf{a} := \mathsf{B} \wedge \mathbf{a} + \mathsf{B} \cdot \mathbf{a}$$

for any vector $\mathbf{a}$ and r-vector B with $r > 1$. This should make it possible to prove a good number of other identities. An ideal prover would include adequate definitions for the most popular Clifford algebra operators.

The induction approach is by no means complete. It is even difficult to describe precisely which class of identities such a prover can always succeed in proving. In the case of proving definite identities using our rewrite system, any identity can be proved when the use of coordinates or pseudo-coordinates are allowed. Nevertheless, for proving indefinite identities it is not even clear how coordinates can be applied.

Our induction based approach uses artificial intelligence to imitate human proofs, by incorporating powerful tools of computer algebra and simplification. Despite its limitations, it would be a success if one could use it to prove a large number of Clifford algebra identities automatically.

Acknowledgments. This work has been supported by CEC under Reactive LTR Project 21914 (CUMULI).

Chapter 6

Automated Theorem Proving

Hongbo Li

6.1 Introduction

In modern algebraic methods for automated geometry theorem proving, Wu's characteristic set method (Wu, 1978, 1994; Chou, 1988) and the Gröbner basis method (Buchberger, Collins and Kutzler, 1988; Kutzler and Stifter, 1986; Kapur, 1986) are two basic ones. In these methods, the first step is to set up a coordinate system, and represent the geometric entities and constraints in the hypothesis of a theorem by coordinates and polynomial equations. The second step is to compute a characteristic set or Gröbner basis by algebraic manipulations among the polynomials. The third step is to verify the conclusion of the theorem by using the characteristic set or Gröbner basis.

Coordinate representations do not keep geometric meaning and the algebraic manipulations of polynomials of coordinates are sometimes too complicated to be carried out on modern computers. For this reason, in recent years there has been a trend to use geometric invariants for algebraic representations, combining Wu's method or the Gröbner basis method with an algebra of geometric invariants for algebraic manipulations. Among these methods are those of Crapo and Richter-Gebert (1995) which integrate Grassmann-Cayley algebra with the Gröbner basis method, the method of Mourrain (1999) which combine Grassmann-Cayley algebra with Wu's method, and the area method (Chou, Gao and Zhang, 1994; Yang, Gao, Chou and Zhang, 1996) which combine a set of high-level geometric invariants with Wu's method. Proofs based on these methods are often shorter, more readable and have better geometric meaning.

Because of the general applicability of Clifford algebra to geometry, it is natural to consider combining Clifford algebra with Wu's method or the

Academy of Mathematics and Systems Science, Chinese Academy of Sciences, Beijing 100080, P. R. China. This paper is written during the author's visit to the Institut für Informatik und Praktische Mathematik, Christian-Albrechts-Universität, D-24105 Kiel, Germany. It is supported partially by the DFG Foundations and the AvH Foundations of Germany, the Grant NKBRSF of China, the Hundred People Program of the Chinese Academy of Sciences, and the Qiu Shi Science and Technology Foundations of Hong Kong.

Gröbner basis method. In Li (1994, 1996), a general framework is proposed for combining Clifford algebra with Wu's method. Various applications that belong to this framework are carried out in automated theorem proving in Euclidean, non-Euclidean and differential geometries (Li, 1995, 1997a, 1997b, 1999; Li and Cheng, 1997, 1998a, 1998b; Li and Shi, 1997). Research and applications have also been carried out by Wang (1996, 1998) and his group in combining Clifford algebra with the Gröbner basis method and term rewriting techniques.

This chapter is composed of three sections. In the first section, we talk about a general framework of combining Clifford algebra with Wu's method. In the second and third sections, we present examples of automated theorem proving in Euclidean and differential geometries.

6.2 A general Framework for Clifford algebra and Wu's Method

Given a geometric problem, there are different levels of algebraic representations for it. There are corresponding algebraic manipulations for each level of representation. When changing a high-level representation to a low-level one, substitutions, expansions and simplifications are usually enough. However, there is in general no way to reverse this procedure.

For example, given a problem in projective geometry, we can represent it in the Grassmann-Cayley algebra, and use the algebraic operations of "∧" and "∨" for geometric computations. We can also use the projective space of one-dimensional subspaces of a Euclidean space, and use the inner products for algebraic manipulations, although they do not have a projective interpretation. Finally, we can use homogeneous coordinates to represent geometric entities, matrices to represent projective transformations, and homogeneous polynomials for algebraic manipulations. Among these three representations, the first is on the highest level, the third is on the lowest level. From the first representation to the third, there are fewer and fewer geometric invariance, but more and more abundant algebraic manipulations to work with.

In automated theorem proving, if we use a low-level representation at the beginning, for example coordinate representation, then in many cases the polynomials we need to deal with are of tens of variables and thousands of terms. Symbolic computations of such polynomials often fail on present-day computers. On the other hand, if we use a high-level representation at the beginning, then we have only limited algebraic manipulations, and for many theorems we cannot derive the conclusions from the hypotheses.

To improve this situation, it should be possible to represent a geometric problem at a suitably high level at the beginning, and carry out the algebraic manipulations to the hypotheses of the theorem in order to establish the conclusion. If the conclusion fails to be proved at this level, then the level of representation is lowered mechanically for more algebraic manipulations. At the bottom is the coordinate representation. After every change of representation, new algebraic manipulations are carried out to the result of the previous manipulations, so even though the conclusion cannot be proved after algebraic manipulations of a high-level representation, simplification of the hypotheses can often be achieved for later algebraic manipulations.

A general framework realizing the above idea by combining Clifford algebra with Wu's method is proposed by Li (1994, 1996), Li and Cheng (1997). For triangulation in Clifford algebra formalism, a new technique called *vectorial equation-solving* is proposed. For mechanically changing Clifford algebra representations by introducing new geometric parameters, a technique called *parametric equation-solving* is proposed. The method is complete because Wu's method is resorted to at the level of coordinate representation. The method can be described as follows:

Stage 1. Triangulate the hypothesis using substitutions, pseudo-divisions and vectorial equation-solving. The result is called a triangular sequence. Prove the conclusion with the triangular sequence, and continue if the proof fails.

By triangulation, we mean obtaining a set of equations AS from a given set of equations PS such that the leading elements of the equations in AS follow a strictly ascending order with respect to a prescribed order of variables. The equations in AS must satisfy the relation

$$zero(AS/I) \subseteq zero(PS) \subseteq zero(AS), \qquad (2.1)$$

where $zero(AS)$ denotes all complex solutions of AS, I is a set of equations, and

$$zero(AS/I) = zero(AS) - \bigcup_{i \in I} zero(i).$$

The set $\{i \neq 0 | i \in I\}$ is called nondegeneracy conditions.

Stage 2. Triangulate the triangular sequence with vectorial the equation-solving method, the parametric equation-solving method, substitutions and pseudo-divisions. The result is called a *parametric triangular sequence*. Prove the conclusion with the parametric triangular sequence, and continue if the proof fails.

By the parametric equation-solving method, we mean solving equations by allowing the solutions to be in parametric form, the parameters can be either free or constrained.

Stage 3. Select a coordinate system and translate all expressions in the parametric triangular sequence into polynomials of coordinates. Use Wu's method to compute a characteristic set and prove the conclusion with the characteristic set.

A characteristic set AS of a set of polynomials PS is another set of polynomials such that each element of AS is reduced by every other element of AS with respect to a prescribed order of variables, and the relation (2.1) is satisfied.

We illustrate the method with a theorem in solid geometry:

Theorem 6.1.   Let $ABCD$ be a tetrahedron. Let the plane M, N, E, F be defined by the the respective points on the lines AB, AC, DC, DB. If the plane moves in such a way that $MNEF$ is always a parallelogram, then the center O of the parallelogram is always on a fixed straight line.

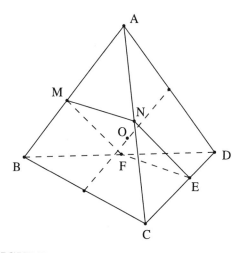

FIGURE 6.1. A theorem in solid geometry.

We embed the space in $\mathcal{R}^4$ as a hyperplane away from the origin. In the Clifford algebra $\mathcal{G}_4$, the hypothesis can be represented by

$$
\begin{cases}
M - N = F - E, & MNEF \text{ is a parallelogram} \\
A \wedge B \wedge M = 0, & A, B, M \text{ are collinear} \\
A \wedge C \wedge N = 0, & A, C, N \text{ are collinear} \\
C \wedge D \wedge E = 0, & C, D, E \text{ are collinear} \\
B \wedge D \wedge F = 0, & B, D, F \text{ are collinear} \\
O = \dfrac{M + E}{2}, & O \text{ is the midpoint of } ME \\
A \wedge B \wedge C \wedge D \neq 0, & A, B, C, D \text{ are not coplanar}
\end{cases}
\qquad (2.2)
$$

The conclusion cannot be algebraized.

The order of the variables for triangulation is: $A \prec B \prec C \prec D \prec M \prec N \prec E \prec F \prec O$.

After triangulation, we get the following triangular sequence:

$$2O - M - E = 0,$$
$$F - E - M + N = 0,$$
$$E(A \wedge B \wedge C \wedge D)^{\sim} - D(A \wedge B \wedge C \wedge D)^{\sim}$$
$$\qquad -D(A \wedge B \wedge D \wedge N)^{\sim} + C(A \wedge B \wedge D \wedge N)^{\sim} = 0,$$
$$N(A \wedge B \wedge C \wedge D)^{\sim} - C(A \wedge B \wedge C \wedge D)^{\sim}$$
$$\qquad -C(B \wedge C \wedge D \wedge M)^{\sim} + A(B \wedge C \wedge D \wedge M)^{\sim} = 0,$$
$$A \wedge B \wedge M = 0.$$

The nondegeneracy condition is $A \wedge B \wedge C \wedge D \neq 0$, which is in the original hypothesis. The conclusion cannot be obtained from the triangular sequence, as M does not have an explicit expression.

After parametric triangulation, we obtain the following parametric triangular sequence:

$$\begin{aligned}
&2O - A - D + \lambda A + \lambda D - \lambda B - \lambda C = 0,\\
&F - D + \lambda D - \lambda B = 0,\\
&E - D + \lambda D - \lambda C = 0, \qquad\qquad\qquad (2.3)\\
&N - A + \lambda A - \lambda C = 0,\\
&M - A + \lambda A - \lambda B = 0,
\end{aligned}$$

where λ is a parameter generated by the parametric equation-solving proceedure.

The nondegeneracy conditions are:

$$A \wedge B \wedge C \wedge D, \quad B - A \neq 0,$$

which are all guaranteed by the original hypothesis.

The conclusion is obvious from the first equality in (2.3):

$$O = (1 - \lambda)\frac{A + D}{2} + \lambda\frac{B + C}{2},$$

i. e., O is on the line passing through the midpoint of the line segment AD and the midpoint of the line segment BC.

6.3   Automated Theorem Proving in Euclidean Geometry and Other Classical Geometries

The method can be used to prove theorems in Euclidean, affine, projective, non-Euclidean, and differential, geometries, and can be used in mechanics and robotics. The proofs produced are often readable because they are short and have geometric interpretations.

We have successfully applied this method to study a conjecture by Erdös et al. around 1994. The original problem is from Erdös, Jackson and Mauldlin:

Let A_{ij}, $1 \leq i < j \leq 5$ be 10 points in the plane. If there are five points A_k, $1 \leq k \leq 5$ in this plane, including points at infinity, such that at least two are distinct and such that A_i, A_j, A_{ij} are collinear for $1 \leq i < j \leq 5$, we say that the five points form a consistent 5-tuple. Now assuming that no three of the A_{ij}'s are collinear, is it true that there are only finitely many consistent 5-tuples?

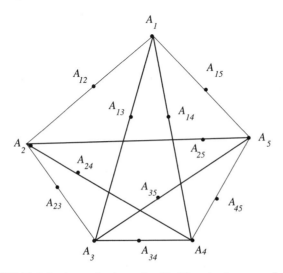

FIGURE 6.2. A conjecture by Erdös et al. around 1994.

Erdös et al. proved that if there are only finitely many solutions, then there are at most 49. They ask Boyer if their theorem prover could solve it. Boyer in turn sent the problem to Chou and Gao, and Chou and Gao sent it to us. Chou and Gao proved that generically there are only finitely many solutions. We proved in Li and Shi (1997), using our method and techniques from algebraic geometry, that

Theorem 6.2. For 10 generic points A_{ij} on the plane, there are at most 6 solutions.

We further proved that

Theorem 6.3. If no three of the 10 points A_{ij} are collinear, and no three of the lines connecting two points are concurrent, except those lines meeting at one of the points, then there are at most 6 solutions.

6.4 Automated Theorem Proving in Differential Geometry

In differential geometry, the local theory of space curves and space surfaces are of fundamental importance. E. Cartan's moving frame method and calculus of exterior differential forms are two important techniques for dealing with these theories. In our method, we integrate both techniques with Wu's method. When applied to theorems in space curve theory, our method can produce proofs that are similar to those given in textbooks. When applied to theorems in space surface theory, our method can often produce proofs that are simpler than those found in textbooks.

Moreover, the method can be used to prove complicated theorems. Below we give an example of a theorem first proposed by E. Cartan and later proved by S. S. Chern (1985).

First, we set up the notation that we will use in the local theory of space surfaces. Consider a sufficiently small piece of the smooth surface M in $\mathbf{E}^3$. Over M, there is a frame of orthonormal fields $\{x; e_1, e_2, e_3\}$ such that for each $x \in M$, e_3 is the unit normal vector at x, and e_1, e_2 are tangent vectors. These fields make up a first-order frame field. If, moreover, e_1 and e_2 are along the principal directions, the fields make up a second-order frame field.

The equations of motion of the first-order frame field are

$$\left\{ \begin{array}{llll} dx = & \omega_1 e_1 & +\omega_2 e_2 & \\ de_1 = & & \omega_{12} e_2 & +\omega_{13} e_3 \\ de_2 = & -\omega_{12} e_1 & & +\omega_{23} e_3 \\ de_3 = & -\omega_{13} e_1 & -\omega_{23} e_2 & \end{array} \right. \tag{4.4}$$

The Gauss-Codazzi equations of a second-order frame field are

$$d\omega_1 = \omega_{12} \wedge \omega_2, \qquad d\omega_2 = \omega_1 \wedge \omega_{12}; \tag{4.5}$$

$$d\omega_{12} = -K\omega_1 \wedge \omega_2; \tag{4.6}$$

$$da = (a - c)(u\omega_1 + *\omega_{12}), \qquad dc = (a - c)(v\omega_2 - *\omega_{12}). \tag{4.7}$$

Here u, v are two scalars, and "$*$" is the Hodge dual operator:

$$*\omega_1 = \omega_2, \quad *\omega_2 = -\omega_1.$$

Theorem 6.4. [Chern's Theorem] A non-trivial family of isometric surfaces having the same principal curvatures is either a family of surfaces of constant mean curvature, or a family of Weingarten surfaces of non-constant mean curvature, assuming that they do not contain umbilics and are c^5.

Let a, c be the principal curvatures at a point of a surface in the family. The non-umbilic assumption is equivalent to $a \neq c$. A Weingarten surface is a surface satisfying $da \wedge dc = 0$ at every point.

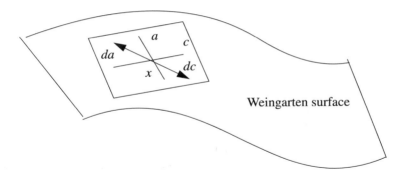

FIGURE 6.3. Chern's Theorem.

Suppose that M^* is a surface which is isometric to M. We shall denote the quantities and equation numbers pertaining to M^* by the same symbols with asterisks. Suppose that τ is the angle between the principal directions of M and M^* at their corresponding points with the same parameters, then

$$\begin{cases} \omega_1^* = \cos\tau\omega_1 - \sin\tau\omega_2 \\ \omega_2^* = \sin\tau\omega_1 + \cos\tau\omega_2 \end{cases}, \tag{4.8}$$

and their principal curvatures are preserved,

$$a^* = a, \quad c^* = c. \tag{4.9}$$

The non-triviality assumption for the family of isometric surfaces is equivalent to the requirement that (4.8) holds for τ over an interval on the τ-axis. We do not have an algebraic representation of this assumption at present.

The hypotheses are (4.8), (4.9) and the above requirement on τ. The conclusion is

$$da \wedge dc = 0, \tag{4.10}$$

since surfaces of constant mean curvature are also Weingarten surfaces.

Now we do triangulation. The input for the triangulation is:

$$(4.5), (4.6), (4.7), (4.5^*), (4.6^*), (4.7^*), (4.8), (4.9).$$

The order of variables for the triangulation is

$$a, c \prec u, v \prec \omega_1, \omega_2 \prec \omega_{12}, *\omega_{12} \prec du, dv \prec d * \omega_{12} \prec \tau \prec d\tau \prec$$
$$a^*, c^* \prec u^*, v^* \prec \omega_1^*, \omega_2^* \prec \omega_{12}^*, *\omega_{12}^* \prec du^*, dv^* \prec d * \omega_{12}^* \prec \tag{4.11}$$
$$da, dc, d\omega_1, d\omega_2, d\omega_{12}, da^*, dc^*, d\omega_1^*, d\omega_2^*, d\omega_{12}^*.$$

After triangulation, we get the following triangular sequence:
$triseq = (4.5), (4.6), (4.7), (4.8), (4.9),$ and

$$\begin{cases} \omega_{12}^* = \omega_{12} - d\tau \\ u^* = \cos\tau u - \sin\tau v \\ v^* = \sin\tau u + \cos\tau v \\ d\tau = (v\omega_1 + u\omega_2)\sin^2\tau + (-u\omega_1 + v\omega_2)\sin\tau\cos\tau \end{cases}, \tag{4.12}$$

$$\begin{cases} du \wedge \omega_1 + d * \omega_{12} - v\omega_2 \wedge *\omega_{12} + uv\omega_1 \wedge \omega_2 = 0 \\ dv \wedge \omega_2 - d * \omega_{12} - u\omega_1 \wedge *\omega_{12} + uv\omega_1 \wedge \omega_2 = 0 \end{cases}, \qquad (4.13)$$

$$\sin \tau \{ \sin \tau (du \wedge \omega_2 + dv \wedge \omega_1 + (u\omega_2 + v\omega_1) \wedge *\omega_{12}$$
$$- (u^2 + v^2)\omega_1 \wedge \omega_2)$$
$$- \cos \tau (du \wedge \omega_1 - dv \wedge \omega_2 + (u\omega_1 - v\omega_2) \wedge *\omega_{12}) \} = 0. \qquad (4.14)$$

The conclusion cannot be proved by the above triangular sequence. The hypothesis of the non-triviality of family does not have an appropriate algebraic representation, and therefore cannot participate in the triangulation.

The first key to the automatic proof is the algebraization of this hypothesis, based on the triangular sequence obtained so far. Since (4.14) holds for τ over an interval on the τ-axis, the coefficients before $\sin \tau$ and $\cos \tau$ must be zeroes,

$$\begin{cases} du \wedge \omega_2 + dv \wedge \omega_1 + (u\omega_2 + v\omega_1) \wedge *\omega_{12} - (u^2 + v^2)\omega_1 \wedge \omega_2 = 0 \\ du \wedge \omega_1 - dv \wedge \omega_2 + (u\omega_1 - v\omega_2) \wedge *\omega_{12} = 0 \end{cases}.$$

$$(4.15)$$

Thus, we have obtained an algebraic representation of the non-triviality hypothesis.

Now we do triangulation for set of equations

$$(4.5), (4.6), (4.7), (4.8), (4.9), (4.12), (4.13), (4.15).$$

The order of variables is the same as before.

We get the new triangular sequence: $new\text{-}triseq = (4.5), (4.6), (4.7), (4.8), (4.9), (4.12),$ and

$$d*\omega_{12} = 0, \qquad (4.16)$$
$$du \wedge \omega_1 - v\omega_2 \wedge *\omega_{12} + uv\omega_1 \wedge \omega_2 = 0, \qquad (4.17)$$
$$dv \wedge \omega_2 - u\omega_1 \wedge *\omega_{12} + uv\omega_1 \wedge \omega_2 = 0, \qquad (4.18)$$
$$(uvdu + v^2 dv) \wedge \omega_2 + (uvdv + u^2 du) \wedge \omega_1 = 0. \qquad (4.19)$$

The conclusion still cannot be proved by the new triangular sequence.

We have reached the stage of parametric triangulation. The following lemma provides a powerful technique for parametric equation-solving, and is the second key point of our proof:

Lemma 6.1. [Cartan's Lemma] Suppose $\omega_1, \omega_2, \ldots, \omega_r; \theta_1, \theta_2, \ldots, \theta_r$ are 1-forms in the n-dimensional vector space V, and $\omega_1 \wedge \omega_2 \wedge \cdots \wedge \omega_r \neq 0$. If $\sum_{i=1}^{r} \omega_i \wedge \theta_i = 0$, then there exist scalars $a_{ij} = a_{ji}, i, j = 1, 2, \ldots, r$, such that

$$\theta_i = \sum_{j=1}^{r} a_{ij}\omega_j.$$

Applying this lemma to (4.17), (4.18), (4.19) we get that there exist scalars p, q, B such that

$$
\begin{aligned}
\omega_{12} &= p\omega_1 + q\omega_2, & (4.20) \\
udu + vdv &= B(u\omega_1 + v\omega_2). & (4.21)
\end{aligned}
$$

This is the procedure of parameterization.

We are ready to do triangulation again. This time the triangulation is on the set

$$(4.5), (4.6), (4.7), (4.8), (4.9), (4.12),$$
$$(4.16), (4.17), (4.18), (4.19), (4.20), (4.21).$$

The variables $p, q \prec B \prec dp, dq \prec dB$ are inserted before $\omega_{12}, *\omega_{12} \prec du, dv$ in the order of the sequence (4.11).

We get the following parametric triangular sequence: $par\text{-}triseq = (4.5)$, $(4.7), (4.8), (4.9), (4.12), (4.20),$

$$
\begin{cases}
du = (B + v^2 - vp)\omega_1 + (uv - vq)\omega_2 \\
dv = (up - uv)\omega_1 + (B + uq - u^2)\omega_2
\end{cases}, \qquad (4.22)
$$

$$
\begin{cases}
dB = \big((u - 2q)B + u(u^2 + v^2 + ac)\big)\,\omega_1, \\
\qquad + \big((2p - v)B + v(u^2 + v^2 + ac)\big)\,\omega_2 \quad , \\
(u^2 + v^2)dp = f_1\omega_1 + f_2\omega_2
\end{cases} \qquad (4.23)
$$

where f_1, f_2 are polynomials of a, c, u, v, B, p, q, and

$$
pu + qv - uv = 0. \qquad (4.24)
$$

The conclusion is proved using the parametric triangular sequence by simple substitution. This finishes the proof.

6.5 Conclusion

Using Clifford algebra, we are able to design and realize fast computer algorithms which challenge geometric experts by the capability of producing readable computer-generated proofs, discovering new theorems and solving open problems. Clifford algebra proves to be important and efficient in doing scientific research for geometers of various groups.

Part III

Computer Vision

Chapter 7

The Geometry Algebra of Computer Vision

Eduardo Bayro Corrochano and Joan Lasenby

7.1 Introduction

In this chapter we present a mathematical approach for the computation of problems in computer vision which is based on *geometric algebra*. We will show that geometric algebra is a well-founded and elegant language for expressing and implementing those aspects of linear algebra and projective geometry that are useful for computer vision. Since geometric algebra offers both geometric insight and algebraic computational power, it is useful for tasks such as the computation of projective invariants, camera calibration and recovery of shape and motion. We will mainly focus on the geometry of multiple uncalibrated cameras

Geometric algebra [15] is a coordinate-free approach to geometry based on the algebras of Grassmann [10] and Clifford [7]. The algebra is defined on a space spanned with a multivector basis. A multivector is a linear combination of basic geometric objects of different order, e.g. scalars, vectors and bivectors. The system has an associative and fully invertible product called the *geometric product* or *Clifford product*. The existence of such a product gives the system tremendous representational and computational power. For some preliminary applications of geometric algebra in the field of computer vision see [2, 3, 4, 18]. We will show that geometric algebra provides a very natural language for projective geometry and has all the necessary equipment for the tasks which the Grassmann-Cayley algebra is currently used for. The Grassmann-Cayley or double algebra [6] is a mathematical system for computations with subspaces of finite-dimensional vector spaces. While this algebra expresses the ideas of projective geometry, such as the meet and join, very elegantly, it lacks an inner (regressive) product and some other key concepts which are useful both analytically and in reducing the computational cost in calculations.

The next section will give a brief introduction to the 3–D and 4–D geometric algebras. This section is also devoted to the formulation, in the geometric algebra framework, of those aspects of projective geometry relevant for computer vision. The reader can consult [13] for a more complete introduction and for other brief summaries see [3, 5]. Given this background, we will look at the concepts of projective split and projective transformations. Section three presents the algebra of incidence and section four the algebra in projective space of points, lines and planes. The analysis of monocular, binocular and trinocular geometries is given in section five. Conclusions are presented in the final section.

In this chapter vectors will be bold quantities (except for basis vectors) and multivectors will not be slant bold. Lower case is used to denote vectors in 3-D Euclidean space and upper case to denote vectors in 4-D projective space. We will denote a geometric algebra $\mathcal{G}_{p,q,r}$ referring to an n-D geometric algebra in which p basis vectors square to $+1$, q to -1 and r to 0, so that $p + q + r = n$.

7.2   The Geometric Algebras of 3-D and 4-D Spaces

The need for a mathematical framework to understand and process digital camera images of the 3-D world, prompted researchers in the late seventies to use *projective geometry*. Using homogeneous coordinates, we embed the 3-D Euclidean visual space in the projective space P^3 or R^4 and the 2-D Euclidean space of the image plane in the projective space P^2 or R^3. As a result, the inherently non-linear projective transformations from 3-D space to the 2-D image space become linear. In addition, *points* and *directions* are now differentiated instead of being represented by the same quantity. The choice of projective geometry was indeed a step forward; however, there is still the need, [14], for a mathematical system which reconciles projective geometry and multilinear algebra. In most of the computer vision literature we can indeed see that they are considered as divorced mathematical systems. When required, it is also common to resort to other systems; for example, the dual algebra [6] for incidence algebra and the Hamiltonian formulation for motion estimation [23]. Here we suggest the use of a system which offers all of these mathematical facilities. Unlike matrix and tensor algebra, geometric algebra does not obscure the underlying geometry of the problem. We will therefore formulate the main aspects of such problems in geometric algebra, starting with the modelling of 3-D visual space and the 2-D image plane.

7.2.1   3-D space and the 2-D image plane

To introduce the basic geometric models in computer vision, we consider the imaging of a point $\mathbf{X} \in R^4$ into a point $\boldsymbol{x} \in R^3$ assuming that the reader is familiar with the basic concepts of using homogeneous coordinates – these will also be discussed in later sections. The optical centre, C, of the camera may be different from the origin of the world coordinate system, O, as depicted in figure (7.1).

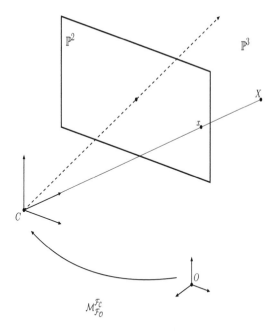

FIGURE 7.1. Pinhole camera model.

In the standard matrix representation, the mapping $P : \mathbf{X} \longrightarrow x$ is expressed by the homogeneous transformation matrix

$$P = \begin{bmatrix} p_{11} & p_{12} & p_{13} & p_{14} \\ p_{21} & p_{22} & p_{23} & p_{14} \\ p_{31} & p_{32} & p_{33} & p_{34} \end{bmatrix} \tag{2.1}$$

which may be decomposed into a product of three matrices

$$P = K P_0 M_0^c. \tag{2.2}$$

where P_0, K and M_0^c will now be defined. P_0 is the 3×4 matrix

$$\begin{bmatrix} 1 & 0 & 0 & 0 \\ 0 & 1 & 0 & 0 \\ 0 & 0 & 1 & 0 \end{bmatrix} \tag{2.3}$$

which simply projects down from 4-D to 3-D, representing a projection from homogeneous coordinates of space to homogeneous coordinates of the image plane.

M_0^c represents the 4×4 matrix containing the rotation and translation which takes the world frame, $\mathcal{F}_0$, to the camera frame $\mathcal{F}_c$ and is given explicitly by

$$M_0^c = \begin{bmatrix} R & t \\ \mathbf{0}^T & 1 \end{bmatrix}. \tag{2.4}$$

This Euclidean transformation is described by the *extrinsic parameters* of rotation (3×3 matrix R) and translation (3×1 vector $\boldsymbol{t}$). Finally, the 3×3 matrix K, expresses the assumed camera model as an *affine transformation* between the camera plane and the image coordinate system, so that K is an upper triangular matrix. In the case of the perspective (or pinhole) camera the matrix K, which we now call K_p, is given by

$$K_p = \begin{bmatrix} \alpha_u & \gamma & u_0 \\ 0 & \alpha_v & v_0 \\ 0 & 0 & 1 \end{bmatrix}. \tag{2.5}$$

The five parameters in K_p represent the camera parameters of scaling, shift and rotation in the camera plane. In this case the distance from the optical centre to the image plane is finite. In later sections we will formulate the *perspective camera* in the geometric algebra framework.

One important task in computer vision is to estimate the matrix of intrinsic camera parameters, K_p, and the rigid motion given in M_0^c, in order to be able to reconstruct 3-D data from image sequences.

7.2.2 The geometric algebra of 3-D Euclidean space

The 3-D space is spanned by three basis vectors $\{\sigma_1, \sigma_2, \sigma_3\}$ (with $\sigma_i^2 = +1$ for all $i = 1, 2, 3$) and the 3–D geometric algebra generated by these basis vectors has $2^3 = 8$ elements given by:

$$\underbrace{1}_{scalar}, \underbrace{\{\sigma_1, \sigma_2, \sigma_3\}}_{vectors}, \underbrace{\{\sigma_1\sigma_2, \sigma_2\sigma_3, \sigma_3\sigma_1\}}_{bivectors}, \underbrace{\{\sigma_1\sigma_2\sigma_3\}}_{trivector} \equiv I. \tag{2.6}$$

Bivectors can be interpreted as oriented areas, trivectors as oriented volumes. Note that we will not use bold for these basis vectors. The highest grade element is a trivector called the unit *pseudoscalar*. It can easily be verified that the pseudoscalar $\sigma_1\sigma_2\sigma_3$ squares to -1 and commutes with all multivectors (a multivector is a general linear combination of any of the elements in the algebra) in the 3-D space. The *unit pseudoscalar I* is crucial when discussing duality. In a space of 3 dimensions we can construct a trivector $\boldsymbol{a} \wedge \boldsymbol{b} \wedge \boldsymbol{c}$, but no 4-vectors exists since there is no possibility of sweeping the volume element $\boldsymbol{a} \wedge \boldsymbol{b} \wedge \boldsymbol{c}$ over a 4th dimension.

The three basis vectors $\{\sigma_i\}$ multiplied by I give the following basis bivectors;

$$I\sigma_1 = \sigma_2\sigma_3 \quad I\sigma_2 = \sigma_3\sigma_1 \quad I\sigma_3 = \sigma_1\sigma_2. \tag{2.7}$$

If we identify the $\boldsymbol{i}$, $\boldsymbol{j}$, $\boldsymbol{k}$ of the quaternion algebra with $\sigma_2\sigma_3$, $-\sigma_3\sigma_1$ and $\sigma_1\sigma_2$, we can recover the famous *Hamilton relations*

$$i^2 = j^2 = k^2 = ijk = -1. \tag{2.8}$$

In geometric algebra a *rotor*, R, is an even-grade element of the algebra which satisfies $R\tilde{R} = 1$. The relation between quaternions and rotors is as follows, if $Q = \{q_0, q_1, q_2, q_3\}$ represents a *quaternion*, then the rotor which performs the same rotation is simply given by

$$R = q_0 + q_1(I\sigma_1) - q_2(I\sigma_2) + q_3(I\sigma_3). \tag{2.9}$$

The quaternion algebra is therefore seen to be a subset of the geometric algebra of 3-space.

7.2.3 *A 4-D geometric algebra for projective space*

For the modelling of the image plane we use $\mathcal{G}_{3,0,0}$ which has the standard *Euclidean signature*. We will show that if we choose to map between projective space and 3-D Euclidean space via the *projective split* (see later), we are then forced to use the 4-D geometric algebra $\mathcal{G}_{1,3,0}$ for $\mathcal{P}^3$. The Lorentzian metric we are using here has no adverse effects in the operations we outline in this chapter, however we will briefly discuss in a later section how a $\{+ + ++\}$ metric for our 4-D space and a different split is being favoured in recent research.

The Lorentzian 4-D algebra has as its vector basis $\gamma_1, \gamma_2, \gamma_3, \gamma_4$, where $\gamma_4^2 = +1$, $\gamma_k^2 = -1$ for $k = 1, 2, 3$. This then generates the following multi-vector basis

$$\underbrace{1}_{scalar}, \underbrace{\gamma_k}_{4\ vectors}, \underbrace{\gamma_2\gamma_3, \gamma_3\gamma_1, \gamma_1\gamma_2, \gamma_4\gamma_1, \gamma_4\gamma_2, \gamma_4\gamma_3}_{6\ bivectors}, \underbrace{I\gamma_k}_{4\ trivectors}, \underbrace{I}_{pseudoscalar} \tag{2.10}$$

The pseudoscalar is $I = \gamma_1\gamma_2\gamma_3\gamma_4$ with

$$I^2 = (\gamma_1\gamma_2\gamma_3\gamma_4)(\gamma_1\gamma_2\gamma_3\gamma_4) = -(\gamma_3\gamma_4)(\gamma_3\gamma_4) = -1. \tag{2.11}$$

The fourth basis vector, γ_4, can also be seen as a selected direction for the *projective split* [3] operation in 4-D. We will see shortly that by taking the geometric product with γ_4 we can associate bivectors of our 4-D space with vectors of our 3-D space. The role and use of the projective split operation will be treated in more detail in a later section.

7.2.4 *Projective transformations*

Historically, the success of *homogeneous coordinates* has partly been due to their ability to represent a general displacement as a single 4×4 matrix and to linearize non-linear transformations [9].

The following indicates how a projective transformation is linearized by going up one dimension in the GA framework. In general a point (x, y, z) in the 3-D space is projected onto the image via a transformation of the form:

$$x' = \frac{\alpha_1 x + \beta_1 y + \delta_1 z + \epsilon_1}{\tilde{\alpha} x + \tilde{\beta} y + \tilde{\delta} z + \tilde{\epsilon}}, \quad y' = \frac{\alpha_2 x + \beta_2 y + \delta_2 z + \epsilon_2}{\tilde{\alpha} x + \tilde{\beta} y + \tilde{\delta} z + \tilde{\epsilon}}. \tag{2.12}$$

This transformation, which is expressed as the ratio of two linear transformations, is indeed non-linear. In order to convert this non–linear transformation in $\mathcal{E}^3$ into a linear transformation in R^4 we define a linear function $\underline{f}_p$ mapping vectors onto vectors in R^4 such that the action of $\underline{f}_p$ on the basis vectors $\{\gamma_i\}$ is given by

$$
\begin{aligned}
\underline{f}_p(\gamma_1) &= \alpha_1 \gamma_1 + \alpha_2 \gamma_2 + \alpha_3 \gamma_3 + \tilde{\alpha} \gamma_4 \\
\underline{f}_p(\gamma_2) &= \beta_1 \gamma_1 + \beta_2 \gamma_2 + \beta_3 \gamma_3 + \tilde{\beta} \gamma_4 \\
\underline{f}_p(\gamma_3) &= \delta_1 \gamma_1 + \delta_2 \gamma_2 + \delta_3 \gamma_3 + \tilde{\delta} \gamma_4 \\
\underline{f}_p(\gamma_4) &= \epsilon_1 \gamma_1 + \epsilon_2 \gamma_2 + \epsilon_3 \gamma_3 + \tilde{\epsilon} \gamma_4
\end{aligned} \tag{2.13}
$$

When we use homogeneous coordinates a general point P in $\mathcal{E}^3$ given by $\boldsymbol{x} = x\sigma_1 + y\sigma_2 + z\sigma_3$ becomes the point $\mathbf{X} = (X\gamma_1 + Y\gamma_2 + Z\gamma_3 + W\gamma_4)$ in R^4, where $x = X/W$, $y = Y/W$, $z = Z/W$. Now using $\underline{f}_p$ the linear map of $\mathbf{X}$ onto $\mathbf{X'}$ is given by

$$\mathbf{X'} = \sum_{i=1}^{3}\{(\alpha_i X + \beta_i Y + \delta_i Z + \epsilon_i W)\gamma_i\} + (\tilde{\alpha} X + \tilde{\beta} Y + \tilde{\delta} Z + \tilde{\epsilon} W)\gamma_4 \tag{2.14}$$

The coordinates of the vector $\boldsymbol{x'} = x'\sigma_1 + y'\sigma_2 + z'\sigma_3$ in $\mathcal{E}^3$ which correspond to $\mathbf{X'}$ are given by

$$x' = \frac{\alpha_1 X + \beta_1 Y + \delta_1 Z + \epsilon_1 W}{\tilde{\alpha} X + \tilde{\beta} Y + \tilde{\delta} Z + \tilde{\epsilon} W} = \frac{\alpha_1 x + \beta_1 y + \delta_1 z + \epsilon_1}{\tilde{\alpha} x + \tilde{\beta} y + \tilde{\delta} z + \tilde{\epsilon}}, \tag{2.15}$$

and similarly

$$y' = \frac{\alpha_2 x + \beta_2 y + \delta_2 z + \epsilon_2}{\tilde{\alpha} x + \tilde{\beta} y + \tilde{\delta} z + \tilde{\epsilon}}, \quad z' = \frac{\alpha_3 x + \beta_3 y + \delta_3 z + \epsilon_3}{\tilde{\alpha} x + \tilde{\beta} y + \tilde{\delta} z + \tilde{\epsilon}}. \tag{2.16}$$

If the above represents projection from the world onto a camera image plane, we should take into account the focal length of the camera. This would require $\alpha_3 = f\tilde{\alpha}$, $\beta_3 = f\tilde{\beta}$ etc., thus we can define $z' = f$ (focal length) independent of the point chosen. The non-linear transformation in $\mathcal{E}^3$ then becomes a linear transformation, $\underline{f}_p$, in R^4. The linear function $\underline{f}_p$ can be used to prove the invariant nature of various quantities under projective transformations, [5].

7.2.5   The projective split

The idea of the *projective split* was introduced by Hestenes [14] in order to connect *projective geometry* and *metric geometry*. This is done by associating the even subalgebra of $\mathcal{G}_{n+1}$ with the geometric algebra of one dimension less, $\mathcal{G}_n$. One can define a mapping between the spaces by choosing a preferred direction in $\mathcal{G}_{n+1}$, γ_{n+1}. If we then take the geometric product of a vector $\mathbf{X} \in \mathcal{G}_{n+1}$ and γ_{n+1}

$$\mathbf{X}\gamma_{n+1} = \mathbf{X}\cdot\gamma_{n+1} + \mathbf{X}\wedge\gamma_{n+1} = \mathbf{X}\cdot\gamma_{n+1}(1 + \frac{\mathbf{X}\wedge\gamma_{n+1}}{\mathbf{X}\cdot\gamma_{n+1}}) \qquad (2.17)$$

we can associate the vector $\boldsymbol{x} \in \mathcal{G}_n$ with the bivector $\frac{\mathbf{X}\wedge\gamma_{n+1}}{\mathbf{X}\cdot\gamma_{n+1}} \in \mathcal{G}_{n+1}$. This result can be projectively interpreted as the pencil of all lines passing though the point γ_{n+1}. In physics the projective split is called the *space time split* which relates the spacetime system $\mathcal{G}_4$ with Minkowski metric, to the observable system $\mathcal{G}_3$ with Euclidean metric.

In computer vision, we are interested in relating elements of projective space with their associated elements in the Euclidean space of the image plane. Optical rays (bivectors) are mapped to points (vectors), optical planes (tricectors) are mapped to lines (bivectors) and optical volumes (4-vectors) to planes (trivector or pseudoscalar).

Suppose we choose γ_4 as a selected direction in R^4, we can then define a mapping which associates the bivectors $\gamma_i\gamma_4$, $i = 1, 2, 3$, in R^4 with the vectors σ_i, $i = 1, 2, 3$, in $\mathcal{E}^3$;

$$\sigma_1 \equiv \gamma_1\gamma_4, \quad \sigma_2 \equiv \gamma_2\gamma_4, \quad \sigma_3 \equiv \gamma_3\gamma_4. \qquad (2.18)$$

Note that in order to preserve the Euclidean structure of the spatial vectors $\{\sigma_i\}$ (i.e. $\sigma_i^2 = +1$) we are forced to choose a non-Euclidean metric for the basis vectors in R^4. That is why we select the basis $\gamma_4^2 = +1$, $\gamma_i = -1$, $i = 1, 2, 3$ for $\mathcal{G}_{1,3,0}$. This is precisely the metric structure of Lorentzian spacetime used in studies of relativistic physics. We note here that although we have chosen here to relate our spaces via the projective split, it is possible to use a Euclidean metric $\{+ + ++\}$ for our 4D space and define the split using reciprocal vectors [20]. It is becoming apparent that this is the preferred procedure and generalizes nicely to splits from higher dimensional spaces. However, for the areas discussed in this chapter, we encounter no problems by using the projective split.

Let us now see how we associate points via the projective split. For a vector $\mathbf{X} = X_1\gamma_1 + X_2\gamma_2 + X_3\gamma_3 + X_4\gamma_4$ in R^4 the projective split is obtained by taking the geometric product of $\mathbf{X}$ and γ_4;

$$\mathbf{X}\gamma_4 = \mathbf{X}\cdot\gamma_4 + \mathbf{X}\wedge\gamma_4 = X_4\left(1 + \frac{\mathbf{X}\wedge\gamma_4}{X_4}\right) \equiv X_4(1 + \boldsymbol{x}). \qquad (2.19)$$

According to equation (2.18) we can associate $\mathbf{X}\wedge\gamma_4/X_4$ in R^4 with the vector $\boldsymbol{x}$ in $\mathcal{E}^3$. Similarly, if we start with a vector $\boldsymbol{x} = x_1\sigma_1 + x_2\sigma_2 + x_3\sigma_3$

in $\mathcal{E}^3$, we represent it in R^4 by the vector $\mathbf{X} = X_1\gamma_1 + X_2\gamma_2 + X_3\gamma_3 + X_4\gamma_4$ such that

$$
\begin{aligned}
\boldsymbol{x} &= \frac{\mathbf{X}\wedge\gamma_4}{X_4} = \frac{X_1}{X_4}\gamma_1\gamma_4 + \frac{X_2}{X_4}\gamma_2\gamma_4 + \frac{X_3}{X_4}\gamma_3\gamma_4 \\
&= \frac{X_1}{X_4}\sigma_1 + \frac{X_2}{X_4}\sigma_2 + \frac{X_3}{X_4}\sigma_3,
\end{aligned}
\tag{2.20}
$$

which implies $x_i = \frac{X_i}{X_4}$, for $i = 1, 2, 3$. The approach of representing $\boldsymbol{x}$ in a higher dimensional space can therefore be seen to be equivalent to using *homogeneous coordinates*, $\mathbf{X}$, for $\boldsymbol{x}$.

Let us now look at the representation of a line L in R^4; a line is given by the outer product of two vectors:

$$
\begin{aligned}
L &= \boldsymbol{A}\wedge\boldsymbol{B} \\
&= (L^{14}\gamma_1\gamma_4 + L^{24}\gamma_2\gamma_4 + L^{34}\gamma_3\gamma_4) + (L^{23}\gamma_2\gamma_3 + L^{31}\gamma_3\gamma_1 + L^{12}\gamma_1\gamma_2) \\
&= (L^{14}\gamma_1\gamma_4 + L^{24}\gamma_2\gamma_4 + L^{34}\gamma_3\gamma_4) - I(L^{23}\gamma_1\gamma_4 + L^{31}\gamma_2\gamma_4 + L^{12}\gamma_3\gamma_4) \\
&= \boldsymbol{n} - I\boldsymbol{m},
\end{aligned}
\tag{2.21}
$$

the six quantities $\{n_i, m_i\}$ $i = 1, 2, 3$ are precisely the Plücker coordinates of the line. $\{L^{14}, L^{24}, L^{34}\}$ are the coefficients of the *spatial part* of the bivector which represents the line direction $\boldsymbol{n}$. $\{L^{23}, L^{31}, L^{12}\}$ are the coefficients of the *non-spatial part* of the bivector which represents the *moment* of the line $\boldsymbol{m}$.

Let us now see how we can related this line representation to an $\mathcal{E}^3$ representation via the projective split. We take a line , L, joining points $\mathbf{A}$ and $\mathbf{B}$

$$
L = \mathbf{A}\wedge\mathbf{B} = \langle \mathbf{AB}\rangle_2 = \langle \mathbf{A}\gamma_4\gamma_4\mathbf{B}\rangle_2
\tag{2.22}
$$

here, the notation $\langle M\rangle_k$, tells us to take the grade k part of the multivector M. Now, using our previous expansions of $\mathbf{X}\gamma_4$ in the projective split for vectors, we can write

$$
L = (\mathbf{A}\cdot\gamma_4)(\mathbf{B}\cdot\gamma_4)\langle(1 + \boldsymbol{a})(1 - \boldsymbol{b})\rangle_2
\tag{2.23}
$$

where $\boldsymbol{a} = \frac{\mathbf{A}\wedge\gamma_4}{\mathbf{A}\gamma_4}$ and $\boldsymbol{b} = \frac{\mathbf{B}\wedge\gamma_4}{\mathbf{B}\gamma_4}$ are the $\mathcal{E}^3$ representations of $\mathbf{A}$ and $\mathbf{B}$. Writing $A_4 = \mathbf{A}\cdot\gamma_4$ and $B_4 = \mathbf{B}\cdot\gamma_4$ then gives us

$$
\begin{aligned}
L &= A_4 B_4\langle 1 + (\boldsymbol{a} - \boldsymbol{b}) - \boldsymbol{ab}\rangle_2 \\
&= A_4 B_4\{(\boldsymbol{a} - \boldsymbol{b}) + \boldsymbol{a}\wedge\boldsymbol{b}\}.
\end{aligned}
\tag{2.24}
$$

Let us now 'normalize' the spatial and non-spacial parts of above bivector

$$
L' = \frac{L}{A_4 B_4|\boldsymbol{a} - \boldsymbol{b}|} = \frac{(\boldsymbol{a} - \boldsymbol{b})}{|\boldsymbol{a} - \boldsymbol{b}|} + \frac{(\boldsymbol{a}\wedge\boldsymbol{b})}{|\boldsymbol{a} - \boldsymbol{b}|}
\tag{2.25}
$$

$$= (n_x\sigma_1 + n_y\sigma_2 + n_z\sigma_3) + (m_x\sigma_2\sigma_3 + m_y\sigma_3\sigma_1 + m_z\sigma_1\sigma_2)$$

$$= (n_x\sigma_1 + n_y\sigma_2 + n_z\sigma_3) + I_3(m_x\sigma_1 + m_y\sigma_2 + m_z\sigma_3)$$

$$= \boldsymbol{n}' + I_3\boldsymbol{m}'. \tag{2.26}$$

Here $I_3 = \sigma_1\sigma_2\sigma_3 \equiv I_4$. Note that in $\mathcal{E}^3$ the line has two components, a vector representing the direction of the line and the dual of a vector (bivector) representing the moment of the line. This completely encodes the position of the line in 3–D space by specifying the plane in which the line lies and the perpendicular distance of the line from the origin.

7.3   The Algebra of Incidence

This section will discuss the use of geometric algebra for the *algebra of incidence* [16]. Firstly we will define the bracket and consider the duality principle. We will define the important concept of the *bracket*, discuss *duality* and then show that the basic projective geometry operations of meet and join can be expressed easily in terms of standard operations within the geometric algebra. We also briefly discuss the linear algebra framework in GA indicating how one will be able to use this within projective geometry. One of the main reasons for moving to a projective space is so that lines, planes etc have representations as real geometric objects and so that operations of intersection etc., can be performed by simple manipulations (instead of via solutions of sets of equations, as in $\mathcal{E}^3$).

7.3.1   The bracket

In a nD space any pseudoscalar will span a hypervolume of dimension n. Since, up to scale, there can only be one such hypervolume, all pseudoscalars, P, are multiples of the unit pseudoscalar, I, $P = \alpha I$, with α a scalar. We compute this scalar multiple by multiplying the pseudoscalar, P, with the inverse of I

$$PI^{-1} = \alpha II^{-1} = \alpha \equiv [P]. \tag{3.27}$$

Thus, the *bracket* of the pseudoscalar P, $[P]$, is its magnitude, arrived at by multiplication on the right by I^{-1}. This bracket is precisely the bracket of the Grassmann-Cayley algebra. The sign of the bracket does not depend on the signature of the space and as such it has been a useful quantity for the non-metrical applications of projective geometry.

The bracket of n vectors $\{\boldsymbol{x}_i\}$ is

$$\begin{aligned} [\boldsymbol{x}_1\boldsymbol{x}_2\boldsymbol{x}_3...\boldsymbol{x}_n] &= [\boldsymbol{x}_1\wedge\boldsymbol{x}_2\wedge\boldsymbol{x}_3\wedge...\wedge\boldsymbol{x}_n] \\ &= (\boldsymbol{x}_1\wedge\boldsymbol{x}_2\wedge\boldsymbol{x}_3\wedge...\wedge\boldsymbol{x}_n)I^{-1} \end{aligned}$$

It can also be shown that this is equivalent to the definition of the determinant of the matrix whose row vectors are the vectors $\boldsymbol{x}_i$.

To understand how we can express a bracket in projective space in terms of vectors in Euclidean space we can expand a pseudoscalar P using the projective split for vectors:

$$\begin{aligned}
P &= \mathbf{X}_1 \wedge \mathbf{X}_2 \wedge \mathbf{X}_3 \wedge \mathbf{X}_4 = \langle \mathbf{X}_1 \gamma_4 \gamma_4 \mathbf{X}_2 \mathbf{X}_3 \gamma_4 \gamma_4 \mathbf{X}_4 \rangle_4 \\
&= W_1 W_2 W_3 W_4 \langle (1 + \boldsymbol{x}_1)(1 - \boldsymbol{x}_2)(1 + \boldsymbol{x}_3)(1 - \boldsymbol{x}_4) \rangle_4
\end{aligned}$$

where $W_i = \mathbf{X}_i \gamma_4$ from equation (2.19). A pseudoscalar part is produced by taking the product of three spatial vectors (there are no (spatial bivector) $\times$ (spatial vector) terms), i.e.

$$\begin{aligned}
P &= W_1 W_2 W_3 W_4 \langle -\boldsymbol{x}_1 \boldsymbol{x}_2 \boldsymbol{x}_3 - \boldsymbol{x}_1 \boldsymbol{x}_3 \boldsymbol{x}_4 + \boldsymbol{x}_1 \boldsymbol{x}_2 \boldsymbol{x}_4 + \boldsymbol{x}_2 \boldsymbol{x}_3 \boldsymbol{x}_4 \rangle_4 \\
&= W_1 W_2 W_3 W_4 \langle (\boldsymbol{x}_2 - \boldsymbol{x}_1)(\boldsymbol{x}_3 - \boldsymbol{x}_1)(\boldsymbol{x}_4 - \boldsymbol{x}_1) \rangle_4 \qquad (3.28) \\
&= W_1 W_2 W_3 W_4 \{ (\boldsymbol{x}_2 - \boldsymbol{x}_1) \wedge (\boldsymbol{x}_3 - \boldsymbol{x}_1) \wedge (\boldsymbol{x}_4 - \boldsymbol{x}_1) \}.
\end{aligned}$$

If the $W_i = 1$, we can summarize the above relationships between the brackets of 4 points in R^4 and $\mathcal{E}^3$ as follows

$$\begin{aligned}
[\mathbf{X}_1 \mathbf{X}_2 \mathbf{X}_3 \mathbf{X}_4] &= (\mathbf{X}_1 \wedge \mathbf{X}_2 \wedge \mathbf{X}_3 \wedge \mathbf{X}_4) I_4^{-1} \\
&\equiv \{ (\boldsymbol{x}_2 - \boldsymbol{x}_1) \wedge (\boldsymbol{x}_3 - \boldsymbol{x}_1) \wedge (\boldsymbol{x}_4 - \boldsymbol{x}_1) \} I_3^{-1}. \quad (3.29)
\end{aligned}$$

7.3.2 The duality principle and the meet and join operations

In order to introduce the concepts of *duality* which are so important in projective geometry, we should firstly define the dual A^* of an r-vector A as

$$A^* = A I^{-1}. \qquad (3.30)$$

This notation A^* relates the ideas of duality to the notion of a *Hodge dual* in differential geometry. Note that in general I^{-1} may not necessarily commute with $\boldsymbol{A}$.

We see therefore that the dual of an r-vector is an $(n - r)$-vector, for example in 3-D space the dual of a vector $(r = 1)$ is a plane or bivector $(n - r = 3 - 1 = 2$).

Using the ideas of duality we are able to relate the inner product to, incidence operators and we will see this in what follows. In an n-D space suppose we have an r-vector A and an s-vector B, where B has dual $B^* = B I^{-1} \equiv B \cdot I^{-1}$. Here, since $B I^{-1} = B \cdot I^{-1} + B \wedge I^{-1}$ we can replace the geometric product by the inner product as the outer product gives zero (there can be no $(n + 1)$-D vector). Now, using the identity

$$A_r \cdot (B_s \cdot C_t) = (A_r \wedge B_s) \cdot C_t \quad \text{for} \quad r + s \leq t, \qquad (3.31)$$

we can write

$$A \cdot (BI^{-1}) = A \cdot (B \cdot I^{-1}) = (A \wedge B) \cdot I^{-1} = (A \wedge B)I^{-1}. \qquad (3.32)$$

This expression can be rewritten using the definition of the dual as follows

$$A \cdot B^* = (A \wedge B)^*. \qquad (3.33)$$

This equation shows the relationship between the inner and outer products in terms of the duality operator. Now, if $r + s = n$, then $A \wedge B$ is of grade n and is therefore a pseudoscalar. Using equation (3.27) it follows that

$$\begin{aligned} A \cdot B^* &= (A \wedge B)^* = (A \wedge B)I^{-1} = ([A \wedge B]I)I^{-1} \\ &= [A \wedge B]. \end{aligned} \qquad (3.34)$$

We see therefore that the bracket relates the inner and outer products to non-metric quantities. It is via this route that the inner product, normally associated with a metric, can be used in a non-metric theory such as projective geometry. It is also interesting to note that since duality is expressed as a simple multiplication by an element of the algebra, there is no need to introduce any special operators or any concept of a different space.

Now, when we work with lines and planes, it will clearly be necessary to have operations for computing the intersections or *joins* of such geometric objects. We require a means of performing the set- theory operations of intersection, $\cap$, and union, $\cup$.

If in an n-dimensional geometric algebra the r-vector A and the s-vector B have no common subspace, one can define the *join* of both vectors as follows

$$J = A \wedge B. \qquad (3.35)$$

So that the join is simply the outer product (an $r + s$ vector) of the two vectors. However, if A and B have no common subspace , the join would not simply be given by the wedge but by the subspace they span. The operation join J can be interpreted as a *common dividend of lowest grade* and is defined up to a scale factor. The join gives the pseudoscalar if $(r + s) \geq n$. We will use $\bigwedge$ for the join only when the blades A and B have a common subspace, otherwise the ordinary exterior product $\wedge$ will be used.

If there exists a k-vector C such that for A and B, we can write $A = A'C$ and $B = B'C$ for some A' and B', then we can define the *intersection* or *meet* using the duality principle as follows

$$(A \vee B)^* = A^* \wedge B^*. \qquad (3.36)$$

A beautiful result telling us that the dual of the meet is given by the join of the duals. Since the dual of $A \vee B$ will be taken with respect to the *join* of A and B, we must be careful to specify what space we use for the dual in equation (3.36). However, in most cases of practical interest this join will indeed be the whole space and the meet and we are therefore able to obtain a more useful expression for the meet using equation (3.33)

$$A \vee B = ((A \vee B)^*)^* = (A^* \wedge B^*)I = (A^* \wedge B^*)(I^{-1}I)I = (A^* \cdot B) \quad (3.37)$$

The above concepts are discussed further in [16].

7.3.3 Linear algebra

This section presents the geometric algebra approach to the basic concepts of linear algebra – it is presented here for completeness. Although it will not be discussed in this chapter, the treatment of invariants [5] uses linear algebra and projective geometry to create geometric entities which are invariant under projective transformations.

A linear function f maps vectors to vectors in the same space, the extension of f to act linearly on multivectors is possible via the so called *outermorphism*, $\underline{f}$, defining the action of $\underline{f}$ on r–blades by

$$\underline{f}(a_1 \wedge a_2 \wedge \ldots \wedge a_r) = \underline{f}(a_1) \wedge \underline{f}(a_2) \wedge \ldots \wedge \underline{f}(a_r). \quad (3.38)$$

$\underline{f}$ is called an outermorphism, because $\underline{f}$ preserves the grade of any r-vector it acts on. The action of $\underline{f}$ on general multivectors is then defined through linearity. $\underline{f}$ must therefore satisfy the following conditions

$$\begin{aligned}
\underline{f}(a_1 \wedge a_2) &= \underline{f}(a_1) \wedge \underline{f}(a_2) \\
\underline{f}(A_r) &= \langle \underline{f}(A_r) \rangle_r \\
\underline{f}(\alpha_1 a_1 + \alpha_2 a_2) &= \alpha_1 \underline{f}(a_1) + \alpha_2 \underline{f}(a_2).
\end{aligned} \quad (3.39)$$

Accordingly, the outermorphism of a product of two linear functions is the product of the outermorphisms, i.e. if $f(a) = f_2(f_1(a))$ we write $\underline{f} = \underline{f}_2 \underline{f}_1$. The *adjoint* $\overline{f}$ of a linear function f acting on the vectors a and b can be defined by the property

$$\underline{f}(a) \cdot b = a \cdot \overline{f}(b). \quad (3.40)$$

If $\underline{f} = \overline{f}$, the function is *self-adjoint* and can be represented by a symmetric matrix, F ($F = F^T$).

Since the outermorphism preserves grade, the unit pseudoscalar must be mapped onto some multiple of itself – this multiple is the *determinant* of $\underline{f}$;

$$\underline{f}(I) = \det(\underline{f})I. \quad (3.41)$$

This is a particularly simple definition of the determinant from which many properties determinants follow straightforwardly.

7.4   Algebra in Projective Space

Having introduced duality, defined the operations of meet and join and given the geometric approach to linear algebra, we are now ready to carry out geometric computations using the algebra of incidence.

Consider three non-collinear points, P_1, P_2, P_3, represented by vectors x_1, x_2, x_3 in $\mathcal{E}^3$ and by vectors $\mathbf{X}_1$, $\mathbf{X}_2$, $\mathbf{X}_3$ in R^4. The line L_{12} joining points P_1 and P_2 can be expressed in R^4 by the bivector

$$L_{12} = \mathbf{X}_1 \wedge \mathbf{X}_2. \tag{4.42}$$

Any point P, represented in R^4 by $\mathbf{X}$, on the line through P_1 and P_2, will satisfy

$$\mathbf{X} \wedge L_{12} = \mathbf{X} \wedge \mathbf{X}_1 \wedge \mathbf{X}_2 = 0. \tag{4.43}$$

This is therefore the equation of the line in R^4. In general such an equation is telling us that $\mathbf{X}$ belongs to the subspace spanned by $\mathbf{X}_1$ and $\mathbf{X}_2$, i.e. that

$$\mathbf{X} = \alpha_1 \mathbf{X}_1 + \alpha_2 \mathbf{X}_2 \tag{4.44}$$

for some α_1, α_2. In computer vision we can use this as a geometric constraint to test whether a point $\mathbf{X}$ lies on L_{12}.

The plane Φ_{123} passing through points P_1, P_2, P_3 is expressed by the following trivector in R^4

$$\Phi_{123} = \mathbf{X}_1 \wedge \mathbf{X}_2 \wedge \mathbf{X}_3. \tag{4.45}$$

In 3-D space there are generally three types of intersections we wish to consider; the intersection of a line and a plane, a plane and a plane, and a line and a line. To compute these intersections we will make use of the following general formula [15], giving the inner product of an r-blade, $A_r = a_1 \wedge a_2 \wedge \ldots \wedge a_r$, and an s-blade, $B_s = b_1 \wedge b_2 \wedge \ldots \wedge b_s$ (for $s \leq r$)

$$B_s \cdot (a_1 \wedge a_2 \wedge \ldots \wedge a_r) = \tag{4.46}$$
$$\sum_{j} \epsilon(j_1 j_2 \ldots j_r) B_s \cdot (a_{j_1} \wedge a_{j_2} \wedge \ldots \wedge a_{j_s}) a_{j_s+1} \wedge \ldots \wedge a_{j_r}$$

where we sum over all combinations $j = (j_1, j_2, \ldots, j_r)$ such that no two j_k's are the same. $\epsilon(j_1 j_2 \ldots j_r) = +1$ if j is an even permutation of $(1, 2, 3, \ldots, r)$ and -1 if it is an odd permutation.

7.4.1   Intersection of a line and a plane

In the space R^4 consider the line $A = \mathbf{X}_1 \wedge \mathbf{X}_2$ intersecting the plane $\Phi = \mathbf{Y}_1 \wedge \mathbf{Y}_2 \wedge \mathbf{Y}_3$. We can compute the intersection point using the meet operation as follows

$$A \vee \Phi = (\mathbf{X}_1 \wedge \mathbf{X}_2) \vee (\mathbf{Y}_1 \wedge \mathbf{Y}_2 \wedge \mathbf{Y}_3) = A \vee \Phi = A^* \cdot \Phi. \qquad (4.47)$$

where we have used equation (3.37) and the fact that in this case the join is the whole space.

Note that the pseudoscalar, I_4 in $\mathcal{G}_{1,3,0}$ for R^4, squares to -1, commutes with bivectors but anticommutes with vectors and trivectors and has inverse $I_4^{-1} = -I_4$. This therefore leads to

$$A^* \cdot \Phi = (AI^{-1}) \cdot \Phi = -(AI) \cdot \Phi. \qquad (4.48)$$

Now using the equation (4.47) we can then expand the meet as

$$A \vee \Phi = -(AI) \cdot (\mathbf{Y}_1 \wedge \mathbf{Y}_2 \wedge \mathbf{Y}_3) = -\{(AI) \cdot (\mathbf{Y}_2 \wedge \mathbf{Y}_3)\} \mathbf{Y}_1 +$$
$$+\{(AI) \cdot (\mathbf{Y}_3 \wedge \mathbf{Y}_1)\} \mathbf{Y}_2 + \{(AI) \cdot (\mathbf{Y}_1 \wedge \mathbf{Y}_2)\} \mathbf{Y}_3 \qquad (4.49)$$

Noting that $(AI) \cdot (\mathbf{Y}_i \wedge \mathbf{Y}_j)$ is a scalar, we can evaluate the above by taking scalar parts. For example, $(AI) \cdot (\mathbf{Y}_2 \wedge \mathbf{Y}_3) = \langle I(\mathbf{X}_1 \wedge \mathbf{X}_2)(\mathbf{Y}_2 \wedge \mathbf{Y}_3)\rangle = I(\mathbf{X}_1 \wedge \mathbf{X}_2 \wedge \mathbf{Y}_2 \wedge \mathbf{Y}_3)$. From the definition of the bracket given earlier, we can see that if $P = \mathbf{X}_1 \wedge \mathbf{X}_2 \wedge \mathbf{Y}_2 \wedge \mathbf{Y}_3$, then $[P] = (\mathbf{X}_1 \wedge \mathbf{X}_2 \wedge \mathbf{Y}_2 \wedge \mathbf{Y}_3)I_4^{-1}$. If we therefore write $[\mathbf{A}_1\mathbf{A}_2\mathbf{A}_3\mathbf{A}_4]$ as a shorthand for the magnitude of the pseudoscalar formed from the four vectors, then we can readily see that the meet reduces to

$$A \vee \Phi = [\mathbf{X}_1\mathbf{X}_2\mathbf{Y}_2\mathbf{Y}_3]\mathbf{Y}_1 + [\mathbf{X}_1\mathbf{X}_2\mathbf{Y}_3\mathbf{Y}_1]\mathbf{Y}_2 + [\mathbf{X}_1\mathbf{X}_2\mathbf{Y}_1\mathbf{Y}_2]\mathbf{Y}_3 \qquad (4.50)$$

giving the intersection point (vector in R^4).

7.4.2 Intersection of two planes

The *line of intersection of two planes*, $\Phi_1 = \mathbf{X}_1 \wedge \mathbf{X}_2 \wedge \mathbf{X}_3$ and $\Phi_2 = \mathbf{Y}_1 \wedge \mathbf{Y}_2 \wedge \mathbf{Y}_3$, can be computed via the meet of Φ_1 and Φ_2

$$\Phi_1 \vee \Phi_2 = (\mathbf{X}_1 \wedge \mathbf{X}_2 \wedge \mathbf{X}_3) \vee (\mathbf{Y}_1 \wedge \mathbf{Y}_2 \wedge \mathbf{Y}_3). \qquad (4.51)$$

As in previous section, this can be expanded as

$$\begin{aligned}
\Phi_1 \vee \Phi_2 &= \Phi_1^* \cdot (\mathbf{Y}_1 \wedge \mathbf{Y}_2 \wedge \mathbf{Y}_3) \\
&= -\{(\Phi_1 I) \cdot \mathbf{Y}_1\}(\mathbf{Y}_2 \wedge \mathbf{Y}_3) + \{(\Phi_1 I) \cdot \mathbf{Y}_2\}(\mathbf{Y}_3 \wedge \mathbf{Y}_1) + \\
&\quad +\{(\Phi_1 I) \cdot \mathbf{Y}_3\}(\mathbf{Y}_1 \wedge \mathbf{Y}_2). \qquad (4.52)
\end{aligned}$$

Again, the join is the whole space and so the dual is easily formed. Following the arguments of the previous section we can show that $(\Phi_1 I) \cdot \mathbf{Y}_i \equiv -[\mathbf{X}_1\mathbf{X}_2\mathbf{X}_3\mathbf{Y}_i]$, so that the meet is

$$\begin{aligned}
\Phi_1 \vee \Phi_2 &= [\mathbf{X}_1\mathbf{X}_2\mathbf{X}_3\mathbf{Y}_1](\mathbf{Y}_2 \wedge \mathbf{Y}_3) + [\mathbf{X}_1\mathbf{X}_2\mathbf{X}_3\mathbf{Y}_2](\mathbf{Y}_3 \wedge \mathbf{Y}_1) + \\
&\quad +[\mathbf{X}_1\mathbf{X}_2\mathbf{X}_3\mathbf{Y}_3](\mathbf{Y}_1 \wedge \mathbf{Y}_2), \qquad (4.53)
\end{aligned}$$

producing a line of intersection or bivector in R^4.

7.4.3 Intersection of two lines

Two lines will intersect only if they are coplanar. This means that their representations in R^4, $A = \mathbf{X}_1 \wedge \mathbf{X}_2$, and $B = \mathbf{Y}_1 \wedge \mathbf{Y}_2$ will satisfy

$$A \wedge B = 0. \tag{4.54}$$

This fact suggests that the computation of the intersection should be carried out in the 2-D Euclidean space which has an associated 3-D projective counterpart, R^3. In this plane the intersection point is given by

$$
\begin{aligned}
A \vee B &= A^* \cdot B = -(AI_3) \cdot (\mathbf{Y}_1 \wedge \mathbf{Y}_2) \\
&= -\{((AI_3) \cdot \mathbf{Y}_1)\mathbf{Y}_2 - ((AI_3) \cdot \mathbf{Y}_2)\mathbf{Y}_1\}
\end{aligned} \tag{4.55}
$$

where I_3 is the pseudoscalar for R^3. Once again we evaluate $((AI_3) \cdot \mathbf{Y}_i)$ by taking scalar parts

$$(AI_3) \cdot \mathbf{Y}_i = \langle \mathbf{X}_1 \mathbf{X}_2 I_3 \mathbf{Y}_i \rangle = I_3 \mathbf{X}_1 \mathbf{X}_2 \mathbf{Y}_i = -[\mathbf{X}_1 \mathbf{X}_2 \mathbf{Y}_i]. \tag{4.56}$$

The meet can therefore be written as

$$A \vee B = [\mathbf{X}_1 \mathbf{X}_2 \mathbf{Y}_1]\mathbf{Y}_2 - [\mathbf{X}_1 \mathbf{X}_2 \mathbf{Y}_2]\mathbf{Y}_1 \tag{4.57}$$

where the bracket $[\mathbf{A}_1 \mathbf{A}_2 \mathbf{A}_3]$ in R^3 is understood to mean $(\mathbf{A}_1 \wedge \mathbf{A}_2 \wedge \mathbf{A}_3)I_3^{-1}$. The above is often an impractical means of performing the intersection of two lines – see [20] for a method which creates a plane and intersects one of the lines with this plane. See also [8] for a discussion of what information can be gained when the lines do not intersect. See also Chapter 13 for a complete treatment of the incidence relations between points, lines and planes in the n–affine plane.

7.4.4 Implementation of the algebra

In order to implement the expressions and procedures outlined so far in this chapter we have used a computer algebra package written for MAPLE. The program originates from [17] which works with geometric algebras of $\mathcal{G}_{1,3,0}$ and $\mathcal{G}_{3,0,0}$; a more general version of this program, which works with a user-defined metric on an n-D algebra is available on [1]. Using these packages we are easily able to simulate the situation of several cameras (or one moving camera) looking at a world scene and to do so entirely in projective (4D) space. Much of the work described in subsequent sections has been tested in MAPLE.

7.5 Visual Geometry of n Uncalibrated Cameras

This section will give an analysis of the constraints relating the geometry of n uncalibrated cameras. Firstly the pinhole camera model for one view will be defined in terms of lines and planes. In two and three views the epipolar geometry is defined in terms of bilinear and trilinear constraints. Since the constraints are based on the coplanarity of lines, we can only find relationships expressed by a single tensor for up to four cameras. For more than four cameras the constraints are linear combinations of bilinearities, trilinearities and quadrilinearities.

7.5.1   Geometry of one view

We begin with the monocular case depicted in Figure 7.2. Here the image plane is defined by a vector basis of three arbitrary non–collinear points $\mathbf{A}_1$, $\mathbf{A}_2$ and $\mathbf{A}_3$ with the optical center given by $\mathbf{A}_0$ (all vectors in R^4). Thus, $\{\mathbf{A}_i\}$ can be used as a coordinate basis for the image plane $\Phi_A = \mathbf{A}_1 \wedge \mathbf{A}_2 \wedge \mathbf{A}_3$, so that any point $\mathbf{A}'$ lying in Φ_A can be written as

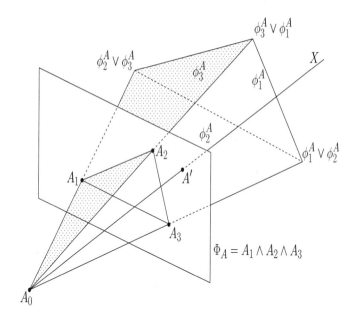

FIGURE 7.2. Sketch of projection into a single camera – the monocular case.

$$\mathbf{A}' = \alpha_1 \mathbf{A}_1 + \alpha_2 \mathbf{A}_2 + \alpha_3 \mathbf{A}_3. \tag{5.58}$$

We are also able to define a bivector basis of the image plane, $\{L_i^A\}$, spanning the lines in Φ_A;

$$L_1^A = \mathbf{A}_2 \wedge \mathbf{A}_3 \qquad L_2^A = \mathbf{A}_3 \wedge \mathbf{A}_1 \qquad L_3^A = \mathbf{A}_1 \wedge \mathbf{A}_2 \tag{5.59}$$

The bivectors $\{L_i^A\}$ together with the optical center allow us to define three planes, ϕ_i^A, as follows;

$$\begin{aligned}
\phi_1^A &= \mathbf{A}_0 \wedge \mathbf{A}_2 \wedge \mathbf{A}_3 = \mathbf{A}_0 \wedge L_1^A \\
\phi_2^A &= \mathbf{A}_0 \wedge \mathbf{A}_3 \wedge \mathbf{A}_1 = \mathbf{A}_0 \wedge L_2^A \qquad (5.60) \\
\phi_3^A &= \mathbf{A}_0 \wedge \mathbf{A}_1 \wedge \mathbf{A}_2 = \mathbf{A}_0 \wedge L_3^A.
\end{aligned}$$

We will call the planes, ϕ_j^A, *optical planes*. Clearly each is a trivector and can be written as

$$\phi_j^A = t_{j1}(I\gamma_1) + t_{j2}(I\gamma_2) + t_{j3}(I\gamma_3) + t_{j4}(I\gamma_4) \equiv t_{jk}(I\gamma_k) \qquad (5.61)$$

since there are 4 basis trivectors in our 4-D space. These optical planes clearly intersect the image plane in the lines $\{L_j^A\}$. Furthermore, the intersections of the optical planes also define a bivector basis which spans the pencil of *optical rays* (rays passing through the optical centre of the camera) in R^4;

$$\begin{aligned}
L_{A1} &= \phi_2 \vee \phi_3 \equiv \mathbf{A}_0 \wedge \mathbf{A}_1 \\
L_{A2} &= \phi_3 \vee \phi_1 \equiv \mathbf{A}_0 \wedge \mathbf{A}_2 \qquad (5.62) \\
L_{A3} &= \phi_1 \vee \phi_2 \equiv \mathbf{A}_0 \wedge \mathbf{A}_3,
\end{aligned}$$

so that any optical ray resulting from projecting a world point $\mathbf{X}$ onto the image plane can be written as

$$\mathbf{A}_0 \wedge \mathbf{X} = x_j L_{Aj}.$$

We can now interpret the camera matrices, used so widely in computer vision applications, in terms of the quantities defined in this section.

The projection of any world point, $\mathbf{X}$, onto the image plane is x and is given by the intersection of line $\mathbf{A}_0 \wedge \mathbf{X}$ with the plane Φ_A

$$x = (\mathbf{A}_0 \wedge \mathbf{X}) \vee (\mathbf{A}_1 \wedge \mathbf{A}_2 \wedge \mathbf{A}_3) = X_\mu \{(\mathbf{A}_0 \wedge \gamma_\mu) \vee (\mathbf{A}_1 \wedge \mathbf{A}_2 \wedge \mathbf{A}_3)\}$$

where μ is summed over 1 to 4. We can now expand the above meet to give

$$\begin{aligned}
x &= X_j \{[\mathbf{A}_0 \wedge \gamma_j \wedge \mathbf{A}_2 \wedge \mathbf{A}_3] \mathbf{A}_1 + [\mathbf{A}_0 \wedge \gamma_j \wedge \mathbf{A}_3 \wedge \mathbf{A}_1] \mathbf{A}_2 + \\
&\quad + [\mathbf{A}_0 \wedge \gamma_j \wedge \mathbf{A}_1 \wedge \mathbf{A}_2] \mathbf{A}_3 \}. \qquad (5.63)
\end{aligned}$$

Since $x = x^k \mathbf{A}_k$, the above implies $x = X_j P_{jk} \mathbf{A}_k$ and therefore that

$$x^k = P_{jk} X_j$$

where

$$P_{jk} = [\mathbf{A}_0 \wedge \gamma_j \wedge L_k^A] \equiv [\phi_k \wedge \gamma_j] = -t_{kj} \qquad (5.64)$$

since $I\gamma_j \wedge \gamma_k = -I\delta_{jk}$. The matrix P takes $\mathbf{X}$ to $\mathbf{x}$ and is therefore the standard camera projection matrix. If we define a set of vectors $\{\phi_A^j\}$, $j = 1, 2, 3$, which are the duals of the planes $\{\phi_j^A\}$, i.e. $\phi_A^j = \phi_j^A I^{-1}$, it is then simple to see that

$$\phi_A^j = -\phi_j^A I = I\phi_j^A = -[t_{j1}\gamma_1 + t_{j2}\gamma_2 + t_{j3}\gamma_3 + t_{j4}\gamma_4]. \qquad (5.65)$$

Thus, we see that the projected point $\mathbf{x} = x^j \mathbf{A}_j$ is then given by

$$x^j = \mathbf{X} \cdot \phi_A^j \qquad \text{or} \qquad \mathbf{x} = (\mathbf{X} \cdot \phi_A^j) \mathbf{A}_j \qquad (5.66)$$

i.e. the coefficients in the image plane are formed by projecting $\mathbf{X}$ onto the vectors formed by taking the duals of the optical planes. This is, of course, equivalent to the matrix formulation

$$\mathbf{x} = \begin{bmatrix} x_1 \\ x_2 \\ x_3 \end{bmatrix} = \begin{bmatrix} \phi_A^1 \\ \phi_A^2 \\ \phi_A^3 \end{bmatrix} \mathbf{X} = \begin{bmatrix} t_{11} & t_{12} & t_{13} & t_{14} \\ t_{21} & t_{22} & t_{23} & t_{24} \\ t_{31} & t_{32} & t_{33} & t_{34} \end{bmatrix} \begin{bmatrix} X_1 \\ X_2 \\ X_3 \\ X_4 \end{bmatrix}$$

$$\equiv \mathbf{PX}. \qquad (5.67)$$

The elements of the camera matrix are therefore simply the coefficients of each optical plane in the coordinate frame of the world point. They encode the intrinsic and extrinsic camera parameters as given in equation (2.2).

Next we consider the projection of world lines in R^4 onto the image plane. Suppose we have a world line $L = \mathbf{X}_1 \wedge \mathbf{X}_2$ joining the points $\mathbf{X}_1$ and $\mathbf{X}_2$. If $\mathbf{x}_1 = (\mathbf{A}_0 \wedge \mathbf{X}_1) \vee \Phi_A$ and $\mathbf{x}_2 = (\mathbf{A}_0 \wedge \mathbf{X}_2) \vee \Phi_A$ (i.e. the intersections of the optical rays with the image plane) then the projected line in the image plane is clearly given by

$$l = \mathbf{x}_1 \wedge \mathbf{x}_2$$

As we can express l in the bivector basis for the plane, we have

$$l = l^j L_j^A$$

where $L_1^A = \mathbf{A}_2 \wedge \mathbf{A}_3$ etc. as before. From our previous expressions for projections given in equation (5.66), we see that we can also write l as follows

$$l = \mathbf{x}_1 \wedge \mathbf{x}_2 = (\mathbf{X}_1 \cdot \phi_A^j)(\mathbf{X}_2 \cdot \phi_A^k) \mathbf{A}_j \wedge \mathbf{A}_k \equiv l^p L_p^A \qquad (5.68)$$

which tells us that the *line coefficients*, $\{l^j\}$, are

$$\begin{aligned} l^1 &= (\mathbf{X}_1 \cdot \phi_A^2)(\mathbf{X}_2 \cdot \phi_A^3) - (\mathbf{X}_1 \cdot \phi_A^3)(\mathbf{X}_2 \cdot \phi_A^2) \\ l^2 &= (\mathbf{X}_1 \cdot \phi_A^3)(\mathbf{X}_2 \cdot \phi_A^1) - (\mathbf{X}_1 \cdot \phi_A^1)(\mathbf{X}_2 \cdot \phi_A^3) \\ l^3 &= (\mathbf{X}_1 \cdot \phi_A^1)(\mathbf{X}_2 \cdot \phi_A^2) - (\mathbf{X}_1 \cdot \phi_A^2)(\mathbf{X}_2 \cdot \phi_A^1). \end{aligned} \qquad (5.69)$$

Using the identity in equation (3.36) we are then able to deduce identities of the following form for each of the l^j

$$l^1 = (\mathbf{X}_1 \wedge \mathbf{X}_2) \cdot (\phi_A^2 \wedge \phi_A^3) = (\mathbf{X}_1 \wedge \mathbf{X}_2) \cdot (\phi_2^A \vee \phi_3^A)^* = L \cdot L_1^{A*}$$

using the fact that the join of the duals is the dual of the meet. We therefore have the general result

$$l^j = L \cdot L_j^{A*} \equiv L \cdot L_A^j \tag{5.70}$$

where we have defined L_A^j to be the dual of L_A^A. Thus, we have again expressed the projection of a line L onto the image plane by contracting L with the set of lines dual to those formed by intersecting the optical planes.

Below we summarize the two results derived here for the projections of points ($\mathbf{X}_1$ and $\mathbf{X}_2$) and lines ($L = \mathbf{X}_1 \wedge \mathbf{X}_2$) onto the image plane:

$$\boldsymbol{x}_1 = (\mathbf{X}_1 \cdot \phi_A^j) \mathbf{A}_j \qquad \boldsymbol{x}_2 = (\mathbf{X}_2 \cdot \phi_A^j) \mathbf{A}_j$$
$$l = (L \cdot L_A^j) L_j^A \equiv l^k L_k^A \tag{5.71}$$
$$\tag{5.72}$$

Having formed the sets of dual planes, $\{\phi_A^j\}$, and dual lines, $\mathbf{L}_A^j\}$, for a given image plane, it is then conceptually very straightforward to project any point or line onto that plane.

If we express the world and image lines as bivectors, $L = \alpha_j \sigma_j + \tilde{\alpha}_j I \sigma_j$ and $L_A^P = \beta_j \sigma_j + \tilde{\beta}_j I \sigma_j$, we can write equation (5.72) as a matrix equation:

$$l = \begin{bmatrix} l^1 \\ l^2 \\ l^3 \end{bmatrix} = \begin{bmatrix} u_{11} & u_{12} & u_{13} & u_{14} & u_{15} & u_{16} \\ u_{21} & u_{22} & u_{23} & u_{24} & u_{25} & u_{26} \\ u_{31} & u_{32} & u_{33} & u_{34} & u_{35} & u_{36} \end{bmatrix} \begin{bmatrix} \alpha_1 \\ \alpha_2 \\ \alpha_3 \\ \tilde{\alpha}_1 \\ \tilde{\alpha}_2 \\ \tilde{\alpha}_3 \end{bmatrix} \equiv P_L \bar{l} \tag{5.73}$$

where $\bar{l}$ is the vector of *Plücker coordinates*, $[\alpha_1, \alpha_2, \alpha_3, \tilde{\alpha}_1, \tilde{\alpha}_2, \tilde{\alpha}_3]$ and the matrix P_L contains the β and *beta*'s – i.e. information about the camera configuration.

When we back-project a point, $\boldsymbol{x}$, or line, l, in the image plane we produce their duals, i.e. a line, l_x, or a plane, ϕ_l, respectively. These back-projected lines and planes are given by the following expressions

$$l_x = \mathbf{A}_0 \wedge \boldsymbol{x} = (\mathbf{X} \cdot \phi_A^j) \mathbf{A}_0 \wedge \mathbf{A}_j = (\mathbf{X} \cdot \phi_A^j) \mathbf{L}_j^A \tag{5.74}$$
$$\phi_l = \mathbf{A}_0 \wedge l = (L \cdot L_A^j) \mathbf{A}_0 \wedge L_j^A = (L \cdot L_A^j) \phi_j^A. \tag{5.75}$$

7.5.2 Geometry of two views

In this and following sections we will work in projective space, R^4, although returning to 3-D Euclidean space will be necessary when discussing invariants in terms of image coordinates; this will be done via the projective split. Figure 7.3.a shows a world point $\mathbf{X}$ projecting onto points $\mathbf{A}'$ and $\mathbf{B}'$ in the two image planes ϕ_A and ϕ_B respectively.

The so called epipoles $\mathbf{E}_{AB}$ and $\mathbf{E}_{BA}$ correspond to the intersections of the line joining the optical centres with the image planes. Since the points $\mathbf{A}_0, \mathbf{B}_0, \mathbf{A}', \mathbf{B}'$ are coplanar, we can formulate the bilinear constraint using the fact that the outer product of these four vectors must vanish:

$$\mathbf{A}_0 \wedge \mathbf{B}_0 \wedge \mathbf{A}' \wedge \mathbf{B}' = 0. \tag{5.76}$$

Now, if we let $\mathbf{A}' = \alpha_i \mathbf{A}_i$ and $\mathbf{B}' = \beta_j \mathbf{B}_j$, then equation (5.76) can be written as

$$\alpha_i \beta_j \{ \mathbf{A}_0 \wedge \mathbf{B}_0 \wedge \mathbf{A}_i \wedge \mathbf{B}_j \} = 0. \tag{5.77}$$

Defining $\tilde{F}_{ij} = \{ \mathbf{A}_0 \wedge \mathbf{B}_0 \wedge \mathbf{A}_i \wedge \mathbf{B}_j \} I^{-1} \equiv [\mathbf{A}_0 \mathbf{B}_0 \mathbf{A}_i \mathbf{B}_j]$ gives us

$$\tilde{F}_{ij} \alpha_i \beta_j = 0 \tag{5.78}$$

which corresponds in R^4 to the well-known relationship between the components of the *fundamental matrix* or *bilinear constraint* in E^3, F, and the image coordinates [19]. This suggests that $\tilde{F}$ can be seen as a linear function mapping two vectors onto a scalar:

$$\tilde{F}(\mathbf{A}, \mathbf{B}) = \{ \mathbf{A}_0 \wedge \mathbf{B}_0 \wedge \mathbf{A} \wedge \mathbf{B} \} I^{-1} \tag{5.79}$$

So that $\tilde{F}_{ij} = \tilde{F}(\mathbf{A}_i, \mathbf{B}_j)$. Note that viewing the fundamental matrix as a linear function means that we have a coordinate-independent description. Now, if we use the projective split to associate our point $\mathbf{A}' = \alpha_i \mathbf{A}_i$ in the image plane with its $\mathcal{E}^3$ representation $\boldsymbol{a}' = \delta_i \boldsymbol{a}_i$, where $\boldsymbol{a}_i = \frac{\mathbf{A}_i \wedge \gamma_4}{\mathbf{A}_i \cdot \gamma_4}$, it is not difficult to see that the coefficients are related by

$$\alpha_i = \frac{\mathbf{A}' \cdot \gamma_4}{\mathbf{A}_i \cdot \gamma_4} \delta_i \tag{5.80}$$

Thus, we are able to relate our 4-D fundamental matrix, $\tilde{F}$ to an *observed* fundamental matrix F by

$$\tilde{F}_{kl} = (\mathbf{A}_k \cdot \gamma_4)(\mathbf{B}_l \cdot \gamma_4) F_{kl} \tag{5.81}$$

so that

$$\alpha_k \tilde{F}_{kl} \beta_l = (\mathbf{A}' \cdot \gamma_4)(\mathbf{B}' \cdot \gamma_4) \delta_k F_{kl} \epsilon_l \tag{5.82}$$

where $\boldsymbol{b}' = \epsilon_i \boldsymbol{b}_i$, with $\boldsymbol{b}_i = \frac{\mathbf{B}_i \wedge \gamma_4}{\mathbf{B}_i \cdot \gamma_4}$. F is the standard fundamental matrix that we would form from observations.

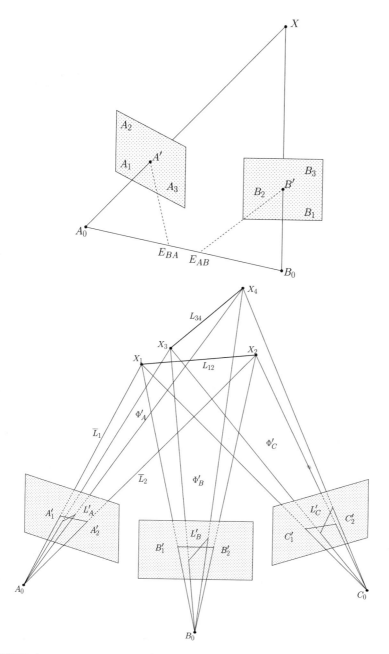

FIGURE 7.3. Sketch of a) binocular projection of a world point, b) trinocular projection.

7.5.3   Geometry of three views

The so called trilinear constraint captures the geometric relationships existing between points and lines in three camera views. Figure 7.3.b shows three image planes ϕ_A, ϕ_B and ϕ_C with bases $\{\mathbf{A}_i\}$, $\{\mathbf{B}_i\}$ and $\{\mathbf{C}_i\}$ and optical centres $\mathbf{A}_0, \mathbf{B}_0, \mathbf{C}_0$. Intersections of two world points $\mathbf{X}_i$ with the planes occur at points $\mathbf{A}_i'$, $\mathbf{B}_i'$, $\mathbf{C}_i'$, $i = 1, 2$. The line joining the world points is $L_{12} = \mathbf{X}_1 \wedge \mathbf{X}_2$, and the projected lines are denoted by L_A', L_B' and L_C'.

We first define three planes

$$\Phi_A' = \mathbf{A}_0 \wedge \mathbf{A}_1' \wedge \mathbf{A}_2', \quad \Phi_B' = \mathbf{B}_0 \wedge \mathbf{B}_1' \wedge \mathbf{B}_2', \quad \Phi_C' = \mathbf{C}_0 \wedge \mathbf{C}_1' \wedge \mathbf{C}_2'. \quad (5.83)$$

It is clear that L_{12} can be formed by intersecting Φ_B' and Φ_C',

$$L_{12} = \Phi_B' \vee \Phi_C' = (\mathbf{B}_0 \wedge L_B') \vee (\mathbf{C}_0 \wedge L_C'). \quad (5.84)$$

If $L_{A_1} = \mathbf{A}_0 \wedge \mathbf{A}_1'$ and $L_{A_2} = \mathbf{A}_0 \wedge \mathbf{A}_2'$, then we can easily see that L_1 and L_2 intersect with L_{12} at $\mathbf{X}_1$ and $\mathbf{X}_2$ respectively. We therefore have

$$L_1 \wedge L_{12} = 0 \quad \text{and} \quad L_2 \wedge L_{12} = 0 \quad (5.85)$$

which can then be written as

$$(\mathbf{A}_0 \wedge \mathbf{A}_i') \wedge \{(\mathbf{B}_0 \wedge L_B') \vee (\mathbf{C}_0 \wedge L_C')\} = 0 \quad \text{for } i = 1, 2. \quad (5.86)$$

This therefore suggests that we should define a linear function T which maps a point and two lines onto a scalar:

$$T(\mathbf{A}', L_B', L_C') = (\mathbf{A}_0 \wedge \mathbf{A}') \wedge \{(\mathbf{B}_0 \wedge L_B') \vee (\mathbf{C}_0 \wedge L_C')\}. \quad (5.87)$$

Now, using the line bases of the planes B and C similar as the ones of the plane A in equation (5.59), we can write

$$\mathbf{A}' = \alpha_i \mathbf{A}_i, \quad L_B' = l_j^B L_j^B \quad L_C' = l_k^C L_k^C. \quad (5.88)$$

If we define the components of a tensor as $T_{ijk} = T(\mathbf{A}_i, L_j^B, L_k^C)$, then if $\mathbf{A}', L_B', L_C'$ are all derived from projections of the same two world points, equation (5.86) tells us that we can write

$$T_{ijk} \alpha_i l_j^B l_k^C = 0. \quad (5.89)$$

T is the *trifocal tensor* and equation (5.89) is the *trilinear constraint*. In in [11, 21] this was arrived at by consideration of camera matrices, here, however, equation (5.89) is arrived at from purely geometric considerations, namely that two planes intersect in a line which in turn intersects with another line. To see how we relate the three projected *lines*, we express the line in image plane ϕ_A joining $\mathbf{A}'_1$ and $\mathbf{A}'_2$ as the intersection of the plane joining $\mathbf{A}_0$ to the world line L_{12} with the image plane $\Phi_A = \mathbf{A}_1 \wedge \mathbf{A}_2 \wedge \mathbf{A}_3$

$$L'_A = \mathbf{A}'_1 \wedge \mathbf{A}'_2 = (\mathbf{A}_0 \wedge L_{12}) \vee \Phi_A. \tag{5.90}$$

Considering L_{12} as the meet of the planes $\Phi'_B \vee \Phi'_C$ and using the expansions of L'_A, L'_B, L'_C given in equation (5.88), we can rewrite this equation as

$$l^A_i \mathbf{L}^A_i = \left((\mathbf{A}_0 \wedge \mathbf{A}_i) \wedge l^B_j l^C_k \{(\mathbf{B}_0 \wedge L^B_j) \vee (\mathbf{C}_0 \wedge L^C_k)\} \right) \vee \Phi_A. \tag{5.91}$$

Using the expansion of the meet given in equation (4.53) we have

$$l^A_i \mathbf{L}^A_i = [(\mathbf{A}_0 \wedge \mathbf{A}_i) \wedge l^B_j l^C_k \{(\mathbf{B}_0 \wedge L^B_j) \vee (\mathbf{C}_0 \wedge L^C_k)\}] L^A_i \tag{5.92}$$

which, when we equate coefficients, gives

$$l^A_i = T_{ijk} l^B_j l^C_k. \tag{5.93}$$

Thus we obtain the familiar equation which relates the projected lines in the three views.

7.5.4   Geometry of n-views

If we have n-views, let us choose 4 of these views and denote them by A, B, C and N. As before, we assume that $\{\mathbf{A}_j\}$, $\{\mathbf{B}_j\}$ etc. ... $j = 1, 2, 3$ define the image planes.

Let $\Phi_{Ai} = \mathbf{A}_0 \wedge \mathbf{A}_i \wedge \mathbf{A}'$, $\Phi_{Bi} = \mathbf{B}_0 \wedge \mathbf{B}_i \wedge \mathbf{B}'$ etc. where $\mathbf{A}'$, $\mathbf{B}'$ etc are the projections of a world point P onto the image planes. $\Phi_{Aj} \vee \Phi_{Bk}$ gives a line passing through the world point P as does $\Phi_{Cl} \vee \Phi_{Nm}$. Since these two lines intersect we have the condition

$$\{\Phi_{Aj} \vee \Phi_{Bk}\} \wedge \{\Phi_{Cl} \vee \Phi_{Nm}\} = 0. \tag{5.94}$$

Consider also the world line $L = \mathbf{X}_1 \wedge \mathbf{X}_2$ which projects down to l_a, l_b, l_c, l_n in the four image planes. We know from the previous sections that it is possible to write L in terms of these image lines as the meet of two planes in several ways

$$L = (\mathbf{A}_0 \wedge l_a) \vee (\mathbf{B}_0 \wedge l_b) \tag{5.95}$$
$$L = (\mathbf{C}_0 \wedge l_c) \vee (\mathbf{N}_0 \wedge l_n) \tag{5.96}$$

Now, since $L \wedge L = 0$ and taking $l_a = \ell^i_a L^A_i$ etc, we can write

$$\ell^i_a \ell^j_b \ell^k_c \ell^m_n [(\mathbf{A}_0 \wedge L^A_i) \vee (\mathbf{B}_0 \wedge L^B_j)] \wedge [(\mathbf{C}_0 \wedge L^C_k) \vee (\mathbf{N}_0 \wedge L^N_m)] = 0 \tag{5.97}$$

which can be written as

$$\ell^i_a \ell^j_b \ell^k_c \ell^m_n Q_{ijkm} = 0. \tag{5.98}$$

Here, Q is the so-called *quadrifocal tensor* recently discussed in [12]. The above constraint in terms of lines is straightforward but it is also possible to find a relationship between point coordinates and Q. To do this we expand equation (5.94) as follows

$$\alpha_r \beta_s \delta_t \eta_u \{[(\mathbf{A}_0 \wedge L^A_{jr}) \vee (\mathbf{B}_0 \wedge L^B_{ks})] \wedge [(\mathbf{C}_0 \wedge L^C_{lt}) \vee (\mathbf{N}_0 \wedge L^N_{mu})]\} = 0. \quad (5.99)$$

where we have used the notation $L^A_{jr} = \mathbf{A}_j \wedge \mathbf{A}_r \equiv \epsilon_{ijr} L^A_i$. Thus we can also write the above equation as

$$\alpha_r \beta_s \delta_t \eta_u \epsilon_{i_1 jr} \epsilon_{i_2 ks} \epsilon_{i_3 lt} \epsilon_{i_4 mu} Q_{i_1 i_2 i_3 i_4} = 0, \quad (5.100)$$

for any $\{i, j, k, m\}$.

7.6   Conclusions

This chapter has outlined the use of geometric algebra as a framework for analysis and computation in computer vision. In particular, the framework for projective geometry was described and the analysis of tensorial relations between multiple camera views was presented in a wholly geometric fashion. The projective geometry operations of meet and join are easily expressed analytically and easily computed in geometric algebra. Indeed it is the ease with which we can perform the algebra of incidence (intersections of lines planes etc.) that simplifies many of the otherwise complex tensorial relations. The concept of duality has been discussed and used specifically in projecting down from the world to image planes – in geometric algebra, duality is a particularly simple concept and one in which the non-metric properties of the inner product becomes apparent.

Acknowledgments

Eduardo Bayro-Corrochano was supported by the project SO-201 of the Deutsche Forschungsgemeinschat. Joan Lasenby is supported by the Royal Society of London.

Chapter 8

Using Geometric Algebra for Optical Motion Capture

Joan Lasenby and Adam Stevenson

8.1 Introduction

Optical motion capture refers to the process by which accurate 3D data from a moving subject is reconstructed from the images in two or more cameras. In order to achieve this reconstruction it is necessary to know how the cameras are placed relative to each other, the internal characteristics of each camera and the matching points in each image. The goal is to carry out this process as automatically as possible. In this paper we will outline a series of calibration techniques which use all of the available data simultaneously and produce accurate reconstructions with no complicated calibration equipment or procedures. These techniques rely on the use of geometric algebra and the ability therein to differentiate with respect to multivectors and linear functions.

Optical motion capture involves the use of multiple cameras to observe a moving subject. From the 2D data in each camera the goal is to obtain a moving 3D reconstruction of our subject. This process has applications in medicine, biomechanics, sports training and animation. The whole motion capture process starts by *calibrating* the cameras – i.e. determining their relative positions and orientations and the internal camera characteristics. In any practical system, we require this process to be easy to accomplish and the results to be accurate. This paper will look in detail at this initial stage of the motion capture process, in particular the determination of the relative orientations and positions of any number of cameras given no special calibration object. The algorithms developed for this purpose involve the use of geometric algebra and result in an iterative scheme which does not require any non-linear minimization stage. There are already many examples of the use of geometric algebra in other computer vision applications a few of which are given in [1, 8, 9]. During the m-camera calibration process we shall see that two very useful algorithms emerge: firstly, a straightforward, analytic means of estimating the relative translations between cameras (not simply their directions) given that the relative rotations are known, is presented. Secondly, given any number of cameras and their relative rotations and translations, we show how to produce a robust, optimal (in a least squares sense) estimate of the world coordinates. Both techniques could be useful in a variety of applications and are each programmed in just a few lines of code.

The setup we use consists of three 50Hz monochrome CCD cameras each connected to the inputs of a framegrabber card located in a PC – a synch signal is fed into the cameras so that the digitised data comes from simultaneous frames, see figure 8.1. The system will shortly be extended to 6 cameras.

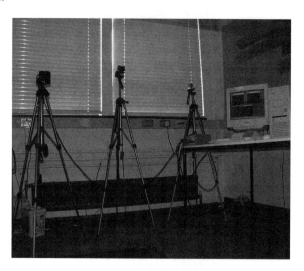

FIGURE 8.1. 3-camera motion capture system.

Retroreflective markers are placed on the moving subject and these are illuminated with IR radiation directed from each of the cameras. Image sequences of bright blobs are then captured – one for each camera. Storing only the locations of the bright blobs dispenses with the need for expensive frame-stores. In the subsequent processing, the bright blobs in each frame are reduced to single points by an algorithm which attempts to find the 'centre of mass' of each blob. We are therefore left with a list of the pixel coordinates for the points seen in each frame for each image. Assuming we are able to reconstruct 3D data from matched image points, it is essential that we are able to track and match the points through the sequences. For complicated motions, tracking can be the hardest part of the whole process; points crossing, being occluded, performing abrupt changes of direction, all add to the difficulties. Experience has shown that one reliable means of tracking is to track the points in space, i.e. to track the 3D motion – this enables one to use rigidity and length constraints (i.e. information from a model) in a simple fashion to improve the prediction process. Therefore, for reliable tracking it is very important that we have a good initial calibration of the system, otherwise the reconstructions will be poor and the tracking may experience problems. This is one of the main incentives for developing an accurate user-friendly means of obtaining the system calibration parameters. The following section will explain what the calibration parameters are and how we can estimate these using geometric algebra (GA) techniques. This will be followed by some results showing the accuracy of the calibrations via simulations and tests on real data. Throughout the paper we will assume that the readers are familiar with basic GA manipulations – for simple introductions to GA see [4, 7, 3, 5]. In this paper we will use the convention that where indices are repeated in the contravariant and covariant positions, i.e. $a^i b_i$, they are summed over unless explicitly stated otherwise.

8.2   External and Internal Calibration

In this section we will explain what is meant by external and internal calibration and show how we can use GA techniques to determine the unknown calibration parameters.

8.2.1   External calibration

Suppose that we have m cameras which we label 1 to m – these cameras are placed about the field of view. The aim is to place the cameras such that at any point in the image sequence, any given world point will always be visible in at least two of the cameras – this may not always be possible, but the tracking software can often make sensible predictions based on the rest of the tracked sequence when no prediction from the data is possible. Let us take the first camera, 1, as our reference camera. Then the position and orientation of camera j will be completely specified by a rotor R_j and a translation t_j as shown in figure 8.2.

Part of the calibration process will therefore be to determine, as accurately as possible, the $m - 1$ rotors and the $m - 1$ translations.

8.2.2   Internal calibration

A world point $\boldsymbol{X} = (X, Y, Z)$ is projected onto an image plane to give an image point $\boldsymbol{x} = (x, y, f)$ where f is the focal length of the camera (pinhole camera model), see figure 8.3

However, from the image we will measure pixel coordinates $\boldsymbol{u} = (u, v, 1)$. In order to move between pixel and image coordinates it is easy to show that there exists a 3×3 matrix, C which takes $\boldsymbol{x}$ to $\boldsymbol{u}$:

$$\boldsymbol{u} = C(\boldsymbol{x}/f), \qquad \boldsymbol{x}/f = C^{-1}\boldsymbol{u}$$

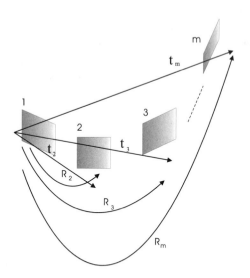

FIGURE 8.2. Rotations and translations of cameras relative to reference, chosen as camera 1.

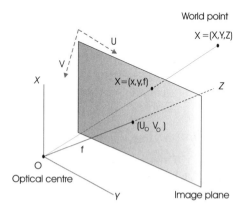

FIGURE 8.3. Factors determining the internal calibration parameters.

where C is of the form

$$C = \begin{pmatrix} \alpha & \beta & u_0 \\ \gamma & \delta & v_0 \\ 0 & 0 & 1 \end{pmatrix}$$

(u_0, v_0) is known as the principal point – it is where the optical axis of the camera cuts the image plane. $\alpha, \beta, \gamma, \delta$ depend upon the possible scaling and skewing of the pixel axes and f is the focal length (distance along the optical axis from the optical centre to the image plane).

The remainder of the calibration process will therefore be to determine the internal camera parameters. The internal parameters can be found via a variety of techniques and, once found, are unlikely to vary over reasonable timescales. In this paper we will mainly focus on how to accurately estimate the external parameters given knowledge of the internal parameters (in this case we say we are working with *calibrated cameras*, although a later section will indicate how we can include estimation of the internal parameters in the estimation procedure.)

8.3   Estimating the External Parameters

Suppose first that we know internal calibration matrices C_j for each camera, $j = 1, .., m$. Let the N world points that we observe with our cameras be $\mathbf{X}_i, i = 1, .., N$, and define an *occlusion field* O_{ij} such that $O_{ij} = 1$ if $\mathbf{X}_i$ is visible in camera j and 0 if it is not visible in camera j. In practice, we would like to be able to do this external calibration without having to track points (recall the tracking *uses* the calibration information). This is done by waving a single marker or light source over the viewing area (usually a volume of around 2m^3 should be covered for adequate calibration). In this way each camera will see no points or only one point and there is no tracking or matching problem. It is of course possible that some cameras will see more than one point due to the presence of spurious sources – if this occurs the frame is not used in the calibration process.

Let $\mathbf{u}_{ij}$ be the observed pixel coordinates (of the form $(u, v, 1)$) of the projection of world point $\mathbf{X}_i$ in camera j. Since we know the internal calibration parameters of each camera, we can recover the image coordinates, $\mathbf{x}_{ij}$, for this point via $\mathbf{x}_{ij} = C_j^{-1}\mathbf{u}_{ij}$ (from hereon we will take it that $\mathbf{x}_{ij} \equiv \mathbf{x}_{ij}/f$ to reduce the complication). If R_j and $\mathbf{t}_j$ are the rotor and translation which relate the frame and position of camera j to the reference frame of camera 1 then the following relation holds

$$\mathbf{X}_{ij} = \tilde{R}_j(\mathbf{X}_i - \mathbf{t}_j)R_j \tag{3.1}$$

where $\mathbf{X}_{ij}$ is world point i in the coordinate frame of camera j, see [7].

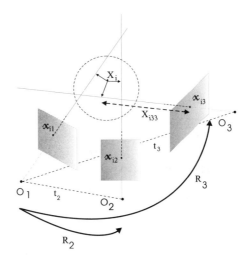

FIGURE 8.4. Geometric depiction of the meaning of cost function S_2.

If the noise occurs in the image planes, we might expect that our estimates of the Rs and ts would best be found via minimization of the following cost function

$$S_1 = \sum_{j=1}^{m} \sum_{i=1}^{N} \left[\boldsymbol{x}_{ij} - \frac{\tilde{R}_j (\boldsymbol{X}_i - \boldsymbol{t}_j) R_j}{[\tilde{R}_j (\boldsymbol{X}_i - \boldsymbol{t}_j) R_j] \cdot \boldsymbol{e}_3} \right]^2 O_{ij} \qquad (3.2)$$

This is effectively minimizing the sum of the squared distances in the image planes between the observed image points and the projected points. We should note here that $R_1 \equiv \mathcal{I}$ (the identity), $\boldsymbol{t}_1 \equiv \boldsymbol{0}$, and the presence of the O_{ij} ensures that if the point $\boldsymbol{X}_i$ is not visible in camera j then there is no contribution from this term. However, we can see immediately that the presence of the parameters we are trying to estimate in the denominator of the right-hand term makes this equation a difficult one – we would certainly have to find the minimum via some non-linear optimization technique.

Now, suppose that instead we consider the following cost function:

$$S_2 = \sum_{j=1}^{m} \sum_{i=1}^{N} \left[X_{ij3} \boldsymbol{x}_{ij} - \tilde{R}_j (\boldsymbol{X}_i - \boldsymbol{t}_j) R_j \right]^2 O_{ij} \qquad (3.3)$$

Here X_{ij3} is the distance we have to move out along the ray joining the optical centre of camera j to image point $\boldsymbol{x}_{ij}$ in order to minimize the distance between the world point $\boldsymbol{X}_i$ and the point $X_{ij3} \boldsymbol{x}_{ij}$. The above cost function is therefore the sum of squared distances between the world points and their closest points on the camera rays projecting out from the observed image points. Thus, while S_1 represents a cost function in the image planes, S_2 represents a cost function in the world, see figure (8.4).

Our observations are the image points in the m cameras, therefore the noise on our observations occurs in these image planes – if one assumes Gaussian noise one might therefore want to minimize the cost function S_1. However, it is also true that minimizing S_1 does not ensure that the reconstructed world points are in some way 'as close as possible to the observed rays – which one might also deem desirable. In fact, both cost functions are likely to give good results and we choose to optimize S_2 in order to obtain avoid a non-linear minimization.

Now, let us try to optimize S_2 over our parameters R_j, t_j, X_{ij3}, X_i. Although for external calibration purposes we are only interested in the relative rotations and translations of the cameras, here we shall adopt a maximum likelihood approach and differentiate with respect to all of our unknown parameters. We shall show in the following sections that it is possible to obtain an iterative solution to this minimization problem and that this procedure converges reliably provided the data is not very poor. This differentiation will involve differentiation with respect to scalars, vectors and rotors.

In the following sections we will frequently use the quantities defined below:

$$n_j = \sum_{i=1}^{N} O_{ij} \qquad \text{for} \qquad j = 1, 2, .., m \qquad (3.4)$$

$$m_i = \sum_{j=1}^{m} O_{ij} \qquad \text{for} \qquad i = 1, 2, .., N \qquad (3.5)$$

Here, n_j is the number of points visible in camera j and m_i is the number of cameras that can see world point i.

8.3.1 Differentiation w.r.t. t_k

When we take the derivative, ∂_a, with respect to (w.r.t.) a vector quantity a we use the fact that the differential operator ∂_a can be written (in terms of a basis $\{e_i\}$) as

$$\partial_a = e^i \frac{\partial}{\partial a^i} \qquad \text{where} \qquad a = a^i e_i \qquad (3.6)$$

Here $\{e^i\}$ is the **reciprocal frame** to $\{e_i\}$, and is defined by $e_i \cdot e^j = \delta_i^j$, for $i, j = 1, 2, 3$. Note that we do not write vectors in bold when they appear as subscripts in the vector derivative. We now want to differentiate S_2 w.r.t. t_k, where k can take values $2, 3, ..., m$. Consider first differentiating a vector squared, x^2, w.r.t. $t = t^j e_j$. Taking out a factor of e^j on the left and using the fact that $uv + vu$ is equivalent to the inner product of the two vectors, we have that

$$\partial_t(\boldsymbol{x}\boldsymbol{x}) = e^j \frac{\partial \boldsymbol{x}}{\partial t^j}\boldsymbol{x} + e^j \boldsymbol{x}\frac{\partial \boldsymbol{x}}{\partial t^j}$$

$$= 2e^j \left\{\frac{\partial \boldsymbol{x}}{\partial t^j}\cdot\boldsymbol{x}\right\} \tag{3.7}$$

Thus, if we let $\{X_{ik3}\boldsymbol{x}_{ik} - \tilde{R}_k(\boldsymbol{X}_i - \boldsymbol{t}_k)R_k\} \equiv \boldsymbol{Y}_{ik}$, then $\partial_{t_k}S_2 = 0$ gives

$$\partial_{t_k}S_2 = 2\sum_{i=1}^{N} e^j \frac{\partial}{\partial t_k^j}\{\tilde{R}_k\boldsymbol{t}_k R_k\}\cdot\boldsymbol{Y}_{ik}O_{ik}$$

$$= 2\sum_{i=1}^{N} e^j[(\tilde{R}_k e_j R_k)\cdot\boldsymbol{Y}_{ik}]O_{ik}$$

$$= 2\sum_{i=1}^{N} e^j \left(e_j\cdot R_k\boldsymbol{Y}_{ik}\tilde{R}_k\right)$$

$$= 2\sum_{i=1}^{N} R_k\boldsymbol{Y}_{ik}\tilde{R}_k = 2R_k\left[\sum_{i=1}^{N}\boldsymbol{Y}_{ik}\right]\tilde{R}_k = 0$$

$$\Rightarrow \qquad \sum_{i=1}^{N}\boldsymbol{Y}_{ik} = 0 \tag{3.8}$$

where we have used the fact that $(Ra\tilde{R})\cdot\boldsymbol{b} = \boldsymbol{a}\cdot(\tilde{R}\boldsymbol{b}R)$. Since $\sum_{i=1}^{N}\boldsymbol{Y}_{ik}$ is linear in $\boldsymbol{t}_k$, it is straightforward to solve equation (3.8) for $\boldsymbol{t}_k$ to give

$$\boldsymbol{t}_k = \frac{1}{n_k}\sum_{i=1}^{N}\left[\boldsymbol{X}_i - X_{ik3}R_k\boldsymbol{x}_{ik}\tilde{R}_k\right]O_{ik} \tag{3.9}$$

We have $m - 1$ such equations as k goes from 2 to m. Thus, if we have the data and have estimates for the world points, the rotors and the X_{ik3} values, we can solve for each of the translations.

8.3.2   Differentiation w.r.t. R_k

In geometric algebra we can differentiate w.r.t. any element of the algebra (for more details on multivector differentiation see [7, 6, 3]) and therefore w.r.t. rotors. Let us write

$$\partial_{R_k}S_2 = \partial_{R_k}\sum_{i=1}^{N}(\boldsymbol{v}_{ik} - \tilde{R}_k\boldsymbol{u}_{ik}R_k)^2 O_{ik} \tag{3.10}$$

where $v_{ik} = X_{ik3}x_{ik}$ and $u_{ik} = X_i - t_k$. The RHS has now been put in a standard form for which the solution (see [7] for details) is as follows

$$\partial_{R_k} S_2 = 4\tilde{R}_k \sum_{i=1}^{N} v_{ik} \wedge (\tilde{R}_k u_{ik} R_k) O_{ik} \tag{3.11}$$

For a minimum we require $\partial_{R_k} S_2 = 0$, and therefore the R_k must satisfy

$$\sum_{i=1}^{N} v_{ik} \wedge (\tilde{R}_k u_{ik} R_k) O_{ik} = \sum_{i=1}^{N} \{X_{ik3}x_{ik} \wedge \tilde{R}_k (X_i - t_k) R_k\} O_{ik} = 0 \tag{3.12}$$

or, substituting for t_k from equation (3.9)

$$\begin{aligned}
\text{LHS} &= \sum_{i=1}^{N} \{X_{ik3}x_{ik} \wedge \tilde{R}_k \\
&\qquad \left[X_i - \frac{1}{n_k} \sum_{j=1}^{N} \left[X_j - X_{jk3} R_k x_{jk} \tilde{R}_k \right] O_{jk} \right] R_k \} O_{ik} \\
&= \sum_{i=1}^{N} \{X_{ik3}x_{ik} \wedge \tilde{R}_k \left(X_i - \frac{1}{n_k} \sum_{j=1}^{N} X_j O_{jk} \right) R_k \} O_{ik} \\
&\equiv \sum_{i=1}^{N} (\tilde{v}_{ik} \wedge \tilde{R}_k \tilde{u}_{ik} R_k) = 0
\end{aligned} \tag{3.13}$$

where we now have $\tilde{v}_{ik} = X_{ik3}x_{ik}O_{ik}$ and $\tilde{u}_{ik} = X_i - \frac{1}{n_k} \sum_{j=1}^{N} X_j O_{jk}$. The second line in the set of equations (3.13) is obtained by noting that

$$\sum_{i=1}^{N} [O_{ik}X_{ik3}x_{ik}] \wedge \frac{1}{n_k} \sum_{j=1}^{N} X_{jk3}x_{jk}O_{jk} = 0.$$

We can now solve for R_k via SVD as outlined in [7] – i.e.

$$\tilde{R}_k = V U^T \quad \text{where} \quad F^k = U S V^T$$

$$\text{with} \quad F^k_{\alpha\beta} = \sum_{i=1}^{N} (e_\alpha \cdot \tilde{u}_{ik})(e_\beta \cdot \tilde{v}_{ik}) \tag{3.14}$$

This can be done for each k. Thus, we see from the above that provided we have the data, the world points and the X_{ij3} values, we can make an estimate of the rotations using the maximum likelihood estimator for the translations.

8.3.3 Differentiation w.r.t. the X_{pq3}

Next we would like to differentiate w.r.t. the scalars X_{pq3} – recall these represent the distance along the ray we have to move to bring us 'as close as possible' to the world point.

For each X_{pq3} we have

$$
\begin{aligned}
\partial_{X_{pq3}} S_2 &= \partial_{X_{pq3}} \left\{ X_{pq3} \boldsymbol{x}_{pq} - \tilde{R}_q (\mathbf{X}_p - \boldsymbol{t}_q) R_q \right\}^2 O_{pq} \\
&= 2 \left\{ X_{pq3} \boldsymbol{x}_{pq} - \tilde{R}_q (\mathbf{X}_p - \boldsymbol{t}_q) R_q \right\} \cdot \boldsymbol{x}_{pq} O_{pq} = 0
\end{aligned}
$$
(3.15)

For $O_{pq} \neq 0$ we therefore have

$$
X_{pq3} = \frac{[\tilde{R}_q (\mathbf{X}_p - \boldsymbol{t}_q) R_q] \cdot \boldsymbol{x}_{pq}}{\boldsymbol{x}_{pq}^2} \equiv \frac{(\mathbf{X}_p - \boldsymbol{t}_q) \cdot [R_q \boldsymbol{x}_{pq} \tilde{R}_q]}{\boldsymbol{x}_{pq}^2}
$$
(3.16)

This equation tells us how to estimate the values of the $\{X_{ij3}\}$ given we know the data, world points, rotations and translations.

8.3.4 Differentiation w.r.t. the $\mathbf{X}_k$

If we expand S_2 it is easy to see that the derivative w.r.t. $\mathbf{X}_k$ (for k from 1 to N) is given by

$$
\begin{aligned}
\partial_{X_k} S_2 &= \partial_{X_k} \sum_{j=1}^{m} \left\{ -2 X_{kj3} (R_j \boldsymbol{x}_{kj} \tilde{R}_j) \cdot \mathbf{X}_k + (\mathbf{X}_k - \boldsymbol{t}_j)^2 \right\} O_{kj} \\
&= 2 \sum_{j=1}^{m} \left[-X_{kj3} R_j \boldsymbol{x}_{kj} \tilde{R}_j + (\mathbf{X}_k - \boldsymbol{t}_j) \right] O_{kj} = 0
\end{aligned}
$$
(3.17)

where we have used the fact that $\partial_a (\boldsymbol{a} \cdot \boldsymbol{b}) = \boldsymbol{b}$. The above expression can then be rearranged to give

$$
\mathbf{X}_k = \frac{1}{m_k} \sum_{j=1}^{m} [\boldsymbol{t}_j + X_{kj3} R_j \boldsymbol{x}_{kj} \tilde{R}_j] O_{kj}
$$
(3.18)

if $m_k \neq 0$. k can take the values 1 to N. Thus, we are able to estimate the world points given values of the rotations, translations and the $\{X_{ij3}\}$.

8.3.5 Refining the estimates of $\boldsymbol{t}_j$ and $\mathbf{X}_k$

From our data (consisting of one point in many frames viewed by each camera) it is relatively straightforward to obtain an initial guess at the R_j

– this can be done by taking two cameras at a time and applying some standard algorithm (e.g. decomposing the Essential matrix [10], Weng et al's algorithm [12], etc.). Of course, this will not give a *consistent* set of rotations (e.g. $R_{23}R_2 \neq R_3$, where R_{23} is the rotor which takes the frame at camera 2 to the frame at camera 3), but it will give a reasonable starting point for the algorithm. Now, it would then be nice if we were able to estimate a consistent set of translations from these rotations and the data – but currently equation (3.9) gives t in terms of the other unknown parameters as well as the rotations. In addition, for reconstruction purposes, we would like to have an expression for the world points, $\{X_k\}$, in terms of just the rotations and translations. This is clearly also going to be essential when we have calibrated our cameras and we are wanting to reconstruct in an optimal fashion, points in the world from all of our m-camera data. We will deal with the case of reconstruction first.

8.3.6   Optimal reconstruction from calibrated data

If we substitute equation (3.16) into equation (3.18) to eliminate the $\{X_{pq3}\}$ values, we have

$$X_k = \frac{1}{m_k} \sum_{j=1}^{m} \left[t_j + \frac{1}{x_{kj}^2} \left\{ (X_k - t_j) \cdot R_j x_{kj} \tilde{R}_j \right\} R_j x_{kj} \tilde{R}_j \right] O_{kj} \qquad (3.19)$$

To simplify the notation we write $w_{ij} = R_j x_{ij} \tilde{R}_j$. If we take the inner product of the above equation with e_i, $i = 1, 2, 3$, we can rearrange to give

$$X_k \cdot \left[e_i - \frac{1}{m_k} \sum_{j=1}^{m} \frac{1}{x_{kj}^2} (w_{kj} \cdot e_i) w_{kj} O_{kj} \right] =$$
$$\frac{1}{m_k} \sum_{j=1}^{m} \left[t_j \cdot e_i - \frac{1}{x_{kj}^2} (w_{kj} \cdot t_j)(w_{kj} \cdot e_i) \right] O_{kj} \qquad (3.20)$$

We have $3 \times N$ such equations ($k = 1, .., N$ and $i = 1, 2, 3$). For each k we can construct a matrix equation for X_k

$$A_k X_k = b_k \qquad \Rightarrow \qquad X_k = A_k^{-1} b_k \qquad (3.21)$$

where the matrix A_k and vector b_k are given by

$$A_{ip}^k = \delta_{ip} - \frac{1}{m_k} \sum_{j=1}^{m} \frac{1}{x_{kj}^2} (w_{kj} \cdot e_i)(w_{kj} \cdot e_p) O_{kj} \qquad (3.22)$$

$$b_k \cdot e_i = b_i^k = \frac{1}{m_k} \sum_{j=1}^{m} t_j \cdot \left[e_i - \frac{1}{x_{kj}^2} (w_{kj} \cdot e_i) w_{kj} \right] O_{kj} \qquad (3.23)$$

Thus, if we have a knowledge of the calibration (Rs and ts), we see that via equation (3.21) we can very quickly reconstruct the 3D world points with a method that uses *all* of the available data at once in a sensible way. More generally the SVD can be used to solve $A_k X_k = b_k$ to avoid possible degeneracy.

8.3.7 An initial estimate for the translations

Suppose we substitute for X_{ik3} from equation (3.16) into equation (3.9) (still using $w_{ij} = R_j x_{ij} \tilde{R}_j$)

$$t_k = \frac{1}{n_k} \sum_{i=1}^{N} \left[X_i - \frac{1}{x_{ik}^2} \{ (X_i - t_k) \cdot w_{ik} \} w_{ik} \right] O_{ik} \tag{3.24}$$

for $n_k \neq 0$. If we now take the inner product of the above equation with e_j we have

$$t_k \cdot \left[e_j - \frac{1}{n_k} \sum_{i=1}^{N} y_{ijk} O_{ik} \right] = \frac{1}{n_k} \sum_{i=1}^{N} X_i \cdot \left[e_j - y_{ijk} \right] O_{ik} \tag{3.25}$$

with $y_{ijk} = \frac{1}{x_{ik}^2} (w_{ik} \cdot e_j) w_{ik}$. Now, writing $a_{ijk} = [e_j - y_{ijk}] O_{ik}$ and $p_{jk} = e_j - \frac{1}{n_k} \sum_{i=1}^{N} y_{ijk} O_{ik}$ the above equation can be written more concisely as

$$t_k \cdot p_{jk} = \frac{1}{n_k} \sum_{i=1}^{N} X_i \cdot a_{ijk} \tag{3.26}$$

Recall from the previous section that we can write $X_i = A_i^{-1} b_i$ where A_i is a matrix which is a function of the Rs only and b_i is a vector which is a function of both the Rs and the ts. Let us therefore write $X_i = \underline{f}_i(b_i)$, where $\underline{f}_i$ is the linear function corresponding to A_i^{-1}. Using the fact that $\underline{f}(c) \cdot d = c \cdot \overline{f}(d)$, we can now rewrite equation (3.26) as

$$t_k \cdot p_{jk} = \frac{1}{n_k} \sum_{i=1}^{N} b_i \cdot \overline{f}_i(a_{ijk}) \tag{3.27}$$

The next step is to note that we can write equation (3.23) as

$$b_k \cdot e_i = \frac{1}{m_k} \sum_{j=1}^{m} t_j \cdot a_{kij} \tag{3.28}$$

Letting $\overline{f}_i(a_{ijk}) = \tilde{a}_{ijk}^s e_s$ we can write $b_i \cdot \overline{f}_i(a_{ijk})$ as

$$b_i \cdot \overline{f}_i(a_{ijk}) = \frac{1}{m_i} \tilde{a}_{ijk}^s \sum_{l=1}^{m} (t_l \cdot a_{isl}) \tag{3.29}$$

From this equation we can see that it will now be possible to use equation (3.27) in order to form a linear equation in the ts. With some manipulation is it possible to obtain, for given j and k, the following expression

$$\sum_{l=1}^{m} t_l \cdot \left\{ \frac{1}{n_k} \sum_{i=1}^{N} \frac{1}{m_i} \tilde{a}_{ijk}^{s} O_{ik} \boldsymbol{a}_{isl} - \boldsymbol{p}_{jk} \delta_{lk} \right\} = 0 \qquad (3.30)$$

This can be written as a matrix equation of the form $Q\boldsymbol{T} = 0$ where $\boldsymbol{T} = [t_{21}, t_{22}, t_{23}, t_{31}, \ldots, t_{m3}]^T$ (since $\boldsymbol{t}_1 = 0$) and can therefore be solved by assigning to $\boldsymbol{T}$ the eigenvector corresponding to the smallest eigenvalue of the matrix $Q^T Q$ (alternatively use SVD). Thus, given only an estimate of the Rs we have been able to formulate an estimate of the ts – again, using all of the available data simultaneously.

8.3.8 *The iterative calibration scheme*

Having worked out all of the necessary steps in the previous sections, we are now in a position to outline the iterative scheme by which the external calibration is carried out.

1. Guess an initial set of Rs given only the data (use standard 2-camera algorithms)

2. Estimate a set of ts given these Rs

3. Estimate the world points $\{\mathbf{X}_i\}$ given these Rs and ts

4. Estimate the $\{X_{pq3}\}$s given all of the above

5. Obtain a new estimate of the Rs using values from (2),(3),(4) and start the next iteration by returning to step (2).

In practice each step of the procedure can be performed quickly and convergence is achieved within a few tens of iterations. In estimating the ts we should note that we are only able to do this up to scale. One may therefore set a value to unity (say $\boldsymbol{t}_2 \cdot \boldsymbol{e}_3$) and evaluate the other values relative to this – when doing this however, checks must be made that the signs of the estimated ts do not produce negative depths (if they do, we will need to take $\boldsymbol{t}_2 \cdot \boldsymbol{e}_3 = $ -1).

The above external calibration routine requires a very simple initial data gathering stage (waving a single point over a volume representative of where the world points will be) and utilises all of the image data simultaneously in order to produce optimal estimates of the relative rotations and translations of the cameras. In addition the formula for reconstruction given in equation (3.18) is very simple and gives accurate and robust 3D reconstructions. The value of the cost function (S_2) can also be monitored throughout the iterations; a final value of S_2 which is too large is usually indicative of poor data and a new calibration should be performed.

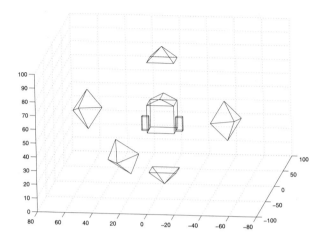

FIGURE 8.5. Wireframe house (26 points) viewed from 5 cameras – the optical centre and 4 defining points of the image planes are shown. The position and orientation of the cameras are such that the house is in view in each camera.

8.4 Examples and Results

In order to illustrate this calibration procedure we will present some results on both simulated and real data. While the procedure is routinely used in the tracking and subsequent reconstruction of *real* motion capture data, a quantitative evaluation of its behaviour is more easily obtained from simulations. The real data presented attempts to evaluate the performance of the calibration by checking that rays from the image planes, from which we reconstruct the world point, do indeed cross approximately at a single point in space. Real multiple camera data together with example reconstructions can be downloaded from http://www.sig-proc.eng.cam.ac.uk/vision.

We use 5 cameras, the first camera placed with its optical centre at the origin $[0, 0, 0]$ and with its optical axis along the z-direction, viewing a wireframe house which is placed about 50 units in z away from the origin. Cameras 2 to 4 are rotated and translated from camera 1 as shown in figure 8.5.

R_j is the rotor which takes the frame at camera 1 to the frame at camera j, and the axes, $\hat{n}_j$, and angles, θ_j, which characterise R_j (since $R_j = \exp{-\frac{I\hat{n}_j\theta_j}{2}}$) are given in table 1.

Rotor	$\hat{n}$	θ	t
R_2	$[0.7071, -0.7071, 0]$	$51.498°$	$[40, 40, 5]$
R_3	$[0.7593, -0.6508, 0]$	$83.810°$	$[60, 70, 40]$
R_4	$[-0.7071, 0.7071, 0]$	$96.721°$	$[-60, -60, 60]$
R_5	$[-1, 0, 0]$	$180°$	$[0, 0, 100]$

Table 1 Table showing true values of the rotations and translations of the cameras.

In order to calibrate the cameras we used 30 points generated at random from a cube centred at $[0, 0, 50]$ with side length 30 – these points simulated the calibration process whereby one bright marker is moved around the scene over a number of frames. Here we will assume that each of the 30 points is visible in all cameras. The 30 points were projected into the 5 cameras and the image points from each image plane were the only data given to the calibration routines. In the image planes Gaussian noise was added. Three different levels of noise were tested having standard deviation, $\sigma = 0.001, 0.005, 0.01$ – with the image plane coordinates ranging roughly from -0.45 to +0.45, at a resolution of 1000×1000 this would correspond to standard deviations ranging from 1 pixel to 10 pixels. To initialise the algorithm, an initial set of Rs and ts were found by taking two cameras at a time and performing some simple method to determine the parameters, e.g. the algorithm of Weng *et al.* [12] – call these R_k^0 and t_k^0. 20 iterations of the algorithm were allowed in each case, although generally fewer were needed to achieve adequate convergence. Let the final estimated values be R_k^f and t_k^f.

Using R_k^f and t_k^f we can then reconstruct the wireframe house. We use a realistic set of data which consists of the image points in each camera of those points from the house that were visible in that camera (i.e. we include the relevant occlusion field). For the case depicted in figure (8.5), we can see, for example, that the uppermost camera will not see any of the vertices on the lower side of the house. For these simulations it was the case that every vertex was visible in at least two cameras. Also the same data and occlusion field were used to perform the 3D reconstruction using the initial guesses R_k^0 and t_k^0. The 3D reconstruction was carried out using equation (3.21) in both cases.

Figure (8.6) shows 6 different 3D views of the true wireframe house – the azimuth and elevation ([az,el], in degrees) of the viewpoint for each of the views is as follows (from top left to bottom right)

$$[-38, 30], \ [-15, 5], \ [-110, 20], \ [80, -25], \ [-90, 90], \ [-90, 0]$$

Figure (8.7) shows the reconstructions obtained for the case of added noise, $\sigma = 0.001$ – the left column shows the results from the iterative scheme

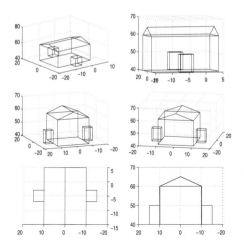

FIGURE 8.6. Six views of the true vertices of the simulated house.

(20 iterations), while the right column shows the results for reconstruction from the two-camera estimates. The top, middle and bottom views have azimuth and elevation as for the left column of figure (8.6). Figures (8.8) and (8.9) show similar plots for $\sigma = 0.005$ and $\sigma = 0.01$. We see that with little noise the reconstruction is very good for both cases. However, as the noise gets more severe, we see that the iterative scheme tends to give better reconstructions. Even under higher noise levels the reconstruction remains acceptable.

As well as comparing the reconstructions it is also instructive to see how the estimated rotors compare with the true rotors in each of the above cases. If a rotor R, is written as $R = \exp(-I\hat{n}\theta/2)$, then the bivector describing the rotation is $I\hat{n}\theta/2$, so that a good way of comparing rotors is to compare the bivector components: i.e. $n_1\theta$, $n_2\theta$, $n_3\theta$, with $n_i = \hat{n}\cdot e_i$. Figure (8.10) compares these components for the true rotors, and the two sets of rotors described above for four noise values, $\sigma = 0.001, 0.005, 0.007, 0.01$. Similar comparisons for the translations are shown in figure (8.11).

In order to show the performance of the calibration algorithms on real data we used three cameras to take a sequence of 300 frames of a person performing a golf swing, with markers placed on shoulders, elbows and wrists. The cameras were calibrated prior to taking the data by waving a single bright marker over a representative volume and applying the algorithms outlined in section 1.3. Figure 8.12 shows an example of the reconstruction by showing the linked points for frame 3 of the sequence. Although this plot does not tell us much without detailed information of the real subject, figure 8.13 gives some idea of the accuracy of the calibration by plotting the rays from the matching image points (four such points were taken) through the optical centres of the cameras. The positions of the cameras are obtained from the calibration. If the calibration is good, we would ex-

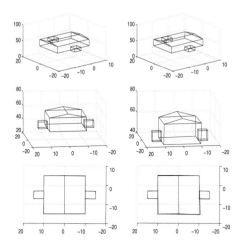

FIGURE 8.7. Results of the reconstruction with $\sigma = 0.001$. The left column shows results of iterative algorithm; right column shows results from taking two-camera estimates.

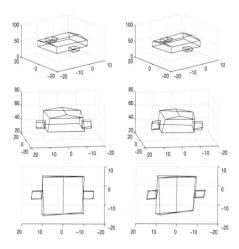

FIGURE 8.8. Results of the reconstruction with $\sigma = 0.005$. The left column shows results of iterative algorithm; right column shows results from taking two-camera estimates.

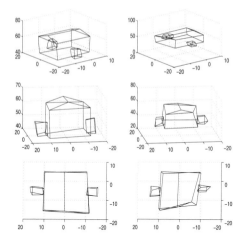

FIGURE 8.9. Results of the reconstruction with $\sigma = 0.01$. **The left column shows results of iterative algorithm; right column shows results from taking two-camera estimates.**

pect all matching image points to intersect more or less at a single point in space. From figure 8.12 we can see that this is indeed the case for the particular frame chosen, and is also the case throughout the rest of the sequence.

We can see that on the whole, the iterative algorithm described in this paper produces good estimates of the bivectors and of the translations over a wide range of noise cases. The two-camera estimates that we have compared the algorithm with are, of course, not something that would be routinely used in practice. However, most calibration schemes would start with some such estimate and generally proceed via non-linear minimization. Such minimizations use gradient descent methods and as such are crucially dependent on the initial guess as they will tend to find the local minimum in the vicinity of this initial guess. Other methods of calibration involve building up the external calibration parameters camera by camera; such methods have to ensure that the final estimates are independent of the particular order of estimation and form a self-consistent set. Some calibration schemes in the literature are given in [11, 2], however, code is generally not available to compare such algorithms with those discussed here.

8.5   Extending to Include Internal Calibration

The discussion in this paper has assumed that we have the internal calibration of the cameras. Typically, for the motion capture system, this is done every few weeks or so, and the values are assumed not to change sig-

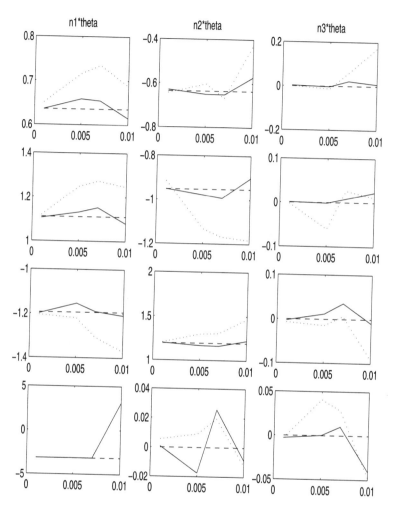

FIGURE 8.10. Moving left to right in columns shows results for $n_1\theta$, $n_2\theta$, $n_3\theta$, while moving down rows from top to bottom shows results for R_2, R_3, R_4, R_5. In each plot the dashed line gives the true value of the bivector component, the solid line gives the bivector component from the iterative algorithm and the dotted line gives that from the two-camera estimate. The x-axis in each case gives the standard deviation of the Gaussian noise added.

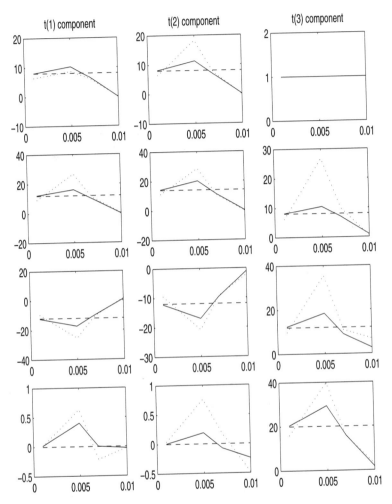

FIGURE 8.11. Moving left to right in columns shows results for the first, second and third components of the translation vectors, while moving down rows from top to bottom shows results for cameras 2 to 4. In each plot the dashed line gives the true value of the translation component, the solid line gives the component from the iterative algorithm and the dotted line gives that from the two-camera estimate. The x-axis in each case gives the standard deviation of the Gaussian noise added. Note that the translations are normalised so that $t_2 \cdot e_3 = 1$, hence the graph in the upper right.

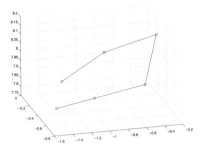

FIGURE 8.12. 3D snapshot at frame 3 of the shoulders, elbows and wrists of the golfer.

nificantly on this timescale. However, it is possible to adjust the algorithm presented here to include determination of the internal parameters. If we return to equation 3.3, but replace x_{ij} by u_{ij},

$$u_{ij} = f_j(x_{ij}) \equiv C_j x_{ij} \tag{5.31}$$

where the linear function f_j represents the 3×3 camera matrix C_j, our cost function S_2 in terms of the observations u and the internal calibration, becomes

$$S_2 = \sum_{j=1}^{m} \sum_{i=1}^{N} \left[X_{ij3} f_{cj}(u_{ij}) - \tilde{R}_j(X_i - t_j)R_j \right]^2 O_{ij} \tag{5.32}$$

where $f_{cj} \equiv f_j^{-1}$. We can now minimize over the $\{f_{cj}\}$ as well as the other parameters using the ability in GA to carry out functional differentiation. One must note, however, that the f_cs take a particular form (which can be made equivalent to an upper triangular matrix), so this constraint must be allowed for. A detailed description of this self-calibration procedure will be presented elsewhere.

8.6   Conclusions

A means of determining the external calibration parameters (relative rotations and translations) for any number of cameras observing a scene has been presented. Using geometric algebra to differentiate with respect to the the unknowns in the problem, we are able to build up an iterative estimation scheme. In the process, we also produce an efficient and robust reconstruction algorithm which can be used for estimating the world points once the calibration has been achieved. The method is essentially a maximum likelihood technique in which we substitute maximum likelihood estimators in order to eliminate the parameters we do not want to estimate

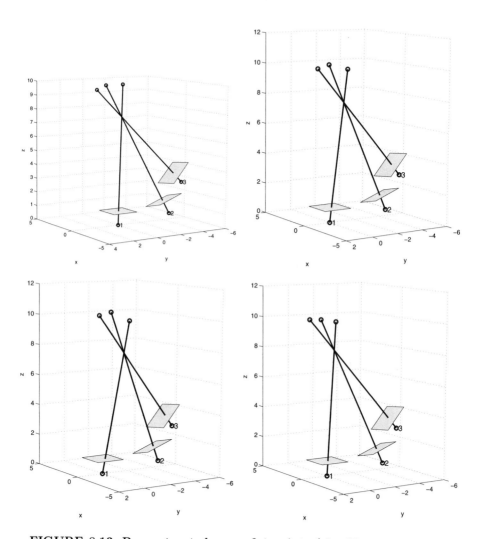

FIGURE 8.13. Reconstructed rays of 4 points (shoulders, one elbow and one wrist) in randomly chosen frames from a 300 frame sequence of a golf swing. The reconstruction was carried out using calibration data determined by the iterative scheme described in section 1.3.

(e.g. the world points and the $\{X_{ij3}\}$s). Another technique which can be employed is a Bayesian approach, which marginalises over these parameters (nuisance parameters) prior to estimating the Rs and ts – a review of the geometric algebra approach to this procedure is given in this volume 9. Indeed, if the parameters in question have a multivariate Gaussian distribution then the two techniques should give the same results. Preliminary tests indicate that, even though the noise is unlikely to be multivariate Gaussian in real data, the two approaches produce very similar results on good data.

The calibration scheme presented here is currently used on an optical motion capture system. The algorithms are used with data from a single moving marker to produce the external calibration. This calibration is then used in the tracking and reconstruction of subsequent data taken from the subject. The algorithm is relatively quick, robust and is easily effected, meaning that the cameras can be moved and the system speedily recalibrated.

In summary, we have presented a technique for external camera calibration which used the ease of expressing geometric entities in geometric algebra and the ability to differentiate with respect to any element of the algebra. Using rotors provides a very efficient way of optimizing over a rotation manifold; it is a minimally parameterized system, does not have the singularities associated with Euler angles and is less cumbersome and more easily extendable than quaternions (in the sense that rotors can rotate any geometric object, not just vectors and have the same form in any dimension). The results presented here can be used alone or used to initialize algorithms which employ minimization techniques and different cost functions. The intermediate steps of determining the best estimate of the world points from known data points and given calibration, and of determining the relative translations between cameras given the rotations and data points, are also useful in many reconstruction and tracking scenarios.

Acknowledgments

JL would like to thank the Royal Society of London for their support in the form of a University Research Fellowship. AXSS acknowledges the support of the EPSRC in the form of a Studentship. The authors would like to thank Ken Yam and Anthony Lasenby for useful comments on the text, Maurice Ringer for the tracking of the golf data and the referees for helpful suggestions.

Chapter 9

Bayesian Inference and Geometric Algebra: An Application to Camera Localization

Chris Doran

9.1 Introduction

Geometric algebra is an extremely powerful language for solving complex geometric problems in engineering [4, 8]. Its advantages are particularly clear in the treatment of rotations. Rotations of a vector are performed by the double-sided application of a *rotor*, which is formed from the *geometric product* of an even number of unit vectors. In three dimensions a rotor is simply a normalised element of the even subalgebra of $\mathcal{G}_3$, the geometric algebra of three dimensional space. In this paper we are solely interested in rotations in space, and henceforth all reference to rotors can be assumed to refer to the 3-d case. Rotors have a number of useful features. They can be easily parameterised in terms of the bivector representing the plane of rotation. Their product is a very efficient way of computing the effect of compound rotations, and is numerically very stable.

Rotors are normalised elements in a 4-d algebra (the even subalgebra of $\mathcal{G}_3$), so they can be represented by points on the unit sphere in 4-d. This is called a 3-sphere, and is the rotor *group manifold* [2, 5]. The simple structure of this manifold makes it very easy to extrapolate between rotations, which is useful in many fields including finite element analysis and rigid body dynamics. The extrapolation method can be easily understood in terms of relaxing the normalisation constraint and working with unnormalised rotors, and normalising the result at the end of a computation. This is also the key to simplifying the problem of differentiating with respect to rotations. Ordinarily, a function of a rotation is viewed as taking its value on the group manifold. Derivatives of this function take their values in the tangent space to the group manifold. This is mathematically rigorous, but rather cumbersome computationally. A better idea is to move off the group manifold and work in the 4-d linear space, where the rules of calculus are much simpler [3, 4, 8]. Used properly, this trick can significantly simplify optimisation problems involving rotations.

The main applications considered here are to variations of the camera localization problem in computer vision [7, 8, 10, 11, 13, 14]. Suppose that a number of cameras are placed in unknown positions and they observe the same scene. In order to reconstruct the scene, we need to determine the relative positions and orientations of the cameras. Given a sufficient number of point matches between the cameras, this information can be accurately recovered without any external measurements. For most cases this problem can be reduced to a least squares minimisation over a set of rotations and translations, and this can be simplified considerably using the techniques of rotor calculus. The least squares likelihood functions used here are derived from a simple Bayesian probabilistic model, which helps to expose some of the underlying assumptions in the choice of likelihood function [12]. This is useful in pointing the way to constructing improved models. In this paper we assume a projective camera model, and will further assume that the internal camera parameters are all known. A preliminary discussion of how geometric algebra can be used to estimate these internal parameters is contained in [9]. The basic techniques described here can be generalised in a number of ways to deal with more complex situations and at various points we discuss how one might exploit this. In particular, the extension beyond two cameras is straightforward. This is an area where more traditional tensor-based approaches run into difficulties.

9.2 Geometric Algebra in Three Dimensions

The geometric algebra of three-dimensional space is generated by a right-handed orthonormal set of vectors $\{e_1, e_2, e_3\}$. Their geometric product satisfies

$$e_i e_j = \delta_{ij} + I\epsilon_{ijk}e_k \qquad (2.1)$$

where I is the pseudoscalar

$$I = e_1 \wedge e_2 \wedge e_3 = e_1 e_2 e_3. \qquad (2.2)$$

The full algebra is spanned by

$$\begin{array}{cccc} 1 & \{e_i\} & \{Ie_i\} & I \\ \text{1 scalar} & \text{3 vectors} & \text{3 bivectors} & \text{1 trivector.} \end{array} \qquad (2.3)$$

The dot and wedge symbols have their usual meaning as inner and outer products, and for vectors

$$a \cdot b = \frac{1}{2}(ab + ba) \qquad a \wedge b = \frac{1}{2}(ab - ba). \qquad (2.4)$$

The geometric product for general multivectors is denoted simply by juxtaposition, and throughout inner and outer products take precedence over geometric products. Angled brackets $\langle\rangle_n$ are used for the projection onto grade operation, and the scalar part of a multivector A is denoted simply by $\langle A \rangle$. The scalar part satisfies the cyclic reordering property

$$\langle AB \cdots C \rangle = \langle B \cdots CA \rangle. \tag{2.5}$$

The *reverse* of a multivector is formed by reversing the order of geometric products of vectors in the multivector and is denoted with a tilde. An arbitrary multivector M can be decomposed as

$$M = \alpha + a + B + \beta I, \tag{2.6}$$

where α and β are scalars, a is a vector and B is a bivector. The reverse of M is

$$\tilde{M} = \alpha + a - B - \beta I. \tag{2.7}$$

9.3   Rotors and Rotations

A *rotor* is a normalised element of the even subalgebra,

$$R = \alpha + B, \tag{3.8}$$

where α is a scalar and B is a bivector. The normalisation condition is that

$$R\tilde{R} = \tilde{R}R = \alpha^2 - B^2 = 1. \tag{3.9}$$

Rotors generate rotations of vectors via the double-sided transformation law

$$a \mapsto a' = Ra\tilde{R}. \tag{3.10}$$

This same law holds for bivectors, since

$$
\begin{aligned}
(Ra\tilde{R}) \wedge (Rb\tilde{R}) &= \tfrac{1}{2}\left(Ra\tilde{R}Rb\tilde{R} - Rb\tilde{R}Ra\tilde{R}\right) \\
&= \tfrac{1}{2}R(ab - ba)\tilde{R} \\
&= Ra \wedge b\,\tilde{R}.
\end{aligned} \tag{3.11}
$$

It is also simple to check that rotors leave inner products invariant,

$$a' \cdot b' = \langle Ra\tilde{R}Rb\tilde{R} \rangle = \langle ab \rangle = a \cdot b. \tag{3.12}$$

The rotor transformation law $a \mapsto Ra\tilde{R}$ also leaves trivectors invariant, so has determinant $+1$ and must be a rotation.

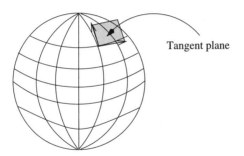

FIGURE 9.1. *Tangent Space.* **At each point on the sphere one can attach a tangent plane.**

Rotors can be parameterised directly in terms of the plane of rotation by writing

$$R = \exp(-B/2). \tag{3.13}$$

The rotor R now generates a rotation through an angle $|B|$ in the plane specified by B, with the same orientation as B. In three dimensions we can also write

$$R = \exp(-\theta I\hat{n}/2) \tag{3.14}$$

where $\theta = |B|$, and $\hat{n} = -IB/|B|$ is the unit vector representing the rotation axis. The map between a vector n and the bivector In is called a *duality transformation*. Bivectors can only be dualised to vectors in three dimensions, so the concept of an axis of rotation only exists for three-dimensional space.

9.3.1   The group manifold

Rotors are elements of a four-dimensional space, normalised to 1. They can be represented as points on a *3-sphere* — the set of unit vectors in four dimensions. This is the rotor *group manifold*. At any point on the manifold, the *tangent space* is three-dimensional. This is the analog of the tangent plane to a sphere in three dimensions (see Figure 9.1).

Rotors require three parameters to specify them uniquely. One common parameterisation is in terms of the *Euler angles* (θ, ϕ, ψ),

$$R = \exp(-e_1e_2\phi/2)\exp(-e_2e_3\theta/2)\exp(-e_1e_2\psi/2). \tag{3.15}$$

But often it is more convenient to use the set of bivector generators, with

$$|B^2| \leq \pi. \tag{3.16}$$

The rotors R and $-R$ generate the *same* rotation, because of their double-sided action. It follows that the *rotation* group manifold is more complicated than the rotor group manifold — it is a projective 3-sphere with points R and $-R$ identified. This is one reason why it is usually easier to work with rotors.

9.3.2 Extrapolating between rotations

Suppose we are given two estimates of a rotation, R_0 and R_1, how do we find the mid-point? With rotors this is remarkably easy! We first make sure sure they have smallest angle between them in four dimensions. This is done by ensuring that

$$\langle R_0 \tilde{R}_1 \rangle = \cos\theta > 0. \tag{3.17}$$

If this inequality is not satisfied, then the sign of one of the rotors should be flipped. The 'shortest' path between the rotors on the group manifold is defined by

$$R(\lambda) = R_0 \exp(\lambda B), \tag{3.18}$$

where

$$R(0) = R_0, \quad R(1) = R_1. \tag{3.19}$$

It follows that we can find B from

$$\exp(B) = \tilde{R}_0 R_1. \tag{3.20}$$

The path defined by $\exp(\lambda B)$ is an invariant construct. If both endpoints are transformed, the path transforms in the same way. The midpoint is

$$R_{1/2} = R_0 \exp(B/2), \tag{3.21}$$

which therefore generates the midpoint rotation. This is quite general — it works for any rotor group (or any *Lie group*). For rotations in three dimensions we can do even better. R_0 and R_1 can be viewed as two unit vectors in a four-dimensional space. The path $\exp(\lambda B)$ lies in the plane specified by these vectors (see Figure 9.2).

The rotor path between R_0 and R_1 can be written as

$$R(\lambda) = R_0\big(\cos\lambda\theta + \sin\lambda\theta\,\hat{B}\big), \tag{3.22}$$

where we have used $B = \theta\hat{B}$. But we know that

$$\exp(B) = \cos\theta + \sin\theta\,\hat{B} = \tilde{R}_0 R_1. \tag{3.23}$$

It follows that

$$R(\lambda) = \frac{R_0}{\sin\theta}\big(\sin\theta\cos\lambda\theta + \sin\lambda\theta(\tilde{R}_0 R_1 - \cos\theta)\big) \tag{3.24}$$

$$= \frac{1}{\sin\theta}\big(\sin(1-\lambda)\theta\,R_0 + \sin\lambda\theta\,R_1\big), \tag{3.25}$$

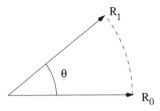

FIGURE 9.2. *The path between two rotors* **The rotors can be treated as unit vectors in four dimensions. The path between them lies entirely in the plane of the two rotors, and therefore defines a segment of a circle.**

which satisfies $R(\lambda)\tilde{R}(\lambda) = 1$ for all λ. The midpoint rotor is therefore simply

$$R_{1/2} = \frac{\sin(\theta/2)}{\sin\theta}(R_0 + R_1). \tag{3.26}$$

This gives us a remarkably simple prescription for finding the midpoint: *add the rotors and normalise the result.* By comparison, the equivalent matrix is quadratic in R, and so is much more difficult to express in terms of the two endpoint rotation matrices.

Suppose now that we have a number of estimates for a rotation and wanted to find the average. Again the answer is simple. First one chooses the sign of the rotors so that they are all in the 'closest' configuration. This will normally be easy if the rotations are all roughly equal. If some of the rotations are quite different then one might have to search around for the closest configuration, though in these cases the average of such rotations is not a useful concept. Once one has all of the rotors chosen, one simply adds them up and normalises the result to obtain the average. This sort of calculation can be useful in computer vision problems where one has a number of estimates of the relative rotations between cameras, and their average is required.

The lesson here is that problems involving rotations can be simplified by working with rotors and relaxing the normalisation criteria. This enables us to work in a four-dimensional linear space and is the basis for a simplified calculus for rotations.

9.4 Rotor Calculus

Any function of a rotation can be viewed as taking its values over the group manifold. In most of what follows we are interested in scalar functions, though there is no reason to restrict to this case. The derivative of the function with respect to a rotor defines a vector in the tangent space at each point on the group manifold. The vector points in the direction of steepest increase of the function. This can all be made mathematically rigorous and is the subject of *differential geometry*. The problem is that much off this is over-complicated for the relatively simple minimisation problems encountered in computer vision. Working intrinsically on the group manifold involves introducing local coordinates (such as the Euler angles) and differentiating with respect to each of these in turn. The resulting calculations can be long and messy and often hide the simplicity of the answer.

Geometric algebra provides us with a more elegant and simpler alternative. We relax the rotor normalisation constraint and replace R by ψ — a general element of the even subalgebra. There is a very simple derivative operator associated with ψ. We first decompose ψ in terms of the $\{e_i\}$ basis as

$$\psi = \psi_0 + \sum_{k=1}^{3} \psi_k I e_k \qquad (4.27)$$

where the $\{\psi_0, \ldots, \psi_3\}$ are a set of scalar components. We now define the *multivector derivative* ∂_ψ by

$$\partial_\psi = \frac{\partial}{\partial \psi_0} - \sum_{k=1}^{3} I e_k \frac{\partial}{\partial \psi_k}. \qquad (4.28)$$

This derivative is independent of the chosen frame. It satisfies the basic result

$$\partial_\psi \langle \psi A \rangle = A \qquad (4.29)$$

where A is a constant, even-grade multivector. All further results for ∂_ψ are built up from this basic result and Leibniz' rule for the derivative of a product.

The basic trick now is to re-write a rotation as

$$R a \tilde{R} = \psi a \psi^{-1}. \qquad (4.30)$$

This works because any even multivector ψ can be written as

$$\psi = \rho^{1/2} R \qquad (4.31)$$

where R is a rotor, $\rho = \psi \tilde{\psi}$ and $\rho = 0$ if and only if $\psi = 0$. The inverse of ψ is then

$$\psi^{-1} = \rho^{-1/2} \tilde{R} \qquad (4.32)$$

so that

$$\psi \psi^{-1} = R \tilde{R} = 1. \qquad (4.33)$$

The equality of equation (4.30) follows immediately. If one imagines a function over a sphere in three dimensions, one can extend this to a function over all space by attaching the same value to all points on each line from the origin. The extension $R \mapsto \psi$ does precisely this, but in a four dimensional space.

We are now able to differentiate functions of the rotation quite simply. The typical application is to a scalar of the type

$$(Ra\tilde{R}) \cdot b = \langle Ra\tilde{R}b \rangle = \langle \psi a \psi^{-1} b \rangle. \tag{4.34}$$

We now have

$$\partial_\psi \langle \psi a \psi^{-1} b \rangle = a\psi^{-1} b + \dot{\partial}_\psi \langle \psi a \dot{\psi}^{-1} b \rangle \tag{4.35}$$

where the overdot denotes the scope of the differential operator (*i.e.* the term being differentiated). We next require a formula for the inverse term. We start by letting M be a constant multivector, and derive

$$0 = \partial_\psi \langle \psi \psi^{-1} M \rangle = \psi^{-1} M + \dot{\partial}_\psi \langle \psi \dot{\psi}^{-1} M \rangle. \tag{4.36}$$

It follows that

$$\dot{\partial}_\psi \langle \dot{\psi}^{-1} M \psi \rangle = -\psi^{-1} M. \tag{4.37}$$

But in this formula we can now let M become a function of ψ, as only the first term, ψ^{-1}, is acted on by the differential operator. We can therefore replace M by $M\psi^{-1}$ to obtain the useful formula

$$\dot{\partial}_\psi \langle \dot{\psi}^{-1} M \rangle = -\psi^{-1} M \psi^{-1}. \tag{4.38}$$

We can now complete the derivation started at (4.35) to find

$$\partial_\psi \langle \psi a \psi^{-1} b \rangle = a\psi^{-1} b - \psi^{-1} b \psi a \psi^{-1}. \tag{4.39}$$

It is convenient to premultiply this expression by ψ to get

$$\psi \partial_\psi \langle \psi a \psi^{-1} b \rangle = \psi a \psi^{-1} b - b \psi a \psi^{-1} = 2(Ra\tilde{R}) \wedge b. \tag{4.40}$$

The fact that the *geometric product* is formed between ψ and ∂_ψ is important. This product is *invertible*, so no information is lost. The fact that a bivector is formed here is sensible. Bivectors belong to a three-dimensional space — the same number of dimensions as the tangent space to the group manifold. The big advantage of the approach used here is that one never leaves the geometric algebra of space, and the resultant bivector is evaluated in the same space, rather than in some abstract tangent space on the group manifold. The result (4.40) is also sensible if one thinks about varying R in $(Ra\tilde{R}) \cdot b$ while keeping the vectors a and b constant. This function clearly has a maximum when $Ra\tilde{R}$ is parallel to b, which is precisely where the derivative vanishes.

This simple derivation turns out to be very useful in a range of applications, including rigid body dynamics and point-particle models for fermions. Here we have chosen to illustrate its use with some applications in computer vision.

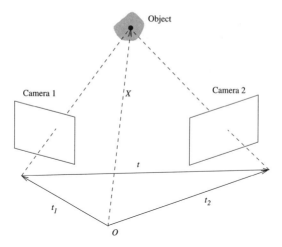

FIGURE 9.3. *The basic two camera setup.* **The same object is viewed from two different directions. The cameras are related by a translation and a rotation. All vectors are expressed relative to some arbitrary origin O. The relative vector between the camera centres, $t = t_2 - t_1$, is independent of the origin.**

9.5   Computer Vision

The main problem of interest in this paper is that of camera localization. Suppose that we have different camera views of the same scene. Given point matches with added noise, we want to find the relative translation and rotation between the cameras. Once the camera geometry has been calculated like this, it is possible to reconstruct the three-dimensional scene. Applications of this basic idea include fields such as motion analysis, reaching and neurocontrol, and robot control. Before studying the more realistic case of a projective camera model (see Section 9.6) we first study a simpler, toy problem whose solution is well known. This is the case where the full 3-d position is measured for the point matches, including the range data. This enables us to introduce some of the tools of Bayesian inference in a simplified setting.

9.5.1   Known range data

Suppose that we know the full three-dimensional coordinates of each point match (which is not very common in practice). The basic solution in this case is well known for the two camera case and has been discussed by many authors [1, 6, 7, 11]. The derivation presented here is slightly different, however, in being based on an underlying probabilistic model for the data, with the rotations and translation recovered via a Bayesian argument. Rel-

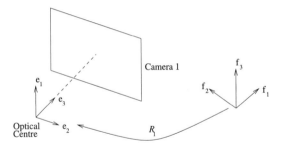

FIGURE 9.4. *The camera frame.* **Each camera has a frame $\{e_i\}$ attached to it, with the 3-axis representing the optical axis. The camera frame is related to an arbitrary global frame $\{f_i\}$ by a rotor, with a separate rotor required for each camera. The rotor taking the camera 1 frame onto the camera 2 frame is then $R_2\tilde{R}_1$, and this is what we aim to find.**

ative to an arbitrary origin, O, the camera centres are located at positions t_1 and t_2, and the point matches at positions X^k (see Figure 9.3). Throughout we use superscript indices to label the point matches, and subscript Latin indices to label frame vectors, $\{e_i\}$, or components of a vector, x_i. Which of these is intended should be obvious, as we only use e_i and f_i for frame vectors. At various points, subscript Greek indices are used to label the cameras.

If we write the two camera frames as $\{e_{1i}\}$ and $\{e_{2i}\}$ respectively, then the data we assume that we can record are a set of coordinates for the point matches,

$$x_{1i}^k \;=\; e_{1i} \cdot (X - t_1) \qquad (5.41)$$
$$x_{2i}^k \;=\; e_{2i} \cdot (X - t_2). \qquad (5.42)$$

We now introduce a third, arbitrary reference frame $\{f_i\}$, which is related to the two camera frames by

$$e_{1i} = R_1 f_i \tilde{R}_1, \quad e_{2i} = R_2 f_i \tilde{R}_2. \qquad (5.43)$$

(See Figure 9.4). The advantage of working with separate rotors for the camera frames, instead of the mutual rotation between them, is that it keeps all formulae symmetric in the choice of frame, and ensures that the equations generalise easily to the n-camera case. This also provides a useful check on the formalism — we should only obtain equations for the mutual rotation between the camera frames, and not the absolute rotations between the camera frames and the $\{f_i\}$. In terms of storing and manipulating the data, everything is done in terms of the $\{f_i\}$ frame, which is usually chosen to coincide with the camera 1 frame. We next define the vectors

$$x_1^k = x_{1i}^k f_i, \quad x_2^k = x_{2i}^k f_i, \qquad (5.44)$$

which should be related by

$$X^k = R_1 x_1^k \tilde{R}_1 + t_1 = R_2 x_2^k \tilde{R}_2 + t_2, \tag{5.45}$$

for all point matches k.

When we measure the position coordinates for a point match the measurements will be subject to various forms of noise due to discretisation (from the conversion to digital pixel coordinates), camera wobble, inexact point matches and many other effects. We will assume that all of this noise can be modeled with a simple Gaussian distribution, centred on the exact value. This is an enormous simplification and is almost certainly incorrect. The main advantage in assuming Gaussian noise is that the various marginalisation integrals can be performed analytically and usually return simple, least squares functions to minimise. The point of adopting a Bayesian framework is that these (often hidden) assumptions are brought out clearly. This in turn suggests various improvements which can lead to more accurate reconstruction.

Our assumed probability density function (pdf) is (ignoring the normalisation)

$$P(x_{1i}^k) \propto \exp\left(\frac{-1}{2\sigma^2}\left(x_{1i}^k - e_{1i} \cdot (X^k - t_1)\right)^2\right) \tag{5.46}$$

$$P(x_{2i}^k) \propto \exp\left(\frac{-1}{2\sigma^2}\left(x_{2i}^k - e_{2i} \cdot (X^k - t_2)\right)^2\right). \tag{5.47}$$

The pdf for the vector x_1^k is therefore simply

$$P(x_1^k) \propto \exp\left(\frac{-1}{2\sigma^2}(R_1 x_1^k \tilde{R}_1 + t_1 - X^k)^2\right), \tag{5.48}$$

with a similar result holding for x_2^k. The full joint probability distribution over all point matches is therefore

$$P(\{x_1^k, x_2^k\}|\{X^k\}, R_1, R_2, t_1, t_2) \propto$$
$$\exp\left(\frac{-1}{2\sigma^2}\sum_k (R_1 x_1^k \tilde{R}_1 + t_1 - X^k)^2 + (R_2 x_2^k \tilde{R}_2 + t_2 - X^k)^2\right). \tag{5.49}$$

Bayes' theorem [12] states that

$$P(X|Y, I) = \frac{P(Y|X, I) \times P(X|I)}{P(Y|I)} \propto P(Y|X, I) \times P(X|I). \tag{5.50}$$

This follows immediately from the product rule of probability theory. The final term $P(X|I)$ is called the *prior* and is chosen to reflect any knowledge we might have about the quantity to be determined prior to any measurements being made. In our case we have no such knowledge, so we assume uniform priors for the camera frames and centres, and for the positions of the point matches. We can therefore use Bayes' theorem to invert our pdf to obtain

$$P(\{x_1^k, x_2^k\}|\{X^k\}, R_1, R_2, t_1, t_2) \propto P(R_1, R_2, t_1, t_2, \{X^k\}|\{x_1^k, x_2^k\}),$$

(5.51)

where we continue to ignore normalisation factors. The next step is to *marginalise* over the actual positions X^k to get the pdf for the rotors R_i and positions t_i in terms of the data. This marginalisation process is performed by simply integrating out the unwanted degrees of freedom,

$$P(R_1, R_2, t_1, t_2|\{x_1^k, x_2^k\})$$

$$\propto \int d^3X^1 \, d^3X^2 \cdots d^3X^n \, P(R_1, R_2, t_1, t_2, \{X^k\}|\{x_1^k, x_2^k\}).$$

(5.52)

The marginalisation integrals are straightforward once one employs the result

$$(X-a)^2 + (X-b)^2 = 2\big(X - \tfrac{1}{2}(a+b)\big)^2 + \tfrac{1}{2}(a-b)^2.$$

(5.53)

All that remains after the integral is therefore

$$P(R_1, R_2, t_1, t_2|\{x_1^k, x_2^k\}) \propto$$

$$\exp\Big(\frac{-1}{2\sigma^2}\sum_k (R_1 x_1^k \tilde{R}_1 - R_2 x_2^k \tilde{R}_2 + t_1 - t_2)^2\Big).$$

(5.54)

Maximising this function therefore reduces to minimising the least squares difference

$$S = \sum_k (R_1 x_1^k \tilde{R}_1 - R_2 x_2^k \tilde{R}_2 + t_1 - t_2)^2,$$

(5.55)

as has been discussed by many authors [1, 6, 7, 11].

9.5.2 Solution

The first point to note is that S of equation (5.55) is a function of $t_1 - t_2$ only, and hence is independent of the absolute origin. This is precisely the behaviour we expect. It follows that minimisation of S with respect to either t_1 or t_2 lead to the same equation, which is simply that

$$t_2 - t_1 = R_1 \bar{x}_1 \tilde{R}_1 - R_2 \bar{x}_2 \tilde{R}_2$$

(5.56)

where

$$\bar{x}_1 = \frac{1}{n}\sum_{k=1}^n x_1^k, \quad \bar{x}_2 = \frac{1}{n}\sum_{k=1}^n x_2^k.$$

(5.57)

The vector $t_2 - t_1$ is simply the difference in the two centroids of the data, and depends on the rotors R_i.

Now that we have found $t_2 - t_1$ we can substitute its value back into S to express S as a function of the rotors only:

$$S = \sum_k \left(R_1(x_1^k - \bar{x}_1)\tilde{R}_1 - R_2(x_2^k - \bar{x}_2)\tilde{R}_2\right)^2. \tag{5.58}$$

On squaring this only the cross terms remain with any rotor dependence, and we are left to maximise

$$S' = \sum_k \langle (x_1^k - \bar{x}_1)\tilde{R}_1 R_2(x_2^k - \bar{x}_2)\tilde{R}_2 R_1\rangle. \tag{5.59}$$

This is a function of the relative rotor $\tilde{R}_1 R_2$ only, again as expected. The same equation is obtained if we differentiate S' with respect to R_1 or R_2. Using the result of equation (4.40) we see that the equation to solve is

$$\sum_k (R_1(x_1^k - \bar{x}_1)\tilde{R}_1) \wedge (R_2(x_2^k - \bar{x}_2)\tilde{R}_2) = 0. \tag{5.60}$$

Taking the inner product with the bivector $e_{1i} \wedge e_{1j}$ produces the equation

$$\mathsf{F}_{ij} - \mathsf{F}_{ji} = 0 \tag{5.61}$$

where

$$\mathsf{F}_{ij} = \sum_k f_i \cdot (x_1^k - \bar{x}_1)\, f_j \cdot (\tilde{R}_1 R_2(x_2^k - \bar{x}_2)\tilde{R}_2 R_1). \tag{5.62}$$

This is easily solved with a singular-value decomposition of F_{ij}, as has been discussed elsewhere [8].

9.5.3   Adding more cameras

The generalisation to n cameras is quite straightforward. Instead of two terms in the pdf of equation (5.49) there are now n of them. The marginalisation integral simply involves completing the square as follows:

$$\sum_{\alpha=1}^{n} (X - a_\alpha)^2 = n\left(X - \frac{1}{n}\sum_{\alpha=1}^{n} a_\alpha\right)^2 + \frac{1}{n}\sum_{\alpha<\beta} (a_\alpha - a_\beta)^2. \tag{5.63}$$

The least squares expression to minimise therefore involves the sum over all $n(n-1)/2$ combinations of different cameras,

$$S = \sum_{\alpha<\beta} \sum_k (R_\alpha x_\alpha^k \tilde{R}_\alpha - R_\beta x_\beta^k \tilde{R}_\beta + t_\alpha - t_\beta)^2, \tag{5.64}$$

where the k sum runs over point matches, and α, β run over the camera pairs. This result is sensible as it is totally symmetric on the camera labels

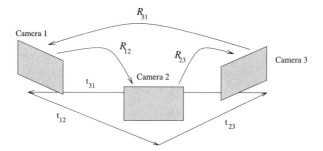

FIGURE 9.5. *The three camera setup.* **The relative vectors between the cameras are given by $t_{ij} = t_j - t_i$. The relative rotations are $R_{ij} = R_j \tilde{R}_i$. These satisfy $t_{12} + t_{23} + t_{31} = 0$ and $R_{31} R_{23} R_{12} = 1$.**

and does not depend on relating everything back to a preferred reference camera.

Minimising S with respect to each of the t_α vectors gives the simple solution for the relative translations

$$t_\alpha - t_\beta = R_\alpha \bar{x}_\alpha \tilde{R}_\alpha - R_\beta \bar{x}_\beta \tilde{R}_\beta. \tag{5.65}$$

Again, the total vector $t_1 + \cdots + t_n$ is unspecified. Substituting the values for the relative vectors into S, we are left with the function

$$S = \sum_{\alpha < \beta} \sum_k \left(R_\alpha (x_\alpha^k - \bar{x}_\alpha) \tilde{R}_\alpha - R_\beta (x_\beta^k - \bar{x}_\beta) \tilde{R}_\beta \right)^2, \tag{5.66}$$

which we want to minimise with respect to the n rotors R_α. As before, one only obtains equations for the relative rotations between two cameras, and not the absolute rotation from the global $\{f_i\}$ frame.

One can get the general feel of this equation structure considering three cameras (Figure 9.5). The three equations from the three rotors reduce to

$$\sum_k \left(R_1(x_1^k - \bar{x}_1)\tilde{R}_1 \right) \wedge \left(R_2(x_2^k - \bar{x}_2)\tilde{R}_2 + R_3(x_3^k - \bar{x}_3)\tilde{R}_3 \right) = 0 \tag{5.67}$$

and

$$\sum_k \left(R_2(x_2^k - \bar{x}_2)\tilde{R}_2 \right) \wedge \left(R_3(x_3^k - \bar{x}_3)\tilde{R}_3 + R_1(x_1^k - \bar{x}_1)\tilde{R}_1 \right) = 0. \tag{5.68}$$

The final equation is just the sum of the first two and contains no further information. Again, this is to be expected as there are always $n - 1$ relative rotations to solve for.

This equation structure is more complicated that the 2-camera case, and cannot by solved simply with a singular-value decomposition. Rather than removing the anti-symmetric component of a single tensor, one has to

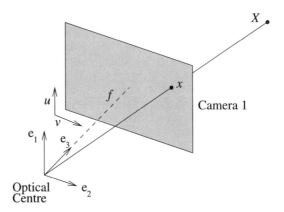

FIGURE 9.6. *Pixel Coordinates.* **In most applications in computer vision one only measures the pixel coordinates of a point in the camera plane. Provided the camera is calibrated, these can be converted to the image coordinates of** *x.*

minimise the anti-symmetric components of 3 independent tensors, using 2 independent rotors. This problem should be numerically quite straightforward to solve, either at the level of the equations, or through direct numerical minimisation of the S of equation (5.66). This latter approach is simplified by the fact that the individual pairwise minimisers for two of the pairs provide good starting points for any minimisation routine.

9.6   Unknown range data

In most computer vision applications we do not have access to the third coordinate giving the direction to a point. Instead what we measure are pixel coordinates in the camera plane (see Figure 9.6). Placing the origin at the camera centre, a world point X has coordinates (X_1, X_2, X_3) expressed in the camera frame. Adopting the projective pinhole camera model, the image point x has coordinates (x_1, x_2, f), where f is the focal length. The pixel coordinates $u = (u_1, u_2, 1)$ are related to the image coordinates by a 3×3 camera matrix C,

$$u = C(x/f), \quad x/f = C^{-1}u. \tag{6.69}$$

(See [9] for more details). Provided the matrix C is known, we can recover the vector x/f. For a projective pinhole camera, the components of this are simply the *homogeneous coordinates* $(X_1/X_3, X_2/X_3, 1)$ of the world point X.

For the 2-camera setup of Figure 9.3, the two coordinates we measure in the Camera 1 system are

$$x_{1i}^k = \frac{e_{1i} \cdot (X^k - t_1)}{e_{13} \cdot (X^k - t_1)}. \quad i = 1, 2. \tag{6.70}$$

A simple model would be to assume is that the observed data is taken from a Gaussian distribution centred on these values. The problem with this is that the resulting marginalisation integral over the X^k cannot be performed analytically. Instead we will use a different model in which the marginalisation integrals can be performed. The result is a likelihood function which can be minimised very quickly and efficiently. The results of this turn out to be reasonable, and geometrically quite sensible.

Our choice of a simplified model, including modeling the combined effects of the various sources of noise with a simple Gaussian distribution, is one of a number of simplifying assumptions we will make in order to find a simple function to minimise. Each of these assumptions can be challenged and modified to construct more realistic models and give better reconstruction. This approach is quite different from the standard alternative, based on the epipolar geometry and the fundamental matrix [10, 14]. In this approach an assortment of least-squares optimisers are considered, none with any underlying justification from a probabilistic model, and an assortment of linear algebra techniques are used to find the mutual translation and rotation. Many of these do not properly account for the structure of the rotation group, which limits their accuracy. They do have some value, however, in providing some fast algorithms to give initial points for the nonlinear schemes developed here.

Our starting point is the pdf of equation (5.49). That is, we start by treating all three coordinates in the same way. Again, we marginalise over the positions X^k to get the 2-camera joint pdf, but this time we view the range data as an unknown parameter and assign it a uniform prior. We therefore arrive at the distribution

$$P(R_1, R_2, t_1, t_2, \{z_1^k, z_2^k\}|\{x_{1i}^k, x_{2i}^k\}) \propto$$
$$\exp\left(\frac{-1}{2\sigma^2} \sum_k (R_1 z_1^k x_1^k \tilde{R}_1 - R_2 z_2^k x_2^k \tilde{R}_2 + t_1 - t_2)^2\right), \tag{6.71}$$

where i runs over the two coordinates in the camera plane, z_α^k is the unknown range (α denotes the camera), and the vectors x_1^k, x_2^k are formed directly from the measured data by

$$x_\alpha^k = \sum_{i=1}^{2} x_{\alpha i}^k f_i + f_3. \tag{6.72}$$

The next step is to marginalise over the unknown ranges z_1 and z_2. Here we make one final simplification by taking the range of the integrals from

$-\infty \ldots \infty$. This allows for points behind the camera to be considered, so is clearly unjustified, but has the advantage that the integrals can be performed analytically. The integral we require has the form

$$I = \int_{-\infty}^{\infty} dz_1 \, dz_2 \exp\left(-(z_1 a_1 - z_2 a_2 + t)^2\right) \tag{6.73}$$

where $a_1 = R_1 x_1^k \tilde{R}_1$, etc. and $t = t_1 - t_2$. To carry out this integral we need the result that

$$\int d^n x \exp\left(- x_i x_j \mathsf{T}_{ij} + 2 x_i b_i\right) = N \exp\left(b_i b_j \mathsf{T}_{ij}^{-1}\right) \tag{6.74}$$

where T_{ij} is an $n \times n$ symmetric matrix, b_i is an n-component vector and N is a normalisation constant. For the integral (6.73) the matrix T_{ij} is given by

$$\mathsf{T}_{ij} = \begin{pmatrix} a_1{}^2 & -a_1 \cdot a_2 \\ -a_1 \cdot a_2 & a_2{}^2 \end{pmatrix}, \tag{6.75}$$

and the vector b_i is

$$b_i = \begin{pmatrix} -a_1 \cdot t \\ a_2 \cdot t \end{pmatrix}. \tag{6.76}$$

It follows that

$$\det \mathsf{T}_{ij} = a_1{}^2 a_2{}^2 - (a_1 \cdot a_2)^2 = -(a_1 \wedge a_2)^2, \tag{6.77}$$

and

$$\mathsf{T}_{ij}^{-1} = -\frac{1}{(a_1 \wedge a_2)^2} \begin{pmatrix} a_2{}^2 & a_1 \cdot a_2 \\ a_1 \cdot a_2 & a_1{}^2 \end{pmatrix}. \tag{6.78}$$

Hence

$$\begin{aligned}
b_i b_j \mathsf{T}_{ij}^{-1} &= -\frac{1}{(a_1 \wedge a_2)^2} \left(a_1{}^2 (a_2 \cdot t)^2 + a_2{}^2 (a_1 \cdot t)^2 - 2 a_1 \cdot a_2 \, a_1 \cdot t \, a_2 \cdot t\right) \\
&= -\frac{1}{(a_1 \wedge a_2)^2} (a_1 \cdot t \, a_2 - a_2 \cdot t \, a_1)^2 \\
&= \left(\frac{t \cdot (a_1 \wedge a_2)}{|a_1 \wedge a_2|}\right)^2,
\end{aligned} \tag{6.79}$$

which assembles into a simple geometric function. Applying these results to the pdf of equation (6.71), and remembering the final $(t_1 - t_2)^2$ term, we arrive at the log-likelihood function

$$S = \sum_{k=1}^{n} \frac{\left((t_1 - t_2) \wedge \left((R_1 x_1^k \tilde{R}_1) \wedge (R_2 x_2^k \tilde{R}_2)\right)\right)^2}{|(R_1 x_1^k \tilde{R}_1) \wedge (R_2 x_2^k \tilde{R}_2)|^2}. \tag{6.80}$$

This is now a simple function of the vectors t_α and the rotors R_α. Again, only the relative translation $(t_1 - t_2)$ enters the problem, and the freedom to choose the f_i reference frame means that one of the rotors is arbitrary.

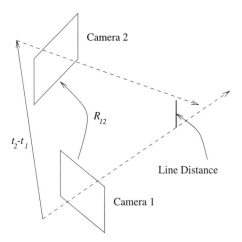

FIGURE 9.7. *Line Distance.* **Given a point match in the two camera planes, the vectors are extended out to three-dimensional space, and the distance between the lines is found. The sum of the squares of these is minimised to find the best fit translation and rotation.**

The function (6.80) has a simple geometric interpretation in terms of the distance between the projective lines for a given point match (see Figure 9.7). Given a point match, the projective lines from the two cameras are extended into space. The function then records the square of the distance between the lines (in units on $|t_1 - t_2|$), and sums these over all point matches. This is certainly a sensible error measure for this problem, and it is instructive to see how it arises from a probabilistic model.

The function (6.80) is scale invariant, since no scale has yet been imposed on the problem. As it stands, therefore, the function is minimised by setting $t_1 - t_2 = 0$. To avoid this we need to impose a scale, which is most simply achieved by setting

$$(t_1 - t_2)^2 = 1. \tag{6.81}$$

This condition is imposed by including a Lagrange multiplier, so the function to minimise becomes

$$S = \sum_{k=1}^{n} \left((t_1 - t_2) \cdot n^k \right)^2 - \lambda\left((t_1 - t_2)^2 - 1 \right), \tag{6.82}$$

where

$$n^k = \frac{I(R_1 x_1^k \tilde{R}_1) \wedge (R_2 x_2^k \tilde{R}_2)}{|(R_1 x_1^k \tilde{R}_1) \wedge (R_2 x_2^k \tilde{R}_2)|}. \tag{6.83}$$

Our final S (6.82) is still quadratic in the relative vector $t = t_1 - t_2$, and minimising gives the simple equation

$$\sum_{k=1}^{n} t \cdot n^k n^k = \lambda t. \tag{6.84}$$

We next construct the symmetric, positive definite function

$$\mathsf{F}(a) = \sum_{k=1}^{n} a \cdot n^k n^k, \tag{6.85}$$

which is a function of the data and the rotation only. The translation t is an eigenvector of this function, with the eigenvalue

$$\lambda = t \cdot \mathsf{F}(t) = \sum_{k=1}^{n} (t \cdot n^k)^2 = S. \tag{6.86}$$

So to minimise the error function S we need to choose t to be the eigenvector with smallest eigenvalue. All we need do, then, is minimise the lowest eigenvalue of F with respect to the rotor R. This is a fairly simple optimisation problem, as we only need to search in the 3-parameter rotor space. Numerical studies of this function reveal that it contains some local minima, but the global minimum lies in a fairly deep valley and it is not hard to find this numerically.

9.7   Extension to three cameras

The Bayesian analysis presented here extends easily to the 3 camera case. A simpler alternative, however, is to take the log-likelihood function of equation (6.80) and sum this function over each of the camera pairs. Incorporating a Lagrange multiplier to impose a suitable constraint, the function we need to minimise is

$$S_3 = \sum_{k=1}^{n} \left((t_1 - t_2) \cdot n_{12}^k \right)^2 + \left((t_2 - t_3) \cdot n_{23}^k \right)^2 + \left((t_3 - t_1) \cdot n_{31}^k \right)^2$$
$$- \lambda \left((t_1 - t_2)^2 + (t_2 - t_3)^2 + (t_3 - t_1)^2 - 1 \right), \tag{7.87}$$

where

$$n_{12}^k = \frac{I(R_1 x_1^k \tilde{R}_1) \wedge (R_2 x_2^k \tilde{R}_2)}{|(R_1 x_1^k \tilde{R}_1) \wedge (R_2 x_2^k \tilde{R}_2)|} \quad etc. \tag{7.88}$$

We only get independent equations from minimising with respect to two of the three translation vectors. Taking these to be t_1 and t_2 the equations we arrive at are

$$\sum_{k=1}^{n} (t_1 - t_2) \cdot n_{12}^k n_{12}^k - (t_3 - t_1) \cdot n_{31}^k n_{31}^k = \lambda(2t_1 - t_2 - t_3) \tag{7.89}$$

$$\sum_{k=1}^{n}(t_2 - t_3) \cdot n_{23}^{k} \, n_{23}^{k} - (t_1 - t_2) \cdot n_{12}^{k} \, n_{12}^{k} = \lambda(2t_2 - t_3 - t_1). \qquad (7.90)$$

If we now set

$$a = 2t_1 - t_2 - t_3, \quad b = 2t_2 - t_3 - t_1, \qquad (7.91)$$

then we recover a 6×6 eigenvalue problem of the form

$$\begin{pmatrix} \mathsf{F}_{12} + 2\mathsf{F}_{31} & \mathsf{F}_{31} - \mathsf{F}_{12} \\ \mathsf{F}_{23} - \mathsf{F}_{12} & 2\mathsf{F}_{23} + \mathsf{F}_{12} \end{pmatrix} \begin{pmatrix} a \\ b \end{pmatrix} = 3\lambda \begin{pmatrix} a \\ b \end{pmatrix} \qquad (7.92)$$

where

$$\mathsf{F}_{12}(a) = \sum_{k=1}^{n} a \cdot n_{12}^{k} \, n_{12}^{k}, \quad etc. \qquad (7.93)$$

As in the 2 camera case, the eigenvalue λ returns the value of S_3 that we are trying to minimise. The minimisation problem therefore reduces to finding a pair of rotors which minimises the lowest eigenvalue of a 6×6 matrix. Numerical implementation of this algorithm will be presented elsewhere.

9.8   Conclusions

Geometric Algebra is an extremely powerful tool for handling rotations in three dimensions. Vectors and the quantities which act on them are united in a single algebra, which has a number of computational advantages. Relaxing the normalisation condition for rotors provides a simplified calculus for rotations which avoids having to work in the tangent space to the group manifold. As a result, many extremisation problems involving rotations can be studied and solved without ever leaving the geometric algebra of 3-d.

The applications to the camera localization problem given here illustrate the various advantages that geometric algebra can provide. This is particularly so when combined with Bayesian inference techniques. The models considered here are highly simplified, though still quite useful. Much work remains in order to construct robust, accurate algorithms to use with real cameras. The effects of the camera matrix must be included, particularly as the cameras often require re-calibrating after they are moved significantly. Similarly, more realistic noise models are required. Discretisation errors, for example, are certainly not well modeled as Gaussian process. In addition, we need to be able to work with arbitrary numbers of cameras, allowing for occlusion effects where point matches may only be shared by a subset of all of the cameras. When tackling each of these problems, however, there seems little doubt that the combination of geometric algebra and Bayesian reasoning advocated here will turn out to be the best way to proceed.

Acknowledgments CD gratefully acknowledges the support of the EPSRC.

Chapter 10

Projective Reconstruction of Shape and Motion Using Invariant Theory

Eduardo Bayro Corrochano and Vladimir Banarer

10.1 Introduction

In this chapter we present a geometric approach for the computation of shape and motion using projective invariants in the geometric algebra framework [6, 7].

In the last years researchers have developed diverse methods to compute projective invariants using n uncalibrated cameras [1, 2, 4, 8]. Different approaches for projective reconstruction have utilized the projective depth [13, 14], projective invariants [4] and factorization methods [11, 15, 16]. The factorization methods require the projective depth. The contribution of this paper is the application of projective invariants depending on the fundamental matrix or trifocal tensor to compute the projective depths. Using these projective depths we initialize the projective reconstruction procedure to compute shape and motion. We also illustrate the application of algebra of incidence for the development of geometric inference rules to complete the 3D data. The experimental part shows projective reconstruction of shape and motion using both simulated and real images.

The organization of the chapter is as follows: section two explains the generation and computation of projective invariants using two and three uncalibrated cameras. We test their performance using both simulated and real images. Section three presents the computation of the projective depth using projective invariants in terms of the trifocal tensor. The treatment of projective reconstruction and the role of the algebra of incidence to complete the 3–D shape is given in section four. The conclusion part follows.

10.2   3-D Projective Invariants from Multiple Views

This section presents the point and line projective invariants computable by means of n uncalibrated cameras. We begin with the generation of geometric invariants using the Plücker–Grassmann quadratic relation. We give a geometric interpretation of the cross–ratio in the 3–D space and in the image plane. We compute then projective invariants using two and three cameras.

10.2.1   Generation of geometric projective invariants

We choose for the visual projective space P^3 the geometric algebra $\mathcal{G}_{1,3,0}$ and for the image or projective plane P^2 the geometric algebra $\mathcal{G}_{3,0,0}$. Any 3D point is written in $\mathcal{G}_{1,3,0}$ as $\mathbf{X}_n = X_n\gamma_1 + Y_n\gamma_2 + Z_n\gamma_3 + W_n\gamma_4$ and its projected image point in $\mathcal{G}_{3,0,0}$ as $\boldsymbol{x}_n = x_n\sigma_1 + y_n\sigma_2 + z_n\sigma_3$, where $x_n = X_n/W_n$, $y_n = Y_n/W_n$, $z_n = Z_n/W_n$. The 3–D projective basis consists of four basis points and a fifth one for normalization: $\boldsymbol{X}_1 = [1, 0, 0, 0]^T$, $\boldsymbol{X}_2 = [0, 1, 0, 0]^T$, $\boldsymbol{X}_3 = [0, 0, 1, 0]^T$, $\boldsymbol{X}_4 = [0, 0, 0, 1]^T$, $\boldsymbol{X}_5 = [1, 1, 1, 1]^T$ and the 2–D projective basis comprises three basis points and one for normalization: $\boldsymbol{x}_1 = [1, 0, 0]^T$, $\boldsymbol{x}_2 = [0, 1, 0]^T$, $\boldsymbol{x}_3 = [0, 0, 1]^T$, $\boldsymbol{x}_4 = [1, 1, 1]^T$. Using them we can express in terms of brackets the 3D projective coordinates X_n, Y_n, Z_n for any 3D point, as well as its 2D projected coordinates x_n, y_n

$$\frac{X_n}{W_n} = \frac{[234n][1235]}{[2345][123n]}, \quad \frac{Y_n}{W_n} = \frac{[134n][1235]}{[1345][123n]}, \quad \frac{Z_n}{W_n} = \frac{[124n][1235]}{[1245][123n]}. \quad (2.1)$$

$$\frac{x_n}{w_n} = \frac{[23n][124]}{[234][12n]}, \quad \frac{y_n}{w_n} = \frac{[13n][124]}{[134][12n]}. \quad (2.2)$$

These equations are projective invariants relations and they can be used for example, to compute the position of a moving camera.

The projective structure and its projection on the 2–D image is related according to the following geometric constraint

$$\begin{pmatrix} 0 & w_5Y_5 & -y_5Z_5 & (y_5 - w_5)W_5 \\ w_5X_5 & 0 & -x_5Z_5 & (x_5 - w_5)W_5 \\ 0 & w_6Y_6 & -y_6Z_6 & (x_5 - w_5)W_5 \\ 0 & w_6Y_6 & -y_6Z_6 & (y_6 - w_6)W_6 \\ w_6X_6 & 0 & -x_6Z_6 & (x_6 - w_6)W_6 \\ 0 & w_7Y_7 & -y_7Z_7 & (y_7 - w_7)W_7 \\ w_7X_7 & 0 & -x_7Z_7 & (x_7 - w_7)W_7 \\ \cdot & \cdot & \cdot & \cdot \\ \cdot & \cdot & \cdot & \cdot \\ \cdot & \cdot & \cdot & \cdot \end{pmatrix} \begin{pmatrix} X_0^{-1} \\ Y_0^{-1} \\ Z_0^{-1} \\ W_0^{-1} \end{pmatrix} = 0, \quad (2.3)$$

where $X_0, Y_0, Z_0\ W_0$ are the coordinates of the view point. Since the matrix is of rank < 4, any determinant of four rows becomes a zero. Considering

$(X_5, Y_5, Z_5 W_5) = (1, 1, 1, 1)$ as a normalizing point and taking the determinant formed by the first four rows of equation (2.3) we get the geometric constraint equation involving six points pointed out by Quan [12]

$$(w_5 y_6 - x_5 y_6)X_6 Z_6 + (x_5 y_6 - x_5 w6)X_6 W_6 + (x_5 w_6 - y_5 w6)X_6 Y_6 +$$
$$+(y_5 x_6 - w_5 x6)Y_6 Z_6 + (y_5 w_6 - y_5 x6)Y_6 W_6 +$$
$$+(w_5 x_6 - w_5 y_6)Z_6 W_6 = 0 \tag{2.4}$$

Carlsson [3] showed that the equation (2.4) can be also derived using the *Plücker–Grassmann relations*. This can be computed as the *Laplace expansion* of the 4×8 rectangular matrix involving the same six points as above

$$[X_1, X_2, X_3, X_4, X_5, X_5, X_6, X_7] = [X_0, X_1, X_2, X_3] \tag{2.5}$$
$$[X_4, X_5, X_6, X_7] - [X_0, X_1, X_2, X_4][X_3, X_5, X_6, X_7] +$$
$$+[X_0, X_1, X_2, X_5][X_3, X_4, X_6, X_7] - [X_0, X_1, X_2, X_6]$$
$$[X_3, X_4, X_5, X_7] + [X_0, X_1, X_2, X_7][X_3, X_4, X_5, X_6] = 0.$$

Using four functions like equation (2.5) in terms of the permutations of six points as indicated by their sub–indices in the table below

X_0	X_1	X_2	X_3	X_4	X_5	X_6	X_7
0	1	5	1	2	3	4	5
0	2	6	1	2	3	4	6
0	3	5	1	2	3	4	5
0	4	6	1	2	3	4	6

we get an expression where the brackets that have two identical points vanish

$$[0152][1345] - [0153][1245] + [0154][1235] = 0,$$
$$[0216][2346] - [0236][1246] + [0246][1236] = 0,$$
$$[0315][2345] + [0325][1345] + [0345][1235] = 0,$$
$$[0416][2346] + [0426][1346] - [0436][1246] = 0. \tag{2.6}$$

It is easy to show that the brackets of image points can be written in the form $[x_i x_j x_k] = w_i w_j w_k [K][X_0 X_i X_j X_k]$, where $[K]$ is the matrix of the intrinsic parameters [10]. Now if we express in equations (2.6) all the brackets which have the point X_0 in terms of the brackets of image points and organize all the bracket products as a 4×4 matrix we get the singular matrix

$$\begin{pmatrix} 0 & [125][1345] & [135][1245] & [145][1235] \\ [216][2346] & 0 & [236][1246] & [246][1236] \\ [315][2345] & [325][1345] & 0 & [345][1235] \\ [416][2346] & [426][1346] & [436][1246] & 0. \end{pmatrix} \tag{2.7}$$

Here the scalars $w_i w_j w_k[K]$ of each matrix entry cancel each other. Now after taking the determinat of this matrix and rearrange the terms conveniently, we obtain the following useful bracket polynomial

$$[125][346]\begin{bmatrix}1236\end{bmatrix}\begin{bmatrix}1246\end{bmatrix}\begin{bmatrix}1345\end{bmatrix}\begin{bmatrix}2345\end{bmatrix} -$$
$$[126][345]\begin{bmatrix}1235\end{bmatrix}\begin{bmatrix}1245\end{bmatrix}\begin{bmatrix}1346\end{bmatrix}\begin{bmatrix}2346\end{bmatrix} +$$
$$[135][246]\begin{bmatrix}1236\end{bmatrix}\begin{bmatrix}1245\end{bmatrix}\begin{bmatrix}1346\end{bmatrix}\begin{bmatrix}2345\end{bmatrix} -$$
$$[136][245]\begin{bmatrix}1235\end{bmatrix}\begin{bmatrix}1246\end{bmatrix}\begin{bmatrix}1345\end{bmatrix}\begin{bmatrix}2346\end{bmatrix} +$$
$$[145][236]\begin{bmatrix}1235\end{bmatrix}\begin{bmatrix}1246\end{bmatrix}\begin{bmatrix}1346\end{bmatrix}\begin{bmatrix}2345\end{bmatrix} -$$
$$[146][235]\begin{bmatrix}1236\end{bmatrix}\begin{bmatrix}1245\end{bmatrix}\begin{bmatrix}1345\end{bmatrix}\begin{bmatrix}2346\end{bmatrix} = 0, \qquad (2.8)$$

Surprisingly this bracket expression is exactly the *shape constraint* for six points given by Quan [12]

$$i_1 I_1 + i_2 I_2 + i_3 I_3 + i_4 I_4 + i_5 I_5 + i_6 I_6 = 0, \qquad (2.9)$$

where $i_1 = [125][346]$, $i_2 = [126][345]$, ..., $i_6 = [146][235]$ and
$I_1 = [1236][1246][1345][2345]$, $I_2 = [1235][1245][1346][2346]$, ...,
$I_6 = [1236][1245][1345][2346]$ are the the relative linear invariants in P^2 and P^3 respectively. Using the shape constraint we are now ready to generate invariants for different purpose.

Let us illustrate this with an example. As shown in the Figure 10.1 there is a configuration of six points which indicates whether or not the end–effector is grasping properly.

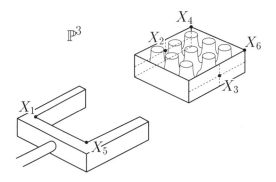

FIGURE 10.1. Grasping a box.

To test this situation we can use an invariant generated from the constraint of equation (2.8). In this particular situation we recognize two planes:

[1235]=0 and [2346]=0. Substituting these six points in equation (2.8) we make some brackets vanish reducing the equation to

$$[125][346]\begin{bmatrix}1236\end{bmatrix}\begin{bmatrix}1246\end{bmatrix}\begin{bmatrix}1345\end{bmatrix}\begin{bmatrix}2345\end{bmatrix} -$$
$$-[135][246]\begin{bmatrix}1236\end{bmatrix}\begin{bmatrix}1245\end{bmatrix}\begin{bmatrix}1346\end{bmatrix}\begin{bmatrix}2345\end{bmatrix} = 0 \qquad (2.10)$$

$$[125][346]\begin{bmatrix}1246\end{bmatrix}\begin{bmatrix}1345\end{bmatrix} - [135][246]\begin{bmatrix}1245\end{bmatrix}\begin{bmatrix}1346\end{bmatrix} = 0 \qquad (2.11)$$

or

$$
\begin{aligned}
Inv &= \frac{(\mathbf{X}_1 \wedge \mathbf{X}_2 \wedge \mathbf{X}_4 \wedge \mathbf{X}_5) I_4^{-1} (\mathbf{X}_1 \wedge \mathbf{X}_3 \wedge \mathbf{X}_4 \wedge \mathbf{X}_6) I_4^{-1}}{(\mathbf{X}_1 \wedge \mathbf{X}_2 \wedge \mathbf{X}_4 \wedge \mathbf{X}_6) I_4^{-1} (\mathbf{X}_1 \wedge \mathbf{X}_3 \wedge \mathbf{X}_4 \wedge \mathbf{X}_5) I_4^{-1}} \\
&= \frac{(\boldsymbol{x}_1 \wedge \boldsymbol{x}_2 \wedge \boldsymbol{x}_5) I_3^{-1} (\boldsymbol{x}_3 \wedge \boldsymbol{x}_4 \wedge \boldsymbol{x}_6) I_3^{-1}}{(\boldsymbol{x}_1 \wedge \boldsymbol{x}_3 \wedge \boldsymbol{x}_5) I_3^{-1} (\boldsymbol{x}_2 \wedge \boldsymbol{x}_4 \wedge \boldsymbol{x}_6) I_3^{-1}}.
\end{aligned} \qquad (2.12)
$$

In this equation any bracket of P^3 after the projective mapping fulfills

$$
\begin{aligned}
(\mathbf{X}_1 \wedge \mathbf{X}_2 \wedge \mathbf{X}_4 \wedge \mathbf{X}_5) I_4^{-1} &\equiv \\
W_1 W_2 W_4 W_5 \{ (\boldsymbol{x}_2 - \boldsymbol{x}_1) &\wedge (\boldsymbol{x}_4 - \boldsymbol{x}_1) \wedge (\boldsymbol{x}_5 - \boldsymbol{x}_1) \} I_3^{-1}, \quad (2.13)
\end{aligned}
$$

The constraint (2.8) makes always sure that the $W_i W_j W_k W_l$ constants are canceled. Furthermore, we can interpret the invariant Inv, the equivalent of the , in P^3 as ratios of volumes and in P^2 as rations of triangle areas

$$Inv = \frac{V_{1245} V_{1346}}{V_{1246} V_{1345}} = \frac{A_{125} A_{346}}{A_{135} A_{246}}. \qquad (2.14)$$

In other words, we can also see this invariant in P^3 as the relation of 4-vectors or volumes built by points lying on a quadric which projected in P^2 represents an invariant build by areas of triangles encircled by conics.

For example utilizing this invariant we can check whether or not the grasper is holding the box correctly. Note that using the observed 3–D points in the image we can compute this invariant and see if the relation of the triangle areas corresponds with the appropriate relation for firm grasping, i.e. if the grasper is away the invariant has a different value from the required value when the points $\mathbf{X}_1$, $\mathbf{X}_5$ of the grasper are near to the objects points $\mathbf{X}_2$, $\mathbf{X}_3$.

10.2.2 Projective invariants using two views

Let us consider a 3–D *projective invariant* derived from the equation (2.8)

$$\boldsymbol{Inv_3} = \frac{[\mathbf{X}_1 \mathbf{X}_2 \mathbf{X}_3 \mathbf{X}_4][\mathbf{X}_4 \mathbf{X}_5 \mathbf{X}_2 \mathbf{X}_6]}{[\mathbf{X}_1 \mathbf{X}_2 \mathbf{X}_4 \mathbf{X}_5][\mathbf{X}_3 \mathbf{X}_4 \mathbf{X}_2 \mathbf{X}_6]}. \qquad (2.15)$$

The computation of the bracket

$$[1234] = (\mathbf{X}_1 \wedge \mathbf{X}_2 \wedge \mathbf{X}_3 \wedge \mathbf{X}_4) \boldsymbol{I}_4^{-1} = ((\mathbf{X}_1 \wedge \mathbf{X}_2) \wedge (\mathbf{X}_3 \wedge \mathbf{X}_4)) \boldsymbol{I}_4^{-1}$$

of four points from R^4, mapped to the cameras with the optical centers $\mathbf{A}_0$ and $\mathbf{B}_0$, suggests to use the binocular model based on incidence algebra as introduced in chapter 7. Defining the lines

$$
\begin{aligned}
\boldsymbol{L}_{12} &= \mathbf{X}_1 \wedge \mathbf{X}_2 = (\mathbf{A}_0 \wedge \boldsymbol{L}_{12}^A) \vee (\mathbf{B}_0 \wedge \boldsymbol{L}_{12}^B) \\
\boldsymbol{L}_{34} &= \mathbf{X}_3 \wedge \mathbf{X}_4 = (\mathbf{A}_0 \wedge \boldsymbol{L}_{34}^A) \vee (\mathbf{B}_0 \wedge \boldsymbol{L}_{34}^B)
\end{aligned}
$$

where lines $\boldsymbol{L}_{ij}^A$ and $\boldsymbol{L}_{ij}^B$ are mappings of the line $\boldsymbol{L}_{ij}$ to the two image planes, results in the following expression for the bracket

$$[1234] = [\mathbf{A}_0 \mathbf{B}_0 \mathbf{A}'_{1234} \mathbf{B}'_{1234}]. \tag{2.16}$$

Here $\mathbf{A}'_{1234}$ and $\mathbf{B}'_{1234}$ are the points of intersection of the lines $\boldsymbol{L}_{12}^A$ and $\boldsymbol{L}_{34}^A$ or $\boldsymbol{L}_{12}^B$ and $\boldsymbol{L}_{34}^B$, respectively. These points, lying in the image planes, can be expanded using the mappings of three points $\mathbf{X}_i$, say $\mathbf{X}_1, \mathbf{X}_2, \mathbf{X}_3$, to the image planes, i.e. $\mathbf{A}_j$ and $\mathbf{B}_j$, $j = 1, 2, 3$, as projective basis, as follows

$$
\begin{aligned}
\mathbf{A}'_{1234} &= \alpha_{1234,1}\mathbf{A}_1 + \alpha_{1234,2}\mathbf{A}_2 + \alpha_{1234,3}\mathbf{A}_3 \\
\mathbf{B}'_{1234} &= \beta_{1234,1}\mathbf{B}_1 + \beta_{1234,2}\mathbf{B}_2 + \beta_{1234,3}\mathbf{B}_3.
\end{aligned}
$$

Then equation (15.73) from chapter 15 follows

$$[1234] = \sum_{i,j=1}^{3} \tilde{F}_{ij} \alpha_{1234,i} \beta_{1234,j} = \boldsymbol{\alpha}_{1234}^T \tilde{F} \boldsymbol{\beta}_{1234}, \tag{2.17}$$

where $\tilde{F}$ is the fundamental matrix given in terms of the projective basis, embedded in R^4 and $\boldsymbol{\alpha}_{1234} = (\alpha_{1234,1}, \alpha_{1234,2}, \alpha_{1234,3})$ and $\boldsymbol{\beta}_{1234} = (\beta_{1234,1}, \beta_{1234,2}, \beta_{1234,3})$ are corresponding points.

The ratio

$$Inv_{3F} = \frac{(\boldsymbol{\alpha}^T{}_{1234} \tilde{F} \boldsymbol{\beta}_{1234})(\boldsymbol{\alpha}^T{}_{4526} \tilde{F} \boldsymbol{\beta}_{4526})}{(\boldsymbol{\alpha}^T{}_{1245} \tilde{F} \boldsymbol{\beta}_{1245})(\boldsymbol{\alpha}^T{}_{3426} \tilde{F} \boldsymbol{\beta}_{3426})} \tag{2.18}$$

is therefore seen to be an invariant using two cameras [2]. Note that equation (2.18) is invariant whatever values of the γ_4 components of the vectors $\mathbf{A}_i, \mathbf{B}_i, \mathbf{X}_i$ etc. are chosen. A confusion arises if we attempt to express the invariant of equation (2.18) in terms of what we actually observe, i.e. the homogeneous Cartesian image coordinates $a_i's$, $b_i's$ and the fundamental matrix F calculated from these image coordinates. In order to avoid that it is necessary to transfer the computations of equation (2.18) carried out in R^4 to R^3. Let us explain now this procedure.

If we define $\tilde{F}$ by

$$\tilde{F}_{kl} = (\mathbf{A}_k \cdot \gamma_4)(\mathbf{B}_l \cdot \gamma_4) F_{kl} \tag{2.19}$$

then using the relationships $\alpha_{ij} = \dfrac{\mathbf{A}'_i \cdot \gamma_4}{\mathbf{A}'_j \cdot \gamma_4} a_{ij}$ and $\beta_{ij} = \dfrac{\mathbf{B}'_i \cdot \gamma_4}{\mathbf{B}'_j \cdot \gamma_4} b_{ij}$ it follows that

$$\alpha_{ik} \tilde{F}_{kl} \beta_{il} = (\mathbf{A}'_i \cdot \gamma_4)(\mathbf{B}'_i \cdot \gamma_4) a_{ik} F_{kl} b_{il}. \tag{2.20}$$

If F is estimated by some method, then an $\tilde{F}$ defined as in equation (2.19) will also act as a *fundamental matrix* or *bilinear constraint* in R^4. Now let us look again at the invariant $\boldsymbol{Inv_{3F}}$. According to the above considerations, we can write the invariant as

$$\boldsymbol{Inv_{3F}} = \frac{(\boldsymbol{a}^T{}_{1234} F \boldsymbol{b}_{1234})(\boldsymbol{a}^T{}_{4526} F \boldsymbol{b}_{4526}) \phi_{1234} \phi_{4526}}{(\boldsymbol{a}^T{}_{1245} F \boldsymbol{b}_{1245})(\boldsymbol{a}^T{}_{3426} F \boldsymbol{b}_{3426}) \phi_{1245} \phi_{3426}} \tag{2.21}$$

where $\phi_{pqrs} = (\mathbf{A}'_{pqrs} \cdot \gamma_4)(\mathbf{B}'_{pqrs} \cdot \gamma_4)$. Therefore we can see that the ratio of the terms $\boldsymbol{a}^T F \boldsymbol{b}$ which resembles the expression for the invariant in R^4 but uses only the observed coordinates and the estimated fundamental matrix will not be an invariant. Instead, we need to include the factors ϕ_{1234} etc., which do not cancel. It is relatively easy to show [1] that these factors can be formed as follows. Since $\boldsymbol{a}'_3$, $\boldsymbol{a}'_4$ and $\boldsymbol{a}'_{1234}$ are collinear, we can write $\boldsymbol{a}'_{1234} = \mu_{1234} \boldsymbol{a}'_4 + (1 - \mu_{1234}) \boldsymbol{a}'_3$. Then, by expressing $\mathbf{A}'_{1234}$ as the intersection of the line joining $\mathbf{A}'_1$ and $\mathbf{A}'_2$ with the plane through $\mathbf{A}_0, \mathbf{A}'_3, \mathbf{A}'_4$ we can use the projective split and equate terms so that they give

$$\frac{(\mathbf{A}'_{1234} \cdot \gamma_4)(\mathbf{A}'_{4526} \cdot \gamma_4)}{(\mathbf{A}'_{3426} \cdot \gamma_4)(\mathbf{A}'_{1245} \cdot \gamma_4)} = \frac{\mu_{1245}(\mu_{3426} - 1)}{\mu_{4526}(\mu_{1234} - 1)}. \tag{2.22}$$

Note that the values of μ are readily obtainable from the images. The factors $\mathbf{B}'_{pqrs} \cdot \gamma_4$ are found in a similar way so that if $\boldsymbol{b}'_{1234} = \lambda_{1234} \boldsymbol{b}'_4 + (1 - \lambda_{1234}) \boldsymbol{b}'_3$ etc., the overall expression for the invariant becomes

$$\begin{aligned} \boldsymbol{Inv_{3F}} &= \frac{(\boldsymbol{a}^T_{1234} F \boldsymbol{b}_{1234})(\boldsymbol{a}^T_{4526} F \boldsymbol{b}_{4526})}{(\boldsymbol{a}^T_{1245} F \boldsymbol{b}_{1245})(\boldsymbol{a}^T_{3426} F \boldsymbol{b}_{3426})} \cdot \\[2mm] &\quad \frac{\mu_{1245}(\mu_{3426} - 1)}{\mu_{4526}(\mu_{1234} - 1)} \frac{\lambda_{1245}(\lambda_{3426} - 1)}{\lambda_{4526}(\lambda_{1234} - 1)}. \end{aligned} \tag{2.23}$$

As conclusion, given the coordinates of a set of 6 corresponding points in two image planes, where these 6 points are projections of arbitrary world points in general position, we can form 3–D projective invariants provided we have some estimate of F.

10.2.3   Projective invariant of points using three views

The technique used to form the 3–D projective invariants for two views can be straightforwardly extended to give expressions for invariants of three views. Considering four world points, $\mathbf{X}_1, \mathbf{X}_2, \mathbf{X}_3, \mathbf{X}_4$, or two lines $\mathbf{X}_1 \wedge \mathbf{X}_2$ and $\mathbf{X}_3 \wedge \mathbf{X}_4$, projected into three camera planes, we can write

$$\begin{aligned} \mathbf{X}_1 \wedge \mathbf{X}_2 &= (\mathbf{A}_0 \wedge \boldsymbol{L}_{12}^A) \vee (\mathbf{B}_0 \wedge \boldsymbol{L}_{12}^B) \\ \mathbf{X}_3 \wedge \mathbf{X}_4 &= (\mathbf{A}_0 \wedge \boldsymbol{L}_{34}^A) \vee (\mathbf{C}_0 \wedge \boldsymbol{L}_{34}^C). \end{aligned}$$

Once again, we can combine the above expressions so that they give to give an equation for the 4-vector $\mathbf{X}_1 \wedge \mathbf{X}_2 \wedge \mathbf{X}_3 \wedge \mathbf{X}_4$,

$$\begin{aligned} \mathbf{X}_1 \wedge \mathbf{X}_2 \wedge \mathbf{X}_3 \wedge \mathbf{X}_4 &= ((\mathbf{A}_0 \wedge \boldsymbol{L}_{12}^A) \vee (\mathbf{B}_0 \wedge \boldsymbol{L}_{12}^B)) \wedge ((\mathbf{A}_0 \wedge \boldsymbol{L}_{34}^A) \vee (\mathbf{C}_0 \wedge \boldsymbol{L}_{34}^C)) \\ &= (\mathbf{A}_0 \wedge \mathbf{A}_{1234}) \wedge ((\mathbf{B}_0 \wedge \boldsymbol{L}_{12}^B) \vee (\mathbf{C}_0 \wedge \boldsymbol{L}_{34}^C)). \quad (2.24) \end{aligned}$$

Writing the lines $\boldsymbol{L}_{12}^B$ and $\boldsymbol{L}_{34}^C$ in terms of the line coordinates we have
$$\boldsymbol{L}_{12}^B = \sum_{j=1}^{3} l_{12,j}^B \boldsymbol{L}_j^B \text{ and } \boldsymbol{L}_{34}^C = \sum_{j=1}^{3} l_{34,j}^C \boldsymbol{L}_j^C.$$

It has been shown in chapter 15 that the components of the *trifocal tensor* (which plays the role of the fundamental matrix for 3 views), can be written in geometric algebra as

$$\tilde{T}_{ijk} = [(\mathbf{A}_0 \wedge \mathbf{A}_i) \wedge ((\mathbf{B}_0 \wedge \boldsymbol{L}_j^B) \vee (\mathbf{C}_0 \wedge \boldsymbol{L}_k^C))] \quad (2.25)$$

so that from equation (2.24) it can be derived:

$$[\mathbf{X}_1 \wedge \mathbf{X}_2 \wedge \mathbf{X}_3 \wedge \mathbf{X}_4] = \sum_{i,j,k=1}^{3} \tilde{T}_{ijk} \alpha_{1234,i} l_{12,j}^B l_{34,k}^C = \tilde{T}(\alpha_{1234}, \boldsymbol{L}_{12}^B, \boldsymbol{L}_{34}^C) \quad (2.26)$$

The invariant $\boldsymbol{Inv}_3$ can then be expressed as

$$\boldsymbol{Inv}_{3T} = \frac{\tilde{T}(\alpha_{1234}, \boldsymbol{L}_{12}^B, \boldsymbol{L}_{34}^C) \tilde{T}(\alpha_{4526}, \boldsymbol{L}_{25}^B, \boldsymbol{L}_{26}^C)}{\tilde{T}(\alpha_{1245}, \boldsymbol{L}_{12}^B, \boldsymbol{L}_{45}^C) \tilde{T}(\alpha_{3426}, \boldsymbol{L}_{34}^B, \boldsymbol{L}_{26}^C)}. \quad (2.27)$$

Note that the factorization must be done so that the same line factorizations occur in both the numerator and denominator. Therefore we have an expression for invariants in three views that is a direct extension of the invariants for two views. Forming the above invariant from observed quantities we note, as before, that some correction factors will be necessary – equation (2.27) is given above in terms of R^4 quantities. Fortunately, this is quite straightforward. Regarding the results of previous section, we can simply consider the $\alpha's$ terms in equation (2.27) as not observable quantities, conversely the line terms like $\boldsymbol{L}_{12}^B, \boldsymbol{L}_{34}^C$ are indeed observed quantities.

As a result, the expression has to be modified using partially the coefficients computed in previous section and for the unique four combinations of three cameras their invariant equations read

$$\boldsymbol{Inv_{3T}} = \frac{T(\boldsymbol{a}_{1234}, l_{12}^B, l_{34}^C) T(\boldsymbol{a}_{4526}, l_{25}^B, l_{26}^C)}{T(\boldsymbol{a}_{1245}, l_{12}^B, l_{45}^C) T(\boldsymbol{a}_{3426}, l_{34}^B, l_{26}^C)} \frac{\mu_{1245}(\mu_{3426} - 1)}{\mu_{4526}(\mu_{1234} - 1)}. \tag{2.28}$$

10.2.4 Comparison of the projective invariants

Invariants using F

0.000	0.590	0.670	0.460
	0	0.515	0.68
		0.59	0
			0.69

0.063	0.650	0.750	0.643
	0.67	0.78	0.687
		0.86	0.145
			0.531

0.148	0.600	0.920	0.724
	0.60	0.96	0.755
		0.71	0.97
			0.596

0.900	0.838	0.690	0.960
	0.276	0.693	0.527
		0.98	0.59
			0.663

Invariants using T

0.000	0.590	0.310	0.630
	0	0.63	0.338
		0.134	0.67
			0.29

0.044	0.590	0.326	0.640
	0	0.63	0.376
		0.192	0.67
			0.389

0.031	0.100	0.352	0.660
	0.031	0.337	0.67
		0.31	0.67
			0.518

0.000	0.640	0.452	0.700
	0.063	0.77	0.545
		0.321	0.63
			0.643

FIGURE 10.2. The distance matrices show the performance of the invariants by increasing Gaussian noise σ: 0.005, 0.015, 0.025 and 0.04.

This section shows simulations with synthetic data and computations using real images. The simulation was implemented in Maple.

The computation of the bilinearity matrix F and the trilinearity focal tensor T was done using a linear method. We believe that for the test purposes these are good enough. Four different sets of six points $S_i = \{\boldsymbol{X}_{i1}, \boldsymbol{X}_{i2}, \boldsymbol{X}_{i3}, \boldsymbol{X}_{i4}, \boldsymbol{X}_{i5}, \boldsymbol{X}_{i6}\}$, where $i = 1, .., 4$, were considered in the simulation and the only three possible invariants were computed for each set $\{I_{1,i}, I_{2,i}, I_{3,i}\}$. Then, the invariants of each set were represented as 3–D vectors ($\mathbf{v}_i = [I_{1,i}, I_{2,i}, I_{3,i}]^T$). We computed four of these vectors that corresponded to four different sets of six points using two images for the F case and three images for the T case (first group of images); and for four of these vectors corresponding to the same point sets we used another two images for the F case or another three images for the T case (second group of images). The comparison of the invariants was done using Euclidean distances of the vectors $d(\mathbf{v}_i, \mathbf{v}_j) = (1 - |\frac{\mathbf{v}_i \cdot \mathbf{v}_j}{||\mathbf{v}_i|| ||\mathbf{v}_j||}|)^{\frac{1}{2}}$; this method was used for the same reason by [5].

Since in $d(\mathbf{v}_i, \mathbf{v}_j)$ we normalize the vectors $\mathbf{v}_i$ and $\mathbf{v}_j$, the distance $d(\mathbf{v}_i, \mathbf{v}_j)$ for any of them does lies between 0 and 1 and it does not vary when $\mathbf{v}_i$ or $\mathbf{v}_j$ is multiplied by a nonzero constant. The figure 10.2 shows a comparison table where each (i, j)-th entry represents the distance computed using $d(\mathbf{v}_i, \mathbf{v}_j)$ between the invariants of set S_i of the points extracted of the first group of images and the set S_j of the points yet using the second group of images. In the ideal case, the diagonal of the distance matrices should be zero, that means that the values of the computed invariants remain constant regardless of which group of images they were used for. The entries off the diagonal mean that we are comparing vectors composed of different coordinates ($\mathbf{v}_i = [I_{1,i}, I_{2,i}, I_{3,i}]^T$), thus they are not parallel and should be bigger than zero and if they are very different the value of $d(\mathbf{v}_i, \mathbf{v}_j)$ should be approximately 1. Now looking at the figure 10.2, we can clearly see that the performance of the invariants based on trilinearities is much better than that of those based on bilinearities, the diagonal values in the T case are in general closer to zero than in the F case and its entries off the diagonal are in general bigger values than in the F case.

FIGURE 10.3. Image sequence taken during navigation by the binocular head of a mobile robot. The upper row shows the left camera images and the lower one shows the right camera ones.

In the case of real images we use a sequence of images taken by a moving robot equipped with a binocular head. The figure 10.3 shows three images of the left eye in the upper row and below these of the right eye respectively. We took image couples, one from the left and one from the right for the invariants using F and two of one eye and one of the other for the invariant using T. From the image we took 38 points semi–automatically and we selected now six sets of points. In each set the points are in general position. Three invariants of each set were computed and the comparison tables were obtained similarly to the previous experiment, see figure 10.4.

using F

0.04	0.79	0.646	0.130	0.679	0.89
	0.023	0.2535	0.278	0.268	0.89
		0.0167	0.723	0.606	0.862
			0.039	0.808	0.91
				0.039	0.808
					0.039

using T

0.021	0.779	0.346	0.930	0.759	0.81
	0.016	0.305	0.378	0.780	0.823
		0.003	0.83	0.678	0.97
			0.02	0.908	0.811
				0.008	0.791
					0.01

FIGURE 10.4. The distance matrices show the performance of the computed invariants using bilinearities (top) and trilinearities (bottom) for the image sequence.

This shows again that the approach to compute the invariants using trilinearities is much more robust than the one using bilinearities, as expected from the theoretical point of view.

10.3   Projective Depth

In a geometric sense the *projective depth* can be seen as the relation between the distance regarding the view center of a 3–D point X_i and the focal distance f as depicted in figure 10.5.

Let us derive the projective depth from a projective mapping. According to the pinhole model explained in chapter 15 the coordinates of a point in

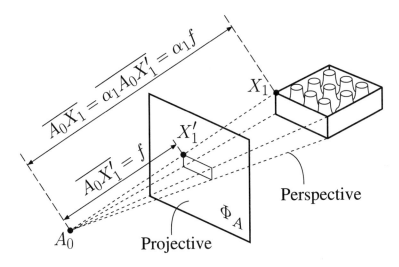

FIGURE 10.5. Geometric interpretation of the projective depth.

the image plane is the result of the projection of the 3–D point to the three optical planes ϕ_A^1, ϕ_A^2, ϕ_A^3. They are spanned by a trivector basis $\gamma_i, \gamma_j, \gamma_k$ and the coefficients t_{ij}. This projective mapping in a matrix representation reads

$$
\lambda x = \begin{bmatrix} x \\ y \\ 1 \end{bmatrix} = \begin{bmatrix} \phi_A^1 \\ \phi_A^2 \\ \phi_A^3 \end{bmatrix} X = \begin{bmatrix} t_{11} & t_{12} & t_{13} & t_{14} \\ t_{21} & t_{22} & t_{23} & t_{24} \\ t_{31} & t_{32} & t_{33} & t_{34} \end{bmatrix} \begin{bmatrix} X \\ Y \\ Z \\ 1 \end{bmatrix}
$$

$$
= \begin{bmatrix} f & 0 & 0 \\ 0 & f & 0 \\ 0 & 0 & 1 \end{bmatrix} \begin{bmatrix} r_{11} & r_{12} & r_{13} & t_x \\ r_{21} & r_{22} & r_{23} & t_y \\ r_{31} & r_{32} & r_{33} & t_z \\ 0 & 0 & 0 & 1 \end{bmatrix} \begin{bmatrix} X \\ Y \\ Z \\ 1 \end{bmatrix} \tag{3.29}
$$

where the projective scale factor is called λ. Note that the projective mapping is further expressed in terms of a f, rotation and translation components. Let us attach the world coordinates to the view center of the camera. The resultant projective mapping becomes

$$
\lambda x = \begin{bmatrix} f & 0 & 0 & 0 \\ 0 & f & 0 & 0 \\ 0 & 0 & 1 & 0 \end{bmatrix} \begin{bmatrix} X \\ Y \\ Z \\ 1 \end{bmatrix} \equiv PX. \tag{3.30}
$$

We can then compute straightforwardly

$$
\lambda = Z. \tag{3.31}
$$

The way how we compute the projective depth ($\equiv \lambda$) of a 3–D point appears simple using invariant theory, namely using equations (2.1). For that we select a basis system taking four 3–D points in general position X_1, X_2, X_3, X_5, the optical center of camera at the new position as the four point X_4, and X_6 as the 3–D point to be reconstructed. This has been depicted in figure 10.6.

Since we use the mapped points, we consider the *epipole* (mapping of the current view center) as the four point and the mapped sixth point as the point with the unknown depth. The other mapped basis points remain constant during the procedure.

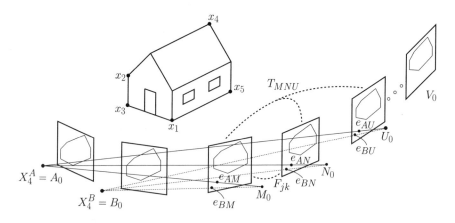

FIGURE 10.6. Computing the projective depths of n cameras.

According to equation (2.1), the tensor based expression for computing the third coordinate or projective depth of a point X_j ($= X_6$) reads

$$\lambda_j = \frac{Z_j}{W_j} = \frac{T(a_{124j}, l_{12}^B, l_{4j}^C)T(a_{1235}, l_{12}^B, l_{35}^C)}{T(a_{1245}, l_{12}^B, l_{45}^C)T(a_{123j}, l_{12}^B, l_{3j}^C)} \cdot \frac{\mu_{1245}\mu_{123j}}{\mu_{124j}\mu_{1235}}. \quad (3.32)$$

In this way we can successively compute the projective depths λ_{ij} of the j–points referred to the i–camera. The λ_{ij} will be used in next section for the 3–D reconstruction using the join image concept and the *singular value decomposition* SVD method.

Since this kind of invariant can be also expressed in terms of the quadrifocal tensor [9], we can compute the projective depth based on four cameras.

10.4   Shape and Motion

The orthographic and paraperspective *factorization method for structure and motion* using the affine camera model was developed by Tomasi, Kanade and Poelman [11, 15]. This method works for cameras viewing small and distance scenes, thus all scale factors of projective depth $\lambda_{ij}=1$. For the case of perspective images the scale factors λ_{ij} are unknown. According to Triggs [16] all λ_{ij} satisfy a set of consistency reconstruction equations of the so–called *join image*. One way to compute λ_{ij} is by using the epipolar constraint. If we use a matrix representation this is given by

$$F_{ik}\lambda_{ij}\boldsymbol{x}_{ij} = \boldsymbol{e}_{ik}\wedge\lambda_{kj}\boldsymbol{x}_{kj} \tag{4.33}$$

which after an inner product gives the relation of projective depths for the j-point between camera i and k

$$\lambda'_{kj} = \frac{\lambda_{kj}}{\lambda_{ij}} = \frac{(\boldsymbol{e}_{ik}\wedge\boldsymbol{x}_{kj})F_{ik}\boldsymbol{x}_{ij}}{||\boldsymbol{e}_{ik}\wedge\boldsymbol{x}_{kj}||^2}. \tag{4.34}$$

Considering the i-camera as reference we can norm the λ_{kj} for all k-cameras and use λ'_{kj} instead. If that is not the case we can norm between neighbor images in a chained relationship [16].

In the previous section we presented a better procedure for the computing of λ_{ij} involving three cameras. The extension of the equation (4.34) in terms of the trifocal or quadrifocal tensor is awkward and unpractical.

10.4.1   The join image

The *join image* $\mathcal{J}$ is nothing else than the intersections of optical rays and planes at the points or lines in the 3–D projective space as depicted in figure (10.7). The interrelated geometry can be linearly expressed by the fundamental matrix and trifocal and quadrifocal tensors. The reader will find more details about these linear constraints in chapter 7.

In order to take into account the interrelated geometry, the *projective reconstruction* procedure should put together all the data of the individual images in a geometrically coherent manner. The way to do that is by considering the observations of the points $\boldsymbol{X}_j$ regarding each i–camera

$$\lambda_{ij}\boldsymbol{x}_{ij} = P_i\boldsymbol{X}_j \tag{4.35}$$

as the i–row of a matrix of rank 4. For m cameras and n points the 3m×n matrix $\mathcal{J}$ of the join image is given by

$$\mathcal{J} = \begin{pmatrix} \lambda_{11}\boldsymbol{x}_{11} & \lambda_{12}\boldsymbol{x}_{12} & \lambda_{13}\boldsymbol{x}_{13} & \cdot & \cdot & \cdot & \lambda_{1n}\boldsymbol{x}_{1n} \\ \lambda_{21}\boldsymbol{x}_{21} & \lambda_{22}\boldsymbol{x}_{22} & \lambda_{23}\boldsymbol{x}_{23} & \cdot & \cdot & \cdot & \lambda_{2n}\boldsymbol{x}_{2n} \\ \lambda_{31}\boldsymbol{x}_{31} & \lambda_{32}\boldsymbol{x}_{32} & \lambda_{33}\boldsymbol{x}_{33} & \cdot & \cdot & \cdot & \lambda_{3n}\boldsymbol{x}_{3n} \\ \cdot & \cdot & \cdot & \cdot & \cdot & \cdot & \cdot \\ \cdot & \cdot & \cdot & \cdot & \cdot & \cdot & \cdot \\ \cdot & \cdot & \cdot & \cdot & \cdot & \cdot & \cdot \\ \lambda_{m1}\boldsymbol{x}_{m1} & \lambda_{m2}\boldsymbol{x}_{m2} & \lambda_{m3}\boldsymbol{x}_{m3} & \cdot & \cdot & \cdot & \lambda_{mn}\boldsymbol{x}_{mn} \end{pmatrix}. \tag{4.36}$$

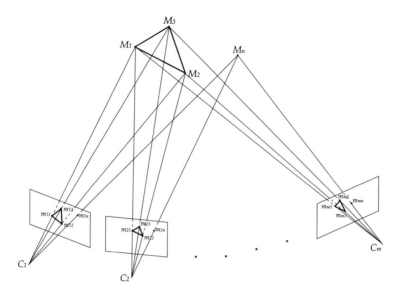

FIGURE 10.7. The geometry of the join image.

For the affine reconstruction procedure the matrix is of rank 3. The matrix $\mathcal{J}$ of the join image is amenable to a singular value decomposition for finding the shape and motion [11, 15].

10.4.2   The SVD method

The application of SVD to $\mathcal{J}$ gives

$$\mathcal{J}_{3m\times n} = U_{3m\times r}S_{r\times r}V_{n\times r}^T, \tag{4.37}$$

where the columns of matrix $V_{n\times r}^T$ and $U_{3m\times r}$ constitute the orthonormal base for the input (co-kernel) and output (range) spaces of $\mathcal{J}$. In order to get a decomposition in motion and shape of the projected point structure, $S_{r\times r}$ can be absorbed into both matrices $V_{n\times r}^T$ and $U_{3m\times r}$ as follows

$$\mathcal{J}_{3m\times n} = (U_{3m\times r}S_{r\times r}^{\frac{1}{2}})(S_{r\times r}^{\frac{1}{2}}V_{n\times r}^T) = \begin{pmatrix} P_1 \\ P_2 \\ P_3 \\ \cdot \\ \cdot \\ \cdot \\ P_m \end{pmatrix}_{3m\times 4} (\boldsymbol{X}_1\boldsymbol{X}_2\boldsymbol{X}_3...\boldsymbol{X}_n)_{4\times n} \tag{4.38}$$

This way to divide $S_{r\times r}$ is not unique. Since the rank of $\mathcal{J}$ is 4 we should take the first four biggest singular values for $S_{r\times r}$. The matrices P_i correspond to the projective mappings or *motion* from the projective space to the individual images and the point structure or *shape* is given by X_j. We test our approach using a simulations program written in Maple. Using the method of section 10.3 firstly we computed the projective depth of the points of a wire house observed with 9 cameras and then using the SVD projective reconstruction method we gained the shape and motion. The reconstructed house after the Euclidean readjustment for the presentation is shown in figure 10.8.

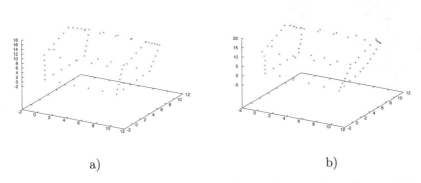

a)   b)

FIGURE 10.8. Reconstructed house using a) noise–free observations and b) noisy observations.

We notice that the reconstruction keeps quite well the original form of the model.

The next section will show how we can improve the shape of the reconstructed model using geometric expressions in terms of the operators of algebra of incidence ∨ (meet) and ∧ (join) and particular tensor based invariants.

10.4.3   Completion of the 3–D shape using geometric invariants

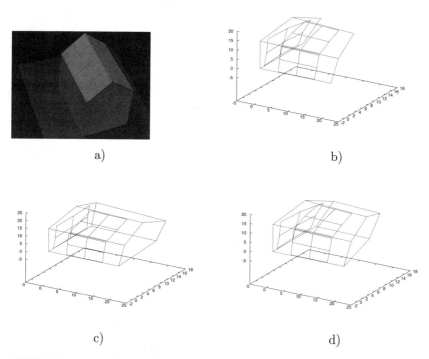

a)   b)

c)   d)

FIGURE 10.9. a) One of the three images, b) reconstructed incomplete house using 3 images c) extending the join image d) completing in the 3–D space.

The projective structure can be improved in two ways: by completing points on the images, by expanding the join image and then by calling the SVD procedure, or, after the reconstruction, by completing points in the 3–D space like the occluded ones. Both approaches can use geometric inference rules based on symmetries or concrete knowledge about the scene. Using three real views of a similar model house with its most right lower corner missing, see figure 10.9.b , we compute in each image the virtual image point of this 3–D point. Then we reconstruct the scene as shown in figure 10.9.c. As opposite, using geometric incidence operations we completed the house employing the space points as depicted in figure 10.9.d. We can see that creating points in the images yields a better reconstruction of the occluded point. Note that in the reconstructed image we transformed the projective shape into an Euclidean one for the presentation of the results.

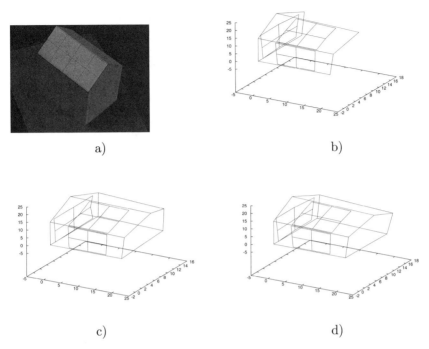

a)   b)

c)   d)

FIGURE 10.10. a) One of the nine images, b) reconstructed incomplete house using 9 images c) extending the join image d) completing in the 3–D space.

We used also lines connecting the reconstructed points only to make visible the house form. Similarly we proceeded using 9 images, as presented in in figure 10.10.a–d.

We can see that the resulting reconstructed point is almost similar in both procedures. As a result we can draw the following conclusion: when we have few views we should extend the join image using virtual image points and in case of several images we should extend the point structure in the 3–D space.

10.5   Conclusions

This chapter focused on the application of projective invariants based on the trifocal tensor. We developed a method to compute the projective depth using this kind of invariants. The resulting projective depths were then used for the initialization of the projective reconstruction of shape and motion.

Furthermore using incidence algebra rules we completed the reconstruction for the case of occluded points.

The main contribution of this paper is that in our geometric method we relate to and extend current approaches regarding projective invariants and their application for reconstruction of shape and motion, as a result the procedures gain geometric transparency and elegance. However, the authors believe that more work have to be done in order to improve the computational algorithms so that the use of projective invariants will be more and more attractive for real systems involving noisy data.

Acknowledgments

Eduardo Bayro-Corrochano was supported by the project SO-201 of the Deutsche Forschungsgemeinschat.

Part IV

Robotics

Chapter 11

Robot Kinematics and Flags

J.M. Selig

11.1 Introduction

In robotics the group of proper rigid transformations of 3-dimensional space
is of central importance. The relevant Clifford algebra in this case is a degenerate one with three generators that square to -1 and a single generator
that squares to 0. The algebra contains a copy of the group's double cover.

In a previous work [10], it was shown that the algebra also contains
representations of the points, lines and planes of space. Moreover, incidence
relations, meets and joins of these linear elements are represented by simple
formulæ in the algebra. Here this work is extended by combining these
linear elements into flags, that is nested sequences of linear elements. These
flags can be used to represent some of the basic joints used in robots. In
particular, a lined plane can be used to represent a prismatic or sliding joint
and either a pointed line or a pointed plane can represent a hinge or revolute
joint. This is because the isotropy group of the flag is the group of rigid body
motions allowed by the joint. This allows us to set up the inverse kinematics
for a serial manipulator with six revolute joints as a problem in the Clifford
algebra. A theorem due to Pieper on the the solubility of such problems is
then fairly straightforward to prove. The theorem states that the inverse
kinematics problem can be solved if any three consecutive joint axes meet
at a point or are parallel. In these cases the methods developed give a
general solution to the inverse kinematics problem. Finally two concrete
examples are given.

11.2 The Clifford Algebra

This work uses a degenerate Clifford algebra, that is the Clifford algebra
associated with a degenerate bilinear form. The algebra has three generators which square to -1, $e_1^2 = e_2^2 = e_3^2 = -1$ and a single generator which
squares to zero $e^2 = 0$. The generators anti-commute in the usual way.

This particular algebra dates back to Clifford himself who concentrated
on its even sub-algebra which he called the 'biquaternions', see [2]. The

reason for looking at this algebra is that is contains the group of rigid body transformations as a sub-group of group of units. More precisely it contains the double cover of the group of rigid transformations. Moreover, as we will see below, it also contains several useful geometric representations of the group, this allows us to turn the geometry of points, lines and planes into Clifford algebra expressions.

11.2.1 The group of rigid body motions

It is well known that the group of rotations in three dimensional space $SO(3)$,can be represented by elements of a Clifford algebra as,

$$r = \cos\frac{\phi}{2} + \sin\frac{\phi}{2}(v_x e_2 e_3 + v_y e_3 e_1 + v_z e_1 e_2)$$

where ϕ is the angle of rotation and the unit vector $(v_x, v_y, v_z)^T$ is the axis of the rotation. It is straightforward to show that these elements form a group under Clifford multiplication with identity 1 and where the inverse of an element r is given by the conjugate r^*. That is, $rr^* = 1$. The action of this group on a point q in $\mathbb{R}^3$ can be written as,

$$q' = rqr^* = r(q_x e_1 + q_y e_2 + q_z e_3)r^*$$

Again it is simple to verify that this corresponds to the standard representation $\mathbf{q}' = R\mathbf{q}$, where $\mathbf{q}$ is the position vector of the point and R is the 3×3 matrix corresponding to a rotation of ϕ about the axis given by $\mathbf{v} = (v_x, v_y, v_z)^T$.

The group product corresponds to the product of rotations. Notice however that both r and $-r$ give the same rotation, so the group found above is not the rotation group itself but its double cover Spin(3).

In robotics we are interested in the group of rigid body motions, the rotations and translations. We are not interested in reflections since no physical machine can reflect an object. So strictly we should refer to the group of proper rigid motions in $\mathbb{R}^3$. In robotics this group is usually denoted $SE(3)$.

In order to incorporate translations we look at Clifford algebra elements of the form,

$$g = r + \frac{1}{2}tre$$

where r is as above and the translation vector $t = t_x e_1 + t_y e_2 + t_z e_3$. We expect the rotations to act on the translation, we know that the group of rigid body motions is a semi-direct product of rotations with translations, $SE(3) = SO(3) \ltimes \mathbb{R}^3$. Combining a pure rotation with a pure translation reveals this action,

$$r(1 + \frac{1}{2}te) = (r + \frac{1}{2}rte) = (r + \frac{1}{2}(rtr^*)re)$$

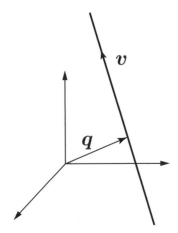

FIGURE 11.1. A line in space.

So the action of rotations on translations is rtr^* as we expect $(R\mathbf{t})$.

As before, the group of elements of this form double cover the rigid body motions as g and $-g$ give the same transformation. Notice that, the group elements g are all members of the even sub-algebra and that $gg^* = 1$. More details can be found in [9, Chap. 9].

11.2.2 Points, lines and planes

In this section we detail the representation of points, lines and planes as elements in the Clifford algebra. Also we show how the group of rigid body motions acts on these elements.

The most well know of these representations is the representation of lines. This is because they are represented by grade 2 elements of the Clifford algebra and hence can be thought of as biquaternions, see for example [5, section 8.2].

Lines in $I\!R^3$ can be specified their Plücker coordinates. Here we think of these as a pair of vectors, a unit vector $\mathbf{v}$, in the direction of the line and a moment vector $\mathbf{u} = \mathbf{q} \times \mathbf{v}$, where $\mathbf{q}$ is the position vector of any point on the line, see figure 11.1. These vectors will thus be orthogonal $\mathbf{v} \cdot \mathbf{u} = 0$. In the Clifford algebra we will represent a line by elements of the form,

$$\ell = (v_x e_2 e_3 + v_y e_3 e_1 + v_z e_1 e_2) + (u_x e_1 e + u_y e_2 e + u_z e_3 e)$$

but satisfying the relation,

$$\ell\ell^* = 1$$

This relation combines the requirements that $\mathbf{v}$ is a unit vector and that $\mathbf{v}$ and $\mathbf{u}$ are orthogonal. These lines are in fact directed lines since $-\ell$ is the same line as ℓ but with the opposite direction.

The effect of a rigid body motion on a line is given in terms of the Plücker coordinates,

$$\mathbf{v}' = R\mathbf{v}, \qquad \mathbf{u}' = R\mathbf{u} + \mathbf{t} \times R\mathbf{v}$$

In the Clifford algebra the effect of a rigid body motion on a line can be represented as,

$$\ell' = g\ell g^*$$

this can be verified by a simple computation.

We can find the group elements which correspond to rotations about a line ℓ as follows. First translate the line so that it passes through the origin using $(1 - (1/2)qe)$, now rotations about a line through the origin are simply rotations about the line's axis $(\cos(\phi/2) + \sin(\phi/2)v)$, finally we put the line back where we found it by translating with $(1 + (1/2)qe)$. The result is a one-parameter subgroup of elements,

$$\begin{aligned} g(\phi) &= (1 + (1/2)qe)(\cos(\phi/2) + \sin(\phi/2)v)(1 - (1/2)qe) \\ &= \cos(\phi/2) + \sin(\phi/2)v + (1/2)\sin(\phi/2)(qv - vq)e \\ &= \cos(\phi/2) + \sin(\phi/2)\ell \end{aligned}$$

Since $qv - vq = 2(q_y v_z - q_z v_y)e_1 + 2(q_z v_x - q_x v_z)e_2 + 2(q_x v_y - q_y v_x)e_3$.

Next we look at planes in $\mathbb{R}^3$, these are represented by grade 1 elements of the Clifford algebra and hence cannot be thought of as biquaternions. A plane can be specified by giving its unit normal vector $\mathbf{n}$ and the perpendicular distance from the origin, see figure 11.2. As usual, the vector equation of the plane is given by,

$$\mathbf{n} \cdot \mathbf{q} = d$$

where $\mathbf{q}$ is any point on the plane. Notice that these are oriented planes since reversing the sign of $\mathbf{n}$ and d will invert the orientation of the plane.

In the Clifford algebra we can represent planes as elements of the form,

$$\pi = n_x e_1 + n_y e_2 + n_z e_3 + de$$

These elements must satisfy the quadratic condition,

$$\pi\pi^* = 1$$

this ensures that the vector $\mathbf{n}$ has unit length. Note that, $\pi^* = -\pi$, hence we could also write the condition as $\pi^2 = -1$. Now if we subject the plane to a rigid body motion the normal vector and distance to the origin will change as follows,

$$\mathbf{n}' = R\mathbf{n}, \qquad d' = d + (R\mathbf{n}) \cdot \mathbf{t}$$

This is most easily seen by considering the effect on the vector equation for the plane above. In the Clifford algebra this can be represented by,

$$\pi' = g\pi g^*$$

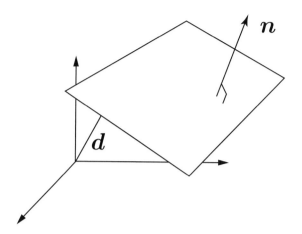

FIGURE 11.2. A plane.

Finally here we look at points in $\mathbb{R}^3$, these will be represented by grade 3 elements of the Clifford algebra of the form,

$$p = e_1 e_2 e_3 + x e_2 e_3 e + y e_3 e_1 e + z e_1 e_2 e$$

The effect of a rigid body motion is given by,

$$p' = gpg^*$$

Notice that these points satisfy the equation $pp^* = 1$, (or $p^2 = 1$, since $p^* = p$) however, they are not the only solutions. There is another $\mathbb{R}^3$ of solutions where the coefficient of $e_1 e_2 e_3$ is -1 instead of $+1$.

This representation is different from the one given in [8] and used in [9]. The representation used there has a slightly different group action and is not homogeneous.

11.2.3 *Some relations*

One of the most useful results of the Clifford algebra outlined above is the fact that many relations between points, lines and planes have simple expressions in terms of this algebra. Here we look at a couple of these, a fuller account may be found in [10].

First we look at the distance from a point to a plane. This is the minimum distance, which will lie along the normal to the plane.

If this distance is l then in the Clifford algebra we have the relation,

$$\frac{1}{2}(\pi p^* + p \pi^*) = l e_1 e_2 e_3 e$$

This enables us to calculate the distance. Moreover, the sign of l tells us which side of the plane the point is.

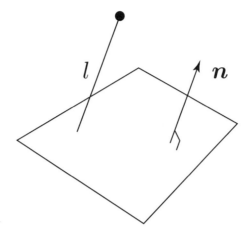

FIGURE 11.3. Distance from a point to a plane.

We also get an incidence relation:

$$\pi p^* + p\pi^* = 0$$

as the condition for the point to lie on the plane.

Next we look at a line and a plane. In general a plane and a line meet at a point. In the Clifford algebra the meeting point can be found from the expression,

$$
\begin{aligned}
\frac{1}{2}(\ell\pi^* + \pi\ell^*) \quad = \quad & \\
& -(n_x v_x + n_y v_y + n_z v_z)e_1 e_2 e_3 - (n_y u_z - n_z u_y + dv_x)e_2 e_3 e \\
& -(n_z u_x - n_x u_z + dv_y)e_3 e_1 e - (n_x u_y - n_y u_x + dv_z)e_1 e_2 e \quad (2.1)
\end{aligned}
$$

dividing by the coefficient of the term $e_1 e_2 e_3$ gives the Clifford algebra element representing the intersection point. If the coefficient vanishes then the line is parallel to the plane and there is no intersection point, unless the line lies in the plane. The line lies in the plane if the whole expression $\ell\pi^* + \pi\ell^*$ vanishes. So we have the incidence relation,

$$\ell\pi^* + \pi\ell^* = 0$$

which implies that the line lies in the plane.

11.3   Flags

In this section we look at how to represent flags in the algebra. A flag is a nested sequence of linear spaces, for example a point contained in a line or a line lying in a plane. The relevance of these figures is that they can be used to represent the joints of a robot. Most industrial robots have revolute or prismatic joints, that is hinges or simple sliding joints. Robots and other mechanisms can have helical or screw joints but the symmetry groups of these joints are not algebraic.

11.3.1   Pointed lines

Our first example is the pointed line, that is a point lying on a line. We may combine the algebra elements found above to represent the flag as,

$$f_{12} = \frac{1}{\sqrt{2}}(p + \ell)$$

Notice that this element is not homogeneous. Not all such non-homogeneous elements represent pointed lines, the line must be a line so we must have $\ell\ell^* = 1$ and, to be a point, the coefficient of $e_1e_2e_3$ must be 1. Finally for the point to lie on the line the point and line must satisfy, $p\ell^* + \ell p^* = 0$. Most of these equations can be written in terms of the flag itself,

$$f_{12}f_{12}^* = 1,$$

Comparing coefficients of the various basis elements gives us all the equations except that we only have that $pp^* + \ell\ell^* = 2$. If we include the relation that the coefficient of $e_1e_2e_3$ must be 1 then we have that $pp^* = 1$ and hence that $\ell\ell^* = 1$. Thus, we see that the space of all pointed lines form an affine algebraic variety, usually called a flag manifold, in this case the $(1,2)$-flag manifold.

The action of the group of rigid motions on these flags is simply,

$$f_{12}' = gf_{12}g^*$$

Since the action is linear, the point and line transform independently. From this we can find the isotropy group of a particular pointed line. Elements of the isotropy group must satisfy,

$$f_{12} = gf_{12}g^*, \quad \text{or equivalently} \quad f_{12}g - gf_{12} = 0$$

That is, elements of the isotropy group must commute with the flag. The only even graded elements which commute with both p and ℓ are linear combinations of 1 and ℓ itself. The result that ℓ and p commute is, of course, a consequence of the fact that p lies on ℓ. The elements of the isotropy group form a line which can be parameterised as,

$$g(\theta) = \cos\frac{\theta}{2} + \sin\frac{\theta}{2}\ell$$

The parameter θ measures the angle turned about the line ℓ, see section 11.2.2.

Another way of looking at this one-parameter group is as the exponential of the line ℓ. That is,

$$\exp(\frac{\theta}{2}\ell) = 1 + \frac{\theta}{2}\ell + \frac{1}{2!}\left(\frac{\theta}{2}\ell\right)^2 + \cdots = \cos\frac{\theta}{2} + \sin\frac{\theta}{2}\ell$$

Recall that $\ell^2 = -1$. This is the exponential map from the Lie algebra of the group to the group itself. In the Clifford algebra the Lie algebra is represented by the elements of grade 2, see [9, section 9.3].

11.3.2   Pointed planes

We can treat pointed planes in much the same way as pointed lines. A general pointed plane will have the form,

$$f_{13} = \frac{1}{\sqrt{2}}(p + \pi)$$

The condition for the point to lie in the plane is $p\pi^* + \pi p^* = 0$ and hence the equations for the $(1,3)$-flag manifold are given by,

$$f_{13}f_{13}^* = 1$$

and also the coefficient of $e_1e_2e_3$ must be 1.

There is a homeomorphism between the two flag manifolds defined above. This can be see by mapping pointed lines to pointed planes and vice versa. Given a pointed line $f_{12} = (1/\sqrt{2})(p + \ell)$ we can find a pointed plane with the same point but where the plane is perpendicular to the original line. These flags are oriented so the direction of the plane (the direction of its normal) will be the same as the direction of the line. In the Clifford algebra we can write the plane perpendicular to the line ℓ and passing through the point p as,

$$\pi^\perp = \frac{1}{2}(p\ell^* - \ell p^*)$$

see [10]. So the mapping is,

$$f_{12} \longrightarrow f_{13}, \qquad \frac{1}{\sqrt{2}}(p + \ell) \longmapsto \frac{1}{\sqrt{2}}(p + \frac{1}{2}(p\ell^* - \ell p^*))$$

The inverse of this map is given by mapping the plane to the line perpendicular to the plane passing through the point. This can be written as,

$$\ell^\perp = \frac{1}{2}(\pi p^* - p\pi^*)$$

It is simple to verify by direct calculation that $\ell^{\perp\perp} = \ell$, if we remember that, $p^2 = -1$, $\ell^* = -\ell$ and also, since p lies on ℓ, we have $p\ell p^* = \ell^*$.

Because of the homeomorphism, the isotropy group of a pointed plane is the same as that for the corresponding pointed line. Again a direct calculation confirms that $\ell = \pi^\perp$ commutes with π. So we can use either a pointed line or a pointed plane to represent a revolute joint. The axis of the joint is given by the line and the point can be any fixed point on the axis.

Finally here, we note another application in robotics for pointed planes, (or pointed lines). In many robotic applications the end-effector is required to trace a path on a smooth surface, an example might be an inspection task. Often there is an axis in the tool which must remain perpendicular to the surface, perhaps the tool is an ultrasonic probe. Now the surface determines a sub-space in the space of all pointed planes, just take each point on the surface together with its tangent plane (the surface lies in $I\!R^3$ and so is orientable). The desired trajectory of the end-effector must be a path in this sub-space. This approach does not seem to have received much attention to date, but see [6].

11.3.3 Lined planes

To represent prismatic joints we use lined planes. In the Clifford algebra these correspond to elements of the form,

$$f_{23} = \frac{1}{\sqrt{2}}(\ell + \pi)$$

The condition for the line to lie on the plane is, $\ell\pi^* + \pi\ell^* = 0$. Hence we see that these elements certainly satisfy,

$$f_{23}f_{23}^* = 1$$

There is an involution on the algebra which we can use, this is sometimes called the *main involution*, it simply reverses the sign of basis elements of odd grade. So for lines we have $\ell^- = \ell$ and $\pi^- = -\pi$ where the superscript $-$ denotes the main involution. This means we can write,

$$(f_{23} + f_{23}^-)^2 = (\sqrt{2}\ell)^2 = -2$$

to ensure that both $\ell\ell^* = 1$ and $\pi\pi^* = 1$.

The isotropy group of a lined plane is simple to find, once again the group elements must commute with the flag,

$$gf_{23} - f_{23}g = 0$$

The linear space of elements which commute with the line are spanned by linear combinations of the set $\{1, \ell, e_1e_2e_3, \ell e_1e_2e_3\}$. The only basis elements which commute with π are 1 and $\ell e_1e_2e_3$,

$$\ell e_1e_2e_3 e\pi - \pi\ell e_1e_2e_3 e = 2(n_x v_x + n_y v_y + n_z v_z)e = 0$$

since the line lies in the plane. Hence a general element of the isotropy group has the form,

$$g(\lambda) = 1 + \frac{1}{2}\lambda \ell e_1 e_2 e_3 e$$

Again, as expected, the exponential of a Lie algebra element:

$$\exp(\frac{\lambda}{2}\ell e_1 e_2 e_3 e) = 1 + \frac{1}{2}\lambda \ell e_1 e_2 e_3 e$$

This time the isotropy group is a one-parameter family of translations, along the axis determined by the line.

11.3.4 Complete flags

Complete flags here comprise a point on a line in a plane,

$$f_{123} = \frac{1}{\sqrt{3}}(p + \ell + \pi)$$

From the discussions above it is not hard to see that the isotropy group of a complete flag is just the trivial group 1, just take the intersection of the isotropy groups of the pointed line and lined plane contained in the complete flag. To each complete flag we can associate a coordinate frame in space, take the point in the flag as the origin, the line as the x-axis and the plane as the xy-plane. The positive x-direction will be determined by the direction of the line and the normal to the plane will determine the positive z-direction. Hence the positive y-direction may be found using the vector cross product. These coordinate frames were studied by Study [12], who called them *soma*. Now there is a 1-to-1 correspondence between elements of the group of rigid body motions $SE(3)$, and the set of all possible frames(somas). To see this, fix a standard or home frame say $f^0 = (e_1 e_2 e_3 + e_2 e_3 + e_3)/\sqrt{3}$, then any particular group element is mapped to the frame obtained by operating on this home element,

$$g \longmapsto g f^0 g^*$$

Remember, the elements g are in fact elements of the double covering of $SE(3)$, both g and $-g$ map to the same frame, but they represent the same element of $SE(3)$.

The inverse of this mapping can be found using linear algebra. Suppose we have a frame f, the corresponding group element satisfies,

$$g f^0 g^* = f, \qquad \text{or equivalently,} \qquad g f^0 - f g = 0$$

This system of equations for g must be solved with the quadratic relation, $g g^* = 1$ and hence we expect two solutions $\pm g$.

In much of the current robotics literature the position and orientation of a rigid body is specified by giving a frame attached to the body. In the following it is more convenient to specify the position and orientation of the body by giving an element of the group of rigid body motions. The above shows that these two views are equivalent.

Finally here, note that spherical (ball-and-socket) joints can be represented by points, since both have an isotropy group which is a copy of $SO(3)$. Planar joints can be represented by planes, they both have isotrpy group $SE(2)$, the group of planar rigid motions. And finally cylindrical joints can be represented by lines, the isotropy group here is $SO(2) \times \mathbb{R}$, rotations about the line and translations along it.

11.4 Robots

11.4.1 Kinematics

Most industrial robot arms consist or six rigid links connected by 6 revolute joints, that is a 6R robot. One of the central problems in robotics is to relate the position and orientation of the robot's last link, end-effector or tool, to the positions of its joints.

The forward or direct kinematic problem is to determine the position and orientation of the tool given the angles of the joints. For a serially connected robot this is relatively straight forward. We begin by choosing a 'home' or standard configuration for the robot. In the home configuration all joint angles will be zero. Now we record the positions of the joint axes in the home configuration, these will be six lines , ℓ_1, $\ell_2 \ldots, \ell_6$. For each of these lines the one-parameter group of rotations about the line is given by,

$$a_i(\theta_i) = \cos\frac{\theta_i}{2} + \sin\frac{\theta_i}{2}\ell_i$$

see section 11.3.1. Now suppose that we set the joint angles to some set of particular values, $\theta_1, \theta_2, \ldots, \theta_6$, what is the rigid body transformation undergone by the end-effector? If we move the last joint first then the tool undergoes a transformation $a_6(\theta_6)$ but the rest of the joints lower down the arm are unchanged. So we can move the 5th joint into position giving an overall transformation $a_5(\theta_5)a_6(\theta_6)$. Continuing in this fashion down the arm it is easy to see that the total transformation is a group element,

$$g = a_1(\theta_1)a_2(\theta_2)a_3(\theta_3)a_4(\theta_4)a_5(\theta_5)a_6(\theta_6)$$

This group element represents the transformation relative to the home configuration. That is, the transformation which would take the end-effector in its home position to the configuration determined by the joint angles $\theta_1, \ldots, \theta_6$.

The inverse kinematic problem for a serial robot is much harder. Given the desired position and orientation for the end-effector, what should we set the joint angles to? Effectively this means solving the above equation for the joint angles given a group element g.

The inverse kinematic problem is often stated in terms of 'tool-frame' coordinate. Imagine a coordinate frame rigidly attached to the tool of a robot, it is often simplest to specify a desired motion of the tool relative to this frame. For instance, if the x-coordinate of this frame is aligned with the axis of the end-effector's gripper jaws say, then we might want to tell the robot to advance some distance in this direction. Suppose that the desired motion is given by a group element h relative to the tool-frame, the corresponding element in the 'world-frame' can be found by conjugation in the group. Let k be the group element which transforms the world-frame to the tool-frame, then in the world-frame the desired motion will be khk^*. It is useful to write the element k as a product $k = bg$. Here b is a constant group element which transforms the world-frame to the tool-frame when the robot is in its home configuration, and g is the group element which takes the end-effector from its home position to the current position. So in the tool-frame the inverse kinematic problem can be written as,

$$a_1(\theta_1)a_2(\theta_2)a_3(\theta_3)a_4(\theta_4)a_5(\theta_5)a_6(\theta_6) = bghg^*b^*$$

That is, the problem is virtually the same as when the desired motion was given in the world-frame. In fact the only difference is that g must be computed and stored at the current position of the robot. The same methods may be used to solve for the joint angles in both cases.

11.4.2   Pieper's theorem

In his Ph.D. thesis, Pieper [7] showed that any 6R robot which has 3 consecutive joint axes meeting at a point has solvable inverse kinematics. Later, Duffy showed [3] that this was also true when any 3 consecutive joints are parallel.

The exact meaning of solvability is not too important here since constructive proofs were given. Clearly if non-solvability results were to be considered the precise meaning of the term 'solvable' would be very important. It would seem that the intention is to use the same concepts as in Galois theory, that is solvable by radicals, but with $\tan\theta_i$ or equivalently $\sin\theta_i$ and $\cos\theta_i$ as the variables.

The demonstration given here roughly follows the work of Pieper, but the computations using the Clifford algebra are simpler and hence the underlying geometry is much clearer. We begin with the kinematic relations for a 6R robot,

$$a_1(\theta_1)a_2(\theta_2)a_3(\theta_3)a_4(\theta_4)a_5(\theta_5)a_6(\theta_6) = g$$

where the a_is are as in the previous section. Now suppose that 3 consecutive joints intersect or are parallel. For the sake of illustration we assume here that it is joints $2, 3$ and 4 which have this property, but it is easy to see how to proceed in other cases.

1. Rearrange the kinematic equation to isolate the 3 intersecting/parallel joints,

$$a_2 a_3 a_4 = a_1^* g a_6^* a_5^*$$

To simplify notation we drop the explicit dependance on the joint angles.

2. If joints $2, 3$ and 4 are intersecting then their common point p will be preservered by a_2, a_3 and a_4, so,

$$a_2 a_3 a_4 p a_4^* a_3^* a_2^* = p = a_1^* g a_6^* a_5^* p a_5 a_6 g^* a_1$$

On the other hand if the joints are parallel there will be a plane π preservered by the joints, any plane perpendicular to the parallel joint axes will do,

$$a_2 a_3 a_4 \pi a_4^* a_3^* a_2^* = \pi = a_1^* g a_6^* a_5^* \pi a_5 a_6 g^* a_1$$

This splits the problem into two pieces, for definiteness we look at the parallel case,

$$
\begin{aligned}
g^* a_1 \pi a_1^* g &= a_6^* a_5^* \pi a_5 a_6 \\
a_2 a_3 a_4 &= a_1 g a_6^* a_5^* = g'
\end{aligned}
$$

3. The first of these equations only involves the joint angles θ_1, θ_5 and θ_6. (Once this equation has been solved, we can evaluate $g' = a_1 g a_6^* a_5^*$ and solve the second equation for the remaining joint angles, θ_2, θ_3 and θ_4.) So our first task is to solve the first of these equations. Notice that this equation is a relation between planes with the general form,

$$\pi_\alpha = a_6^* \pi_\beta a_6$$

If the plane $\pi_\beta = a_5^* \pi a_5$ is rotated about the final joint, the point where this plane meets the axis of the last joint ℓ_6 will remain fixed. This common point is a scalar multiple of the Clifford algebra expression, $\ell_6 \pi_\beta^* + \pi_\beta \ell_6^*$. This allows us to eliminate the last joint angle and write the equation as,

$$g^* a_1 \pi a_1^* g \ell_6^* + \ell_6 g^* a_1 \pi^* a_1^* g = a_5^* \pi a_5 \ell_6^* + \ell_6 a_5^* \pi^* a_5$$

In the case where there are three intersecting joint axes, we will obtain a relation between points. In this case we can find an invariant plane to eliminate another joint angle, this plane passes through the point p given by the equation and is perpendicular to the axis of the joint ℓ. The Clifford algebra expression for such a plane is proportional to $p\ell^* - \ell p^*$. See [10] for more details on these relations.

4. Returning to the parallel joints case we observe that since $a_i = \cos(\theta_i/2) + \sin(\theta_i/2)\ell_i$, the expressions, $a_1\pi a_1^*$ and $a_5\pi a_5^*$ are linear in $\cos\theta_1$, $\sin\theta_1$ and $\cos\theta_5$, $\sin\theta_5$ respectively. There are effectively three linear equations here, the final pair of equations are provided by the trigonometric identities $\cos^2\theta_1 + \sin^2\theta_1 = 1$ and $\cos^2\theta_5 + \sin^2\theta_5 = 1$. Thus, in general, we must solve a pencil of conics. This is a classical problem which is well know to be solvable in terms of radicals and generally has four distinct solutions, see for example [4, chapter 16].

5. Having found θ_1 and θ_5, θ_6 is simple to find using the original equation $g^* a_1\pi a_1^* g = a_6^* a_5^* \pi a_5 a_6$. This system of equations gives essentially two linear equations in the variables $\cos\theta_6$ and $\sin\theta_6$. Hence, we obtain a unique solution for θ_6 given particular values for θ_1 and θ_5.

6. Next we must solve the second of our equations $a_2 a_3 a_4 = g'$, where $g' = a_1 g a_6^* a_5^*$. Notice however, that there are four possible values that g' can take corresponding to the four solutions for the angles θ_1, θ_5 and θ_6. The above is a relation between group elements, so we can eliminate a_4 by acting this group element on the 4th joint axis,

$$a_2 a_3 a_4 \ell_4 a_4^* a_3^* a_2^* = a_2 a_3 \ell_4 a_3^* a_2^* = g' \ell_4 g'^*$$

This is now a relation between lines, if the lines ℓ_2, ℓ_3 and ℓ_4 are intersecting then a_2 can be eliminated using the fact that for a pair of lines ℓ_α, ℓ_β the expression $\ell_\alpha \ell_\beta^* + \ell_\beta \ell_\alpha^*$ is an invariant, with respect to the group of rigid body motions. This leads us to the expression,

$$a_3 \ell_4 a_3^* \ell_2^* + \ell_2 a_3 \ell_4^* a_3^* = g' \ell_4 g'^* \ell_2^* + \ell_2 g' \ell_4^* g'^*$$

Again, this is linear in the variables, $\cos\theta_3$ and $\sin\theta_3$. We expect 2 equations here from the coefficients of 1 and $e_1 e_2 e_3 e$ but the coefficient of $e_1 e_2 e_3 e$ will disappear because the lines are intersecting. Solving the remaining linear equation with the trigonometric identity $\cos^2\theta_3 + \sin^2\theta_3 = 1$, gives two solutions.

On the other hand if the three lines are parallel we can eliminate a_2 using the expression, $(1/2)(\ell_\alpha \ell_\beta^* - \ell_\beta \ell_\alpha^*)$. Because the lines are parallel this will give $s_x e_1 e + s_y e_2 e + s_z e_3 e$ where $\mathbf{s} = (s_x, s_y, s_z)^T$ is a vector from one line to the other, perpendicular to both. The length of this vector $s^2 = s_x^2 + s_y^2 + s_z^2$ is then invariant under an overall rigid motion and will depend only on θ_3. In fact the expression we get will be the cosine rule for the triangle formed by the three lines meeting a perpendicular plane. Hence we obtain two solutions for θ_3 corresponding to the two possible signs for $\sin \theta_3$.

7. In either case we can now retrace our steps and solve,

$$a_2 a_3 \ell_4 a_3^* a_2^* = g' \ell_4 g'^*$$

to get a unique answer for θ_2.

8. Finally we use,

$$a_4 = a_3^* a_2^* g'$$

to recover θ_4.

Notice that we have shown that these robots a maximum of eight distinct solutions for their inverse kinematics. For a general 6R robot, where no three consecutive joint intersect or are parallel, it can be shown that the inverse kinematic problem has 16 solutions.

In the following two sections a pair of examples is given in order to make the procedure more concrete.

11.4.3 Example—the MA2000

This table-top robot arm was designed as a 'home-experiment kit' for an Open University course in robotics. As can be seen from figure 11.4, joints $2, 3$ and 4 are parallel.

We begin with a list of the joint axes in their home configuration,

$$
\begin{aligned}
\ell_1 &= e_1 e_2 \\
\ell_2 &= e_2 e_3 \\
\ell_3 &= e_2 e_3 + l_2 e_2 e \\
\ell_4 &= e_2 e_3 + (l_2 + l_3) e_2 e \\
\ell_5 &= e_1 e_2 - d_4 e_2 e \\
\ell_6 &= e_2 e_3 + (l_2 + l_3 + l_4) e_2 e
\end{aligned}
$$

Here the dimensions l_2, l_3, l_4 and d_4 are constants, sometimes called the design parameters of the robot. A plane perpendicular to joints $2, 3$ and 4 is given by $\pi = e_1$.

The first equation we have to solve is,

$$g^* a_1 \pi a_1^* g \ell_6^* + \ell_6 g^* a_1 \pi^* a_1^* g = a_5^* \pi a_5 \ell_6^* + \ell_6 a_5^* \pi^* a_5$$

Recall that $a_i = \cos(\theta_i/2) + \sin(\theta_i/2)\ell_i$, so after some computation, the right-hand side of the equation becomes,

$$a_5^* \pi a_5 \ell_6^* + \ell_6 a_5^* \pi^* a_5 = -2 \cos \theta_5 e_1 e_2 e_3 - 2 d_4 (\cos \theta_5 - 1) e_2 e_3 e$$

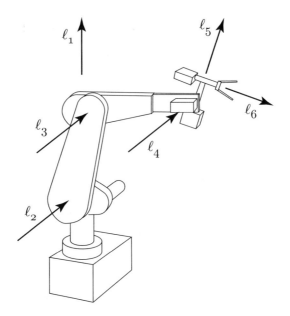

FIGURE 11.4. The telequipment MA200 robot.

$$- \quad 2(l_2 + l_3 + l_4) \cos \theta_5 e_1 e_2 e \qquad (4.2)$$

The left-hand side requires even more computation since we must include a general rigid motion g. Let us write this general motion as a rotation followed by a translation,

$$g = r + \frac{1}{2} tre$$

where

$$r = \cos \frac{\phi}{2} + v_x \sin \frac{\phi}{2} e_2 e_3 + v_y \sin \frac{\phi}{2} e_3 e_1 + v_x \sin \frac{\phi}{2} e_1 e_2$$

and

$$t = t_x e_1 + t_y e_2 + t_z e_3$$

see section 11.2.1. It is useful at this stage to write,

$$g^* a_1 \pi a_1^* g = N_x e_1 + N_y e_2 + N_z e_3 + De$$

where,

$$
\begin{aligned}
N_x &= (\cos \phi + v_x^2 (1 - \cos \phi)) \cos \theta_1 + (v_z \sin \phi + v_x v_y (1 - \cos \phi)) \sin \theta_1 \\
N_y &= (v_x v_y (1 - \cos \phi) - v_z \sin \phi) \cos \theta_1 + (\cos \phi + v_y^2 (1 - \cos \phi)) \sin \theta_1 \\
N_z &= (v_y \sin \phi + v_x v_z (1 - \cos \phi)) \cos \theta_1 + (v_y v_z (1 - \cos \phi) - v_x \sin \phi) \sin \theta_1
\end{aligned}
$$

and

$$D \;=\; N_x t_x + N_y t_y + N_z t_z$$

Now the left-hand side of the equation can be written as,

$$g^* a_1 \pi a_1^* g \ell_6^* + \ell_6 g^* a_1 \pi^* a_1^* g \;=\; -2N_x e_1 e_2 e_3 - 2(D - (l_2 + l_3 + l_4)N_z)e_2 e_3 e$$
$$- \; 2(l_2 + l_3 + l_4)N_x e_1 e_2 e \tag{4.3}$$

Comparing the coefficients of the basis elements gives us just two equations in the first and fifth joint angles,

$$N_x = \cos\theta_5, \qquad \text{and} \qquad D - (l_2 + l_3 + l_4)N_z = d_4(\cos\theta_5 - 1)$$

So we don't have to solve a pair of quadratic equations here we can eliminate $\cos\theta_5$ to get a linear equation in the sine and cosine of θ_1,

$$D - (l_2 + l_3 + l_4)N_z - d_4 N_x + d_4 = 0$$

Solving this with the trigonometric identity $\cos^2\theta_1 + \sin^2\theta_1 = 1$ gives two solutions in general,

$$\cos\theta_1 = \frac{-\alpha\gamma \pm \beta\sqrt{\alpha^2 + \beta^2 - \gamma^2}}{\alpha^2 + \beta^2} \qquad \text{with} \qquad \sin\theta_1 = -(\alpha\cos\theta_1 + \gamma)/\beta$$

The coefficients α, β and γ are functions only of the end-effector's position and orientation,

$$\begin{aligned}
\alpha \;=\; & \big((1 - v_x^2)(t_x - d_4) - v_x v_y t_y - v_x v_z (t_y - l_2 - l_3 - l_4)\big)\cos\phi + \\
& \big(v_y(t_z - l_2 - l_3 - l_4) - v_z t_y\big)\sin\phi + v_x\big(v_x(t_x - d_4) + \\
& v_y t_y - v_z(t_z - l_2 - l_3 - l_4)\big)
\end{aligned}$$

$$\begin{aligned}
\beta \;=\; & \big((1 - v_y^2)t_y - v_x v_y(t_x - d_4) - v_y v_z(t_z - l_2 - l_3 - l_4)\big)\cos\phi + \\
& \big(v_z(t_x - d_4) - v_x(t_z - l_2 - l_3 - l_4)\big)\sin\phi + v_y\big(v_x(t_x - d_4) + \\
& v_y t_y + v_z(t_z - l_2 - l_3 - l_4)\big)
\end{aligned}$$

$$\gamma \;=\; d_4$$

For each of the two solutions for θ_1 we get two solutions for θ_5 given by,

$$\cos\theta_5 = N_x \qquad \text{and} \qquad \sin\theta_5 = \pm\sqrt{1 - N_x^2}$$

To find θ_6 we solve the linear equations,

$$(g^* a_1 \pi a_1^* g) = a_6^*(a_5^* \pi a_5)a_6$$

That is,

$$(N_x e_1 + N_y e_2 + N_z e_3 + De) \;=\;$$

$$= a_6^*(\cos\theta_5 e_1 - \sin\theta_5 e_2 + d_4(\cos\theta_5 - 1)e)a_6$$
$$= \cos\theta_5 e_1 - \cos\theta_6\sin\theta_5 e_2 + \sin\theta_6\sin\theta_5 e_3$$
$$+ (d_4(\cos\theta_5 - 1) + \sin\theta_6\sin\theta_5(l_2 + l_3 + l_3))e \quad (4.4)$$

Comparing coefficients we have that,

$$\cos\theta_6 = N_y/\sin\theta_5, \quad \text{and} \quad \sin\theta_6 = N_z/\sin\theta_5$$

From the results above we can calculate $g' = a_1 g a_6^* a_5^*$, this must be an element of the sub-group generated by a_2, a_3 and a_4 which is the group of motions in the yz-plane. Hence we can write g' as a rotation about the x-axis followed by a translation in the yz-plane,

$$g' = \cos\frac{\phi'}{2} + \sin\frac{\phi'}{2}e_2 e_3 + \frac{1}{2}(t_y'\cos\frac{\phi'}{2} + t_z'\sin\frac{\phi'}{2})e_2 e +$$
$$+ \frac{1}{2}(t_z'\cos\frac{\phi'}{2} - t_y'\sin\frac{\phi'}{2})e_3 e$$

Now we must solve the second part of the problem

$$a_2 a_3 a_4 = g'$$

Since the lines ℓ_2, ℓ_3 and ℓ_4 are parallel, we eliminate a_2 and a_4 by computing,

$$\frac{1}{2}(a_3\ell_4 a_3^*\ell_2^* - \ell_2 a_3\ell_4^* a_3^*) =$$
$$= -l_3\sin\theta_3 e_3 e + (l_2 + l_3\cos\theta_3)e_3 e$$
$$= \frac{1}{2}(g'\ell_4 g'^*\ell_2^* - \ell_2 g'\ell_4^* g'^*) \quad (4.5)$$

so $s^2 = l_2^2 + l_3^2 + 2l_2 l_3\cos\theta_3$. From the right-hand side of the equation we have,

$$s^2 = (l_2 + l_3)^2 + 2(l_2 + l_3)(t_z'\cos\phi' - t_y'\sin\phi') + t_y'^2 + t_z'^2$$

The two solutions for θ_3 are thus,

$$\cos\theta_3 = (s^2 - l_2^2 - l_3^2)/2l_2 l_3, \quad \text{and} \quad \sin\theta_3 = \pm\sqrt{1 - \cos^2\theta_3}$$

Stepping back, we can find θ_2 from,

$$a_2 a_3\ell_4 a_3^* a_2^* = g'\ell_4 g'^*$$

The right-hand side can be written,

$$g'\ell_4 g'^* = e_2 e_3 + ((l_2 + l_3)\cos\phi' + t_z')e_2 e + ((l_2 + l_3)\sin\phi' - t_y')e_3 e$$

which we can abbreviate to $g'\ell_4 g'^* = e_2 e_3 + X e_2 e + Y e_3 e$, with

$$
\begin{aligned}
X &= (l_2 + l_3)\cos\phi' + t'_z \\
Y &= (l_2 + l_3)\sin\phi' - t'_y
\end{aligned}
$$

The left-hand side is,

$$
\begin{aligned}
a_2 a_3 \ell_4 a_3^* a_2^* &= e_2 e_3 + (\cos\theta_2(l_2 + l_3\cos\theta_3) - l_3\sin\theta_2\sin\theta_3)e_2 e + \\
&+ (\sin\theta_2(l_2 + l_3\cos\theta_3) + l_3\cos\theta_2\sin\theta_3)e_3 e \qquad (4.6)
\end{aligned}
$$

So we get a unique solution for θ_2,

$$
\begin{aligned}
\cos\theta_2 &= \big(X(l_2 + l_3\cos\theta_3) + Y l_3\sin\theta_3\big)/(l_2^2 + l_3^2 + 2l_2 l_3\cos\theta_3) \\
\sin\theta_2 &= \big(Y(l_2 + l_3\cos\theta_3) - X l_3\sin\theta_3\big)/(l_2^2 + l_3^2 + 2l_2 l_3\cos\theta_3)
\end{aligned}
$$

To finish the solution we must find θ_4, this can be found from,

$$
a_4 = a_3^* a_2^* g'
$$

By looking at the rotation part of this we get,

$$
\theta_4 = \phi' - \theta_2 - \theta_3
$$

11.4.4   Example—the Intelledex 660

The second example we look at is the Intelledex 660 robot, this is another small robot intended for use in laboratories, see figure 11.5. The design is unusual because the three consecutive intersecting joints are the first three rather than the last three as in a robot with a 3R wrist. Moreover, this robot also has three consecutive parallel joints, that is joints $3, 4$ and 5. This means that we can choose either method to solve the inverse kinematics, we choose to use the fact that the first three joints are intersecting since the other method has already be demonstrated. We will see in a moment that the fact that three of the joints are also parallel makes the problem even easier. As usual we begin by listing the joint axes in their home configuration,

$$
\begin{aligned}
\ell_1 &= e_1 e_2 \\
\ell_2 &= e_2 e_3 \\
\ell_3 &= e_3 e_1 \\
\ell_4 &= e_3 e_1 - l_4 e_1 e \\
\ell_5 &= e_3 e_1 - (l_4 + l_5)e_1 e \\
\ell_6 &= e_1 e_2 + d_5 e_1 e
\end{aligned}
$$

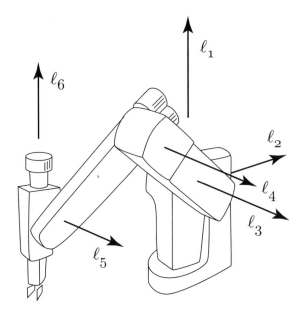

FIGURE 11.5. The intelledex 660 robot.

again l_4, l_5 and d_5 represent design parameters. The point fixed by the first three joints is the origin, $p = e_1e_2e_3$. So we have,

$$a_6 g^* p g a_6^* = a_5^* a_4^* p a_4 a_5$$

Next we eliminate θ_5 by finding the plane through p perpendicular to the fifth joint axis, see section 11.4.2 step 3. In fact, because the fourth and fifth joint axes are parallel the expression, $a_5^* a_4^* p a_4 a_5 l_5^* - l_5 a_5^* a_4^* p^* a_4 a_5$ will be independant of both θ_4 and θ_5. It is not too difficult to see that the expression will be proportional to e_2, the xz-plane. Calculation confirms that it is $2e_2$, so the equation reduces to,

$$a_6 g^* p g a_6^* l_5^* - l_5^* a_6 g^* p^* g a_6^* = 2e_2$$

As usual we write the general rigid motion as $g = r + (1/2)tre$ with $r = \cos(\phi/2) + v_x \sin(\phi/2)e_2e_3 + v_y \sin(\phi/2)e_3e_1 + v_z \sin(\phi/2)e_1e_2$ and $t = t_x e_1 + t_y e_2 + t_z e_3$. This group element transforms the position of the origin to,

$$
\begin{aligned}
g^* p g &= \left(r^* - \frac{1}{2}r^* te\right) e_1 e_2 e_3 \left(r + \frac{1}{2}tre\right) \\
&= e_1 e_2 e_3 + (r^* tr)e_1 e_2 e_3 e \\
&= e_1 e_2 e_3 - t'_x e_2 e_3 e - t'_y e_3 e_1 e - t'_z e_1 e_2 e
\end{aligned}
$$

where

$$t'_x = (v \cdot t)v_x + \left(t_x - (v \cdot t)v_x\right) \cos \phi + \left(t_y v_z - t_z v_y\right) \sin \phi$$

$$\begin{aligned}
t'_y &= (v \cdot t)v_y + \left(t_y - (v \cdot t)v_y\right)\cos\phi + \left(t_z v_x - t_x v_z\right)\sin\phi \\
t'_z &= (v \cdot t)v_z + \left(t_z - (v \cdot t)v_z\right)\cos\phi + \left(t_x v_y - t_y v_x\right)\sin\phi
\end{aligned}$$

This allows us to compute,

$$p' = a_6 g^* pg a_6^* = e_1 e_2 e_3 + X e_2 e_3 e + Y e_3 e_1 e + Z e_1 e_2 e$$

where

$$\begin{aligned}
X &= (d_5\sin\theta_6 + t'_y\sin\theta_6 - t'_x\cos\theta_6) \\
Y &= (d_5(1 - \cos\theta_6) - t'_y\cos\theta_6 - t'_x\sin\theta_6) \\
Z &= -t'_z
\end{aligned}$$

Hence,

$$a_6 g^* pg a_6^* \ell_5^* - \ell_5 a_6 g^* p^* g a_6^* = 2e_2 + 2(d_5(1 - \cos\theta_6) - t'_y\cos\theta_6 - t'_x\sin\theta_6)e$$

Comparing this with the previous calculation gives the linear equation $Y = 0$,

$$d_5(1 - \cos\theta_6) - t'_y\cos\theta_6 - t'_x\sin\theta_6 = 0$$

Solving this with the trigonometric identity $\cos^2\theta_6 + \sin^2\theta_6 = 1$ gives the two solutions,

$$\cos\theta_6 = \frac{d_5(d_5 + t'_y) \pm t'_x\sqrt{t'_x{}^2 + t'_y{}^2 + 2d_5 t'_y}}{(d_5^2 + t'_x{}^2 + t'_y{}^2 + 2d_5 t'_y)}$$

$$\sin\theta_6 = \frac{1}{t'_x}(d_5 - (d_5 + t'_y)\cos\theta_6)$$

Now, for each of the solutions for θ_6 we compute, $p' = a_6 g^* pg a_6^*$. So to find θ_4 and θ_5 we must solve,

$$a_4^* pa_4 = a_5 p' a_5^*$$

This is simply a two joint planar manipulator.

$$\begin{aligned}
e_1 e_2 e_3 + l_4\sin\theta_4 e_2 e_3 e + l_4(1 - \cos\theta_4)e_1 e_2 e &= \\
= e_1 e_2 e_3 + \left(X\cos\theta_5 + Z\sin\theta_5 - (l_4 + l_5)\sin\theta_5\right)e_2 e_3 e &+ \\
+ \left(Z\cos\theta_5 - X\sin\theta_5 + (l_4 + l_5)(1 - \cos\theta_5)\right)e_1 e_2 e
\end{aligned}$$

where we have used the fact that $Y = 0$. We now have two linear equations,

$$\begin{aligned}
l_4\sin\theta_4 &= X\cos\theta_5 + (Z - l_4 - l_5)\sin\theta_5 \\
-l_4\cos\theta_4 &= l_5 + (Z - l_4 - l_5)\cos\theta_5 - X\sin\theta_5
\end{aligned}$$

we can eliminate θ_4 from these equations by squaring and adding them. A little rearrangement using trigonometic identities gives the linear equation,

$$2l_5(Z - l_4 - l_5)\cos\theta_5 - 2l_5 X \sin\theta_5 = l_4^2 - l_5^2 - X^2 - (Z - l_4 - l_5)^2$$

Now we can use the standard solution for a linear equation with the trigonometric identity, $\cos^2\theta_5 + \sin^2\theta_5 = 1$,

$$\cos\theta_5 = \frac{AC \pm B\sqrt{C^2 - B^2 - A^2}}{A^2 + B^2}, \qquad \sin\theta_5 = \frac{1}{B}(C - A\cos\theta_5)$$

where,

$$
\begin{aligned}
A &= 2l_5(Z - l_4 - l_5) \\
B &= -2l_5 X \\
C &= l_4^2 - l_5^2 - X^2 - (Z - l_4 - l_5)^2
\end{aligned}
$$

Having found θ_5, we can find θ_4 immedieatly from the relations,

$$
\begin{aligned}
\cos\theta_4 &= \left(X\sin\theta_5 - (Z - l_4 - l_5)\cos\theta_5 - l_5\right)/l_4 \\
\sin\theta_4 &= \left(X\cos\theta_5 + (Z - l_4 - l_5)\sin\theta_5\right)/l_4
\end{aligned}
$$

The remaining three joint form a spherical mechanism. For each of the four possible solutions found above we compute $g' = ga_6^* a_5^* a_4^*$, and then the equations we must solve are,

$$a_1 a_2 a_3 = g'$$

In order that this equation can be solved we must have that g' is a rotation about the point p, hence we can write,

$$g' = \cos(\phi'/2) + v_x' \sin(\phi'/2)e_2 e_3 + v_y' \sin(\phi'/2)e_3 e_1 + v_z' \sin(\phi'/2)e_1 e_2$$

As mentioned above, we isolate θ_2 using the equation,

$$l_1 a_2 l_3^* a_2^* + a_2 l_3 a_2^* l_1^* = l_1 g' l_3^* g'^* + g' l_3 g'^* l_1^*$$

this gives,

$$\sin\theta_2 = v_x' \sin\phi' + v_y' v_z'(1 - \cos\phi')$$

and hence

$$\cos\theta_2 = \pm\sqrt{1 - \sin^2\theta_2}$$

To find θ_1 we use the equation,

$$a_1 a_2 l_3 a_2^* a_1^* = g' l_3 g'^*$$

expanding this gives two useful equations,

$$\sin\theta_1 \cos\theta_2 = v_z' \sin\phi' - v_x' v_y'(1 - \cos\phi')$$

$$\cos \theta_1 \cos \theta_2 \;\; = \;\; \cos \phi' + v_y'^2 (1 - \cos \phi')$$

In general, when $\cos \theta_2 \neq 0$, this will give a unique solution for θ_1.

Finally we must find θ_3, to do this go back to the equation, $a_1 a_2 a_3 = g'$ and recast it as,

$$a_3 = a_2^* a_1^* g'$$

Comparing coefficients as usual we get,

$$\cos \frac{\theta_3}{2} \;\; = \;\; \cos \frac{\phi'}{2} \cos \frac{\theta_1}{2} \cos \frac{\theta_2}{2} + v_x' \sin \frac{\phi'}{2} \cos \frac{\theta_1}{2} \sin \frac{\theta_2}{2} +$$
$$v_y' \sin \frac{\phi'}{2} \sin \frac{\theta_1}{2} \sin \frac{\theta_2}{2} + v_z' \sin \frac{\phi'}{2} \sin \frac{\theta_1}{2} \cos \frac{\theta_2}{2}$$

$$\sin \frac{\theta_3}{2} \;\; = \;\; - \cos \frac{\phi'}{2} \sin \frac{\theta_1}{2} \sin \frac{\theta_2}{2} - v_x' \sin \frac{\phi'}{2} \sin \frac{\theta_1}{2} \cos \frac{\theta_2}{2} +$$
$$v_y' \sin \frac{\phi'}{2} \cos \frac{\theta_1}{2} \cos \frac{\theta_2}{2} + v_z' \sin \frac{\phi'}{2} \cos \frac{\theta_1}{2} \sin \frac{\theta_2}{2}$$

¿From here it is an easy matter find θ_3.

11.5   Concluding Remarks

As mentioned at the beginning of section 11.2 the algebra presented here is very closely related to the original biquaternion algebra introduced by Clifford and used by Blaschke [1]. It is possible to represent points and lines by biquaternions. However, the action of the group of rigid transformations on points is different from that on lines. Hence, flags cannot be simple combinations of the linear elements. Moreover the relations for meets and joins will be more complicated, (for the above algebra, these relations are derived in [10]. So, despite the fact that this algebra has 16 basis elements rather than only 8 for biquaternions, it is much simpler to use for practical examples.

An algorithm has been outlined above which derives the inverse kinematic relations for any 6R robot with three consecutive intersecting or parallel joints. This method may not give the most efficient derivation but this is hardly relevant since for any robot the derivation will only be performed once. The computations in this work were done by hand but the Clifford algebra is ideally suited to automation using a symbolic algebra computer program. For example, it would be a simple matter to write a *Mathematica* notebook to check the results. The solutions given above introduce several sets of intermediate variables. This reduces the size of the equations and so makes the problem tractable for hand calculation. From the results given above it would not be too difficult to write numerical programs to find the inverse kinematics for these machines, values would have to be assigned to the design parameters of course. Care should be taken, however, because in the above we have not addressed the problem of singularities. This is not too difficult for these examples since we only need to look for divisors vanishing or the discriminant of a quadratic dissappearing.

Although in general we have to solve a pencil of conics, in the examples given above, the problem was in fact simpler in the end, we only had to solve quadratic equations. The 'special geometry' for which these simplifications happen has been extensively studied by Smith [11].

Can the methods outlined above be used for other types of robots? An obvious application would be to robots containing prismatic joints. From the above it is reasonably clear what to do, we must look for sets of consecutive joints which form sub-groups of the group of rigid body motions. If these sub-groups fix a point, a line or a plane then we can eliminate the corresponding joint angles from the kinematic equation an hence simplify it.

Suppose we have a robot with no three consecutive joints parallel or intersecting, it may still be possible to simplify the kinematics and produce a 'semi-analytic' solution. Unfortunately there are good reasons why such designs do not make good practical robots.

Finally, the algebra and methods described above are directly applicable to the theory of mechanisms, in particular spatial mechanisms. Like robots, mechanisms consist of links and joints. However mechanisms usually contain kinematic loops, that is, a ring of links each connected to their neighbours by a joint. We could characterise the joints by flags as above and then write down the relationship between pairs of joints at either end of a link using the Clifford algebra. This would be a simple way to find the algebraic expressions characterising the configuration space of the mechanism. We could also consider fixing a point, line or more generally a flag in the coupler link and then set up equations to determine whether or not this flag could be brought into coincidence with a similar flag fixed in space.

Chapter 12

The Clifford Algebra and the Optimization of Robot Design

Shawn G. Ahlers and John Michael McCarthy

12.1 Introduction

The goal of this chapter is a computer aided design environment that assists the inventor to formulate a task and evaluate candidate devices. The task trajectory of a robot is specified as a set of homogeneous transforms that define key frames for a desired end-effector trajectory. These key frames are converted to double quaternions and interpolated by generalizing well known techniques for Bezier interpolation of quaternions. The result is an efficient interpolation algorithm.

Our focus here is the design of a five degree of freedom TS robot that reaches the given task trajectory. The TS robot is constructed by connecting a pair of revolute joints perpendicular to each other as the base pivot to a spherical (S) joint by a fixed distance, see Figure 12.1. The pair of revolute joints is also known as a gimbal (T) or universal joint. The set of reachable positions and orientations of this device is its workspace which may not include the entire specified trajectory. Our goal is to find the TS robot minimizes the local error between its workspace and this task trajectory.

12.2 Literature Review

Bezier interpolation is used in computer drawing systems to generate curves through specified points (Farin [5]). Shoemake [13] shows that this technique can be used to interpolate rotation key frames specified by quaternion coordinates (Hamilton [9]); the result is an efficient animation algorithm. Ge and Ravani [8] generalize Shoemake's results to spatial displacements using double quaternions (Clifford [1]). These results were refined to ensure smooth transitions at each key frame by Ge and Kang [7].

In this chapter, we apply the results of Ge and Kang to double quaternion interpolation. Etzel and McCarthy [4] show how spatial displacements can be transformed to 4×4 rotations in E^4, and then to double quaternions. A

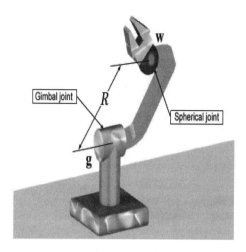

FIGURE 12.1. TS robot.

benefit of this approach is that the interpolation algorithm can be applied to the quaternion components separately.

The robot design problem seeks the dimensions of the device that satisfy geometric constraints (Suh and Radcliffe [14]). The structure of the TS robot requires the wrist **w** to lie on a sphere about the fixed gimbal joint **g**. Innocenti [10] presents a design algorithm that yields as many as 20 TS chains that reach seven arbitrary positions. Our goal is to find the TS robot that fits our end-effector trajectory with arbitrarily many positions.

12.3   Overview of the Design Algorithm

The design algorithm begins with the specification of the task. The task is defined by the N+1 user-specified key frames. These key frames are converted from their representation as homogeneous transforms to double quaternions. These double quaternions are interpolated to define the task trajectory of a desired robot. To compute a TS robot, the frames of this task trajectory are converted back to their homogeneous transforms. By using four position synthesis, the parameters of a TS robot are computed from four frames of the task trajectory. The synthesis procedure is repeated for all combinations of four frames of the task trajectory. The optimization procedure begins by calculating the closest positions and orientations reachable by a designed TS robot to the remaining frames of the trajectory. These new reachable frames are converted to double quaternions. The local error between a frame from task trajectory and the reachable frame is calculated as the magnitude of the difference of these double quaternions. This local error is summed for each frame on the task trajectory and divided by the

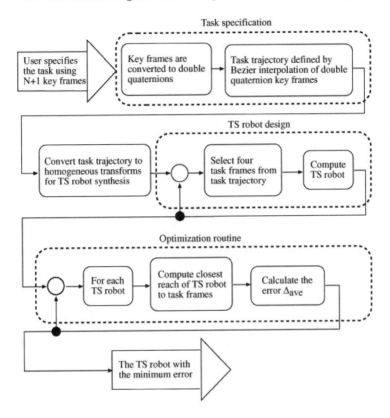

FIGURE 12.2. TS robot design flowchart.

total number of frames to obtain the error. The optimization procedure is repeated for each TS robot obtained from the synthesis procedure. The TS robot with the minimum error is the optimum fit of the robot to the task trajectory, see Figure 12.2. If the optimum TS robot is not satisfactory, the user may alter the key frames and the design process is repeated to obtain another robot candidate.

12.4   Double Quaternions

12.4.1   Homogeneous transforms

The transformation equation for a spatial displacement is not a linear transformation. A spatial displacement consists of a 3×3 rotation matrix and a 3×1 displacement vector. A standard strategy to adjust for this inhomogeneity is to add a fourth component to our position vectors that will always equal 1, then we can introduce the 4×4 *homogeneous transform*

$$\left\{ \begin{array}{c} \mathbf{y} \\ 1 \end{array} \right\} = \left[\begin{array}{cc} A & \mathbf{d} \\ 0\,0\,0 & 1 \end{array} \right] \left\{ \begin{array}{c} \mathbf{x} \\ 1 \end{array} \right\} \qquad (4.1)$$

which we write as

$$\mathbf{y} = [H]\mathbf{x}. \qquad (4.2)$$

The rotation-translation pairs $[H] = [A, \mathbf{d}]$ represent all the spatial positions of M relative to F, known as the special Euclidean group, $SE(3)$.

12.4.2   The Clifford algebra on E^4

A set of hypercomplex numbers called *double quaternions* may be obtained from the even Clifford Algebra of four dimensional Euclidean space E^4. Let $\mathbf{e}_i, i = 1, \ldots, 4$ be the natural coordinate vectors of E^4, then we can construct the multilinear algebra of points in E^4. Introduce the Clifford product

$$\mathbf{e}_i \mathbf{e}_j + \mathbf{e}_j \mathbf{e}_i = -2\mathbf{e}_i \cdot \mathbf{e}_j, \qquad (4.3)$$

where the dot denotes the usual Euclidean scalar product. The even subalgebra $C^+(E^4)$ is of the rank 8, and a typical element can be written as

$$\tilde{\mathbf{Q}} = \mathbf{G} + \omega \mathbf{H}, \qquad (4.4)$$

where $\mathbf{G}$ and $\mathbf{H}$ are Hamilton's quaternions and $\omega = \mathbf{e}_1 \mathbf{e}_2 \mathbf{e}_3 \mathbf{e}_4$ satisfies the identity $\omega^2 = 1$.

Clifford shows that depending on the definition of the scalar product in equation (4.3) we can also obtain dual quaternions, $\omega^2 = 0$, and complex quaternions, $\omega^2 = -1$. Using the double quaternion algebra, $\omega^2 = 1$, we now introduce the symbols $\xi = (1 - \omega)/2$ and $\eta = (1 + \omega)/2$, and construct the double quaternion

$$\tilde{\mathbf{Q}} = (\mathbf{G} - \mathbf{H})\xi + (\mathbf{G} + \mathbf{H})\eta \qquad (4.5)$$

Notice that $\xi^2 = \xi$, $\eta^2 = \eta$, and $\xi\eta = 0$. These identities provide a complete separation of the operations on the quaternions $(\mathbf{G} - \mathbf{H})$ and $(\mathbf{G} + \mathbf{H})$. For example, for any two double quaternions $\tilde{\mathbf{P}} = \mathbf{P}_1 \xi + \mathbf{P}_2 \eta$ and $\tilde{\mathbf{R}} = \mathbf{R}_1 \xi + \mathbf{R}_2 \eta$, we have

$$\tilde{\mathbf{P}}\tilde{\mathbf{R}} = \mathbf{P}_1 \mathbf{R}_1 \xi + \mathbf{P}_2 \mathbf{R}_2 \eta. \qquad (4.6)$$

Since operations on the quaternions may be done independently, the interpolation technique defined for a single quaternion may be utilized for the individual quaternions of the double quaternions. As we will show in a later section, this will allow us to interpolate the quaternions independently.

12.4.3   *Homogeneous transformations as a rotations in E^4*

The general 4×4 homogeneous transform for a spatial displacement can be written as

$$[H] = [A, \mathbf{d}] = \begin{bmatrix} c\theta c\psi - s\theta s\phi s\psi & -c\theta s\phi - s\theta s\phi c\psi & s\theta c\phi & d_x \\ c\phi s\psi & c\phi c\psi & s\phi & d_y \\ -s\theta c\psi - c\theta s\phi s\psi & s\theta s\psi - c\theta s\phi c\psi & c\theta c\phi & d_z \\ 0 & 0 & 0 & 1 \end{bmatrix}, \quad (4.7)$$

where the angles θ, ϕ, and ψ are the longitude, latitude, and roll angles defining the orientation of the displaced frame, respectively, and c and s represent the cosine and sine functions.

Now define the angles α, β, and γ be defined such that

$$\alpha = \frac{d_x}{R}, \quad \beta = \frac{d_y}{R}, \quad \text{and} \quad \gamma = \frac{d_z}{R}, \quad (4.8)$$

where R is the radius of the hypersphere to which the translational elements are computed. We can compute the 4×4 rotation matrix $[J]$ composed of successive rotations of α in the W-X plane, β in the W-Y plane, and γ in the W-Z plane to obtain

$$[J] = \begin{bmatrix} c\alpha & 0 & 0 & s\alpha \\ -s\beta s\alpha & c\beta & 0 & s\beta c\alpha \\ -s\gamma c\beta s\alpha & -s\gamma s\beta & c\gamma & s\gamma c\beta c\alpha \\ -c\gamma c\beta s\alpha & -s\beta c\gamma & -s\gamma & c\gamma c\beta c\alpha \end{bmatrix}. \quad (4.9)$$

If we let $A(\theta, \phi, \psi)$ be the upper left 3×3 submatrix of the 4×4 matrix $[K]$ and keep a 1 in the fourth diagonal location, we may express a general rotation in four dimensional space, E^4, as the product of two 4×4 rotation matrices $[D]=[J(\alpha,\beta,\gamma)][K(\theta,\phi,\psi)]$. Explicitly written

$$= \begin{bmatrix} c\alpha & 0 & 0 & s\alpha \\ -s\beta s\alpha & c\beta & 0 & s\beta c\alpha \\ -s\gamma c\beta s\alpha & -s\gamma s\beta & c\gamma & s\gamma c\beta c\alpha \\ -c\gamma c\beta s\alpha & -s\beta c\gamma & -s\gamma & c\gamma c\beta c\alpha \end{bmatrix}$$

$$\begin{bmatrix} c\theta c\psi - s\theta s\phi s\psi & -c\theta s\phi - s\theta s\phi c\psi & s\theta c\phi & 0 \\ c\phi s\psi & c\phi c\psi & s\phi & 0 \\ -s\theta c\psi - c\theta s\phi s\psi & s\theta s\psi - c\theta s\phi c\psi & c\theta c\phi & 0 \\ 0 & 0 & 0 & 1 \end{bmatrix}.$$

If we assume that the angles α, β, and γ are small, $\cos\alpha = \cos\beta = \cos\gamma = 1$, and

$$\sin\alpha = \frac{d_x}{R}, \quad \sin\beta = \frac{d_y}{R}, \quad \text{and} \quad \sin\gamma = \frac{d_z}{R}. \quad (4.11)$$

Then the 4×4 rotation matrix becomes

$$
= \begin{bmatrix} 1 & 0 & 0 & \frac{d_x}{R} \\ 0 & 1 & 0 & \frac{d_y}{R} \\ 0 & 0 & 1 & \frac{d_z}{R} \\ -\frac{d_x}{R} & -\frac{d_y}{R} & -\frac{d_z}{R} & 1 \end{bmatrix}
$$

$$
\begin{bmatrix} c\theta c\psi - s\theta s\phi s\psi & -c\theta s\phi - s\theta s\phi c\psi & s\theta c\phi & 0 \\ c\phi s\psi & c\phi c\psi & s\phi & 0 \\ -s\theta c\psi - c\theta s\phi s\psi & s\theta s\psi - c\theta s\phi c\psi & c\theta c\phi & 0 \\ 0 & 0 & 0 & 1 \end{bmatrix}.
$$

If we shift the coordinate frame to $W = R$ to cancel the $1/R$ terms, the result is an approximation to a spatial displacement of order $O(1/R^2)$.

The parameter R is identified by specifying a maximum length L for the problem, then the error of this approximation is $\varepsilon \leq (L/R)^2$. Specify ε and solve for R in order to define the rotation in E^4 that approximates a given spatial displacement.

12.4.4   Double quaternion for a spatial displacement

In this subsection, we will reformulate $[H]$ in terms of double quaternions. After we have converted a spatial displacement $[H] = [A, \mathbf{d}]$ to a 4×4 rotation $[D]$, we may use Cayley's formula (Bottema and Roth, 1979), to obtain the skew symmetric matrix

$$
[B] = [D - I][D + I]^{-1} = \begin{bmatrix} 0 & -u_3 & u_2 & v_1 \\ u_3 & 0 & -u_1 & v_2 \\ -u_2 & u_1 & 0 & v_3 \\ -v_1 & -v_2 & -v_3 & 0 \end{bmatrix}. \tag{4.13}
$$

We now define the matrix $[B']$ by interchanging the u_i and v_i terms, in order to obtain the matrices (Etzel and McCarthy, [4])

$$
k_1[S] = \frac{[B] + [B']}{2} = k_1 \begin{bmatrix} 0 & -s_3 & s_2 & s_1 \\ s_3 & 0 & -s_1 & s_2 \\ -s_2 & s_1 & 0 & s_3 \\ -s_1 & -s_2 & -s_3 & 0 \end{bmatrix} \tag{4.14}
$$

and

$$
k_2[T] = \frac{[B] - [B']}{2} = k_2 \begin{bmatrix} 0 & -t_3 & t_2 & -t_1 \\ t_3 & 0 & -t_1 & -t_2 \\ -t_2 & t_1 & 0 & -t_3 \\ t_1 & t_2 & t_3 & 0 \end{bmatrix} \tag{4.15}
$$

where $[B] = k_1[S] + k_2[T]$ and $\sum s_i^2 = \sum t_i^2 = 1$. We can then compute μ and ν by the equations

$$
\begin{aligned} \mu &= \arctan(k_1 + k_2) + \arctan(k_1 - k_2) \\ \nu &= \arctan(k_1 + k_2) - \arctan(k_1 - k_2). \end{aligned} \tag{4.16}
$$

The double quaternion is now given by $\tilde{\mathbf{G}} = \mathbf{G}_1\xi + \mathbf{G}_2\eta$, where

$$\mathbf{G}_1 = \left\{ \begin{array}{c} \sin\mu s \\ \cos\mu \end{array} \right\} \quad \text{and} \quad \mathbf{G}_2 = \left\{ \begin{array}{c} \sin\nu \mathbf{t} \\ \cos\nu \end{array} \right\}. \tag{4.17}$$

where $\mathbf{s} = (s_1, s_2, s_3)^T$ and $\mathbf{t} = (t_1, t_2, t_3)^T$ define the axes to which the angles μ and ν are to be rotated, respectively. Again, notice that each 4×4 rotation matrix defines a double quaternion, that can be separated into a pair of quaternions that multiply separately.

12.4.5 Spatial displacement from a double quaternion

Assuming we have a double quaternion $\tilde{\mathbf{G}}$ of the form of equation (4.17), we compute the associated spatial displacement as follows. Note that a 4×4 rotation matrix can be written in exponential form

$$[D] = e^{[M]} \tag{4.18}$$

where $[M]$ is a 4×4 skew symmetric matrix. The matrix $[M]$ has the form $[M] = \mu[S] + \nu[T]$ (Ge, [6]). Thus

$$[D] = e^{(\mu[S] + \nu[T])} = e^{\mu[S]}e^{\nu[T]}. \tag{4.19}$$

The series expansion of $e^{\mu[S]}$ and $e^{\nu[T]}$ and the identities $[S]^2 = [T]^2 = -[I]$ yield the formulas

$$e^{\mu[S]} = \begin{bmatrix} \cos\mu & -s_3\sin\mu & s_2\sin\mu & s_1\sin\mu \\ s_3\sin\mu & \cos\mu & -s_1\sin\mu & s_2\sin\mu \\ -s_2\sin\mu & s_1\sin\mu & \cos\mu & s_3\sin\mu \\ -s_1\sin\mu & -s_2\sin\mu & -s_3\sin\mu & \cos\mu \end{bmatrix}, \tag{4.20}$$

and

$$e^{\nu[T]} = \begin{bmatrix} \cos\nu & -t_3\sin\nu & t_2\sin\nu & -t_1\sin\nu \\ t_3\sin\nu & \cos\nu & -t_1\sin\nu & -t_2\sin\nu \\ -t_2\sin\nu & t_1\sin\nu & \cos\nu & -t_3\sin\nu \\ t_1\sin\nu & t_2\sin\nu & t_3\sin\nu & \cos\nu \end{bmatrix}. \tag{4.21}$$

The result is the 4×4 rotation matrix $[D]$ defined by $\tilde{\mathbf{G}} = \mathbf{G}_1\xi + \mathbf{G}_2\eta$.

The 4×4 homogeneous transform approximating the rotation is $[H] = [A, \mathbf{d}]$, where A is the upper left 3×3 rotation matrix and the translation vector $\mathbf{d} = (d_x, d_y, d_z)$ is given by

$$d_x = d_{14}R, \quad d_y = d_{24}R, \quad d_z = d_{34}R, \tag{4.22}$$

where d_{ij} is the ijth element of the $[D]$ matrix. The longitude, latitude, and roll angles are

$$\begin{array}{l} \theta = \arctan(d_{13}/d_{33}), \\ \phi = \arctan(d_{23}\cos\theta/d_{33}) = \arctan(d_{23}\sin\theta/d_{13}), \\ \psi = \arctan(d_{21}/d_{22}). \end{array} \tag{4.23}$$

Thus, a homogeneous transformation may be computed from a double quaternion.

12.5   The Task Trajectory

The task of the robot is defined in terms of the trajectory of the end-effector. In order to define this task, we specify a set of key frames and use Bezier interpolation to generate the trajectory. Consider $N+1$ key frames defined by the homogenous transforms $[H_k], k = 0, \ldots, N$. The double quaternions associated with each key frame are denoted as $\tilde{\mathbf{P}}_k = \mathbf{P}_{k,1}\xi + \mathbf{P}_{k,2}\eta$. We use the Bezier interpolation of double quaternions developed by Ge and Kang [7] to create the task trajectory.

Bezier interpolation for double quaternions follows the principles of Bezier interpolation for curves, see Farin [5]. There are two main features the generation of a curve segment between two key frames using the deCasteljau [2, 3] algorithm, and the joining of these segments together to maintain G^1 and G^2 continuity. These continuity conditions ensure a smooth movement of the body along the trajectory.

12.5.1   The DeCasteljau algorithm

In order to generate a trajectory segment between the two key frames $\tilde{\mathbf{P}}_i$ and $\tilde{\mathbf{P}}_{i+1}$, we need a Bezier polygon $\tilde{\mathbf{B}}_{3i}, \tilde{\mathbf{B}}_{3i+1}, \tilde{\mathbf{B}}_{3i+2}$, and $\tilde{\mathbf{B}}_{3i+3}$. The first and last double quaternions of the Bezier polygon are identified with the two key frames,

$$\tilde{\mathbf{B}}_{3i} = \tilde{\mathbf{P}}_i \quad \text{and} \quad \tilde{\mathbf{B}}_{3i+3} = \tilde{\mathbf{P}}_{i+1}. \tag{5.24}$$

The intermediate Bezier double quaternions $\tilde{\mathbf{B}}_{3i+1}$ and $\tilde{\mathbf{B}}_{3i+2}$ are calculated to provide the desired continuity conditions when the complete trajectory is assembled. We show in the next section how this is done.

Here we show how the DeCasteljau algorithm is used to generate positions along the trajectory between two key frames for a given Bezier polygon. The central feature of the algorithm is an interpolation formula between two double quaternions $\tilde{\mathbf{P}}_i$ and $\tilde{\mathbf{P}}_{i+1}$, which is a generalization of Shoemake's [13] original results. Let $\tilde{\mathbf{V}} = \xi\mathbf{V}_1 + \eta\mathbf{V}_2$ and $\tilde{\mathbf{W}} = \xi\mathbf{W}_1 + \eta\mathbf{W}_2$ be two unit double quaternions, then the great circular arc $\tilde{\mathbf{L}}(t)$ between them is defined by the formula

$$\tilde{\mathbf{L}}(t) = \frac{\sin(1-t)\tilde{\rho}}{\sin \tilde{\rho}}\tilde{\mathbf{V}} + \frac{\sin t\tilde{\rho}}{\sin \tilde{\rho}}\tilde{\mathbf{W}} \tag{5.25}$$

where $\cos \tilde{\rho} = \tilde{\mathbf{V}} \cdot \tilde{\mathbf{W}}$.

Expanding the double quaternions in this equation, we obtain

$$\tilde{\mathbf{L}}(t) = \xi\left(\frac{\sin(1-t)\rho_1}{\sin\rho_1}\mathbf{V}_1 + \frac{\sin t\rho_1}{\sin\rho_1}\mathbf{W}_1\right) + \eta\left(\frac{\sin(1-t)\rho_2}{\sin\rho_2}\mathbf{V}_2 + \frac{\sin t\rho_2}{\sin\rho_2}\mathbf{W}_2\right).$$
$$(5.26)$$

Notice that this equation separates to define the interpolation of the quaternion components of $\tilde{\mathbf{V}}$ and $\tilde{\mathbf{W}}$, individually. Thus our formalism simply requires us to apply Shoemake's interpolation formula twice. In fact, all of our algorithms handle the components of the double quaternions independently.

For a particular value of the parameter t we now seek the double quaternion $\tilde{\mathbf{D}}(t)$ along the Bezier curve segment. The DeCasteljau algorithm uses equation (5.25) to generate circular arcs between each of the Bezier double quaternions $\tilde{\mathbf{B}}_{3i}, \tilde{\mathbf{B}}_{3i+1}, \tilde{\mathbf{B}}_{3i+2}$, and $\tilde{\mathbf{B}}_{3i+3}$ associated with this ith segment. To do this we first compute the double quaternions $\tilde{\mathbf{X}}_0, \tilde{\mathbf{X}}_1$, and $\tilde{\mathbf{X}}_2$ on each of these arcs by the formula

$$\tilde{\mathbf{X}}_m(t) = \frac{\sin(1-t)\tilde{\rho}_m}{\sin\tilde{\rho}_m}\tilde{\mathbf{B}}_{3i+m} + \frac{\sin t\tilde{\rho}_m}{\sin\tilde{\rho}_m}\tilde{\mathbf{B}}_{3i+(m+1)}, m = 0,1,2. \quad (5.27)$$

where m denotes the arc connecting the Bezier double quaternions $\tilde{\mathbf{B}}_{3i+m}$ and $\tilde{\mathbf{B}}_{3i+m+1}$. Next repeat this process in order to define the double quaternions $\tilde{\mathbf{Y}}_0$ and $\tilde{\mathbf{Y}}_1$ on the arcs joining $\tilde{\mathbf{X}}_0, \tilde{\mathbf{X}}_1$ and $\tilde{\mathbf{X}}_1, \tilde{\mathbf{X}}_2$, defined by the formula

$$\tilde{\mathbf{Y}}_n(t) = \frac{\sin(1-t)\tilde{\sigma}_n}{\sin\tilde{\sigma}_n}\tilde{\mathbf{X}}_n + \frac{\sin t\tilde{\sigma}_n}{\sin\tilde{\sigma}_n}\tilde{\mathbf{X}}_{n+1}, n = 0,1. \quad (5.28)$$

Finally, we obtain the frame $\tilde{\mathbf{D}}(t)$ as

$$\tilde{\mathbf{D}}_i(t) = \frac{\sin(1-t)\tilde{\tau}}{\sin\tilde{\tau}}\tilde{\mathbf{Y}}_0 + \frac{\sin t\tilde{\tau}}{\sin\tilde{\tau}}\tilde{\mathbf{Y}}_1. \quad (5.29)$$

As the parameter t varies from 0 to 1, $\tilde{\mathbf{D}}(t)$ will define the trajectory from $\tilde{\mathbf{P}}_i$ to $\tilde{\mathbf{P}}_{i+1}$. This procedure can be generalized for Bezier polygons with more intermediate vertices, see Farin [5].

12.5.2   Bezier interpolation

To define the entire task trajectory, we must compute the Bezier polygon for each of the N segments. The intermediate Bezier double quaternions, $\tilde{\mathbf{B}}_{3i+1}$ and $\tilde{\mathbf{B}}_{3i+2}$ are determined to ensure continuity at each junction.

To ensure G^1 continuity, $\tilde{\mathbf{B}}_{3i-1}$ and $\tilde{\mathbf{B}}_{3i+1}$ and the key frame $\tilde{\mathbf{P}}_i$ must lie the same arc, see Figure 12.3. Therefore, $\tilde{\mathbf{P}}_i$ is related to $\tilde{\mathbf{B}}_{3i+1}$ and $\tilde{\mathbf{B}}_{3i-1}$ by

$$\sin\tilde{\phi}_i\tilde{\mathbf{P}}_i = \sin\tilde{v}_i\tilde{\phi}_i\tilde{\mathbf{B}}_{3i-1} + \sin\tilde{u}_i\tilde{\phi}_i\tilde{\mathbf{B}}_{3i+1}. \quad (5.30)$$

The parameter $\tilde{\phi}_i = \arccos(\tilde{\mathbf{B}}_{3i-1}\cdot\tilde{\mathbf{B}}_{3i+1})$ is the double arclength between $\tilde{\mathbf{B}}_{3i-1}$ and $\tilde{\mathbf{B}}_{3i+1}$. The parameters $\tilde{u}_i$ and $\tilde{v}_i$ locate $\tilde{\mathbf{P}}_i$ on this arc, such

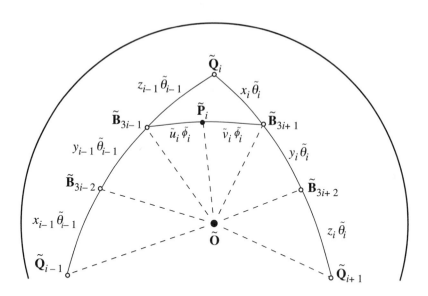

FIGURE 12.3. Construction of Bezier double quaternions.

that $\tilde{u}_i\tilde{\phi}_i = \arccos(\tilde{\mathbf{B}}_{3i-1} \cdot \tilde{\mathbf{P}}_i), \tilde{v}_i\tilde{\phi}_i = \arccos(\tilde{\mathbf{P}}_i \cdot \tilde{\mathbf{B}}_{3i+1})$. These parameters satisfy the relation

$$\tilde{u}_i + \tilde{v}_i = 1. \tag{5.31}$$

To ensure G^2 continuity, the five double quaternions $\tilde{\mathbf{B}}_{3i-2}$, $\tilde{\mathbf{B}}_{3i-1}$, $\tilde{\mathbf{P}}_i$, $\tilde{\mathbf{B}}_{3i+1}$, and $\tilde{\mathbf{B}}_{3i+2}$ must lie on the same great sphere (see Ge and Kang [7]). To do this, we introduce the control double quaternion $\tilde{\mathbf{Q}}_i$ that is defined to be the intersection of the arcs through by $\tilde{\mathbf{B}}_{3i-2}$, $\tilde{\mathbf{B}}_{3i-1}$, and $\tilde{\mathbf{B}}_{3i+1}$, $\tilde{\mathbf{B}}_{3i+1}$. These Bezier double quaternions lie on arcs through $\tilde{\mathbf{Q}}_{i-1}$ and $\tilde{\mathbf{Q}}_i$, and $\tilde{\mathbf{Q}}_i$ and $\tilde{\mathbf{Q}}_{i+1}$. They are located by the parameters x_i, y_i, z_i, so that

$$\sin\tilde{\theta}_{i-1}\tilde{\mathbf{B}}_{3i-1} = \sin z_{i-1}\tilde{\theta}_{i-1}\tilde{\mathbf{Q}}_{i-1} + \sin(x_{i-1} + y_{i-1})\tilde{\theta}_{i-1}\tilde{\mathbf{Q}}_i, \\ \text{for} \quad i = 2, \ldots, N-1, \tag{5.32}$$

and

$$\sin\tilde{\theta}_i\tilde{\mathbf{B}}_{3i+1} = \sin(y_i + z_i)\tilde{\theta}_i\tilde{\mathbf{Q}}_i + \sin x_i\tilde{\theta}_i\tilde{\mathbf{Q}}_{i+1}, \\ \text{for} \quad i = 1, \ldots, N-2, \tag{5.33}$$

where the angle $\tilde{\theta}_i = \arccos(\tilde{\mathbf{Q}}_i \cdot \tilde{\mathbf{Q}}_{i+1})$. Note, the x_i, y_i, z_i are greater than zero and satisfy the constraint

$$x_i + y_i + z_i = 1 \quad \text{for} \quad 1 \leq i \leq N-2. \tag{5.34}$$

At the endpoints of the trajectory $x_0 = 0$ and $z_{N-1} = 0$.

To complete the condition for G^2 continuity, we require the parameters $\tilde{u}_i$ and $\tilde{v}_i$ to satisfy the constraint derived in [7],

$$\frac{\tilde{v}_i \sin \tilde{v}_i \tilde{\phi}_i}{\tilde{u}_i \sin \tilde{u}_i \tilde{\phi}_i} = \frac{y_{i-1} \tilde{\theta}_{i-1} \sin x_i \tilde{\theta}_i}{y_i \tilde{\theta}_i \sin z_{i-1} \tilde{\theta}_{i-1}}. \tag{5.35}$$

This equation together with equation (5.31) can be solved to determine the parameters $\tilde{u}_i$ and $\tilde{v}_i$ in terms of the angles $\tilde{\theta}_i$ and $\tilde{\phi}_i$. These angles are computed from the locations of $\tilde{\mathbf{Q}}_i$ and $\tilde{\mathbf{B}}_{3i\pm1}$. The control double quaternions $\tilde{\mathbf{Q}}_i$ must be found to fit the user-specified $\tilde{\mathbf{P}}_i$. The parameters $\tilde{u}_i$ and $\tilde{v}_i$ are determined by solving equations (5.35) and (5.31) numerically.

Substitute equations (5.32) and (5.33) into (5.30) to define the key frame double quaternions $\tilde{\mathbf{P}}_i$ directly in terms of the control quaternions $\tilde{\mathbf{Q}}_i$

$$\tilde{\mathbf{P}}_i = a_i \tilde{\mathbf{Q}}_{i-1} + b_i \tilde{\mathbf{Q}}_i + c_i \tilde{\mathbf{Q}}_{i+1} \tag{5.36}$$

where

$$\tilde{a}_i = \frac{\sin z_{i-1} \tilde{\theta}_{i-1} \sin \tilde{v}_i \tilde{\phi}_i}{\sin \tilde{\theta}_{i-1} \sin \tilde{\phi}_i},$$

$$\tilde{b}_i = \frac{\sin(x_{i-1} + y_{i-1}) \tilde{\theta}_{i-1} \sin \tilde{v}_i \tilde{\phi}_i}{\sin \tilde{\theta}_{i-1} \sin \tilde{\phi}_i} + \frac{\sin(y_i + z_i) \tilde{\theta}_i \sin \tilde{u}_i \tilde{\phi}_i}{\sin \tilde{\theta}_i \sin \tilde{\phi}_i},$$

$$\tilde{c}_i = \frac{\sin x_i \tilde{\theta}_i \sin \tilde{u}_i \tilde{\phi}_i}{\sin \tilde{\theta}_i \sin \tilde{\phi}_i}.$$

This relationship can be written in matrix form as

$$\begin{bmatrix} \tilde{a}_1, & \tilde{b}_1, & \tilde{c}_1, & 0, & \cdots & 0, & 0, & 0 \\ 0, & \tilde{a}_2, & \tilde{b}_2, & \tilde{c}_2, & \cdots & 0, & 0, & 0 \\ \vdots & \vdots & \vdots & \vdots & & \vdots & \vdots & \vdots \\ 0, & 0, & 0, & 0, & \cdots & \tilde{a}_{N-1}, & \tilde{b}_{N-1}, & \tilde{c}_{N-1} \end{bmatrix} \left\{ \begin{array}{c} \mathbf{Q}_1 \\ \vdots \\ \mathbf{Q}_{N-1} \end{array} \right\} =$$

$$= \left\{ \begin{array}{c} \mathbf{P}_1 \\ \vdots \\ \mathbf{P}_{N-1} \end{array} \right\}, \tag{5.38}$$

or

$$[\mathrm{M}]\tilde{\mathbf{Q}} = \tilde{\mathbf{P}}, \tag{5.39}$$

where $[M]$ is defined as the coefficient matrix, $\vec{\mathbf{Q}} = (\tilde{\mathbf{Q}}_1, \ldots, \tilde{\mathbf{Q}}_{N-1})^T$, and $\vec{\mathbf{P}} = (\tilde{\mathbf{P}}_1, \ldots, \tilde{\mathbf{P}}_{N-1})^T$. Note, the coefficients $\tilde{a}_i, \tilde{b}_i$, and $\tilde{c}_i$ are dependent on the angles $\tilde{\theta}_i$ and $\tilde{\phi}_i$ which, in turn, are dependent on the control quaternions $\tilde{\mathbf{Q}}_i$.

Given an estimate for the angles $\tilde{\theta}_i$ and $\tilde{\phi}_i$ and the variables $\tilde{u}_i$ and $\tilde{v}_i$, we can compute $[M^j]$ and solve equation (5.39) for $\vec{\mathbf{Q}}^{j+1}$ such that

$$\vec{\mathbf{Q}}^{j+1} = [\mathrm{M}^j]^{-1}\vec{\mathbf{P}}. \tag{5.40}$$

From $\vec{\mathbf{Q}}^{j+1}$, we calculate $\tilde{\mathbf{B}}_{3i-1}$ and $\tilde{\mathbf{B}}_{3i+1}$ from equations (5.32, 5.33). At this point, we correct the estimates for θ_i, $\tilde{\phi}_i$, $\tilde{u}_i$, and $\tilde{v}_i$ and recompute $\tilde{\mathbf{Q}}_i$. The process stops when equation (5.30) is satisfied.

The Bezier interpolation procedure is as follows

Step 1. The special cases of the endpoints are handled by defining the control quaternions at the boundaries, that is, $\tilde{\mathbf{Q}}_{-1} = \tilde{\mathbf{B}}_0 = \tilde{\mathbf{P}}_0$ and $\tilde{\mathbf{Q}}_{N+1} = \tilde{\mathbf{B}}_{3N} = \tilde{\mathbf{P}}_N$. The adjacent Bezier double quaternions are defined as $\tilde{\mathbf{B}}_1 = \tilde{\mathbf{Q}}_0$ and $\tilde{\mathbf{B}}_{3N-1} = \tilde{\mathbf{Q}}_N$. However, the choices for $\tilde{\mathbf{Q}}_0$ and $\tilde{\mathbf{Q}}_N$ are arbitrary. We choose them to lie one-tenth $(t = 0.1)$ of the way from the first and last key frames on the arc-segments passing through $\tilde{\mathbf{P}}_0$, $\tilde{\mathbf{P}}_1$, and $\tilde{\mathbf{P}}_N$, $\tilde{\mathbf{P}}_{N-1}$, such that

$$\tilde{\mathbf{Q}}_0 = \frac{\sin(1 - 0.1)\tilde{\rho}_0}{\sin \tilde{\rho}_0}\tilde{\mathbf{P}}_0 + \frac{\sin 0.1\tilde{\rho}_0}{\sin \tilde{\rho}_0}\tilde{\mathbf{P}}_1 \qquad (5.41)$$

and

$$\tilde{\mathbf{Q}}_N = \frac{\sin(1 - 0.1)\tilde{\rho}_N}{\sin \tilde{\rho}_N}\tilde{\mathbf{P}}_N + \frac{\sin 0.1\tilde{\rho}_N}{\sin \tilde{\rho}_N}\tilde{\mathbf{P}}_{N-1} \qquad (5.42)$$

where $\tilde{\rho}_0 = \arccos(\tilde{\mathbf{P}}_0 \cdot \tilde{\mathbf{P}}_1)$ and $\tilde{\rho}_N = \arccos(\tilde{\mathbf{P}}_N \cdot \tilde{\mathbf{P}}_{N-1})$.

We also set the variables $x_i = y_i = z_i = \frac{1}{3}$ for $i = 1, \ldots, N - 2$; and near the ends of the trajectory, we select $y_0 = y_{N-1} = 0.6$ and $z_0 = x_{N-1} = 0.4$ for equation (5.34). Recall, $x_0 = z_{N-1} = 0$.

Step 2. Determine the initial $[M^0]$ such that $j = 0$ in equation (5.40).

- Let the initial $\tilde{u}_i$ and $\tilde{v}_i$ be defined by

$$\tilde{u}_i = \frac{(x_i y_{i-1})^{1/2}}{(x_i y_{i-1})^{1/2} + (y_i z_{i-1})^{1/2}}, \qquad (5.43)$$

$$\tilde{v}_i = \frac{(y_i z_{i-1})^{1/2}}{(x_i y_{i-1})^{1/2} + (y_i z_{i-1})^{1/2}}. \qquad (5.44)$$

- Compute initial values for the matrix components $[M^0]$ by

$$\begin{aligned} \tilde{a}_i &= z_{i-1} v_i, \\ \tilde{b}_i &= (x_{i-1} + y_{i-1}) v_i + (y_i + z_i) u_i, \qquad (5.45) \\ \tilde{c}_i &= x_i u_i. \quad \text{for} \quad i = 1, \ldots, N - 1. \end{aligned}$$

Step 3. Solve (5.40) to determine the control double quaternions $\vec{\mathbf{Q}}^{j+1}$. Normalize each $\tilde{\mathbf{Q}}_i$ and compute $\theta_i = \arccos(\tilde{\mathbf{Q}}_i \cdot \tilde{\mathbf{Q}}_{i+1})$.

Step 4. Compute the Bezier quaternions $\tilde{\mathbf{B}}_{3i\pm1}$ from equations (5.32) and (5.33). Determine $\tilde{\phi}_i = \arccos(\tilde{\mathbf{B}}_{3i-1} \cdot \tilde{\mathbf{B}}_{3i+1})$, then calculate $\tilde{\mathbf{P}}_i^j$ from the equation (5.30).

FIGURE 12.4. Double Quaternion Interpolated Path.

Step 5. Compare the computed key frames $\tilde{\mathbf{P}}_i^j$ to the actual key frames $\tilde{\mathbf{P}}_i$ by calculating the angle $\tilde{\omega}_i = \arccos(\tilde{\mathbf{P}}_i^j \cdot \tilde{\mathbf{P}}_i)$. We define the error E to be the sum

$$E = \sum_{i=0}^{N} \tilde{\omega}_i^2. \qquad (5.46)$$

The iterative procedure stops when $E \leq \delta$, where δ is the tolerance for convergence, in our case $\delta = 10^{-5}$.

If $E > \delta$,

- Calculate parameters $\tilde{u}_i$ and $\tilde{v}_i$ from the G^2 continuity equations (5.35) and (5.31).

- Compute the new components of the matrix $[M^j]$ using equation (5.37) for the next iteration, and return to step 3.

The result of this procedure is the set of Bezier polygons for each segment of the entire trajectory. DeCasteljau's algorithm is used to determine the frames along each segment.

12.5.3   Example of double quaternion interpolation

To illustrate the double quaternion interpolation procedure, we interpolate the E^3 key frames listed in Table 12.1. The double quaternions associated with these key frames is listed in Table 12.2. The resulting interpolated trajectory is shown in Figure 12.4 where the white end-effectors correspond

	x	y	z	θ	ϕ	ψ
M_1	0.0	0.0	0.0	90°	-45°	0°
M_2	2.0	0.0	2.0	0°	0°	0°
M_3	3.5	1.0	4.0	0°	45°	0°
M_4	5.0	3.0	3.0	20°	20°	22.5°
M_5	6.5	3.0	2.0	45°	0°	45°
M_6	8.0	2.0	0.0	90°	30°	0°

TABLE 12.1. The key frame data for end-effector trajectory.

	P_1				P_2			
M_1	0.271	0.653	-0.271	0.653	0.271	0.653	-0.271	0.653
M_2	0.006	0.000	0.006	0.999	-0.006	0.000	-0.006	0.999
M_3	-0.373	-0.002	0.013	0.928	-0.393	0.002	-0.013	0.920
M_4	-0.119	0.205	0.232	0.943	-0.149	0.197	0.205	0.947
M_5	0.165	0.355	0.364	0.845	0.128	0.352	0.342	0.862
M_6	-0.165	0.683	0.201	0.683	-0.201	0.683	0.165	0.683

TABLE 12.2. Double quaternion key displacements.

to the key frames and the black end-effectors are the interpolated frames. The interpolation procedure converged in four interations.

12.6 The Design of the TS Robot

The TS robot is a five-degree of freedom mechanism that has as its base a gimbal joint and is connected to a spherical joint by a rigid link, see Figure 12.1.

Let $\mathbf{v}$ be the coordinate vector in E^3 of the wrist center in a frame M attached to a workpiece. The TS chain constrains a the point $\mathbf{v}$ to lie on a sphere of radius R about the shoulder joint $\mathbf{g}$. The point, $\mathbf{v}$, in the moving frame, M, takes the position $\mathbf{w}^i = [H_i]\mathbf{v}$ in the fixed frame, F. We have the constraint equation

$$(\mathbf{w} - \mathbf{g}) \cdot (\mathbf{w} - \mathbf{g}) = R^2. \tag{6.47}$$

For a given $\mathbf{w}$ and $\mathbf{g}$, the set of all transformations, $[H]$, that satisfies the equation (6.47) defines the workspace of the TS robot.

We use this constraint equation to define a TS robot for a given trajectory. Choose $v = (0, 0, 0)^T$ to be the origin of the moving frame, M. This reduces the non-linear design problem to a set of linear equations.

If we choose a reference position M_1 and subtract equation 1 from the rest of the equations of the form (6.47), we obtain.

$$\mathbf{w}^i \cdot \mathbf{w}^i - \mathbf{w}^1 \cdot \mathbf{w}^1 - 2\mathbf{w}^i \cdot \mathbf{g} + 2\mathbf{w}^1 \cdot \mathbf{g}, \quad i = 2, \ldots, n. \tag{6.48}$$

Because their are six unknown parameters, $\mathbf{g} = (x, y, z)^T$ and those of $\mathbf{w}^1 = (u, v, w)^T$, in general, we may obtain exact solutions for up to seven arbitrary positions (Innocenti [10] and McCarthy and Liao [11]).

For this chapter, we specify $\mathbf{w}^1$ and solve for the ground pivot $\mathbf{g}$. This yields a system of linear equations that has a unique solution for four specified spatial positions ($n = 4$). Writing the three constraint equations in matrix form, we have the system

$$\begin{bmatrix} (\mathbf{w}^2 - \mathbf{w}^1)^T \\ (\mathbf{w}^3 - \mathbf{w}^1)^T \\ (\mathbf{w}^4 - \mathbf{w}^1)^T \end{bmatrix} \mathbf{g} = \frac{1}{2} \left\{ \begin{array}{c} \mathbf{w}^2 \cdot \mathbf{w}^2 - \mathbf{w}^1 \cdot \mathbf{w}^1 \\ \mathbf{w}^3 \cdot \mathbf{w}^3 - \mathbf{w}^1 \cdot \mathbf{w}^1 \\ \mathbf{w}^4 \cdot \mathbf{w}^4 - \mathbf{w}^1 \cdot \mathbf{w}^1 \end{array} \right\} \tag{6.49}$$

or

$$[W]\mathbf{g} = \mathbf{c}. \tag{6.50}$$

The solution $\mathbf{g}$ may be obtained by inverting the matrix $[W]$. This unique solution $\mathbf{g} = (x, y, z)^T$ is the center of the sphere that passes through the four points $\mathbf{w}^i, i = 1, 2, 3, 4$. The calculation of the armlength R of the TS chain can be computed from equation (6.47),

$$R = \sqrt{(x - u)^2 + (y - v)^2 + (z - w)^2}. \tag{6.51}$$

By calculating the base location $\mathbf{g}$ and the armlength R, the workspace for a specific TS chain is defined. Now, we design an optimimal TS chain such that the workspace of the chain attempts to satisfy the task trajectory.

12.7   The Optimum TS Robot

At this point, we are able to create a task trajectory from user-defined key positions and orientations. We also can determine a TS robot from four specified positions. The goal now is to find the best fit of a TS robot to our task trajectory.

Select four frames from the task trajectory. These four positions become $\mathbf{w}^i$, $i = 1, 2, 3, 4$ in the design equations (6.49) of the TS chain. The physical parameters of the robot $\mathbf{g}$ and R are then calculated. The TS robot will pass through four positions and orientations of our task trajectory and approximate the rest of the task frames. To get the closest point $\mathbf{w}$ of the robot's workspace to an arbitrary pose $\mathbf{a}$ from the task trajectory, we define

Best Fixed Pivot $\mathbf{g}$	(4.49, 0.44, 0.10)
Arm Length R	3.84
Error Δ_{ave}	0.001

TABLE 12.3. TS robot design parameters.

the unit direction vector $\mathbf{v}$ from the fixed pivot $\mathbf{g} = (x, y, z)$ to the point $\mathbf{a} = (a, b, c)$ as

$$\mathbf{v} = \frac{\mathbf{a} - \mathbf{g}}{|\mathbf{a} - \mathbf{g}|} \qquad (7.52)$$

The point $\mathbf{w} = R\mathbf{v}$ is the closest point of the TS robot workspace to the task frame $\mathbf{a}$. The end-effector of the TS robot can attain the exact orientation of the task frame. This position and orientation is converted to the double quaternion $\tilde{\mathbf{W}} = \xi\mathbf{W}_1 + \eta\mathbf{W}_2$. Let the double quaternion $\tilde{\mathbf{A}} = \xi\mathbf{A}_1 + \eta\mathbf{A}_2$ define the task frame. The local error is defined as the magnitude of the eight-vector

$$\Delta(\tilde{\mathbf{A}}, \tilde{\mathbf{W}}) = \sqrt{(\mathbf{A}_1 - \mathbf{W}_1)^2 + (\mathbf{A}_2 - \mathbf{W}_2)^2}. \qquad (7.53)$$

The total number of task frames of a task trajectory is given by

$$T = (N)(s + 1) + 1 \qquad (7.54)$$

where $N + 1$ is the total number of key frames and s is the number of interpolations between any two key frames. The error is defined as the summation of the local errors between the TS robot workspace and each frame of the task trajectory divided by the total number of task frames for specific TS design

$$\Delta_{ave} = \frac{\sum_{k=0}^{T-1} \Delta(\tilde{\mathbf{A}}_k, \tilde{\mathbf{W}}_k)}{T}. \qquad (7.55)$$

This error value is one value of the cost function which must be minimized. A new set of four task frames is selected and the process is repeated. An exhaustive search of all combinations of four task frames in the task trajectory is utilized. The TS robot with minimum error is the optimum fit to the task trajectory.

12.7.1   The optimum TS robot and the task trajectory

The TS robot synthesized to fit the task trajectory obtained from the previous example is shown in Figure 12.4. The fixed pivot $\mathbf{g}$ and the arm length R are listed in Table 12.3. Figure 12.5 shows the sphere reachable by the wrist of the TS robot. The grey end-effectors show the closest positions to the task frames. The center of the sphere is the location of fixed pivot, $\mathbf{g}$.

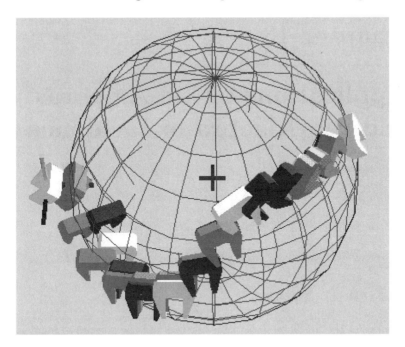

FIGURE 12.5. The TS robot and the task trajectory.

12.8 Conclusion

In this chapter, we present a method to design a TS robot that reaches a specified task trajectory. The task trajectory is defined by interpolating a set of key frames selected by the designer. The interpolation is done using a double quaternion representation of the specified key frames to obtain an efficient formulation. The TS robot that best fits this trajectory is determined by minimizing local error between the workspace and the task frames. An example of this design algorithm is presented.

This procedure allows a user to design a TS robot to accomplish a desired task. If the designed robot is not satisfactory, the user alters the task and the procedure is repeated. This interactive process helps to formulate a task as a set of spatial positions and orientations, scan and evaluate candidate devices, including assessment of range of motion and mechanical advantage, and, finally, select a TS robot to achieve the desired performance.

Chapter 13

Applications of Lie Algebras and the Algebra of Incidence

Eduardo Bayro Corrochano and Garret Sobczyk

13.1 Introduction

We present the fundamentals of Lie algebra and the algebra of incidence in the n–dimensional affine plane. The difference between our approach and previous contributions, [5, 4, 2] is twofold. First, our approach is easily accessible to the reader because there is a direct translation of the familiar matrix representations to our representation using bivectors from the appropriate geometric algebra. Second, our "hands on" approach provides examples from robotics and image analysis so that the reader can become familiar with the computational aspects of the problems involved. This chapter is to some extent complimentary to the above mentioned references. Lie group theory is the appropriate tool for the study and analysis of the action of a group on a manifold. Geometric algebra makes it possible to carry out computations in a coordinate-free manner by using a bivector representation of the most important Lie algebras [5]. Using the bivector representation of a Lie operator, we can easily compute a variety of invariants useful in robotics and image analysis. In our study of rigid motion in the n–dimensional affine plane, we use both the structure of the Lie algebra alongside the operations of meet and join from incidence algebra.

The organization of this chapter is as follows. Section two examines the basic properties of the general linear group from the perspective of geometric algebra. Section three presents the algebra of incidence in the n–dimensional affine plane. Section four studies rigid motion in the affine plane. Section five carries out computations for three typical problems in robotics, using the incidence relations developed in section 3. Section six uses the bivector algebra in an experiment involving real and simulated images for the recognition of visual invariants. Concluding remarks are given in section seven.

13.2   The General Linear Group

The *General Linear Group* $GL(\mathcal{N})$ is defined to be the subset of all $f \in End(\mathcal{N})$ with the property that $f \in GL(\mathcal{N})$ if and only if $\det(f) \neq 0$, [10]. The determinant of f is defined in the algebra $\mathcal{G}_N$ by

$$f(e_1) \wedge f(e_2) \wedge \ldots \wedge f(e_n) = \det(\mathcal{F}) e_1 \wedge e_2 \wedge \ldots \wedge e_n,$$

where $\det(\mathcal{F})$ is just the ordinary determinant of the matrix of f with respect to the basis $\{e\}$. Choosing the basis $\{e\}$ makes explicit the isomorphism between the general linear group $GL(\mathcal{N})$ and $GL(n, \mathcal{C})$ the general linear group of all complex $n \times n$ matrices $\mathcal{F}$ with $\det \mathcal{F} \neq 0$. The theory of Lie groups and their corresponding Lie algebras can be considered largely to be the study of the group-manifold $GL(n, \mathcal{C})$, since any Lie group is isomorphic to a subgroup of $GL(n, \mathcal{C})$, [3, pp.501].

Since we have referred to $GL(\mathcal{N})$ as a *manifold*, we must be careful to give it the structure of an n^2–dimensional topological metric space. We define the inner product $< f, g >$ of $f, g \in GL(\mathcal{N})$ to be the usual hermitian positive definite inner product

$$< f, g > = \sum_{j=1}^{n} \sum_{i=1}^{n} \overline{f_{ij}} g_{ij},$$

where $f_{ij}, g_{ij} \in \mathcal{C}$ are the components of the matrices $\mathcal{F}$ and $\mathcal{G}$ of f and g, respectively, with respect to the basis $\{e\}$. The positive definite norm $|f|$ of $f \in GL(\mathcal{N})$ is defined by

$$|f|^2 = < f, f > = \sum_{j=1}^{n} \sum_{i=1}^{n} \overline{f_{ij}} f_{ij};$$

and is clearly zero if and only if $f = 0$.

The crucial relationship between a Lie group and its corresponding Lie algebra is almost an immediate consequence of the properties of the exponential of a linear operator $f \in End(\mathcal{N})$. The *exponential mapping* may be directly defined by the usual Taylor series

$$e^f = \sum_{i=0}^{\infty} \frac{f^i}{i!},$$

where convergence is with respect to the norm $|f|$. Note that $f^0 = 1$ is the identity operator on $\mathcal{N}$, and f^k is the composition of f with itself k times.

The logarithm of a linear operator, $\theta_f = \log(f)$, exists and is well defined for any $f \in GL(\mathcal{N})$. The logarithm can also be defined in terms of an infinite series, or more directly in terms of the *spectral form* of the f, [20]. Since the logarithm is the inverse function of the exponential function, we can write $f = e^{\theta_f}$ for any $f \in GL(\mathcal{N})$. The logarithmic form $f = e^{\theta_f}$ of $f \in GL(\mathcal{N})$ is useful for defining the *one parameter group* $\{f_t\}$ of the operator $f \in GL(\mathcal{N})$,

$$f_t(x) = e^{t\theta_f} x.$$

The one parameter group $\{f_t\}$ is *continuously connected to the identity* in the sense that $f_0(x) = x$, and $f_1(x) = f(x)$. Note that

$$f_0'(x) = \theta_f e^{t\theta_f}|_{t=0}(x) = \theta_f(x), \qquad (2.1)$$

so θ_f is *tangent* to f_t at the identity. The reason why $\{f_t\}$ is called a one parameter group is because it satisfies the basic additive property that

$$f_s \ f_t = e^{t\theta_f} e^{s\theta_f} = e^{(s+t)\theta_f} = f_{s+t}.$$

We can now define the *general linear Lie algebra* $gl(\mathcal{N})$ of the general linear Lie group $GL(\mathcal{N})$. As a set, $gl(\mathcal{N}) \equiv End(\mathcal{N})$, which is just the set of all *tangent operators* $\theta_f = \log(f) \in End(\mathcal{N})$ to the one parameter groups $f_t = e^{t\theta_f}$ defined for each $f \in GL(\mathcal{N})$. To complete the definition of $gl(\mathcal{N})$, we must specify the algebraic operations of addition and multiplication which makes $End(\mathcal{N})$ into the Lie algebra $gl(\mathcal{N})$. Addition is just the ordinary addition of linear operators, and multiplication is defined by the *Lie bracket* $[\theta_f, \theta_g]$ for $\theta_f, \theta_g \in gl(\mathcal{N})$. An analytic definition of the Lie bracket, which directly ties it to the group structure of $GL(\mathcal{N})$, is given by

$$[\theta_f, \theta_g] = \frac{d}{d(t^2)} f_t g_t f_{-t} g_{-t}|_{t=0} = \frac{1}{2t} \frac{d}{dt} f_t g_t f_{-t} g_{-t}|_{t=0},$$

[18, pp.3].
    Evaluating the Lie bracket, we find that

$$
\begin{aligned}
[\theta_f, \theta_g] &= \frac{1}{2t} \frac{d}{dt} (f_t g_t f_{-t} g_{-t})|_{t=0} \\
&= \frac{1}{2t} (\theta_f f_t g_t f_{-t} g_{-t} + f_t \theta_g g_t f_{-t} g_{-t} - f_t g_t \theta_f f_{-t} g_{-t} - \\
&\qquad - f_t g_t f_{-t} \theta_g g_{-t}) \\
&= (\frac{1}{2t} f_t (\theta_f g_t - g_t \theta_f) f_{-t} g_{-t})|_{t=0} + (\frac{1}{2t} f_t g_t (\theta_g f_{-t} - \\
&\qquad - f_{-t} \theta_g) g_{-t})|_{t=0} \\
&= \frac{1}{2} (\theta_f \theta_g - \theta_g \theta_f) + \frac{1}{2} (-\theta_g \theta_f + \theta_f \theta_g) \\
&= \theta_f \theta_g - \theta_g \theta_f \qquad (2.2)
\end{aligned}
$$

where we have used the Taylor series expansions

$$g_t = 1 + t\theta_g + \ldots, \quad \text{and} \quad f_{-t} = 1 - t\theta_f + \ldots.$$

We have thus shown that the Lie bracket, defined analytically above, reduces to the *commutator product* of the linear operators θ_f and θ_g in $gl(\mathcal{N})$. As such, it is not difficult to show that they satisfy the famous *Jacobi identity*, which is equivalent to the distributive law

$$[\theta_f, [\theta_g, \theta_h]] = [[\theta_f, \theta_g], \theta_h] + [\theta_g, [\theta_f, \theta_h]].$$

When we choose a particular basis $\{e\}$ of $\mathcal{N}$, the isomorphism between the general linear Lie algebra $gl(n, \mathcal{C})$ and $gl(\mathcal{N})$ becomes explicit and the Lie bracket of linear operators just becomes the Lie bracket of $n \times n$ complex matrices (2.26). Alternatively, using the bivector representation (2.23), the Lie bracket of linear operators is expressed in terms of the Lie bracket of the bivectors of the operators (2.25).

13.2.1 The orthogonal groups

The most simple example of an *orthogonal group* is $SO(2)$, which is a subgroup of the general linear group $GL(\mathcal{N}^2)$. As a matrix group it is generated by all 2×2 matrices of the form

$$X_\theta = \begin{pmatrix} \cos\theta & -\sin\theta \\ \sin\theta & \cos\theta \end{pmatrix}.$$

The matrix X_θ generates a counterclockwise rotation in the xy-plane through the angle θ. Using (2.23), we get the corresponding bivector representation

$$\boldsymbol{X}_\theta = \cos(\theta)e_1 \wedge \bar{e}_1 - \sin(\theta)e_1 \wedge \bar{e}_2 + \sin(\theta)e_2 \wedge \bar{e}_1 + \cos(\theta)e_2 \wedge \bar{e}_2$$

For matrices $X_{\theta_1}, X_{\theta_2} \in SO(2)$ the group operation is ordinary matrix multiplication, $X_{\theta_2} X_{\theta_1} = X_{\theta_1 + \theta_2}$. For the bivector representation $\boldsymbol{X}_{\theta_1}, \boldsymbol{X}_{\theta_2} \in SO(2)$, the group operation is defined by the *generalized dot product*, $\boldsymbol{X}_{\theta_1} : \boldsymbol{X}_{\theta_2}$, that is for $\boldsymbol{x} \in \mathcal{N}^2$,

$$(\boldsymbol{X}_{\theta_1} \boldsymbol{x} : \boldsymbol{X}_{\theta_2}) \equiv \boldsymbol{X}_{\theta_2} \cdot (\boldsymbol{X}_{\theta_1} \cdot \boldsymbol{x}) = \boldsymbol{X}_{\theta_1 + \theta_2} \cdot \boldsymbol{x}. \tag{2.3}$$

Note that the bivectors $\boldsymbol{X}_\theta$ are in $\mathcal{G}^2_{n,n}$.

Taking the derivatives of X_θ and $\boldsymbol{X}_\theta$, with respect to θ and evaluating at $\theta = 0$, gives the corresponding generators of the associated Lie algebra $so(2)$. As a matrix Lie algebra, under the bracket operation of matricies, we find the generator

$$\frac{dX_\theta}{d\theta}\Big|_{\theta \to 0} = \begin{pmatrix} 0 & -1 \\ 1 & 0 \end{pmatrix}.$$

As a *bivector Lie algebra*, under the bracket operation of bivectors, we find the bivector generator

$$\boldsymbol{B} = \frac{d\boldsymbol{X}_\theta}{d\theta}\Big|_{\theta \to 0} = -e_1 \wedge \bar{e}_2 + e_2 \wedge \bar{e}_1 = -\sigma_{12} + \eta_{12}. \tag{2.4}$$

The bivector representation of the Lie group $SO(2)$, as a subgroup of the larger Lie group $SO(2,2)$, makes possible the *spinor* representations of these groups discussed in chapter 3. The *spinor group* $Spin(2)$ is defined by taking the exponential of the bivector (2.4),

$$Spin(2) = \{\exp(\frac{1}{2}\theta B)|\ \theta \in I\!R\}$$

The exponential $\exp(\frac{1}{2}\theta B)$ can be calculated by noting that the bivector B satisfies the *minimal polynomial*

$$B^3 + 4B = B(B - 2i)(B + 2i) = 0,$$

which implies the decomposition

$$B = 0p_1 + 2ip_2 - 2ip_3$$

where the mutually annihiliating idempotents are defined by

$$p_1 = \frac{B^2 + 4}{4}, \quad p_2 = -\frac{1}{8}B(B + 2i), \quad p_3 = -\frac{1}{8}B(B - 2i)$$

Using this decomposition, we find that

$$
\begin{aligned}
\exp(\frac{1}{2}\theta B) &= \exp(\frac{0 \cdot \theta}{2})p_1 + \exp(i\theta)p_2 + \exp(-i\theta)p_3 \\
&= p_1 + \cos(\theta)(p_2 + p_3) + \sin(\theta)i(p_2 - p_3) \\
&= p_1 + \cos(\theta)(p_2 + p_3) + \frac{B}{2}\sin(\theta) \quad\quad (2.5)
\end{aligned}
$$

By using Theorem 2.27 from chapter 2, the group action is given by

$$x' = \exp(\frac{1}{2}\theta B)x \exp(-\frac{1}{2}\theta B)$$

where $x = \{e\}x_{\{e\}} = x_1 e_1 + x_2 e_2$. We say that $Spin(2)$ is a *double covering of the orthogonal group* $SO(2)$, because the spinors $\pm \exp(\frac{1}{2}\theta B)$ represent the same group element. Note that we now have the easy composition rule for the composition of two group elements $\exp(\frac{1}{2}\theta_1 B)$ and $\exp(\frac{1}{2}\theta_2 B)$,

$$\exp(\frac{1}{2}\theta_1 B)\exp(\frac{1}{2}\theta_2 B) = \exp(\frac{1}{2}(\theta_1 + \theta_2)B)$$

If we are solely interested in the group $SO(2)$, a more natural place to carry out the calculations is in the Euclidean space $I\!R^2$. We project the null cone $\mathcal{N}^2$ down to $I\!R^2$ by using the reciprocal elements I_2 and $\overline{I_2}$, defined in chapter 2 by

$$I_2 = \sigma_1 \sigma_2 \quad \text{and} \quad \overline{I_2} = (2 - \sqrt{2})^2(\overline{e}_2 + \sigma_2)(\overline{e}_1 + \sigma_1)$$

Thus, for $x = \{e\}x_{\{e\}} = x_1 e_1 + x_2 e_2 \in \mathcal{N}^2$, the projection $x' = P_I(x)$ gives

$$x' = P_I(x) = (x \cdot \bar{I}) \cdot I = x_1 \sigma_1 + x_2 \sigma_2 \in \mathbb{R}^2.$$

As noted in chapter 2, this projection is invertible in the sense that we can find $P_{I'}$ such that $x = P_{I'}(x')$. The projection $P_{I'}$ is specified by

$$x = P_{I'}(x') = (x' \cdot \bar{I}) \cdot I' = x_1 e_1 + x_2 e_2$$

where $\bar{I}$ is defined as before and where $I' = e_1 e_2$.

In $\mathbb{R}^2$, the generator of rotations is the simple bivector $\sigma_2 \sigma_1$. This bivector can be obtained from the bivector (2.4) in $spin(2,2)$ by the projection

$$I_2^{-1} I_2 \cdot B = \sigma_2 \sigma_1 = -I_2$$

onto the Lie algebra $so(2)$. For $x' = x_1 \sigma_1 + x_2 \sigma_2 \in \mathbb{R}^2$, the equivalent rotation is given by

$$y' = \exp(\frac{1}{2}\theta\sigma_2\sigma_1)x' \exp(-\frac{1}{2}\theta\sigma_2\sigma_1)$$

The above ideas can be immediately generalized to the general Lie group $GL(\mathcal{N}^n)$ of null cone $\mathcal{N}^n$, and the orthogonal subgroups $SO(p,q)$, where $p + q = n$. The orthogonal group $SO(p,q)$ acts on the space $\mathbb{R}^{p,q}$. Thus, if we wish to work in this Lie group or the corresponding Lie algebra, we first project the null cone $\mathcal{N}^n$ onto $\mathbb{R}^{p,q}$ by using the reciprocal elements given in chapter 2, carry out the rotation, and then return to the null cone by using the inverse projection.

13.2.2   The Lie group and Lie algebra
of the affine plane

The Lie algebra of the neutral affine plane $\mathcal{A}_{e_3}(\mathcal{N}^2)$ is useful in analysis of visual invariants (section 13.6), so we will begin with its treatment here. The well-known matrix representation of the Lie group of affine transformations in the plane has six independent parameters or degrees of freedom, and consists of all matrices of the form

$$g(A, \vec{v}) = \begin{bmatrix} a_{11} & a_{12} & a \\ a_{21} & a_{22} & b \\ 0 & 0 & 1 \end{bmatrix}, \tag{2.6}$$

where $\det g(A, \vec{v}) = \det A \neq 0$.

The one-parameter sub-groups are generated by the matrices

$$T_x = \begin{bmatrix} 1 & 0 & x \\ 0 & 1 & 0 \\ 0 & 0 & 1 \end{bmatrix}, \qquad T_y = \begin{bmatrix} 1 & 0 & 0 \\ 0 & 1 & y \\ 0 & 0 & 1 \end{bmatrix},$$

$$D_u = \begin{bmatrix} e^u & 0 & 0 \\ 0 & e^u & 0 \\ 0 & 0 & 1 \end{bmatrix}, \qquad R_\theta = \begin{bmatrix} cos(\theta) & -sin(\theta) & 0 \\ sin(\theta) & cos(\theta) & 0 \\ 0 & 0 & 1 \end{bmatrix}, \qquad (2.7)$$

$$S_v = \begin{bmatrix} e^v & 0 & 0 \\ 0 & e^{-v} & 0 \\ 0 & 0 & 1 \end{bmatrix}, \qquad H_\phi = \begin{bmatrix} cosh(\phi) & sinh(\phi) & 0 \\ sinh(\phi) & cosh(\phi) & 0 \\ 0 & 0 & 1 \end{bmatrix}.$$

Using equation (2.1), the matrix representation of the Lie algebra basis generators are obtained by taking the derivative of the equation (2.7), and evaluating the parameter at zero

$$\mathcal{L}_x = \begin{pmatrix} 0 & 0 & 1 \\ 0 & 0 & 0 \\ 0 & 0 & 0 \end{pmatrix}, \qquad \mathcal{L}_y = \begin{pmatrix} 0 & 0 & 0 \\ 0 & 0 & 1 \\ 0 & 0 & 0 \end{pmatrix},$$

$$\mathcal{L}_u = \begin{pmatrix} 1 & 0 & 0 \\ 0 & 1 & 0 \\ 0 & 0 & 0 \end{pmatrix}, \qquad \mathcal{L}_\theta = \begin{pmatrix} 0 & -1 & 0 \\ 1 & 0 & 0 \\ 0 & 0 & 0 \end{pmatrix}, \qquad (2.8)$$

$$\mathcal{L}_v = \begin{pmatrix} 1 & 0 & 0 \\ 0 & -1 & 0 \\ 0 & 0 & 0 \end{pmatrix}, \qquad \mathcal{L}_\phi = \begin{pmatrix} 0 & 1 & 0 \\ 1 & 0 & 0 \\ 0 & 0 & 0 \end{pmatrix}.$$

The above matrix Lie group and matrix Lie algebra can be directly translated into the corresponding Lie group and Lie algebra of the affine plane $\mathcal{A}_{e_3}(\mathcal{N}^2)$. Each of the matrix generators in (2.7) and (2.8) can be replaced by its corresponding bivector representation (2.23). The bivector representations of the generators of this Lie algebra are

$$\mathcal{L}_x = bivector(\mathcal{L}_x) = e_1 \wedge \bar{e}_3, \quad \mathcal{L}_y = bivector(\mathcal{L}_y) = e_2 \wedge \bar{e}_3.$$

$$\mathcal{L}_u = bivector(\mathcal{L}_u) = e_1 \wedge \bar{e}_1 + e_2 \wedge \bar{e}_2, \quad \mathcal{L}_\theta = bivector(\mathcal{L}_\theta) = e_2 \wedge \bar{e}_1 - e_1 \wedge \bar{e}_2.$$

$$\mathcal{L}_v = bivector(\mathcal{L}_v) = e_1 \wedge \bar{e}_1 - e_2 \wedge \bar{e}_2, \quad \mathcal{L}_\phi = bivector(\mathcal{L}_\phi) = e_1 \wedge \bar{e}_2 + e_2 \wedge \bar{e}_1.$$

Expanding these bivector generators in the standard basis (2.6), we get

$$\begin{array}{rcl} \mathcal{L}_x & = & \frac{1}{2}\sigma_1\sigma_3 - \frac{1}{2}\sigma_1\eta_3 - \frac{1}{2}\sigma_3\eta_1 - \frac{1}{2}\eta_1\eta_3, \\ \mathcal{L}_y & = & \frac{1}{2}\sigma_2\sigma_3 - \frac{1}{2}\sigma_2\eta_3 - \frac{1}{2}a_3\eta_2 - \frac{1}{2}\eta_2\eta_3, \\ \mathcal{L}_u & = & -\sigma_1\eta_1 - \sigma_2\eta_2, \\ \mathcal{L}_\theta & = & -\sigma_1\sigma_2 + \eta_1\eta_2, \\ \mathcal{L}_v & = & -\sigma_1\eta_1 + \sigma_2\eta_2, \end{array}$$

$$\mathcal{L}_\phi = -\sigma_1 \eta_2 - \sigma_2 \eta_1. \tag{2.9}$$

Let us see how the Lie algebra of the affine plane can be represented as a Lie algebra of vector fields over the null cone $\mathcal{N}^3$. The *vector derivative* or *gradient* $\overline{\partial}_x = \frac{\partial}{\partial x}$ at the point $x = x e_1 + y e_2 + z e_3 \in \mathcal{N}^3$ is defined by requiring that $a \cdot \overline{\partial}_x$ is the *directional derivative* in the direction of a. It follows that $a \cdot \overline{\partial}_x x = a$ and $e_i \cdot \overline{\partial}_x \equiv \frac{\partial}{\partial x_i}$. We also have

$$\overline{\partial}_x x = \overline{\partial}_x \cdot x + \overline{\partial}_x \wedge x = 3 + \sum_{i=1}^{3} \overline{e}_i \wedge e_i,$$

where $\{e\}$ and $\{\overline{e}\}$ are reciprocal basis for the reciprocal null cones $\mathcal{N}^3$ and $\overline{\mathcal{N}}^3$. In terms of the reciprocal basis $\{\overline{e}\}$, $\overline{\partial}_x = \sum_i \overline{e}_i \frac{\partial}{\partial x_i}$.

Now let $a = a(x)$ and $b = b(x)$ be vector fields in $\mathcal{N}^3$. The Lie bracket $[a, b]$ is defined by

$$[a, b] = a \cdot \overline{\partial}_x b - b \cdot \overline{\partial}_x a$$

Since in $\mathcal{N}^3$, $\overline{\partial}_x \wedge \overline{\partial}_x = 0$, we have the important integrability condition that

$$(a \wedge b) \cdot (\overline{\partial}_x \wedge \overline{\partial}_x) = [a, b] \cdot \overline{\partial}_x - [a \cdot \overline{\partial}_x, b \cdot \overline{\partial}_x] = 0$$

where

$$[a \cdot \overline{\partial}_x, b \cdot \overline{\partial}_x] = a \cdot \overline{\partial}_x b \cdot \overline{\partial}_x - b \cdot \overline{\partial}_x a \cdot \overline{\partial}_x$$

is the Lie bracket or commutator product of the partial derivatives $a \cdot \overline{\partial}_x$ and $b \cdot \overline{\partial}_x$. It follows from this identity that

$$[a, b] \cdot \overline{\partial}_x = [a \cdot \overline{\partial}_x, b \cdot \overline{\partial}_x]$$

relating the Lie bracket of the vector fields $[a, b]$ to the standard Lie bracket of the partial derivatives $[a \cdot \overline{\partial}_x, b \cdot \overline{\partial}_x]$.

Let us consider in detail the translation of the Lie Algebra of the affine plane to the null vector formulation in the null cone $\mathcal{N}^2$. Recall that the two dimensional affine plane $\mathcal{A}_e(\mathcal{N}^2)$ in $\mathcal{N}^3$ defined by

$$\mathcal{A}_e(\mathcal{N}) = \{ x \in \mathcal{N}^3 | \ x = x e_1 + y e_2 + e_3 \}. \tag{2.10}$$

We have already seen that the Lie algebra of the affine plane can be defined by a Lie algebra of matricies, or by an equivalent Lie algebra of bivectors. We now define this same Lie algebra as a Lie algebra of partial derivatives, and as a Lie algebra of vector fields. We have the following correspondence:

$$\mathcal{L}_x = \frac{\partial}{\partial x} = e_1 \cdot \overline{\partial}_x = L_x \cdot (x \wedge \overline{\partial}_x) \leftrightarrow \mathcal{L}_x x = L_x \cdot x = e_1 = L_x \tag{2.11}$$

where $L_x = e_1 \wedge \overline{e}_3$,

$$\mathcal{L}_y = \frac{\partial}{\partial y} = e_2 \cdot \overline{\partial}_x = L_y \cdot (x \wedge \overline{\partial}_x) \leftrightarrow \mathcal{L}_y x = L_y \cdot x = e_2 = L_y \tag{2.12}$$

where $\boldsymbol{L}_y = e_2 \wedge \bar{e}_3$,

$$\begin{aligned} \mathcal{L}_s &= x\frac{\partial}{\partial x} + y\frac{\partial}{\partial y} = (\boldsymbol{x} - e_3) \cdot \bar{\partial}_x = \boldsymbol{L}_s \cdot (\boldsymbol{x} \wedge \bar{\partial}_x) \\ &\leftrightarrow \mathcal{L}_s \boldsymbol{x} = \boldsymbol{L}_s \cdot \boldsymbol{x} = xe_1 + ye_2 = \boldsymbol{x} - e_3 = \boldsymbol{L}_s \end{aligned} \tag{2.13}$$

where $\boldsymbol{L}_s = e_1 \wedge \bar{e}_1 + e_2 \wedge \bar{e}_2$,

$$\mathcal{L}_r = -y\frac{\partial}{\partial x} + x\frac{\partial}{\partial y} = \boldsymbol{L}_r \cdot (\boldsymbol{x} \wedge \bar{\partial}_x) \leftrightarrow \mathcal{L}_r \boldsymbol{x} = \boldsymbol{L}_r \cdot \boldsymbol{x} = \boldsymbol{L}_r \tag{2.14}$$

where $\boldsymbol{L}_r = e_2 \wedge \bar{e}_1 - e_1 \wedge \bar{e}_2$,

$$\mathcal{L}_b = x\frac{\partial}{\partial x} - y\frac{\partial}{\partial y} = \boldsymbol{L}_b \cdot (\boldsymbol{x} \wedge \bar{\partial}_x) \leftrightarrow \mathcal{L}_b \boldsymbol{x} = \boldsymbol{L}_b \cdot \boldsymbol{x} = \boldsymbol{L}_b \tag{2.15}$$

where $\boldsymbol{L}_b = e_1 \wedge \bar{e}_1 - e_2 \wedge \bar{e}_2$,

$$\mathcal{L}_B = y\frac{\partial}{\partial x} + x\frac{\partial}{\partial y} = \boldsymbol{L}_B \cdot (\boldsymbol{x} \wedge \bar{\partial}_x) \leftrightarrow \mathcal{L}_B \boldsymbol{x} = \boldsymbol{L}_B \cdot \boldsymbol{x} = \boldsymbol{L}_B \tag{2.16}$$

where $\boldsymbol{L}_B = e_1 \wedge \bar{e}_2 + e_2 \wedge \bar{e}_1$.

Thus, the Lie algebra of the affine plane is generated by the bivectors

$$\mathcal{M} = \{\boldsymbol{L}_x, \boldsymbol{L}_y, \boldsymbol{L}_s, \boldsymbol{L}_r, \boldsymbol{L}_b, \boldsymbol{L}_B\}, \tag{2.17}$$

or, equivalently, by the vector fields of the form $\boldsymbol{L} \cdot \boldsymbol{x}$ for $\boldsymbol{L} \in \mathcal{M}$. The Lie bracket $[\boldsymbol{L}_1 \cdot \boldsymbol{x}, \boldsymbol{L}_2 \cdot \boldsymbol{x}]$ is given by

$$[\boldsymbol{L}_1 \cdot \boldsymbol{x}, \boldsymbol{L}_2 \cdot \boldsymbol{x}] = \boldsymbol{L}_2 \cdot (\boldsymbol{L}_1 \cdot \boldsymbol{x}) - \boldsymbol{L}_1 \cdot (\boldsymbol{L}_2 \cdot \boldsymbol{x}) = (\boldsymbol{L}_2 \times \boldsymbol{L}_1) \cdot \boldsymbol{x}$$

where $\boldsymbol{L}_1 \times \boldsymbol{L}_2 = \frac{1}{2}(\boldsymbol{L}_1 \boldsymbol{L}_2 - \boldsymbol{L}_2 \boldsymbol{L}_1)$ is the commutator product of the bivectors $\boldsymbol{L}_1, \boldsymbol{L}_2 \in \mathcal{M}$.

The Lie algebra of the affine plane is useful for the analysis of motion in the image plane. The vector fields of this Lie algebra are tangent to their flows, or integral curves of their group action on the manifold, and are presented in Figure 13.1 as images.

We have found the generators

$$\begin{array}{|c|c|c|} \hline \mathcal{L}_x = \frac{\partial}{\partial x} & \mathcal{L}_r = -y\frac{\partial}{\partial x} + x\frac{\partial}{\partial y} & \mathcal{L}_b = x\frac{\partial}{\partial x} - y\frac{\partial}{\partial y} \\ \hline \mathcal{L}_y = \frac{\partial}{\partial y} & \mathcal{L}_s = x\frac{\partial}{\partial x} + y\frac{\partial}{\partial y} & \mathcal{L}_B = y\frac{\partial}{\partial x} + x\frac{\partial}{\partial y} \\ \hline \end{array} \tag{2.18}$$

of the Lie Algebra of the affine plane $\mathcal{A}_{e_3}(\mathcal{N}^2)$ as vector fields along integral curves. Taking the commutator products of these infinitesimal differential generators, gives the multiplication table 13.2.2 for this Lie algebra.

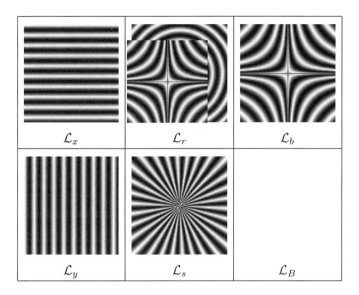

FIGURE 13.1. Lie Algebra basis in the form of images.

$[\cdot,\cdot]$	$\mathcal{L}_x$	$\mathcal{L}_y$	$\mathcal{L}_s$	$\mathcal{L}_r$	$\mathcal{L}_b$	$\mathcal{L}_B$
$\mathcal{L}_x$	0	0	$\mathcal{L}_x$	$\mathcal{L}_y$	$\mathcal{L}_x$	$\mathcal{L}_y$
$\mathcal{L}_y$	0	0	$\mathcal{L}_y$	$-\mathcal{L}_x$	$-\mathcal{L}_y$	$\mathcal{L}_x$
$\mathcal{L}_s$	$-\mathcal{L}_x$	$-\mathcal{L}_y$	0	0	0	0
$\mathcal{L}_r$	$-\mathcal{L}_y$	$\mathcal{L}_x$	0	0	$-2\mathcal{L}_B$	$2\mathcal{L}_b$
$\mathcal{L}_b$	$-\mathcal{L}_x$	$\mathcal{L}_y$	0	$2\mathcal{L}_B$	0	$2\mathcal{L}_r$
$\mathcal{L}_B$	$-\mathcal{L}_y$	$-\mathcal{L}_x$	0	$-2\mathcal{L}_b$	$-2\mathcal{L}_r$	0

FIGURE 13.2. The Lie algebra of the affine plane.

Using the table, we can verify the Jacobi identity for $\mathcal{L}_x, \mathcal{L}_s$ and $\mathcal{L}_b$, getting

$$
\begin{array}{ccccccc}
[\mathcal{L}_x[\mathcal{L}_s\mathcal{L}_b]] & + & [\mathcal{L}_s[\mathcal{L}_b\mathcal{L}_x]] & + & [\mathcal{L}_b[\mathcal{L}_x\mathcal{L}_s]] & = & \\
[\mathcal{L}_x 0] & - & [\mathcal{L}_s\mathcal{L}_y] & + & [\mathcal{L}_b\mathcal{L}_x] & = & \qquad (2.19)\\
0 & + & \mathcal{L}_y & - & \mathcal{L}_y & = & 0.
\end{array}
$$

Or, equivalently, using CLICAL and the bivector representation for $\boldsymbol{L}_x, \boldsymbol{L}_r$ and $\boldsymbol{L}_b$, we calculate

$$
\begin{aligned}
[L_x[L_r L_b]] &+ [L_r[L_b L_x]] &+ [L_b[L_x L_r]] &= \\
2[L_x L_B] &- [L_r L_y] &+ [L_b L_y] &= \\
2L_x &- L_x &- L_x &= 0.
\end{aligned}
\tag{2.20}
$$

13.3   Algebra of Incidence

In various applications in robotics, image analysis and computer vision, projective geometry and the algebra of incidence are very useful. Fortunately, both of these mathematical systems can be efficiently handled in the geometric algebra framework. In this section, we show how to apply the algebra of incidence to problems in robotics.

In Chapter 2, the *meet* and *join* operations in Π^n were characterized in terms of the *intersection* and *union* of the subspaces in $\mathcal{N}^{n+1}$ which name the corresponding objects in Π^n. Since each k-subspace can be associated with a nonzero k-blade of the geometric algebra $\mathcal{G}(\mathcal{N})$, it follows that the corresponding $(k-1)$-plane in Π^n can be named by the k-direction of a k-blade A_k.

Suppose that r points $a_1, a_2, \ldots, a_r \in \Pi^n$ are given in general position (linearly independent vectors in $\mathcal{N}^{n+1}$). Then an $(r-1)$-plane in Π^n is specified by the r-blade

$$
A_r = a_1 \wedge a_2 \wedge \ldots \wedge a_r \neq 0.
$$

Similarly, an $(s-1)$-plane in Π^n is specified by the s-blade

$$
B_s = b_1 \wedge b_2 \wedge \ldots \wedge b_s \neq 0
$$

determined by the s points b_i in general position in Π^n. Considering the a's and b's to be the basis elements of respective subspaces $\mathcal{A}_r$ and $\mathcal{B}_s$, they can be sorted in such a way that

$$
\mathcal{A}^r \cup \mathcal{B}^s = span\{a_1, a_2, \ldots a_s, b_{\lambda_1}, \ldots, b_{\lambda_k}\}
$$

Supposing that

$$
B_s = b_{\lambda_1} \wedge \ldots \wedge b_{\lambda_k} \wedge b_{\alpha_1} \wedge \ldots \wedge b_{\alpha_{s-k}},
$$

it follows that

$$
A_r \cup B_s = A_r \wedge b_{\lambda_1} \wedge \ldots \wedge b_{\lambda_k}
$$

and

$$
\mathcal{A}^r \cap \mathcal{B}^s = span\{b_{\alpha_1}, \ldots, b_{\alpha_{s-k}}\}.
$$

The problem of "meet" and "join" of the r-blade A_r and s-blade B_s has thus been solved.

Defining the reciprocal pseudoscalar element $\overline{I}_{A \cup B}$ of the join $\mathcal{A}^r \cup \mathcal{B}^s$,

$$\overline{I}_{A \cup B} = \overline{a}_1 \wedge \overline{a}_2 \wedge \ldots \wedge \overline{a}_s \wedge \overline{b}_{\lambda_1} \wedge \ldots \wedge \overline{b}_{\lambda_k},$$

the $(s-k)$-blade of the meet can be expressed in terms of the r- and s-blades A_r and B_s,

$$A_r \cap B_s = A_r \cdot (B_s \cdot \overline{I}_{A \cup B}). \tag{3.21}$$

A more complete discussion of these ideas can be found in chapters 2 and 3.

13.3.1 Incidence relations in the affine n-plane

In this subsection, we present very useful incidence relations between points, lines and planes, and higher dimensional k-planes in the affine n-plane $\mathcal{A}_e^n = \mathcal{A}_e(\mathbb{R}^n) \subset \mathbb{R}^{n+1,1}$. Recall that

$$\mathcal{A}_e^n = \{x_h = x + e \mid x \in \mathbb{R}^n\}$$

where $e = \frac{1}{2}(\sigma_{n+1} + \eta_{n+1})$, and the reciprocal element $\overline{e} = \sigma_n - \eta_n$.

Suppose that we are given k-points $a_1^h, a_2^h, \ldots, a_k^h \in \mathcal{A}_e^n$ where each $a_i^h = a_i + e$ for $a_i \in \mathbb{R}^n$. Taking the outer product of these points, we get

$$a_1^h \wedge a_2^h \wedge \ldots \wedge a_k^h = a_1^h \wedge (a_2^h - a_1^h) \wedge a_3^h \wedge \ldots \wedge a_k^h = \ldots$$

$$= a_1^h \wedge (a_2^h - a_1^h) \wedge (a_3^h - a_2^h) \wedge \ldots \wedge (a_k^h - a_{k-1}^h)$$

$$= a_1^h \wedge (a_2 - a_1) \wedge a_3 - a_2) \wedge \ldots \wedge (a_k - a_{k-1})$$

Projectively speaking, this tells us that the $(k-1)$-plane A^h in Π^n, which is the join of the these points, can be expressed in the form

$$A^h = a_1^h \wedge a_2^h \wedge \ldots \wedge a_k^h = a_1 \wedge a_2 \wedge \ldots \wedge a_k +$$
$$+ e \wedge (a_2 - a_1) \wedge (a_3 - a_2) \wedge \ldots \wedge (a_k - a_{k-1}). \tag{3.22}$$

Whereas (3.22) represents a $(k-1)$-plane in Π^n, it also belongs to the affine n-plane $\mathcal{A}_e^n$, and thus contains important metrical information. Dotting this equation with $\overline{e}$, we find that

$$\overline{e} \cdot A^h = \overline{e} \cdot (a_1^h \wedge a_2^h \wedge \ldots \wedge a_k^h) = (a_2 - a_1) \wedge (a_3 - a_2) \wedge \ldots \wedge (a_k - a_{k-1}).$$

This result motivates the following: the directed content of the $(k-1)$-simplex $A^h = a_1^h \wedge a_2^h \wedge \ldots \wedge a_k^h$ in the affine n-plane is given by

$$\frac{\overline{e} \cdot A^h}{(k-1)!} = \frac{\overline{e} \cdot (a_1^h \wedge a_2^h \wedge \ldots \wedge a_k^h)}{(k-1)!} = \frac{(a_2 - a_1) \wedge (a_3 - a_2) \wedge \ldots \wedge (a_k - a_{k-1})}{(k-1)!}$$

We shall now give a number of useful results in the affine plane that have both projective and metrical content.

$$d[a_1^h \wedge \ldots a_k^h, b^h] \equiv \tag{3.23}$$
$$[\{\bar{e} \cdot (a_1^h \wedge \ldots \wedge a_k^h)\} \ (\bar{e} \cdot b^h)]^{-1}[\bar{e} \cdot (a_1^h \wedge \ldots \wedge a_k^h \wedge b^h)]$$
$$= [a_2 - a_1) \wedge \ldots \wedge (a_k - a_{k-1})]^{-1}[(a_2 - a_1) \wedge \ldots \wedge (a_k - a_{k-1}) \wedge (b - a_k)]$$

represents the *directed distance* from the $(k-1)$-plane $a_1^h \wedge \ldots \wedge a_k^h$ to the point b^h.

$$d[a_1^h \wedge a_2^h, b_1^h \wedge b_2^h] \equiv [\{\bar{e} \cdot (a_1^h \wedge a_2^h)\} \wedge \{\bar{e} \cdot (b_1^h \wedge b_2^h)\}]^{-1}[\bar{e} \cdot (a_1^h \wedge a_2^h \wedge b_1^h \wedge b_2^h)]$$

$$= [(a_2 - a_1) \wedge (b_2 - b_1)]^{-1}[(a_2 - a_1) \wedge (b_1 - a_2) \wedge (b_2 - b_1)]$$

represents the *directed distance* between the two lines $a_1^h \wedge a_2^h$ and $b_1^h \wedge b_2^h$ in the affine n-plane. More generally,

$$d[a_1^h \wedge \ldots \wedge a_r^h, b_1^h \wedge \ldots \wedge b_s^h] \equiv \tag{3.24}$$
$$[\{\bar{e} \cdot (a_1^h \wedge \ldots \wedge a_r^h)\} \wedge \{\bar{e} \cdot (b_1^h \wedge \ldots \wedge b_s^h)\}]^{-1}[\bar{e} \cdot (a_1^h \wedge \ldots \wedge a_r^h \wedge b_1^h \wedge \ldots \wedge b_s^h)]$$
$$= [(a_2 - a_1) \wedge \ldots \wedge (a_r - a_{r-1}) \wedge (b_2 - b_1) \wedge \ldots \wedge (b_s - b_{s-1})]^{-1}$$
$$[(a_2 - a_1) \wedge \ldots \wedge (a_r - a_{r-1}) \wedge (b_1 - a_r) \wedge (b_2 - b_1) \wedge \ldots \wedge (b_s - b_{s-1})]$$

represents the *directed distance* between the $(r-1)$-plane $A^h = a_1^h \wedge \ldots \wedge a_r^h$ and the $(s-1)$-plane $B^h = b_1^h \wedge \ldots \wedge b_s^h$ in the affine n-plane.

If $A^h \wedge B^h = 0$, the directed distance may or may not be equal to zero! If

$$(a_1^h \wedge \ldots \wedge a_r^h) \wedge (b_1^h \wedge \ldots \wedge b_{s-1}^h) \neq 0,$$

we can calculate the meet between the $(r-1)$-plane A^h and $(s-1)$-plane B^h,

$$p = (a_1^h \wedge \ldots \wedge a_r^h) \cap (b_1^h \wedge \ldots \wedge b_s^h)$$

$$= (a_1^h \wedge \ldots \wedge a_r^h) \cdot [(b_1^h \wedge \ldots \wedge b_s^h) \cdot \bar{I}_{A \cup B}]$$

where

$$\bar{I}_{A \cup B} = \{\bar{e} \cdot [(a_1^h \wedge \ldots \wedge a_r^h) \wedge (b_1^h \wedge \ldots \wedge b_{s-1}^h)]\} \wedge \bar{e}.$$

The point $p = A^h \cap B^h$ may not be in the affine n-plane, but the *normalized* point $p^h = \frac{p}{\bar{e} \cdot p}$ will either be in the affine plane or will be undefined. Oftentimes in calculations it is not necessary to find the "normalized point", but it is always necessary when the metric is important, or as an indicator of parallel hyperplanes.

13.3.2 Incidence relations in the affine 3-plane

We give incidence relations for the 3D Euclidean space in the affine 3-plane $\mathcal{A}_e^3$, having the pseudoscalar $I = \sigma_{123}e$ and the reciprocal pseudoscalar

$\bar{I} = \bar{e}\sigma_{321}$ which satisfy $I \cdot \bar{I} = 1$. Similar incidence relations were given by Blaschke [11] using dual quaternions, and later by Selig using the 4D degenerate geometric algebra $\mathcal{G}_{3,0,1}$ (see chapter 11). Unlike the formulas given by these authors, our formulas are generally valid in any dimension and are expressed completely in terms of the meet and join operations in the affine plane. Blaschke, Selig, and previously Bayro [1], were not able to exploit the meet and join operations because they are using a geometric algebra with a degenerate metric. We give here the incidence relations for the 3D Euclidean space which we will need later.

The distance of a point b^h to the line $L^h = a_1^h \wedge a_2^h$ is the *magnitude* or *norm* of their directed distance

$$|d| = \left\| \left[\{\bar{e} \cdot (a_1^h \wedge a_2^h)\} \wedge \{(\bar{e} \cdot (b^h))\} \right]^{-1} \left[\bar{e} \cdot (a_1^h \wedge a_2^h \wedge b^h) \right] \right\| \quad (3.25)$$

The distance of a point b^h to the plane $A^h = a_1^h \wedge a_2^h \wedge a_3^h$ is

$$|d| = \left\| \left[\{\bar{e} \cdot (a_1^h \wedge a_2^h \wedge a_3^h)\} \wedge \{(\bar{e} \cdot (b^h))\} \right]^{-1} \left[\bar{e} \cdot (a_1^h \wedge a_2^h \wedge a_3^h \wedge b^h) \right] \right\| \quad (3.26)$$

The incidence relation between the lines $L_1^h = a_1^h \wedge a_2^h$ and $L_2^h = b_1^h \wedge b_2^h$ is completely determined by their join $I_{L_1^h \cup L_2^h} = L_1^h \cup L_2^h$. If $I_{L_1^h \cup L_2^h}$ is a bivector, the lines coincide and $L_1^h = tL_2^h$ for some $t \in \mathbb{R}$. If $I_{L_1^h \cup L_2^h}$ is a 3-vector, the lines are either parallel or intersect in a common point. In this case the meet

$$p = L_1^h \cap L_2^h = L_1^h \cdot [(L_2^h \cdot \bar{I}_{L_1^h \cup L_2^h})]. \quad (3.27)$$

If $\bar{e} \cdot p = 0$ the lines are parallel, otherwise they intersect at the point $p_h = \frac{p}{\bar{e} \cdot p}$ in the affine 3-space $\mathcal{A}_e^3$. Finally, if $I_{L_1^h \cup L_2^h}$ is a 4-vector, the lines are skew. In this case the distance is given by (13.3.1).

The incidence relation between a line $L^h = a_1^h \wedge a_2^h$ and a plane $B^h = b_1^h \wedge b_2^h \wedge b_3^h$ is also determined by their join $L^h \cup B^h$. Clearly, if the join is a trivector, the line L^h lies in the plane B^h. The only other possibility is that their join is the pseudoscalar $I = \sigma_{123}e$. In this case, we calculate the meet

$$p = L^h \cap B^h = L^h \cdot [(B^h \cdot \bar{I}). \quad (3.28)$$

If $\bar{e} \cdot p = 0$, the line is parallel to the plane with the directed distance determined by (3.25). Otherwise, $p_h = \frac{p}{\bar{e} \cdot p}$ is their point of intersection in the affine plane.

Two planes $A^h = a_1^h \wedge a_2^h \wedge a_3^h$ and $B^h = b_1^h \wedge b_2^h \wedge b_3^h$ in the affine plane $\mathcal{A}_e^3$ are either parallel, intersect in a line, or coincide. If their join is a trivector, i.e., $A^h = tB^h$ for some $t \in \mathbb{R}^*$, they obviously coincide. If they do not coincide, then their join is the pseudoscalar $I = \sigma_{123}e$. In this case, we calculate the meet

$$L = A^h \cap B^h = [(\bar{I}) \cdot A^h] \cdot B^h. \tag{3.29}$$

If $\bar{e} \cdot L = 0$ the planes are parallel with the directed distance determined by (3.25). Otherwise, L represents the line of intersection in the affine plane having the direction $\bar{e} \cdot L$.

The equivalent of the above incidence relations were given by Blaschke [11] in terms the dual quaternions, and by Selig [17] in a special 4 dimensional singular algebra. Whereas Blaschke uses only pure quaternions (bivectors) for his representation, Selig uses trivectors for points and vectors for planes. In contrast, in the affine 3-plane points are always represented by vectors, lines by bivectors, and planes by trivectors. This offers a comprehensive and consistent interpretation, which greatly simplifies the underlying conceptual framework.

$$equation\ (3.25) \quad \equiv \quad \frac{1}{2}(\tilde{p}l + \tilde{l}p) \tag{3.30}$$

$$equation\ (3.26) \quad \equiv \quad \frac{1}{2}(\tilde{p}\pi + \tilde{\pi}p) \tag{3.31}$$

$$equation\ (3.28) \quad \equiv \quad \frac{1}{2}(\tilde{l}\pi + \tilde{\pi}l), \tag{3.32}$$

The right sides of the equations gives the equivalent expressions used by Blaschke and Selig.

13.3.3 Geometric constraints as indicators

It is often required to check a geometric configuration during a rigid motion in Euclidean space. Simple geometric incidence relations can be used for this purpose. For example, a point p is on a line L if and only if

$$p \wedge L = 0. \tag{3.33}$$

Similarly, a point p is on a plane A iff

$$p \wedge A = 0. \tag{3.34}$$

A line L will lie in the plane A iff

$$L \cap A = A. \tag{3.35}$$

Alternatively, the line L can meet the plane A in a single point p, in which case

$$L \cap A = p,$$

or, if the line L is parallel to the plane A, $L \cap A = 0$.

13.4   Rigid Motion in the Affine Plane

A rotation in the affine n-plane $\mathcal{A}_e^n = \mathcal{A}_e(\mathbb{R}^n)$, just as in the Euclidean space $\mathbb{R}^n$, is the product of two reflections through two intersecting hyperplanes. If the normal unit vectors to these hyperplanes are m and n, respectively, then the versor of the rotation is given by

$$R = mn = e^{\frac{\theta}{2}B} = \cos(\frac{\theta}{2}) + B\sin(\frac{\theta}{2}), \qquad (4.36)$$

where B is the unit bivector defining the plane of the rotation.

A translation of the vector $x_h \in \mathcal{A}_e^n$, along the vector $t \in \mathbb{R}^n$, to the vector $x_h' = x_h + t \in \mathcal{A}_e^n$, is effected by the versor

$$T = \exp(\frac{1}{2}t\bar{e}) = 1 + \frac{1}{2}t\bar{e}$$

when it is followed by the projection $P_A(x') \equiv (x \wedge \bar{e}) \cdot e$. Thus for $x_h \in \mathcal{A}_e^n$, we get

$$x' = TxT^{-1} = \exp(\frac{1}{2}t\bar{e})x_h \exp(-\frac{1}{2}t\bar{e})$$

$$= (1 + \frac{1}{2}t\bar{e})x_h(1 - \frac{1}{2}t\bar{e}) = x_h + \frac{1}{2}t\bar{e}x_h - \frac{1}{2}x_h t\bar{e} - \frac{1}{4}t\bar{e}x_h t\bar{e}$$

$$= x_h + t + t \cdot (\bar{e} \wedge x^h) - \frac{1}{2}t^2\bar{e} = x_h + t - (t \cdot x_h + \frac{1}{2}t^2)\bar{e}. \qquad (4.37)$$

Applying P_A to this result, we get the expected translated vector

$$x_h' = P_A(x') = P_A[x_h + t - (t \cdot x_h + \frac{1}{2}t^2)\bar{e}] = x_h + t \qquad (4.38)$$

The above calculation shows the close relationship between a translation in the affine plane, and its representation in the horosphere as presented in other chapters. The advantage of carrying out translations in the affine plane rather than in the horosphere, is that the affine plane is still a linear model of Euclidean space, whereas the horosphere is a more complicated *non-linear* model.

Combining the versors for a rotation and a translation, we get the expression for the versor $M = TR$ of a rigid motion. For $x_h \in \mathcal{A}_e^n$, we find

$$x_h' = P_A[Mx_hM^{-1}] = P_A[TRx_hR^{-1}T^{-1}]. \qquad (4.39)$$

Equivalently, we will often write $M^{-1} \equiv \widetilde{M}$, expressing M^{-1} in terms of operation of *conjugation*. Whenever a calculation involves a translation, we must always apply the projection P_A to guarantee that our end result will be in the affine plane. In order to keep our notation as simple as possible, we will assume that whenever a translation is carried out, a projection P_A back to the affine plane is aways carried out, even if not always explicitly stated. The above calculations can be checked with CLICAL 4.0, [8]. Comparisons can also be made to the corresponding calculations made by Hestenes and Li [13] on the horosphere.

Note that all of our computations in the affine n-plane are carried out in a unified manner, regardless of whether we are calculating incidence relations among points and planes, or calculating rigid motions of points and planes. In contrast, other authors, using the 4D degenerated algebra $\mathcal{G}_{3,0,1}$ represent points with trivectors and planes with vectors [17], but when using the motor algebra $\mathcal{G}_{3,0,1}^+$ points, lines and planes are represented solely in terms of quaternion bivectors, [1].

13.5   Application to Robotics

This section carries out computations in affine 3D space for three problems in robotics. The difference between our approach and other approaches used in [17, 1] is that all our calculations, including calculations involving the meet and join operations, are carried out in the affine plane. Note that we will always assume that the projection P_A back to the affine plane carried out following each translation, even if not explicitly mentioned.

13.5.1   Inverse kinematic computing

We illustrate the power of incidence computations in affine 3D space by computing the inverse kinematics for a robot manipulator. Robot manipulators are designed to satisfy certain maneuvering constraints. In carrying out computations, it is highly desirable to use a mathematical framework in which the computations are as simple as possible and clearly reflect the underlying geometry. We claim that the affine 3D space meets these objectives.

The transformation M_t of a robot manipulator which takes the end–effector from its home position to a configuration determined by the n–degrees of freedoms of the joint angles $\theta_1, \theta_2, ..., \theta_n$, is given by

$$M_t = M_1 M_2 M_3 ... M_n, \tag{5.40}$$

where the screw versor of a joint $M_i = T_i R_i$ is dependent on the angle θ_i.

The *inverse kinematics* problem is the task of calculating the angles θ_i for a given final configuration of the the end–effector. Robot manipulators are equiped with a parallel revoluted axis and with some intersecting ones. The latter can be at the end–effector or at the home position. Two typical configurations are illustrated in Figure 13.3.a–b. The mechanical characteristics of the robot manipulators can be used to simplify the computations by considering the invariant plane ϕ^h, in the case of three parallel revoluted line axis Figure 13.3.a , or an invariant point p^h in the case of a intersecting revoluted line axis Figure 13.3.b.

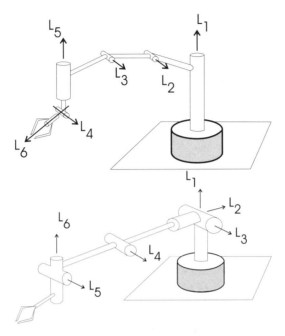

FIGURE 13.3. Manipulators: a) (top) intersect revoluted line axis at the end effector, b) and at the home position.

We can solve the inverse kinematics problem by breaking the problem up into a series of separate equations, using the strategy suggested by Selig [17] (chapter 11). We will illustrate the procedure for a robot with 6 degrees of freedom. First we rearrange the terms of the equation (5.40)

$$M_2 M_3 M_4 = \widetilde{M}_1 M_t \widetilde{M}_6 \widetilde{M}_5 \tag{5.41}$$

In the case of three parallel joints, we can isolate them by considering the common perpendicular plane ϕ^h, which satisfies

$$\phi^h = M_2 M_3 M_4 \, \phi^h \, \widetilde{M}_4 \widetilde{M}_3 \widetilde{M}_2 = \widetilde{M}_1 M_t (\widetilde{M}_6 (\widetilde{M}_5 \, \phi^h \, M_5) M_6) \widetilde{M}_t M_1. \tag{5.42}$$

If they meet at the point p^h, we can isolate the 3 coincident joints by

$$p^h = M_2 M_3 M_4 \, p^h \, \widetilde{M}_4 \widetilde{M}_3 \widetilde{M}_2 = \widetilde{M}_1 M_t \widetilde{M}_6 \widetilde{M}_5 \, p^h \, M_5 \widetilde{M}_6 \widetilde{M}_t M_1. \tag{5.43}$$

In this way, we have separated the problem into the system of two equations

$$\widetilde{M}_t M_1 \, \phi^h \, \widetilde{M}_1 M_t = \widetilde{M}_6 \widetilde{M}_5 \, \phi^h \, M_5 M_6 \tag{5.44}$$

or

$$\widetilde{M_t M_1} \, p^h \, \widetilde{M_1 M_t} = \widetilde{M_6 M_5} \, p^h \, M_5 M_6, \tag{5.45}$$

$$M_2 M_3 M_4 = M_1 M_t \widetilde{M_6 M_5} = M'. \tag{5.46}$$

We can first compute for θ_1, θ_5, θ_6, with the help of either equation (5.44) (Fig. 13.3.a) or equation (5.45) (Fig. 13.3.b). Then, using these results and equation (5.46), we can solve for θ_2, θ_3, θ_4.

Let us see how the procedure works for the case of the three intersecting revoluted joint axes in the common plane at the end–effector in (Figure 13.3.a). When the plane ϕ_b^h (perpendicular to the line axes l_2, l_3 and l_4) is rotated about the end joint, the point p_i^h on the line axis of the revoluted end–joint l_6^h remains invariant. Using the operation of *meet* and equation (5.44), the angle θ_6 can be eliminated,

$$\begin{aligned} p_i^h &= (\widetilde{M_t M_1} \, \phi^h \, \widetilde{M_1 M_t}) \cap l_6^h = (\widetilde{M_6 M_5} \, \phi^h \, M_5 M_6) \cap l_6^h \\ &= (\widetilde{M_5} \, \phi^h \, M_5) \cap l_6^h. \end{aligned} \tag{5.47}$$

In the case of the robot manipulator of Figure 13.3.b, the revoluted joint axes is the manipulator base. Equation (5.45) shows that the point p^h is an invariant for the fourth parallel and fifth line axis. Thus, we can use the equation

$$M_6 \widetilde{M_t} \, p^h \, M_t \widetilde{M_6} = \widetilde{M_5 M_4} \, p^h \, M_4 M_5 \tag{5.48}$$

to solve for the angles θ_4 and θ_5. Using the line l_5 and p^h, we get the invariant plane

$$\phi_i^h = M_6 \widetilde{M_t} \, p^h \, M_t \widetilde{M_6} \wedge l_5. \tag{5.49}$$

The 3D coordinates of this plane correspond to the x, z-plane e_{32}, and thus this equation allows us to solve for the angle θ_6. Having determined θ_6, and using the equations (5.48), we can easily complete the calculations of θ_4 and θ_5.

Consider now the three coincident line axis l_1^h, l_2^h, l_3^h, given in Figure 13.3.b. We can isolate the angle θ_2 by considering the invariant relation based on the meet of two of these lines

$$(M_2 \, l_3^h \, \widetilde{M_2}) \cap l_1^h = (M' \, l_3^h \, \widetilde{M'}) \cap l_1^h = p_0^h \tag{5.50}$$

where $M' = M_1 M_2 M_3$ and p_0^h is the invariant intersecting point. When the lines are parallel, as shown in Figure 13.3.a, we can use the same invariant relation by considering the intersecting point to be at infinity, giving $M' = M_2 M_3 M_4$.

13.5.2   Robot manipulation guidance

Consider a robot arm laser welder, see Figure 13.4.a. The welding distance has to be kept constant and the end–effector should follow a line on the surface. Again, we will carry out all computations in the affine 3D space.

Since the laser has to be kept at a constant distance from the surface for proper welding, we need to check if a given point p_d^h at the end of the laser cannon is always at the distance d from the welding surface $\phi^h = a_1^h \wedge a_2^h \wedge a_3^h$. We compute

$$
\begin{aligned}
|d^h| &= \left| \left[\{\bar{e} \cdot (a_1^h \wedge a_2^h \wedge a_3^h)\} \{(\bar{e} \cdot p_d^h)\} \right]^{-1} \left[\bar{e} \cdot (a_1^h \wedge a_2^h \wedge a_3^h \wedge p_d^h) \right] \right| \\
&= [a_2 - a_1) \wedge (a_3 - a_2)]^{-1} [(a_2 - a_1) \wedge (a_3 - a_2) \wedge (p - a_3)]. (5.51)
\end{aligned}
$$

Note that we use the simple equation (3.24) for computing this distance. The point of intersection p_i^h on the line l^h, aligned with the moving laser beam and the work surface ϕ^h, is given by

$$
\begin{aligned}
p_i^h &= \left(M l^h \widetilde{M} \right) \cap \phi^h = \left((M l^h \widetilde{M} \wedge \bar{e}) e \right) \cap \phi^h \\
&= \left((M l^h \widetilde{M} \wedge \bar{e}) e \right) \cdot (\phi^h \cdot \bar{I}). \qquad (5.52)
\end{aligned}
$$

In order to follow the welding line l_w on the surface, which is parallel to the welding curve, the robot arm should fulfill the point constrain

$$
\begin{aligned}
p_i^h \cap l_w^h &= \left((M l^h \widetilde{M}) \cap \phi^h \right) \cap l_w^h \\
&= \{ \left((M l^h \widetilde{M} \wedge \bar{e}) e \right) \cdot (\phi^h \cdot \bar{I}) \} \cdot (l_w^h \cdot \bar{I}) = 0. \qquad (5.53)
\end{aligned}
$$

13.5.3   Checking for a critical configuration

The control of the movement of a robot arm often requires a direct test to determine whether it has arrived at a prohibited configuration. This can be computed in a straightforward manner by using a determinant function of lines. The six lines are operated on by the screw versors $M_i = T_i R_i$. After the lines have reached their new position, they have in the affine 3D space the form

$$
\begin{aligned}
l_1^h &= M_1 (x'^h_1 \wedge x'^h_2) \widetilde{M}_1 = x_1^h \wedge x_2^h, \\
l_2^h &= M_2 (y'^h_1 \wedge y'^h_2) \widetilde{M}_2 = y_1^h \wedge y_2^h, ..., \\
l_6^h &= M_6 (v'^h_1 \wedge v'^h_2) \widetilde{M}_6 = v_1^h \wedge v_2^h.
\end{aligned}
$$

We compute the determinant function of these six lines, called the *super-bracket*, and get

$$
[l_1^h, l_2^h, l_3^h, l_4^h, l_5^h, l_6^h] = [x_1^h \wedge x_2^h, y_1^h \wedge y_2^h, z_1^h \wedge z_2^h, w_1^h \wedge w_2^h, u_1^h \wedge u_2^h, v_1^h \wedge v_2^h]
$$

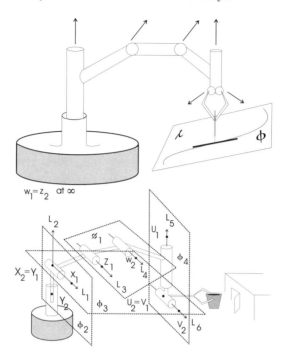

FIGURE 13.4. a) Laser welding (top) b) guidance using a critical configuration constraint.

$$
\begin{aligned}
= \ & [x_1^h x_2^h y_1^h y_2^h][z_1^h z_2^h w_1^h u_1^h][w_2^h u_2^h v_1^h v_2^h] \ + \\
& + [x_1^h x_2^h y_1^h z_1^h][y_2^h z_2^h w_1^h w_2^h][u_1^h u_2^h v_1^h v_2^h] \ - \\
& - [x_1^h x_2^h y_1^h z_1^h][y_2^h w_1^h w_2^h u_1^h][z_2^h u_2^h v_1^h v_2^h] \ + \\
& + [x_1^h x_2^h y_1^h w_1^h][y_2^h z_1^h z_2^h u_1^h][w_2^h u_2^h v_1^h v_2^h].
\end{aligned}
\tag{5.54}
$$

Details about bracket algebra is given in chapter 2. The decomposition of the superbracket, in terms of the bracket polynomial given above, was done by McMillan and Withe [16].

Let us apply the superbracket to identify a critical configuration of the six–revoluted–joint robot arm depicted in Figure 13.4.b, where the revoluted joints are represented with cylinders. The axis of each joint is determined by any two distinct points lying on it. The base line of the stereo system should always be parallel to the target line. Another condition that must be satisfied is that the plane of the end–effector and the base line should not be parallel with the plane spanned by the third and fourth axes. Also, the arm should not move below a given minimun height, or the end–effector could be damaged. All these conditions can be simultaneously tested by using the superbracket. Simplifying the superbracket (5.54) for the Figure 13.4.b, we get the final expression

$$
\begin{aligned}
[l_1^h, l_2^h, l_3^h, l_4^h, l_5^h, l_6^h] \;=\; & -[x_1^h x_2^h y_2^h z_1^h][x_2^h z_2^h u_2^h v_2^h][z_2^h u_1^h u_2^h v_2^h] \\
& +[x_1^h x_2^h y_2^h z_2^h][x_2^h z_1^h z_2^h u_2^h][w_2^h u_1^h u_2^h v_2^h] \\
=\; & [z_1^h x_1^h x_2^h y_2^h][w_2^h x_2^h z_2^h v_2^h][z_2^h u_1^h u_2^h v_2^h]. \qquad (5.55)
\end{aligned}
$$

This equation is just the meet of the four planes, given by

$$
\phi_1^h \cap \phi_2^h \cap \phi_3^h \cap \phi_4^h = \qquad\qquad\qquad\qquad\quad (5.56)
$$
$$
(z_1^h \wedge w_2^h \wedge z_2^h) \cap (x_1^h \wedge x_2^h \wedge y_2^h) \cap (x_2^h \wedge z_2^h \wedge u_2^h) \cap (u_1^h \wedge u_2^h \wedge v_2^h).
$$

A critical configuration is reached when at least one of the brackets in the equation (5.55) is zero. Geometrically, this means that one or more of the resulting planes have become degenerated, or that the resulting planes have a nonempty intersection. For example, the superbracket becomes zero for the Figure 13.4.b, when the third and forth joints and the base line and the target line lie in the same plane, or whenever the position of the end–effector is below the minimum height.

13.6   Application II: Image Analysis

This section carries out the computations in the affine plane $\mathcal{A}_{e_3}(\mathcal{N}^2)$ for two experiments in image analysis. The first experiment utilizes the Lie algebra of the affine plane in the design of an image filter. The second experiment uses the properties of Lie operators for the recognition of hand gestures. The third experiment shows the meet operation applied to image filters.

13.6.1   The design of an image filter

In the experiment we used simulated images of the *optical flow* for two motions: Figure 13.5.a shows a rotational and a translational motion, and Figure 13.6.a shows a dilation and a translational motion. The experiment uses only bivector computations to determine the type of motion, the axis of rotation, and/or the center of the dilation.

To study the motions in the affine plane, we used the Lie algebra of bivectors in the geometric algebra $\mathcal{A}_{e_3}(\mathcal{N}^2)$. The computations were carried out with the help of a computer program which we wrote in C^{++}. Each *flow vector* at any point $\mathbf{x}$ of the image was coded $\mathbf{x} = xe_1 + ye_2 + e_3 \in \mathcal{N}^3$. At each point of the flow image, we applied the commutator product of the six bivectors of the equation (2.17). Using the resultant coefficients of the vectors, the computer program calculated which type of differential invariant or motion was present.

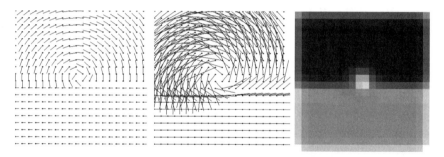

FIGURE 13.5. a) Rotation (L_r) and translational flow (L_x) fields. b) Convolved with a gaussian kernel c) Shows the magnitudes of the convolution.

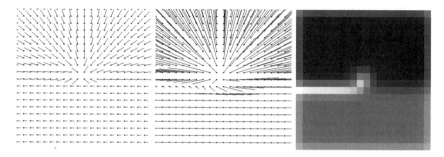

FIGURE 13.6. a) Expansion (L_s) and translational flow (L_x) fields. b) Convolved with a Gaussian kernel c) Shows the magnitudes of the convolution.

In Figure 13.5.b, we depict the result of convolving, via the geometric product, the bivector with a *Gaussian kernel* of size 5×5. Figure 13.5.c presents this result using the output of the kernel. The white center of the image indicates the lowest magnitude. Figure 13.6 shows the results for the case of a flow which is expanding. Comparing the Figure 13.5.c with the Figure 13.6.c, we note the duality of the differential invariants; the center point r of the rotation is invariant, and the invariant of the expansion is a line.

13.6.2   Recognition of hand gestures

Another interesting application, suggested by the seminal paper of Hoffman [12], is to recognize key points of an image by using the previous Lie operators . Figure 13.7.a shows hand gestures given to a robot. Using Lie filters, a robot can interpret whether it should *follow*, *stop* or *move in circles*, see Figure 13.7.b. Table 13.1 gives the firing weights for these filters used when interpreting the various gestures. We can also interpret these Lie filters as the basic elements, or *perceptrons*, of neurocomputing.

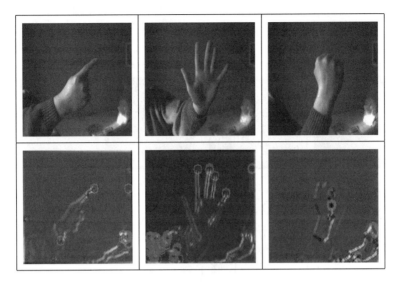

FIGURE 13.7. a) Top images: gestures for robot guidance (follow, stop and explore). b) Lower images: detected gestures by the robot vision system using Lie operators.

Hand gesture	$\mathcal{L}_x$	$\mathcal{L}_y$	$\mathcal{L}_r$	$\mathcal{L}_s$	$\mathcal{L}_b$	$\mathcal{L}_B$	Tolerance
fingertip	0	0	9	−4	11	−9	10%
stop	0	0	−3	1	1	4	10%
fist	0	0	−2	2	2	−1	10%

TABLE 13.1. Firing weights of Lie Operators by key points of hand gestures

13.6.3 The meet filter

The meet operation can be applied to the output of the filters in order to select the relevant points of the image. The meet operation is computed using equation (3.21). The basic idea is that the image is first convolved using two different Lie filters, and then their outputs are combined via the meet operation

$$(\mathcal{F}_a \star I) \cap (\mathcal{F}_b \star I). \tag{6.57}$$

The $\star$ of the filters $\mathcal{F}_a$ and $\mathcal{F}_b$ are entirely expressed in terms of bivectors. The result is the intersection of the filter outputs. Figure 13.8 shows the meet of different Lie filter ouputs. By using a more extensive system of meet filters, we should be able to extract more complicated contours.

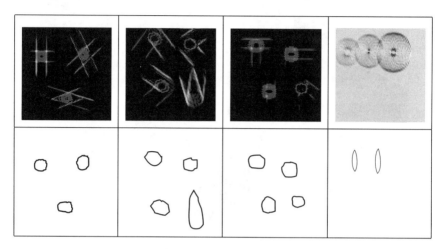

FIGURE 13.8. Top images: meet of Lie filters a) lines b) corners c) joints d) discs. Lower images: detected intersections by the meet of the Lie filter outputs.

13.7 Conclusion

We have shown how geometric algebra can effectively be used to carry out analysis on a manifold which is useful in robotics and image analysis. Geometric algebra offers a clear and concise geometric framework in which calculations can be carried out. Since the elements and operations in geometric algebra are basis free, computations are simpler and geometrically more transparent than in more traditional approaches.

Stereographic projection, and its generalization to the conformal group and projective geometry, have direct applications to image analysis from one or more viewpoints. The key idea is that an image is first represented in the null cone, and is then projected into affine geometries where the image analysis takes place. Since every Lie algebra can be represented by an appropriate bivector algebra in an affine geometry, it follows that a complete motion analysis is possible using their bivector representations. In a novel application in the 3–dimensional affine plane, we have computed rigid motion and applied the algebra of incidence for problems in robotics.

Future work is planned in the reconstruction of 3–D affine motion, in the design of steerable filters, and in the use of bivector algebras in visual robot tracking.

Acknowledgments

Eduardo Bayro Corrochano was supported by the project SO-201 of the Deutsche Forschungsgemeinschaft and CIMAT. Garret Sobczyk gratefully acknowledges the support of INIP of the Universidad de Las Americas-Puebla, and CIMAT-Guanajuato during his Sabbatical, Fall 1999.

Part V

Quantum and Neural Computing, and Wavelets

Chapter 14

Geometric Algebra in Quantum Information Processing by Nuclear Magnetic Resonance

Timothy F. Havel, David G. Cory, Shyamal S. Somaroo, and Ching-Hua Tseng

14.1   Introduction

The relevance of information theoretic concepts to quantum mechanics has been apparent ever since it was realized that the Einstein-Podolsky-Rosen paradox does not violate special relativity because it cannot be used to transmit *information* faster than light [22, 39]. Over the last few years, physicists have begun to systematically apply these concepts to quantum systems. This was initiated by the discovery, due to Benioff [3], Feynman [25] and Deutsch [17], that digital information processing and even universal computation can be performed by finite state quantum systems. Their work was originally motivated by the fact that as computers continue to grow smaller and faster, the day will come when they must be designed with quantum mechanics in mind (as Feynman put it, "there's plenty of room at the bottom"). It has since been found, however, that quantum information processing can accomplish certain cryptographic, communication, and computational feats that are widely believed to be classically impossible [5, 9, 19, 23, 40, 53],as shown for example by the polynomial-time quantum algorithm for integer factorization due to Shor [45]. As a result, the field has now been the subject of numerous popular accounts, including [1, 11, 37, 60]. But despite these remarkable theoretical advances, one outstanding question remains: Can a fully programmable *quantum computer* actually be built?

Most approaches to this problem (*loc. cit.*) have attempted to isolate a single submicroscopic system completely from its environment, so that it can be placed in a known quantum state and coherently controlled, for example by laser light. Although such precise state preparation will certainly be needed to implement a quantum computer that can be scaled to problems beyond the reach of classical computers, it is not an absolute prerequisite for the coherent control and observation of quantum dynamics. The most complex demonstrations of quantum information processing to date have in fact been achieved by liquid-state nuclear magnetic resonance (NMR) spectroscopy, using the spin 1/2 nuclei in macroscopic *ensembles* of molecules at room-temperature [12, 13, 15, 28]. Under these conditions the state of the nuclear spins is almost completely random, but information can nevertheless be stored in their joint statistics. This information is processed by combining the intra-molecular spin Hamiltonian with external radio-frequency fields. These fields are microscopically coherent, and can be engineered so as to act coherently across the entire sample. Special statistical states, called *pseudo-pure states*, can be prepared so that the macroscopic dynamics mirrors the microscopic dynamics of the spins. Finally, the spin degrees of freedom are remarkably well-isolated from the motional and electronic degrees of freedom, so that their decoherence times (i.e. the decay time for a quantum superposition) is typically on the order of seconds in the liquid state. Such decoherence is not only the chief obstacle to performing nontrivial quantum computations by any technology, but is increasingly recognized as playing a fundamental role in how quantum mechanics must be reconciled with classical physics [29].

Nuclear magnetic resonance also provides an experimental paradigm for the study of *multiparticle geometric algebra*, as elegantly developed in [20, 21, 48]. The reason is that the so-called *product operator formalism*, on which the modern theory of NMR spectroscopy is largely based [7, 8, 16, 24, 46, 47, 51, 57], is a nonrelativistic quotient of the multiparticle Dirac (i.e. space-time [33]) algebra. Thus NMR provides a natural and surprisingly easy way to experimentally verify some of the predictions of multiparticle quantum mechanics, as derived by geometric algebra. The existence of a concrete physical application for the theory is also likely to inspire new problems with a more general significance. In addition, NMR is perhaps the most broadly useful form of spectroscopy in existence today, and should greatly benefit from the adoption of the algebraic techniques and geometrical insights afforded by geometric algebra methods. These same benefits have already been shown to apply to the theory of quantum information processing, regardless of its physical realization [49]. The numerous connections between quantum information processing and foundational issues in quantum mechanics, particularly those pertaining to nonlocality and entanglement, bring the circle to a close.

This paper is intended to introduce physicists and mathematicians to the main ideas behind quantum information processing by liquid-state NMR spectroscopy, using the language and techniques of geometric algebra. The first section provides a brief overview of multiparticle geometric algebra, mainly to set the notation and terminology (more complete accounts may be found in the above references). The next section gives a quick introduction to quantum information processing, again referring to the literature for more complete accounts. This is followed by a detailed presentation of the basics of liquid-state NMR spectroscopy, using the product operator formalism, and how NMR can be used to perform universal logical operations on quantum information. The paper concludes with the results of recent experiments which show how geometric algebra can be used to "program" an NMR spectrometer to perform *analog* information processing, i.e. to directly simulate general quantum systems.

14.2   Multiparticle Geometric Algebra

Ever since Hestenes' pioneering work on the applications of geometric algebra to relativistic physics [33], it has been known that the Pauli algebra $\mathcal{G}_3$ is isomorphic to the *even subalgebra* $\mathcal{G}_{1,3}^+$ of the Dirac (space-time) algebra $\mathcal{G}_{1,3}$. This isomorphism is obtained by choosing an inertial frame $[\gamma_0, \gamma_1, \gamma_2, \gamma_3]$, where $\gamma_0^2 = -\gamma_\mu^2 = 1$ and $\gamma_\mu \gamma_\nu = -\gamma_\nu \gamma_\mu$ for all $0 \leq \nu < \mu \leq 3$, and defining the Pauli operators as:

$$\sigma_\mu \equiv \gamma_\mu \gamma_0 \tag{2.1}$$

Note that $\sigma_\mu \sigma_\nu = -\sigma_\nu \sigma_\mu$ $(1 \leq \nu < \mu \leq 3)$ and $\sigma_\mu \sigma_\mu = 1$ $(1 \leq \mu \leq 3)$, thus showing that this mapping gives the desired isomorphism.

The *multiparticle Dirac algebra* $\mathcal{G}_{N,3N}$ [20, 21, 48] is designed to model the internal degrees of freedom of spin 1/2 particles like electrons, protons and the atomic nuclei typically observed by NMR. It is obtained simply by taking a different orthogonal copy of space-time for each of the N distinguishable particles, with bases

$$\left[\gamma_0^k, \gamma_1^k, \gamma_2^k, \gamma_3^k \mid k = 1, \ldots, N\right] , \tag{2.2}$$

and considering the geometric algebra that they generate (note the use of Roman superscripts to label particle spaces). This algebra has dimension 2^{4N}. The subalgebra $(\mathcal{G}_{1,3}^+)^N$ of dimension 2^{3N} generated by the even subalgebras $\mathcal{G}_{1,3}^+$ from each particle space is endowed with a natural tensor product structure, since

$$\sigma_\mu^k \sigma_\nu^\ell = \gamma_\nu^\ell (\gamma_\mu^k \gamma_0^k) \gamma_0^\ell = \sigma_\nu^\ell \sigma_\mu^k \tag{2.3}$$

commutes for all $1 \leq k \leq \ell \leq N$. This plus the fact that it is the algebra, rather than just the underlying vector space, which is physically relevant, explains why the state space of a system of distinguishable particles is the tensor product $(\mathcal{G}_3)^{\otimes N}$ of their individual state spaces $\mathcal{G}_3$.

Nevertheless, this particular tensor product space appears to be larger than is actually needed, since physicists make do with the *complex* tensor product of the Pauli algebras, which has *real* dimension 2^{2N+1}. These superfluous degrees of freedom are due to the fact that the multiparticle geometric algebra contains a different unit pseudo-scalar in every particle space. They may be removed by projecting everything onto the ideal generated by the *correlator*,

$$C \equiv \tfrac{1}{2}(1 - \iota^1 \iota^2) \cdots \tfrac{1}{2}(1 - \iota^1 \iota^N) \,, \tag{2.4}$$

where $\iota^k \equiv \sigma_1^k \sigma_2^k \sigma_3^k$ is the unit pseudo-scalar associated with the k-th particle space. This primitive idempotent is easily seen to commute with the entire multiparticle Pauli algebra, and hence defines a homomorphism into an algebra $(\mathcal{G}_3)^{\otimes N}/C$ of the correct dimension. This C-correlated product of Pauli algebras, in turn, is isomorphic to the algebra of all $2^N \times 2^N$ complex matrices, and so capable of representing all the operations of nonrelativistic multiparticle quantum mechanics.

Interestingly, when restricted to the product of the even subalgebras of the embedded Pauli algebras, $(\mathcal{G}_3^+)^{\otimes N}$, factorization by C acts as an *isomorphism*. This C-correlated even algebra is isomorphic to a real tensor product of N quaternion algebras $(\mathcal{G}_3^+)^{\otimes N}$, and so has dimension 2^{2N} — the same as the real linear space of Hermitian matrices as well as the Lie algebra $\mathsf{u}(2^N)$ of the unitary group. Henceforth, the factor of C in all expressions will be dropped unless there is a specific reason to include it, and the pseudo-scalars from different particle spaces will be identified with the single unit imaginary

$$\iota \equiv \iota^1 C = \cdots = \iota^N C \,. \tag{2.5}$$

One can further define spinor representations of the rotation group $\mathsf{SO}(3)$ within the multiparticle geometric algebra [20, 21, 33, 48]. This relies upon the fact that spinors can be regarded as a minimal left-ideal in the algebra, which is generated by a primitive idempotent E. Including the correlator C, this idempotent may be written in product form as:

$$EC \equiv E_+^1 E_+^2 \cdots E_+^N C \quad (E_+^k \equiv \tfrac{1}{2}(1 + \sigma_3^k), \ k = 1, \ldots, N) \tag{2.6}$$

The left-ideal itself consists of those elements $\Psi \in (\mathcal{G}_3)^{\otimes N}$ such that $\Psi = \Psi\, EC$, which in the Pauli matrix representation of the C-correlated algebra corresponds to matrices with nonzero entries only in their leftmost column. These may be identified with the usual state vectors $|\psi\rangle$ of a (2^N)-dimensional Hilbert space.

Using the relation $E = \sigma_3^k E$ for all k, we can redefine our correlator C in this left-ideal to be

$$D \equiv \tfrac{1}{2}(1 - \iota \sigma_3^1 \iota \sigma_3^2) \cdots \tfrac{1}{2}(1 - \iota \sigma_3^1 \iota \sigma_3^N) \,, \tag{2.7}$$

which will be referred to as the *directional correlator*. It can be shown that, in contrast to C, right-multiplication by D maps the tensor product of quaternion algebras $(\mathcal{G}_3^+)^{\otimes N}$ onto a subalgebra of the correct dimension

2^{N+1}, and that the quaternion algebra in every particle space acts by left-multiplication to give a spinorial representation of the rotation group [21]. Thus this subalgebra provides a covariant parametrization for the space of N-particle states, and for most purposes one can drop the idempotent $\boldsymbol{E}$ and work directly in this subalgebra. Its elements $\psi = \psi \boldsymbol{D}$ are accordingly called *spinors*.

The Pauli operators themselves act on the corresponding one-particle spinors according to

$$\begin{aligned}
\sigma_\mu |\psi\rangle &\leftrightarrow \sigma_\mu \circ \psi \equiv \sigma_\mu \psi \sigma_3 \quad (1 \leq \mu \leq 3) \\
\imath |\psi\rangle &\leftrightarrow \imath \circ \psi \equiv \imath \psi \sigma_3 = \psi \imath \sigma_3 ,
\end{aligned} \tag{2.8}$$

where right-multiplication by σ_3 keeps the results in the Pauli-even subalgebra. This can be viewed as a projection of the geometric product times $\boldsymbol{E}_+$ back into the even subalgebra, since

$$\sigma_\mu \psi \sigma_3 = \sigma_\mu \psi (\boldsymbol{E}_+ - \boldsymbol{E}_-) = \sigma_\mu \psi \boldsymbol{E}_+ + (\widehat{\sigma_\mu \psi \boldsymbol{E}_+}) , \tag{2.9}$$

where the hat "$\widehat{}$" denotes the main involution or parity operation in $\mathcal{G}_3$, which changes the sign of the odd components. This action is readily extended, in a well-defined fashion, to an action of the $\boldsymbol{C}$-correlated products of the Pauli operators on the $\boldsymbol{D}$-correlated products of elements from the even subalgebras of multiple particles.

The multiparticle Dirac algebra is essential to understanding the geometric origin of the tensor product in multi-spin quantum physics, which in turn plays a central role in both quantum computing and NMR (*vide infra*). The remainder of this paper, however, will make direct use of only the non-relativistic quotient algebra. In this regard, it is important to note that the Dirac reverse $\tilde{\boldsymbol{\Gamma}}$ of any $\boldsymbol{\Gamma} \in (\mathcal{G}_{1,3}^+)$ corresponds to the conjugate (i.e. reversion composed with the main involution) in $\mathcal{G}_3$, whereas the Pauli algebra reverse corresponds to the frame-dependent operation $\gamma_0 \tilde{\boldsymbol{\Gamma}} \gamma_0$. Henceforth, the notation $\tilde{\boldsymbol{\Gamma}}$ will be used exclusively for the Pauli algebra reverse. This operation is readily extended to the multiparticle Pauli algebra by defining $(\boldsymbol{\Gamma}^1 \boldsymbol{\Gamma}^2)^\sim \equiv \tilde{\boldsymbol{\Gamma}}^1 \tilde{\boldsymbol{\Gamma}}^2$, and remains well-defined after correlation. In the usual matrix representation, this operation is just the Hermitian conjugate.

14.3 Algorithms for Quantum Computers

Because of the tensor product involved, the exact representation of a collection of finite-state quantum systems on a classical computer takes an amount of memory which grows exponentially with the number of systems. As first noted by Feynman [25], this implies that it may be possible to simulate the evolution of one collection of finite-state quantum systems by another, using only *polynomial* resources (i.e. time and memory). The idea of operating on digital information stored in finite-state quantum systems originated with Benioff [3], and was extended by Deutsch [17] to show that discrete problems can also be solved more rapidly on a quantum computer. At this time, however, very few problems are known which can be solved *exponentially* more rapidly, the most notable being Shor's integer factorization algorithm [45]. A quantum algorithm for solving general search problems with a quadratic speed-up over linear search is available [30], but it is now widely believed that the important class of NP-complete problems [27] cannot be solved in polynomial time even on a quantum computer [4]. The advantages that have been demonstrated are nevertheless significant, and much remains to be learned.

In its standard form, a quantum computer stores binary information in an ordered array of distinguishable two-state quantum systems, e.g. spin $1/2$ particles. These are usually referred to as "qubits". In keeping with their usage, the two orthogonal basis states that represent binary "0" and "1" are denoted by $|0\rangle$ and $|1\rangle$, respectively. Thus a two-bit quantum computer stores the integers 0, 1, 2 and 3 in binary notation as $|00\rangle$, $|01\rangle$, $|10\rangle$ and $|11\rangle$, where

$$|\delta^1\delta^2\rangle \equiv |\delta^1\rangle|\delta^2\rangle \equiv |\delta^1\rangle \otimes |\delta^2\rangle \tag{3.10}$$

($\delta^1, \delta^2 \in \{0,1\}$). This extends in the obvious way to an arbitrary number of qubits N. The interesting feature of qubits is their ability to exist in *superposition* states, $c_0|0\rangle + c_1|1\rangle$ ($c_0, c_1 \in \mathbb{C}$, $|c_0|^2 + |c_1|^2 = 1$). Such a state is not between $|0\rangle$ and $|1\rangle$, as in an analog classical computer with continuous voltages, nor is it really in both states at once, as sometimes stated. It can most accurately be said to be in an indeterminate state, which specifies only the probability $|c_0|^2$ and $|c_1|^2$ with which $|0\rangle$ and $|1\rangle$ will be observed on testing it for this property.

By itself, this is nothing that could not be done on a classical computer with a good random number generator, but things get more interesting when one considers superpositions over multiple qubits, e.g.

$$\begin{aligned} |\psi\rangle &\equiv \tfrac{1}{2}(|00\rangle - |01\rangle + |10\rangle - |11\rangle) \\ &= \tfrac{1}{2}(|0\rangle + |1\rangle)(|0\rangle - |1\rangle) . \end{aligned} \tag{3.11}$$

Let $\boldsymbol{U}_f$ be a unitary transformation of the two qubits, which is defined on the computational basis by

$$\begin{aligned} \boldsymbol{U}_f|00\rangle = |0\rangle|f(0)\rangle , \quad &\boldsymbol{U}_f|01\rangle = |0\rangle|1 - f(0)\rangle , \\ \boldsymbol{U}_f|10\rangle = |1\rangle|f(1)\rangle , \quad &\boldsymbol{U}_f|11\rangle = |1\rangle|1 - f(1)\rangle , \end{aligned} \tag{3.12}$$

where $f : \{0,1\} \to \{0,1\}$ is one of the four possible invertible boolean functions of a single bit, and extended to all superpositions by linearity. This implies that the application of $\boldsymbol{U}_f$ to a superposition over its input (left) qubit effectively computes the value of f on both inputs at once. Applied to the superposition state $|\psi\rangle$ above, where the output (right) qubit is also in a superposition, we obtain after straightforward rearrangements:

$$\boldsymbol{U}_f|\psi\rangle = \tfrac{1}{2}\left((-1)^{f(0)}|0\rangle + (-1)^{f(1)}|1\rangle\right)(|0\rangle - |1\rangle) \qquad (3.13)$$

Now consider a second unitary transformation of the qubits $\boldsymbol{R}_\mathsf{H}$, which is called the *Hadamard transform* and defined on a basis for each bit by

$$\boldsymbol{R}_\mathsf{H}|0\rangle = \tfrac{1}{\sqrt{2}}(|0\rangle + |1\rangle), \quad \boldsymbol{R}_\mathsf{H}|1\rangle = \tfrac{1}{\sqrt{2}}(|0\rangle - |1\rangle) . \qquad (3.14)$$

This is easily seen to transform the above as follows:

$$\boldsymbol{R}_\mathsf{H}\boldsymbol{U}_f|\psi\rangle = \tfrac{1}{\sqrt{2}}\Big(((-1)^{f(0)} + (-1)^{f(1)})|0\rangle + \qquad (3.15)$$
$$((-1)^{f(0)} - (-1)^{f(1)})|1\rangle\Big)|1\rangle$$

Thus if $f(0) = f(1)$ (i.e. f is a constant function), testing the "input" qubit will yield $|0\rangle$ with probability 1, whereas if $f(0) = 1 - f(1)$ (i.e. f is a "balanced" function), it will yield $|1\rangle$ with probability 1. The interesting thing is that this is done with but a single "evaluation" of the function f (via $\boldsymbol{U}_f$), whereas distinguishing these two cases classically would require *two* evaluations. This quantum algorithm is due to Deutsch & Jozsa [18].

The feature of quantum mechanics that makes this possible is the coherent mixing of the basis states by the Hadamard transform, so that those corresponding to the desired solution are amplified and the phase differences among the remainder result in cancellation. Because this can also occur when the state of a qubit is correlated with its spatial coordinate, as in optical diffraction, this is often referred to as interference. By itself, it does not yield an asymptotic reduction in the computation time required, but when combined with the exponential growth in the state space with the number of particles, it becomes possible to cancel exponentially large numbers of possibilities and hence attain exponential speed-ups, as in Shor's algorithm.

It should be noted that *factorizable* states, i.e. those that can written as a product of superpositions over the individual qubits (as in $|\psi\rangle$ above) are effectively parametrized by the coefficients c_0^k and c_1^k of the qubits. Taking the constraints $|c_0^k|^2 + |c_1^k|^2 = 1$ ($1 \le k \le N$) and the fact that there is but a single global phase into account, this implies that the dimension of the manifold of such states increases as $2N + 1$, *not* exponentially. The exponential growth in the dimension thus requires that states can be created which are nonfactorizable, or *entangled*. Entangled states are not only required for efficient quantum computing, but are the source of many quantum "paradoxes" as well [39].

The Hadamard transform is a simple example of a *quantum* logic gate, which maps basis states to superpositions. Unitary transformations like

U_f, on the other hand, constitute logical operations with classical boolean analogues, which must however be *reversible* (since unitary transformations are always invertible). The simplest example is the NOT gate: $N|0\rangle = |1\rangle$ and $N|1\rangle = |0\rangle$. More interesting calculations require feedback, i.e. operation on one qubit conditional on the state of another. Although reversibility precludes operations like AND, which have two inputs but only one output, the XOR gate, with the second qubit passed through unchanged, can be realized as a unitary transformation:

$$S^{1|2}|00\rangle = |00\rangle, \quad S^{1|2}|10\rangle = |10\rangle,$$
$$S^{1|2}|01\rangle = |11\rangle, \quad S^{1|2}|11\rangle = |01\rangle. \tag{3.16}$$

For obvious reasons, this is sometimes called a *controlled-NOT*, or c-NOT, gate. The corresponding three-qubit analog $T^{1|23}$, which NOT's the first qubit if the other two are both $|1\rangle$, is known as the *Toffoli gate* after the person who first realized that it is universal for boolean logic [55]. This follows from the fact that, if one sets the first (target) input bit to 1, the output is the NAND of the other two inputs.

More generally, the c-NOT gates, together with all one-qubit quantum gates, generate the entire unitary group $U(2^N)$ on N qubits [2]. The general problem of "compiling" any given gate U whose generator $\log(U)/(\pi\iota)$ can be factorized into commuting product operators will be solved constructively by geometric algebra below. Nevertheless, the important issue is to characterize those unitary transformations which admit efficient implementations, meaning that the number of "elementary operations" involved grows only polynomially in the number of qubits affected. Such elementary operations are usually required to be "local", in that they involve only a few qubits at a time. The natural Hamiltonians of NMR, for example, have at most two spins in any term, but can only be simulated classically using exponential resources.

14.4   NMR and the Product Operator Formalism

In liquid-state NMR one deals with ensembles of molecules whose spins are in a *mixed state*. A concise description of the relevant statistics is given by the *density operator* [6]. A matrix for the density operator of a *pure* (i.e. single) quantum state is obtained from the corresponding state vector by forming the dyadic product $|\psi\rangle\langle\psi|$ ($\langle\psi|\psi\rangle = 1$). As shown in Refs. [32, 48, 49], the geometric algebra analog of the dyadic product is $\psi E \tilde{\psi}$ ($\langle\psi\tilde{\psi}\rangle = 1$). The density operator of a general mixed state is a convex combination of the density operators of its constituent spin states, namely

$$\rho = 2^N \sum_j p_j \, \psi_j E \, \tilde{\psi}_j, \tag{4.17}$$

where $p_j \geq 0, \sum_j p_j = 1$ can be interpreted as the probabilities of the spin states in the ensemble. Such a representation is, in general, highly redundant. Because the density operator is necessarily Hermitian (reversion symmetric in the product Pauli algebra), a nonredundant, real parametrization can be obtained by expanding it in the *product operator* basis

$$\rho = \sum_{\mu \in \{0,\ldots,3\}^N} \varrho_\mu \, \sigma^1_{\mu^1} \cdots \sigma^N_{\mu^N} , \qquad (4.18)$$

where $\sigma_0 \equiv 1 \equiv 1$.

Evolution of a spinor under a time-independent Hamiltonian H is described by operation with the corresponding propagator as in Eq. 2.8:

$$|\psi\rangle \leftrightarrow \psi \quad \longrightarrow \quad \exp(-t\iota H)|\psi\rangle \leftrightarrow \exp(-t\iota H) \circ \psi \qquad (4.19)$$

Since $\sigma^k_3 E = E = E\sigma^k_3$ for all $1 \leq k \leq N$, it follows that the density operator itself evolves by two-sided multiplication with the propagator and its reverse (i.e. conjugation in the multiplicative group):

$$\rho \quad \longrightarrow \quad \exp(-t\iota H)\rho \exp(t\iota H) = $$
$$2^N \sum_j p_j \, \exp(-t\iota H)\psi_j \, E \, \tilde{\psi}_j \, \exp(t\iota H) \qquad (4.20)$$

Similarly, the expected value of an observable with Hermitian operator A is given by the average of its quantum mechanical expectation values $\langle \psi_j | A | \psi_j \rangle \leftrightarrow 2^N \langle A \psi_j \, E \, \tilde{\psi}_j \rangle$ over the ensemble (where $\langle \cdot \rangle$ denotes the scalar part). It follows that these averages may be obtained directly from the density operator itself as

$$2^N \sum_j p_j \langle A \psi_j E \tilde{\psi}_j \rangle = $$
$$2^N \left\langle A \sum_j p_j \psi_j E \tilde{\psi}_j \right\rangle = \langle A \rho \rangle . \qquad (4.21)$$

It may be seen that the factor of 2^N in our definition of the density operator (Eq. 4.17) compensates for the factor of 2^{-N} in the idempotent E, so that $\langle \rho \rangle = 1$. This normalization of ρ differs from the usual normalization to a trace of unity in a matrix representation, but saves on factors of 2^N when using geometric algebra.

By our remark following Eq. 2.4, it is also possible to represent the Hamiltonians of NMR in product operator notation. The dominant term in these Hamiltonians is the *Zeeman interaction* of the magnetic dipoles of

---

One might hope that one could drop the idempotent in these definitions, as was done previously for spinors, and work with the convex span of products of the form $\psi\tilde{\psi}$. Since these products are even and reversion symmetric, however, they have no bivector part, and thus they do not span enough degrees of freedom to encode for density operators.

the spins with the applied magnetic field $\boldsymbol{B}_0$. Assuming as usual that the field is directed along the z-axis, this term may be written as

$$\boldsymbol{H}_{\mathsf{Z}} \equiv -\tfrac{\hbar}{2}\|\boldsymbol{B}_0\| \sum_{k=0}^{N} \gamma^k (1-\delta^k)\sigma_3^k \equiv -\tfrac{1}{2}\sum_{k=0}^{N} \omega_0^k \sigma_3^k , \quad (4.22)$$

where γ^k here denotes the gyromagnetic ratio of the k-th nucleus, and $\delta^k \ll 1$ is an empirical correction called the *chemical shift* which describes the diamagnetic shielding of the nucleus by the electrons in the molecule. In most of what follows, it will be assumed that we are working with a homonuclear system, wherein $\gamma^k = \gamma^\ell \equiv \gamma$ for all $1 \le k < \ell \le N$.

In accord with the forgoing observations, the density operator of an ensemble of N-spin systems evolves under the Zeeman Hamiltonian as

$$\rho \longrightarrow \exp(-t\iota\boldsymbol{H}_{\mathsf{Z}})\,\rho\exp(t\iota\boldsymbol{H}_{\mathsf{Z}}) =$$

$$\sum_{\mu\in\{0,\dots,3\}^N} \varrho_\mu \exp(-t\iota\omega_0^1\sigma_3^1/2)\,\sigma_{\mu^1}^1 \exp(t\iota\omega_0^1\sigma_3^1/2) \quad (4.23)$$

$$\cdots \exp(-t\iota\omega_0^N\sigma_3^N/2)\,\sigma_{\mu^N}^N \exp(t\iota\omega_0^N\sigma_3^N/2) .$$

Thus the vector given by those terms depending on just a single spin index, e.g. $\varrho_1^k\sigma_1^k + \varrho_2^k\sigma_2^k + \varrho_3^k\sigma_3^k$ ($\varrho_{\nu^k}^k \equiv \varrho_{0\dots\nu^k\dots0}$), precesses about the applied magnetic field at a constant rate ω_0^k. This so-called *Bloch vector* describes the observable macroscopic magnetization due to polarization of the k-th spin over all molecules of the ensemble [24, 26].

In NMR spectroscopy, the spins are controlled by pulses of RF (radiofrequency) radiation about the z axis. The corresponding Hamiltonian

$$\boldsymbol{H}_{\mathsf{RF}} = -\tfrac{1}{2}\sum_{k=1}^{N} \omega_1^k \left(\cos(\omega t)\sigma_1^k + \sin(\omega t)\sigma_2^k\right) \quad (4.24)$$

is *time-dependent*, which normally makes it impossible to give a closed-form solution. Fortunately, in the present case it is possible to transform everything into frame which rotates along with the RF field $\boldsymbol{B}_1$, so that if $\hbar\gamma\|\boldsymbol{B}_1\| \equiv \omega_1^k \gg |\omega_0^k - \omega|$ for all k (i.e. the pulse is strong and hence can be made short enough that the relative precession of the spins over its duration is negligible), we can regard $\boldsymbol{H}_{\mathsf{RF}}$ as a time-independent Hamiltonian which rotates each spin at the constant rate ω_1^k about the x-axis in the rotating frame. By changing the phase of the pulse, one can rotate about any desired axis in the transverse (xy) plane. Henceforth all our transformations will be relative to such a rotating frame (cf. [34]).

The spins, of course, also interact with one another. This paper is exclusively concerned with the NMR of molecules in liquids, where the rapid diffusional motion of the molecules averages the through-space interactions between their nuclear magnetic dipoles to zero much more rapidly than those interactions can have any net effect. Thus the only effective interaction between the nuclei is a through-bond interaction known as *scalar coupling*. Assuming the differences in the resonance frequencies $\omega_0^k - \omega_0^\ell$ of the spins are substantially larger than the scalar coupling constants $J^{k\ell}$ among them, the transverse components $J^{k\ell}(\sigma_1^k\sigma_1^\ell + \sigma_2^k\sigma_2^\ell)$ in each term of this Hamiltonian are similarly averaged to zero by their rapid differential precession. It follows that its effects are well-approximated by the remaining terms parallel to the $\boldsymbol{B}_0$ field, i.e.

$$H_\mathsf{J} \equiv \tfrac{\pi}{2} \sum_{k<\ell} J^{k\ell}\,\sigma_3^k\sigma_3^\ell \tag{4.25}$$

(this is known as the secular, or weak-coupling, approximation). This Hamiltonian transforms the observable "single quantum" (i.e. single Pauli operator) terms according to

$$\begin{aligned}
\sigma_1^k \;\longrightarrow\; & \cos(t\pi J^{k\ell})\sigma_1^k + \sin(t\pi J^{k\ell})\sigma_2^k\sigma_3^\ell \\
= \; & \exp(-\iota t\pi J^{k\ell}\sigma_3^k)\sigma_1^k E_+^\ell + \exp(\iota t\pi J^{k\ell}\sigma_3^k)\sigma_1^k E_-^\ell \;.
\end{aligned} \tag{4.26}$$

In terms of Bloch diagrams (see Figure 14.1), this later form also shows that the magnetization vectors due to spin 1 in those molecules wherein spin 2 is $|0\rangle$ and $|1\rangle$ turn clockwise and counterclockwise in a frame which

In-phase Absorptive	In-phase Dispersive	Anti-phase Absorptive	Anti-phase Dispersive

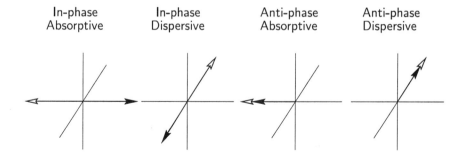

FIGURE 14.1. Bloch diagrams depicting the "single quantum" in-phase absorptive (σ_1^1), dispersive (σ_2^1) and anti-phase absorptive $(\sigma_1^1\sigma_3^2)$, dispersive $(\sigma_2^1\sigma_3^2)$ states of a two-spin system. Vectors with an empty head represent the magnetization from spin 1 in those molecules wherein the second spin is "up" (i.e. $\sigma_\mu^1 E_+^2$, $\mu = 1, 2$) while vectors with a filled head represent the magnetization from spin 1 in those molecules wherein the second spin is "down" (i.e. $\sigma_\mu^1 E_-^2$, $\mu = 1, 2$). Under scalar coupling, these two components of the magnetization counter-rotate at a rate of $2/J$, where J is the scalar coupling strength in Hz, thereby transforming in-phase absorptive into anti-phase dispersive and in-phase dispersive into anti-phase absorptive (see text).

co-rotates with spin 1, respectively, at a rate of $J^{k\ell}/2$ sec^{-1}. It will be shown shortly how this interaction can be used to perform conditional logic operations on the spins.

The final issue to be dealt with is how the density operator and Hamiltonian are manifest in the spectra obtained by NMR. As mentioned above, the precessing magnetic dipole of each spin is described by those components of the density operator which depend on just that spin index. The transverse component of this dipole produces an oscillating signal in the receiver coils, whose Fourier transform contains a peak at the precession frequency ω_0^k of each spin. According to the usual phase conventions of NMR, the peak from $\boldsymbol{\sigma}_1^k$ has an absorptive shape, while that from $\boldsymbol{\sigma}_2^k$ is dispersive (see Figure 14.2). The frequencies of the spins are further modulated by the scalar coupling interactions, which split each peak into a *multiplet* of at most 2^{N-1} peaks at frequencies of $\omega_0^k \pm \pi J^{k1} \pm \cdots \pm \pi J^{kN}$. By multiplying Eq. 4.26 through by $\boldsymbol{\sigma}_3^\ell$ and using the fact that $\boldsymbol{E}_\pm^\ell \boldsymbol{\sigma}_3^\ell = \pm \boldsymbol{E}_\pm^\ell$, it can be shown that transverse-longitudinal correlations (e.g. $\boldsymbol{\sigma}_1^k \boldsymbol{\sigma}_3^\ell$) evolve into observable terms (e.g. $\boldsymbol{\sigma}_2^k$) at frequencies of $\pm \pi J^{k\ell}$, but with *opposite* signs. It follows that the pairs of peaks they generate likewise have opposite sign, or are *anti-phase*, as opposed to *in-phase* peaks with the same signs (see Figure 14.2).

Thus, in effect, an NMR spectrum enables us to directly readout all terms of the density operator with just one transverse component. By collecting spectra following $\pi/2$ readout pulses selective for each spin, it is possible to reconstruct the density operator completely. This kind of measurement contrasts starkly with measurements on single quantum systems, which induce "wave function collapse" to a random eigenstate of the observable so that the density operator can only be reconstructed by collecting statistics over repeated experiments. That wave function collapse does not occur is due to the fact that averages over the ensemble contain insignificant *information* on any one system in it. Such ensemble measurements are sometimes called *weak measurements*, to distinguish them from strong measurements on single quantum systems [39].

14.5   Quantum Computing by Liquid-State NMR

Even at the highest available magnetic fields, the energy of the nuclear Zeeman interaction is at most about 10^{-5} of mean thermal energy per degree of freedom $k_\mathrm{B}T/2$ (where k_B is Boltzmann's constant and T the absolute temperature). Thus in liquid-state NMR the equilibrium state of the spins is almost totally random, so that the probabilities of finding a spin "up" (parallel the field) and "down" are nearly equal. According to the principles of statistical mechanics, these probabilities are given by 2^{-N}

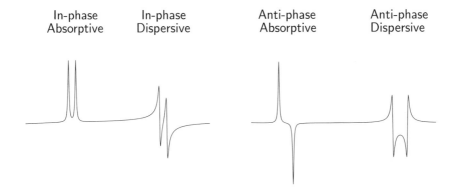

| In-phase | In-phase | Anti-phase | Anti-phase |
| Absorptive | Dispersive | Absorptive | Dispersive |

FIGURE 14.2. Simulated NMR spectra for a weakly coupled two-spin molecule (amplitude of the real part versus frequency). On the left is the spectrum of the spin state $\sigma_1^1 + \sigma_2^2$, which gives a pair of in-phase absorptive peaks for spin 1 (left) and a pair of in-phase dispersive peaks for spin 2 (right). On the right is the spectrum of the spin state $\sigma_1^1 \sigma_3^2 + \sigma_3^1 \sigma_2^2$, which gives a pair of anti-phase absorptive peaks for spin 1 (left) and a pair of anti-phase dispersive peaks for spin 2 (right). Fits to the peak shapes in such spectra after various $\pi/2$ rotations of the individual spins yield sufficient information to uniquely reconstruct the complete density operator.

times the eigenvalues of

$$\rho_{eq} = \frac{\exp(-\boldsymbol{H}_{\mathsf{Z}}/(k_{\mathrm{B}}T))}{\langle \exp(-\boldsymbol{H}_{\mathsf{Z}}/(k_{\mathrm{B}}T)) \rangle} \approx 1 - \boldsymbol{H}_{\mathsf{Z}}/(k_{\mathrm{B}}T) , \qquad (5.27)$$

where the right-hand side is known as the *high-temperature approximation*. Expanding the Zeeman Hamiltonian yields the (high-temperature) equilibrium density operator in product operator notation:

$$\rho_{eq} = 1 + \sum_k \omega_0^k \sigma_3^k / (2k_{\mathrm{B}}T) \qquad (5.28)$$

Since the observables of NMR σ_1^k and σ_2^k have no scalar part, it follows from Eq. 4.21 that the scalar part of any density operator produces no net NMR signal. It also does not evolve under unitary operations, and hence NMR spectroscopists usually drop it altogether. Assuming a homonuclear system (so that $\omega_0^k \approx \omega_0^\ell \equiv \omega_0 \equiv \hbar\gamma\|\boldsymbol{B}_0\|$ for all $1 \leq k, \ell \leq N$), it is also common practice to drop the constant factor $\Delta_0 \equiv \omega_0/(2k_{\mathrm{B}}T)$ in the above. Then the eigenvalues of this "density operator" $\check{\rho}_{eq}$ are given by $\check{\varrho}_i = N - 2\#i$, where $\#i$ is the *Hadamard weight* (number of ones in the binary expansion) of the integer $i = 0, \ldots, 2^N - 1$. Their multiplicities are the binomial coefficients $\binom{N}{\#i}$. For example, the Pauli matrix representation of the homonuclear two-spin equilibrium density operator is:

$$\check{\rho}_{eq} \equiv \sigma_3^1 + \sigma_3^2 \leftrightarrow 2 \begin{bmatrix} 1 & 0 & 0 & 0 \\ 0 & 0 & 0 & 0 \\ 0 & 0 & 0 & 0 \\ 0 & 0 & 0 & -1 \end{bmatrix} \tag{5.29}$$

The obvious way to store binary information in an ensemble of spin systems at equilibrium is to regard each *chemically* distinct type of spin as a "bit" which represents 0 or 1 as its net polarization is up or down, respectively, so that the integers from $i = 0, \ldots, 2^N - 1$ are stored in the states $\pm \sigma_3^1 \pm \cdots \pm \sigma_3^N$. This is, however, a very different thing than storing these integers in the pure states $|i\rangle$ (the Zeeman basis vector obtained by binary expansion of the integer i), the density operators of which are signed sums over all possible products of the form $\sigma_3^{k_1} \cdots \sigma_3^{k_n}$ $(1 \le k_1 < \cdots < k_n \le N)$, as in Eq. 4.18 with $\varrho_\mu = \pm 1$ for all $\mu \in \{0, 3\}$ and 0 otherwise. The problem is that, without these higher-order $(n > 1)$ product terms, it is not possible to perform *conditional* operations on the state, for the simple reason that by linearity these operations act independently on each term of the sum. These higher-order terms are nonnegligible in the equilibrium state only at temperatures approaching absolute zero — which is not an option available in liquid-state NMR!

A class of weakly polarized nonequilibrium states nevertheless exists in which the linear and higher-order terms are all present with equal magnitudes, as they are in pure states. These states, usually known as *pseudo-pure* states [13, 15, 28, 32, 36], are also characterized by having a single non-degenerate eigenvalue in the standard matrix representation, so that they may be written as a trace-preserving rank 1 perturbation on the identity:

$$\begin{aligned} \rho_{pp} &= (1 - \Delta) + 2^N \Delta |\psi\rangle\langle\psi| \\ &\leftrightarrow (1 - \Delta) + 2^N \Delta \psi E \tilde{\psi} \end{aligned} \tag{5.30}$$

The perturbation parameter Δ is restricted by the requirement that the density operator be positive-semidefinite to $-1/(2^N - 1) \le \Delta \le 1$. Assuming that the pseudo-pure state is at equilibrium versus a Hamiltonian of the form $H_0(E - 2^{-N})$, this is related to the polarization $-1 \le \Delta_0 \le 1$ of a single spin versus $H_0 \sigma_z/2$ by $\Delta = \Delta_0/((1 - \Delta_0)2^{N-1} + \Delta_0)$. For example, the two-spin pseudo-pure ground state is given by the above with $|\psi\rangle = |00\rangle \leftrightarrow \psi = 1$, i.e.

$$\begin{aligned} \rho_{00} &= 1 + \Delta\left(2^2 E - 1\right) = 1 + \Delta\left(4E_+^1 E_+^2 - 1\right) \\ &= 1 + \Delta\left(\sigma_3^1 + \sigma_3^2 + \sigma_3^1\sigma_3^2\right). \end{aligned} \tag{5.31}$$

Since the identity 1 commutes with everything, pseudo-pure states are necessarily mapped to pseudo-pure states by unitary operations, and so provide a carrier space for a representation of $\mathsf{SU}(2^N)$ (modulo phase) just like true pure states.

In addition, the fact that pseudo-pure states are realized in the statistics of macroscopic ensembles of identical quantum systems implies that the available measurements are *weak* (section 14.4). Thus NMR measurements on pseudo-pure states actually enable one to directly obtain the expectation value of any observable $\boldsymbol{A}$ relative to the perturbation spinor $\boldsymbol{\psi}$, i.e.

$$\langle \boldsymbol{\rho A} \rangle \;=\; (1-\Delta)\langle \boldsymbol{A} \rangle + 2^N \Delta \langle \boldsymbol{A\psi E\tilde{\psi}} \rangle \;\leftrightarrow\; \Delta \langle \psi \,|\, \boldsymbol{A} \,|\, \psi \rangle \,, \tag{5.32}$$

which follows from the fact that NMR observables have no scalar part. The ensemble nature of NMR also permits certain types of *non*-unitary operations to be performed on the system. Since the eigenvalues of the high-temperature equilibrium and pseudo-pure density operators are different, the preparation of pseudo-pure states necessarily involves such non-unitary operations. There are presently four methods of implementing non-unitary operations in NMR, each of which leads to a physically different (though mathematically equivalent) type of pseudo-pure state.

The conceptually simplest type is a *temporal* pseudo-pure state, which is obtained by averaging the results (signals or spectra) of experiments performed at different times on different states, such that the sum of their density operators is pseudo-pure. This is analogous to *phase-cycling* in NMR [24, 26]. For example, up to a factor of 2/3, the average of the following three two-spin states clearly has the same nonscalar part as the above pseudo-pure ground state:

$$\begin{aligned}
\rho_{\mathsf{A}} &= 1 + \Delta \left(\sigma_3^1 + \sigma_3^2 \right) \\
\rho_{\mathsf{B}} &= 1 + \Delta \left(\sigma_3^1 + \sigma_3^1 \sigma_3^2 \right) \\
\rho_{\mathsf{C}} &= 1 + \Delta \left(\sigma_3^1 \sigma_3^2 + \sigma_3^2 \right)
\end{aligned} \tag{5.33}$$

The first state is the equilibrium state, while the other two may be obtained by permuting the populations in the equilibrium state by the c-NOT gates $\boldsymbol{S}^{2|1}$ and $\boldsymbol{S}^{1|2}$, respectively.

Another way to perform non-unitary operations in NMR relies upon the fact that the observed signal is an integral over the sample volume. Thus if one can create a distribution of states across the sample such that their average is pseudo-pure, one obtains a *spatial* pseudo-pure state. The most straightforward way to do this is to apply a *magnetic field gradient* across the sample, usually a linear gradient along the z-axis parallel to the applied magnetic field $\boldsymbol{B}_0$. This causes the spins to precess at differing rates, depending on their z-coordinates, so that the net transverse magnetization vector perpendicular to the z-axis is wound into a spiral whose average is zero. The transverse phase information thus rendered unobservable is exactly that which would be lost in a strong measurement of the spins along the z-axis, but with the rather striking difference that this phase information can be recovered by inverting the gradient. The next section will present specific RF and gradient pulse sequences for spatial pseudo-pure states.

A rather different approach is to "label" a pseudo-pure subensemble of the spins by a specific state of one or more "ancilla" spins. This approach was used over 20 years ago to demonstrate spinor behavior under rotations by NMR [54], and was first applied to NMR computing by Chuang *et al.* [28, 58]. The correlation with ancilla spin states permits the signal from the pseudo-pure subensemble to be isolated by filtering based on the frequency shifts induced by scalar-coupling. The simplest example of such a *conditional* pseudo-pure state is

$$
\begin{aligned}
\rho \; &= \; 1 + \Delta(\sigma_3^1 + \sigma_3^2 + \sigma_3^1\sigma_3^2)\sigma_3^3 \\
&= \; \left(1 + \Delta(4E_+^1 E_+^2 - 1)\right) E_+^3 + \\
&\quad\; \left(1 - \Delta(4E_+^1 E_+^2 - 1)\right) E_-^3 \; .
\end{aligned}
\tag{5.34}
$$

It can be shown that this state is related to the equilibrium state by unitary c-NOT operations. The latter expression in the equation makes it clear that in the subensemble wherein spin 3 is "up" (i.e. in its ground state E_+^3) the spins 1 & 2 have a population excess in their ground state (assuming $\Delta > 0$). Similarly, in the subensemble with spin 3 "down" spins 1 & 2 have a population deficit in their ground state. Significantly, therefore, on average across the entire ensemble spins 1 and 2 are entirely unpolarized (i.e. random). This can be seen in NMR by *decoupling* spin 3, i.e. by rotating it rapidly with an RF field so that its interactions with spins 1 and 2 are averaged to zero. This effectively removes spin 3 from the system, so that (in the above situation) the spectrum of spins 1 and 2 is reduced to a flat line.

The general operation of "removing" a qubit from a system is known in quantum computing as the *partial trace*. As shown in Ref. [49], this corresponds to dropping all terms which depend upon the spin over which the partial trace is taken in the product operator expansion of the overall density operator. It provides us with our fourth type of pseudo-pure state, which is called a *relative* pseudo-pure state. An example in this case is given by [32]

$$
\begin{aligned}
\check{\rho} \; &= \; 4\left((E_+^1 + E_+^2)E_+^3 E_+^4 + (E_+^1 E_+^2 - E_-^1 E_+^2)E_+^3 E_-^4 + \right.\\
&\quad\; \left.(E_-^1 E_-^2 - E_+^1 E_-^2)E_-^3 E_+^4 - (E_-^1 + E_-^2)E_-^3 E_-^4\right) \\
&= \; 2\left(\sigma_3^3 + \sigma_3^4\right) + \sigma_3^2\left(\sigma_3^3 - \sigma_3^4\right) + \left(\sigma_3^1 + \sigma_3^2 - \sigma_3^1\sigma_3^2\right)\sigma_3^3\sigma_3^4 \\
&\quad\; + \left(\sigma_3^1 + \sigma_3^2 + \sigma_3^1\sigma_3^2\right) \; ,
\end{aligned}
\tag{5.35}
$$

wherein it may be seen that tracing over spins 3 and 4 leaves only the bottom line, which is a two-spin pseudo-pure state. This density operator is again related to the four-spin equilibrium density operator by unitary operations.

The one thing that all these methods of preparing pseudo-pure states from equilibrium states have in common is a rapid loss of signal strength with the number of spins in the resulting pseudo-pure state. This can be understood most simply from the fact that, since the number of Zeeman basis states grows exponentially with the number of spins, at any fixed polarization (specific entropy) the expected population in any one state must likewise decline exponentially [59]. Nevertheless, current methods should be able to prepare usable pseudo-pure states on up to *ca.* 8 − 12 spins.

In addition, it is also at least difficult to study nonlocal effects by NMR, since that would require allowing the spins to interact by scalar coupling through a chemical bond, then rapidly breaking the bond, separating the molecular fragments, and performing further measurements. A more fundamental problem lies in the fact that the microscopic interpretation of experiments on weakly polarized spin systems are always ambiguous, in that there are many different ensembles whose average yields the same overall density operator [10]. Although these issues preclude the use of NMR as a means of studying foundational issues in quantum mechanics involving nonlocality and entanglement, they do not limit its utility as a means of developing the engineering principles needed for quantum information processing [32]. Indeed, the long decoherence times characteristic of nuclear spins, together with the superb coherent control available through modern NMR technology, has enabled demonstrations of many basic features of quantum information processing which had previously existed only in theory. The next section describes how this was done.

14.6   States and Gates by NMR

This section will show how the quantum logic operations introduced in section 14.3 can be represented in the product operator formalism, how they can be implemented in NMR by RF pulse sequences, how they act on density operators in product operator notation, and finally how they can be used together with gradient pulses to generate pseudo-pure states. The simplest logic gate is the **NOT** operation N on a single qubit (spin). This is a rotation by π about a transverse axis, which in the usual phase

---

As further discussed in Ref. [32], there are a number of ways in which this loss can be distributed among the various available resources (i.e. repetitions of the experiment, sample volume and the number of ancillae used), but within the validity of the high-temperature approximation no truly scalable method exists.

conventions is taken as the x-axis:

$$N \;=\; \exp(-(\pi/2)\iota\sigma_1) \;=\; \cos(\pi/2) - \iota\sigma_1\sin(\pi/2) \;=\; -\iota\sigma_1 \quad (6.36)$$

Via the anticommutivity of σ_1 and σ_3, this is readily verified to flip the qubit in question, e.g.

$$
\begin{aligned}
N|0\rangle\langle 0|\tilde{N} \;\leftrightarrow\; & N E_+ \tilde{N} \\
= \;& (-\iota\sigma_1)\tfrac{1}{2}(1+\sigma_3)(\iota\sigma_1) \qquad\qquad (6.37) \\
= \;& \tfrac{1}{2}(1-\sigma_3)(\sigma_1)^2 \;=\; E_- \;\leftrightarrow\; |1\rangle\langle 1| \,.
\end{aligned}
$$

As previously mentioned, such a rotation can be implemented by a single pulse of RF radiation of amplitude $\omega_1 = \hbar\gamma\|\boldsymbol{B}_1\|$ and duration π/ω_1, whose frequency is on-resonance with that of the target spin.

   This can be generalized to a rotation by an arbitrary angle about an arbitrary transverse axis, which implements a one-bit quantum logic gate. The one-bit quantum gate most commonly considered in quantum computing, however, is the Hadamard transform $\boldsymbol{R}_{\mathsf{H}}$ defined in Eq. 3.14. By translating this spinor definition to density operators, it may be seen that this gate acts on the components of the Bloch vector as

$$\boldsymbol{R}_{\mathsf{H}}\,\sigma_1\tilde{\boldsymbol{R}}_{\mathsf{H}} \;=\; \sigma_3 \,, \quad \boldsymbol{R}_{\mathsf{H}}\,\sigma_2\tilde{\boldsymbol{R}}_{\mathsf{H}} \;=\; -\sigma_2 \,, \quad \boldsymbol{R}_{\mathsf{H}}\,\sigma_3\tilde{\boldsymbol{R}}_{\mathsf{H}} \;=\; \sigma_1 \,, \quad (6.38)$$

and so corresponds to a rotation by π about the axis $(\sigma_1 + \sigma_3)/\sqrt{2}$, i.e.

$$\boldsymbol{R}_{\mathsf{H}} \;=\; \exp\!\Big(-(\pi/2)\iota(\sigma_1 + \sigma_3)/\sqrt{2}\Big) \;=\; -\iota(\sigma_1 + \sigma_3)/\sqrt{2} \,. \quad (6.39)$$

Although rotations about non-transverse axes are not easily implemented in most NMR spectrometers, the Hadamard is nevertheless readily obtained from the following sequence of transverse rotations:

$$\boldsymbol{R}_{\mathsf{H}} \;=\; \exp((\pi/4)\iota\sigma_2)\exp(-(\pi/2)\iota\sigma_1)\exp(-(\pi/4)\iota\sigma_2) \quad (6.40)$$

A convenient short-hand (similar to the graphical representation of pulse sequences widely used in NMR) is to just specify the sequence of Hamiltonians applied:

$$\left[\tfrac{\pi}{4}\sigma_2\right] \;\longrightarrow\; \left[\tfrac{\pi}{2}\sigma_1\right] \;\longrightarrow\; \left[-\tfrac{\pi}{4}\sigma_2\right] \quad (6.41)$$

Note that in this sequence, the Hamiltonians are written in left-to-right temporal order, opposite to that in Eq. 6.40.

   Turning now to a two-bit gate, we rewrite the c-NOT defined in Eq. 3.16 as follows:

$$
\begin{aligned}
\boldsymbol{S}^{1|2} \;=\; & |00\rangle\langle 00| + |10\rangle\langle 10| + |01\rangle\langle 11| + |11\rangle\langle 01| \\
= \;& |00\rangle\langle 00| + |10\rangle\langle 10| + \sigma_1^1(|11\rangle\langle 11| + |01\rangle\langle 01|) \qquad (6.42) \\
= \;& (\mathbf{1}\otimes|0\rangle\langle 0|) + \sigma_1^1(\mathbf{1}\otimes|1\rangle\langle 1|) \;\leftrightarrow\; \boldsymbol{E}_+^2 + \sigma_1^1\boldsymbol{E}_-^2
\end{aligned}
$$

This in turn can be expressed in exponential form as

$$
\begin{aligned}
&\exp\big((\pi/2)\iota(1 - \sigma_1^1)E_-^2\big) \\
&= \exp\big((\pi/2)\iota E_-^2\big)\exp\big(-(\pi/2)\iota\sigma_1^1 E_-^2\big) \\
&= \big(E_+^2 + \iota E_-^2\big)\big(E_+^2 - \iota\sigma_1^1 E_-^2\big) \\
&= E_+^2 + \sigma_1^1 E_-^2 = S^{1|2} ,
\end{aligned}
\tag{6.43}
$$

which says that this c-NOT can be regarded as a flip of spin 1 *conditional* on spin 2 being "down". Alternatively, by defining the idempotents $G_\pm^1 \equiv \frac{1}{2}(1 \pm \sigma_1^1)$, we can write this as

$$
S^{1|2} = \exp\big(\pi\iota G_-^1 E_-^2\big) = 1 - 2G_-^1 E_-^2 . \tag{6.44}
$$

This reveals an interesting symmetry: the same c-NOT can also be viewed as inversion of the phase of spin 2 conditional on spin 1 being along $-\sigma_1^1$.

To implement the c-NOT by NMR, it is necessary to use the scalar coupling to induce a conditional phase shift. The pulse sequence can be derived simply by fully expanding the propagator into a product of commuting factors: $\exp\big(\pi\iota G_-^1 E_-^2\big) =$

$$
\exp((\pi/4)\iota)\exp\big(-(\pi/4)\iota\sigma_3^2\big)\exp\big(-(\pi/4)\iota\sigma_1^1\big)\exp\big((\pi/4)\iota\sigma_1^1\sigma_3^2\big) \tag{6.45}
$$

The first factor is just a global phase $\sqrt{\iota}$, which has no effect when a propagator is applied to a density operator and hence can be ignored. The last factor cannot be implemented directly, but can be rotated about σ_2^1 into the scalar coupling Hamiltonian $\exp\big((\pi/4)\iota\sigma_1^1\sigma_3^2\big) =$

$$
\exp\big((\pi/4)\iota\sigma_2^1\big)\exp\big(-(\pi/4)\iota\sigma_3^1\sigma_3^2\big)\exp\big(-(\pi/4)\iota\sigma_2^1\big) . \tag{6.46}
$$

Making this substitution in Eq. 6.45 leaves two transverse rotations of spin 1 adjacent one another, but their product is equivalent to a single transverse rotation and a phase shift:

$$
\begin{aligned}
&\exp\big(-(\pi/4)\iota\sigma_1^1\big)\exp\big((\pi/4)\iota\sigma_2^1\big) \\
&= \exp\big((\pi/4)\iota\sigma_3^1\big)\exp\big(-(\pi/4)\iota\sigma_1^1\big)
\end{aligned}
\tag{6.47}
$$

It follows that the c-NOT may be implemented by the NMR pulse sequence:

$$
\left[\tfrac{\pi}{4}\sigma_2^1\right] \longrightarrow \left[\tfrac{\pi}{4}\sigma_3^1\sigma_3^2\right] \longrightarrow \left[\tfrac{\pi}{4}\sigma_1^1\right] \longrightarrow \left[\tfrac{\pi}{4}\big(\sigma_3^2 - \sigma_3^1\big)\right] \tag{6.48}
$$

Pulse sequences for many other reversible boolean logic gates may be found in Ref. [42].

Even though we are working in a rotating frame, the spins precess at slightly different rates depending on their chemical shifts δ^k (*vide supra*). The c-NOT sequence requires that this differential Zeeman evolution be "turned off" leaving only the coupling Hamiltonian active during the $1/(2J^{12})$ evolution periods. This can be done by inserting "refocusing" π-pulses in the middle and at the end of the period, as follows from:

$$
\begin{aligned}
& \exp\big((\pi/2)\iota(\sigma_1^1 + \sigma_1^2)\big) \exp\big(-(\pi/4)\iota(\sigma_3^1 + \sigma_3^2)\big) \\
& \quad \exp\big(-(\pi/2)\iota(\sigma_1^1 + \sigma_1^2)\big) \exp\big(-(\pi/4)\iota(\sigma_3^1 + \sigma_3^2)\big) \\
=\ & (\iota\sigma_1^1\sigma_1^2) \exp\big(-(\pi/4)\iota(\sigma_3^1 + \sigma_3^2)\big) \\
& \quad (-\iota\sigma_1^1\sigma_1^2) \exp\big(-(\pi/4)\iota(\sigma_3^1 + \sigma_3^2)\big) \\
=\ & \exp\big((\pi/4)\iota(\sigma_3^1 + \sigma_3^2)\big) \big(\sigma_1^1\sigma_1^2\big)^2 \exp\big(-(\pi/4)\iota(\sigma_3^1 + \sigma_3^2)\big) \\
=\ & \exp\big((\pi/4)\iota(\sigma_3^1 + \sigma_3^2)\big) \exp\big(-(\pi/4)\iota(\sigma_3^1 + \sigma_3^2)\big)\ =\ 1
\end{aligned}
\tag{6.49}
$$

It also requires that the scalar coupling evolution be turned off during the Zeeman evolutions at the end of the pulse sequence, which can be done by applying a *selective* π-pulse to just one of the spins while the other evolves, then vice versa, and finally realigning the transmitter phase with that of the spins. This ability to "suspend time" in one part of the system while working on another is an essential component of quantum computing by NMR spectroscopy [35].

Higher-order logic gates can be implemented by analogous sequences. For example, the c^2-NOT or Toffoli gate is:

$$
\begin{aligned}
T^{1|23} &\equiv (1 - E_-^2 E_-^3) + \sigma_1^1 E_-^2 E_-^3 \\
&= 1 - 2G_-^1 E_-^2 E_-^3\ =\ \exp\big(-\pi\iota G_-^1 E_-^2 E_-^3\big)
\end{aligned}
\tag{6.50}
$$

On expanding the propagator as before, one obtains:

$$
\begin{aligned}
& \exp\big(-(\pi/2)\iota E_-^2 E_-^3\big) \exp\big(-(\pi/8)\iota\sigma_1^1\big) \exp\big(-(\pi/8)\iota\sigma_1^1\sigma_3^2\big) \\
& \exp\big(-(\pi/8)\iota\sigma_1^1\sigma_3^3\big) \exp\big(-(\pi/8)\iota\sigma_1^1\sigma_3^2\sigma_3^3\big)
\end{aligned}
\tag{6.51}
$$

The last (left-most) factor in this sequence consists of Zeeman and coupling evolutions, and can be implemented by adjusting their relative rates via refocusing π-pulses. The transverse rotation and "two-body" factors can also be implemented in a fashion similar to that given above for the simple c-NOT gate. The "three-body" factor, on the other hand, must be built-up from successive two-body evolutions (since that is all nature provides us with [56]), for example as $\exp\big(-(\pi/8)\iota\sigma_1^1\sigma_3^2\sigma_3^3\big) =$

$$
\exp\big(-(\pi/4)\iota\sigma_3^1\sigma_3^3\big) \exp\big((\pi/8)\iota\sigma_2^1\sigma_3^3\big) \exp\big((\pi/4)\iota\sigma_3^1\sigma_3^3\big)
\tag{6.52}
$$

Assuming that all the couplings are equal to J, and that the time required for RF pulses is negligible, this sequence requires approximately $2/J$ in time. By neglecting relative phase shifts among the states and allowing multiple simultaneous evolutions, this can be reduced to $3/(4J)$ [15]. A graphical scheme for designing such pulse sequences, and its application to the c^n-NOT for $n \leq 16$, may be found in [41].

In the next section, it will be shown that NMR enables Feynman's idea of simulating one quantum system by another to be demonstrated, using however an ensemble of spins in a pseudo-pure state to simulate a quantum system in a true pure state. This will use the following sequence of RF and gradient (∇) pulses to convert the equilibrium state of a two-spin system into a pseudo-pure ground state (where the propagator for the Hamiltonian over each arrow conjugates the preceding expression to get the next):

$$\sigma_3^1 + \sigma_3^2 \xrightarrow{\left[-\frac{\pi}{8}(\sigma_1^1+\sigma_1^2)\right]} \frac{1}{\sqrt{2}}\left(\sigma_2^1 + \sigma_3^1 + \sigma_2^2 + \sigma_3^2\right)$$

$$\xrightarrow{\left[-\frac{\pi}{4}\sigma_3^1\sigma_3^2\right]} \frac{1}{\sqrt{2}}\left(\sigma_1^1\sigma_3^2 + \sigma_3^1 + \sigma_3^1\sigma_1^2 + \sigma_3^2\right)$$

$$\xrightarrow{\left[\frac{\pi}{12}(\sigma_2^1+\sigma_2^2)\right]} \frac{1}{4}\left(\sqrt{3}\left(\mathbf{E}_+^1\mathbf{E}_+^2 - \frac{1}{4}(1-\sigma_1^1\sigma_1^2)\right) + \frac{1}{2}\left(\sigma_1^1\mathbf{E}_+^2 + \mathbf{E}_+^1\sigma_1^2\right) \right)$$

$$\xrightarrow{\left[\nabla\right]} \frac{\sqrt{3}}{16}\left(\sigma_3^1 + \sigma_3^2 + \sigma_3^1\sigma_3^2 - \sigma_1^1\sigma_1^2 - \sigma_2^1\sigma_2^2\right) \quad (6.53)$$

$$\xrightarrow{\left[-\frac{\pi}{2}\sigma_2^2\right]} \frac{\sqrt{3}}{16}\left(\sigma_3^1 - \sigma_3^2 - \sigma_3^1\sigma_3^2 + \sigma_1^1\sigma_1^2 - \sigma_2^1\sigma_2^2\right)$$

$$\xrightarrow{\left[\nabla\right]} \frac{\sqrt{3}}{16}\left(\sigma_3^1 - \sigma_3^2 - \sigma_3^1\sigma_3^2\right)$$

$$\xrightarrow{\left[\frac{\pi}{2}\sigma_2^2\right]} \frac{\sqrt{3}}{16}\left(\sigma_3^1 + \sigma_3^2 + \sigma_3^1\sigma_3^2\right)$$

It will be observed that the first gradient pulse converts $\sigma_1^1\sigma_1^2$ into the pure zero quantum coherence $\sigma_1^1\sigma_1^2 + \sigma_2^1\sigma_2^2$, by destroying the corresponding double quantum component $\sigma_1^1\sigma_1^2 - \sigma_2^1\sigma_2^2$. This is due to the assumption of a homonuclear system, wherein zero quantum terms have almost no **net** magnetic moment and hence are not rapidly dephased by a gradient. Nevertheless, a π-rotation selective for only one spin converts this back to a double quantum term, which the second gradient wipes out. In a heteronuclear system zero quantum terms *are* rapidly dephased by a gradient, and hence the second gradient would not be **necessary**.

14.7   Quantum Simulation by NMR

This section describes a methodology and proof of concept for the simulation of one quantum system by another, as originally envisioned by Feynman [25] and studied in detail by Lloyd [38]. This will also enable us to illustrate many of the above concepts in quantum information processing.

Unlike the digital quantum computer envisioned by Benioff and Deutsch, however, a quantum simulator is an essentially *analog* device, which maps the state of the simulated system directly onto the joint states of the qubits without discretizing the problem. Such an analog encoding is not only precise in principle, but also more efficient, so that quantum simulations beyond the reach of today's computers could be performed with only 20 to 30 qubits [37, 38]. In addition, since it is usually only the long-term average behavior of quantum systems that is of interest, quantum simulations would be expected to be less sensitive to errors than quantum computations. Finally, the ensemble nature of NMR allows such averages to be observed directly, saving the otherwise requisite repetitions of the same simulation in order to obtain them.

The general scheme used here for quantum simulation is summarized in the following diagram:

$$
\begin{array}{ccc}
|\, s \,\rangle & \xrightarrow{\ \ \boldsymbol{U} = \exp(-T\iota\boldsymbol{H}_{\mathsf{s}})\ \ } & |\, s(T) \,\rangle \\[2pt]
\Big\downarrow {\scriptstyle \phi} & & \Big\downarrow {\scriptstyle \phi} \\[2pt]
|\, p \,\rangle & \xrightarrow{\ \ \boldsymbol{V}_T = \exp(-t_T\iota\bar{\boldsymbol{H}}_{\mathsf{p}})\ \ } & |\, p_T \,\rangle
\end{array}
\tag{7.54}
$$

Here, $|\, s \,\rangle$ and $|\, p \,\rangle$ denote the states of the simulated system and the physical system used to implement the simulation, respectively. The simulated state after a specified amount of time T and the corresponding physical state are denoted by $|\, s(T) \,\rangle$ and $|\, p_T \,\rangle$, respectively (note T is *not* the physical time!). The invertible (generally unitary) linear mapping ϕ encodes the simulated system's states in those of the physical system. Finally, $\boldsymbol{H}_{\mathsf{s}}$ is the simulated Hamiltonian, while $\bar{\boldsymbol{H}}_{\mathsf{p}}$ is the *average* physical Hamiltonian over the time t_T required for the simulation. This average Hamiltonian is obtained by interspersing periods of free evolution under the actual physical Hamiltonian $\boldsymbol{H}_{\mathsf{p}}$ with a sequence of RF pulses which effect unitary operations $\boldsymbol{V}_i$ $(i = 1, \ldots, M)$, so that:

$$
\begin{aligned}
\phi^{-1}\exp(-T\iota\boldsymbol{H}_{\mathsf{s}})\,\phi \ &= \ \exp(-t_T\iota\,\phi^{-1}\boldsymbol{H}_{\mathsf{s}}\,\phi) \\
&= \ \exp(-t_T\,\iota\bar{\boldsymbol{H}}_{\mathsf{p}}) \ = \ \prod_{i=1}^{M}\exp(-t_i\,\iota\boldsymbol{H}_{\mathsf{p}})\,\boldsymbol{V}_i
\end{aligned}
\tag{7.55}
$$

A general methodology has been developed by NMR spectroscopists to permit them to implement arbitrary average Hamiltonians to any desired degree of accuracy [31].

The Hamiltonian to be simulated is often given in canonical form (i.e. in terms of its energy levels). In this case an encoding ϕ which maps the eigenstates of $\boldsymbol{H_s}$ to those of $\bar{\boldsymbol{H}}_p$ is most convenient. Although it is not strictly necessary, the task of implementing the average Hamiltonian by NMR is greatly facilitated by converting it into product operator form. Thus suppose that the simulated Hamiltonian is

$$\boldsymbol{H_s} \equiv \sum_{j=0}^{2^N-1} H_j |j\rangle\langle j| , \tag{7.56}$$

where the energies H_j are arbitrary real numbers. Because no ordering of the energies is assumed, by a choice of indexing every eigenstate encoding can be put in the form

$$|j\rangle \xrightarrow{\phi} |\delta_N^j \cdots \delta_2^j \delta_1^j\rangle \tag{7.57}$$

where $\delta_k^j \in \{0,1\}$ is the k-th bit in the binary expansion of the integer j. In terms of density operators, this becomes

$$|j\rangle\langle j| \longrightarrow \boldsymbol{E_j} \equiv \boldsymbol{E}_{\epsilon_N^j}^N \cdots \boldsymbol{E}_{\epsilon_2^j}^2 \boldsymbol{E}_{\epsilon_1^j}^1 , \tag{7.58}$$

where $\epsilon_k^j = 1 - 2\delta_k^j$ and $\boldsymbol{E}_\epsilon^j$ are the usual idempotents. On expanding these products and regrouping, one obtains:

$$\bar{\boldsymbol{H}}_p = \sum_{j=0}^{2^N-1} H_j \boldsymbol{E_j} \equiv \sum_{j=0}^{2^N-1} \alpha_j \left(\sigma_3^N\right)^{\delta_N^j} \cdots \left(\sigma_3^1\right)^{\delta_1^j} \tag{7.59}$$

Inserting the identity $1 = \sum_{k=0}^{2^N-1} \boldsymbol{E_k}$ and using the relation $\sigma_3^j \boldsymbol{E_k} = \epsilon_k^j \boldsymbol{E_k}$ now yields:

$$\begin{aligned}
\bar{\boldsymbol{H}}_p &= \bar{\boldsymbol{H}}_p \sum_{k=0}^{2^N-1} \boldsymbol{E_k} \\
&= \sum_{j=0}^{2^N-1} \sum_{k=0}^{2^N-1} \alpha_j \left(\epsilon_N^k\right)^{\delta_N^j} \cdots \left(\epsilon_1^k\right)^{\delta_1^j} \boldsymbol{E_k}
\end{aligned} \tag{7.60}$$

Comparison of these two expressions for $\bar{\boldsymbol{H}}_p$ shows that

$$H_k = \sum_{j=0}^{2^N-1} \alpha_j \left(\epsilon_N^k\right)^{\delta_N^j} \cdots \left(\epsilon_1^k\right)^{\delta_1^j} \equiv \boldsymbol{M}^{\otimes N} \boldsymbol{\alpha} , \tag{7.61}$$

where $\boldsymbol{\alpha} = [\alpha_1, \ldots, \alpha_{2^N-1}]^\top$ and $\boldsymbol{M}^{\otimes N}$ a matrix whose jk-th entry is $(-1)^{\#j\&k}$ (with $\#j\&k$ being the Hadamard weight of the AND of j and k).

This linear transformation from the $\boldsymbol{\sigma_3}$ product basis for diagonal operators to eigenstates is known as the *Walsh-Hadamard transform*. As implied by the notation, the matrix $\boldsymbol{M}^{\otimes N}$ is a Kronecker (tensor) power of the 2×2 matrix

$$\boldsymbol{M} \equiv \begin{bmatrix} 1 & 1 \\ 1 & -1 \end{bmatrix} . \tag{7.62}$$

It is easily seen that M and hence $M^{\otimes N}$ is its own inverse up to factors of 2 and 2^N, respectively, and hence this transformation is easily inverted to convert any operator in canonical form into its product operator expansion. Consider, for example, simulating the first 2^N levels of a quantum harmonic oscillator, with $H_j = (2j+1)\Omega/2$ for $j = 0, \ldots, 2^N - 1$. The corresponding product operator form is given by:

$$\bar{H}_\mathsf{p} = \tfrac{1}{2}\Omega\left(2^N - \left(\sigma_3^1 + 2\sigma_3^2 + \cdots + 2^{N-1}\sigma_3^N\right)\right) \qquad (7.63)$$

Significantly, this expansion contains no product terms, so that evolution under it cannot induce new correlations among the qubits. This property depends on the encoding ϕ, however, as may be seen by reordering the (first four) energy levels as $H_0 = \Omega/2$, $H_1 = 3\Omega/2$, $H_2 = 7\Omega/2$ and $H_3 = 5\Omega/2$; this corresponds to a so-called Grey encoding, in which adjacent energy levels differ by single qubit NOT operations. In this case the propagator of the desired average Hamiltonian may be shown to be

$$\begin{aligned} V_T &= \exp\!\left(-T\iota\bar{H}_\mathsf{p}\right) \\ &= \exp\!\left(-T\iota\Omega\!\left(\left(1 + \sigma_3^1/2\right)\sigma_3^2 - 2\right)\right) . \end{aligned} \qquad (7.64)$$

In order to demonstrate these ideas in practice, NMR experiments will now be described which implement the first four levels of a quantum harmonic oscillator in the above Grey encoding [50]. These experiments were done on the molecule $2,3$-dibromothiophene, which contains two weakly coupled hydrogen atoms (see Figure 14.3). Letting $K \equiv (\omega^2 - \omega^1)/(2\pi)$ and placing the receiver on the first spin (i.e. choosing a rotating frame wherein $\omega^1 = 0$), the physical Hamiltonian of this system becomes:

$$H_\mathsf{p} = \pi\left(K + J\sigma_3^1\right)\sigma_3^2 = \pi\left(226.0 + 5.7\sigma_3^1\right)\sigma_3^2 \qquad (7.65)$$

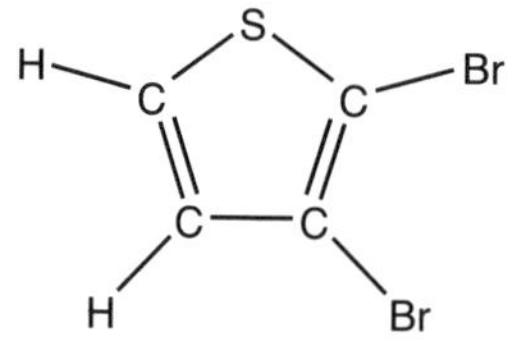

FIGURE 14.3. Chemical diagram of the molecule 2,3-dibromothiophene used for simulation of a quantum harmonic oscillator (see text). The two hydrogen atoms were used as the qubits in an analog representation of the oscillator's first four energy levels.

Up to an overall phase factor, the desired average Hamiltonian is obtained from the following pulse sequence

$$\begin{aligned}
\left[-\pi\left(\sigma_2^1 + \sigma_2^2\right)\right] &\longrightarrow \left[\tau_1/2\sigma_3^1\sigma_3^2\right] \\
\longrightarrow \left[\pi\left(\sigma_2^1 + \sigma_2^2\right)\right] &\longrightarrow \left[(\tau_1/2 + \tau_2)\sigma_3^1\sigma_3^2\right] .
\end{aligned} \tag{7.66}$$

This may be shown by using the fact that $\sigma_2^1\sigma_2^2$ anticommutes with σ_3^2, commutes with $\sigma_3^1\sigma_3^2$ and squares to 1 to rearrange the corresponding sequence of propagators as follows:

$$\begin{aligned}
&\exp\left(-(\tau_1/2 + \tau_2)\iota H_p\right)\exp\left(-(\pi/2)\iota\left(\sigma_2^1 + \sigma_2^2\right)\right) \\
&\exp\left(-(\tau_1/2)\iota H_p\right)\exp\left((\pi/2)\iota\left(\sigma_2^1 + \sigma_2^2\right)\right) \\
=~&\exp\left(-(\tau_1/2 + \tau_2)\iota\pi\left(K + J\sigma_3^1\right)\sigma_3^2\right)(-\iota)\sigma_2^1\sigma_2^2 \\
&\exp\left(-(\tau_1/2)\iota\pi\left(K + J\sigma_3^1\right)\sigma_3^2\right)\iota\sigma_2^1\sigma_2^2 \\
=~&\exp\left(-(\tau_1/2 + \tau_2)\iota\pi\left(K + J\sigma_3^1\right)\sigma_3^2\right) \\
&\exp\left((-\iota)\sigma_2^1\sigma_2^2\left(-(\tau_1/2)\iota\pi\left(K + J\sigma_3^1\right)\sigma_3^2\right)\iota\sigma_2^1\sigma_2^2\right) \\
=~&\exp\left(-(\tau_1/2 + \tau_2)\iota\pi\left(K + J\sigma_3^1\right)\sigma_3^2\right) \\
&\exp\left(-(\tau_1/2)\iota\pi\left(-K + J\sigma_3^1\right)\sigma_3^2\right) \\
=~&\exp\left(-\iota\pi\left(\tau_2 K + (\tau_1 + \tau_2)J\sigma_3^1\right)\sigma_2\right)
\end{aligned} \tag{7.67}$$

Thus the desired propagator V_T at a simulated time T is obtained (up to its overall phase) by setting $\tau_2 = \Omega T/K$ and $\tau_1 = \Omega T/(2J) - \tau_2$.

In order to illustrate the simulation, the spin system was prepared in a pseudo-pure ground state, as described in Eq. 6.53 above. It was then transformed into a double quantum superposition $(|\psi\rangle_{DQ} \equiv |0\rangle + \iota|2\rangle)$, and evolved for a regularly spaced sequence of 64 simulated times T up to one full period Ω^{-1}. For each time T, the corresponding double quantum spin state $(\phi(|\psi\rangle_{DQ}) = |01\rangle + \iota|10\rangle)$ was transformed via a readout pulse selective for a single spin back to a single quantum spin state, which gives rise to a peak in the spectrum whose amplitude could be used to monitor the simulation. A similar set of experiments was also done on the full superposition $(|0\rangle + \cdots + |3\rangle)$ over the first four energy levels of the oscillator. Due to the Grey code used, the single and triple quantum coherences in this case all give rise to observable peaks whose amplitudes could be monitored directly. Figure 14.4 shows these peak amplitudes as a function of simulated time T for each of these cases. Note in particular that a triple-base-frequency oscillation does not occur naturally in a two-spin system, thereby confirming that this simulation involves a nontrivial modification of the system's physical Hamiltonian. The original reference [50] also shows data for the simulation of a driven anharmonic quantum oscillator, which does not rely upon knowledge of the eigenstates, thereby showing that the simulation methodology of Eq. 14.7 is general.

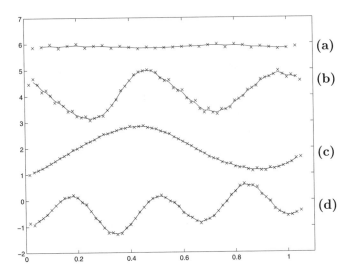

FIGURE 14.4. Plots of selected peak amplitudes (x) versus fraction of harmonic oscillator period simulated for the ground (a), a double quantum (b) and the single (c) and triple (d) quantum coherences in a full superposition over all four energy levels. The solid-lines through each data set were obtained by three-point smoothing.

14.8   Remarks on Foundational Issues

Using a mathematical formalism based on geometric algebra, we have shown how quantum information processing can be performed on small numbers of qubits by liquid-state NMR spectroscopy, where the qubits are physically realized in the joint statistics of a highly mixed ensemble of spin systems. There has nevertheless been considerable controversy over whether or not these experiments are truly "quantum" [10, 43]. The fact that all of quantum mechanics can be done with the multiparticle Dirac algebra, together with the implied geometric interpretation, makes an absolute distinction between "quantum" and "classical" seem a little less profound. Nevertheless, the general consensus now seems to be that liquid-state NMR should be regarded as "quantum" not so much because the measurements that can be made on any one state require the formalism of quantum mechanics for their description, as because the manifold of states and measurement outcomes generated by the available operations do. Thus, even if a highly mixed density operator is expressed as an average over an ensemble of unentangled states, a sequence of RF pulses and evolutions under scalar coupling can always be applied which converts at least some of these states into entangled ones.

Our work also touches upon a number of interesting questions regarding the emergence of classical statistical mechanics from an underlying quantum description of the system and its environment, and at the same time provides a readily accessible experimental system within which these questions can be studied. It is now widely believed that classical statistical mechanics works because the system and its environment become entangled through their mutual interactions [29, 62], so that the partial trace over the environment results in an intrinsically mixed state of the system. If the eigenstates of the resulting density operator are stable under the environmental interactions, the system's dynamics can be described by a classical stochastic process on those eigenstates. This process by which correlations between the selected eigenstates are lost is known as *decoherence*. From this perspective, a pure state is a state of the universe as a whole in which the system and its environment are mutually uncorrelated. Nevertheless, decoherence remains a theory of ensembles; it does not explain what happens in any single system, and hence in particular does not resolve the quantum measurement problem [39].

The potential utility of NMR as a means of exploring some of these issues experimentally is illustrated by our recent demonstration of a *quantum error correcting code* [14]. This extension of the classical theory of error correction to quantum systems was developed in order to control decoherence in quantum computations [44, 52], which would otherwise destroy the coherences on which quantum algorithms depend [11, 60]. Such codes rely upon the fact that the effects of environmental interactions on the system can be completely described by a discrete stochastic process of the form $\rho \longrightarrow \sum_m p_m \boldsymbol{U}_m \rho \tilde{\boldsymbol{U}}_m$, where $p_m \geq 0$ are the probabilities with which the unitary operators $\boldsymbol{U}_m$ are applied to the system. Assuming that this process is known, additional ancillae qubits in a specific state $|0\rangle$ can be added to the system, such that each distinct "error" (i.e. operator $\boldsymbol{U}_m$) maps their joint state into orthogonal subspaces. Thus measurements exist which can determine the error (though *not* the state of the system ρ), enabling it to be corrected. This remarkable ability to intervene in such fundamental processes promises to be useful in characterizing how they occur in nature [32].

The primary experimental question which remains is: How many qubits will we be able to completely observe and control via NMR spectroscopy? The aforementioned signal-to-noise problems associated with preparing pseudo-pure states from high-temperature equilibrium states would appear to impose an upper bound on liquid-state spectroscopy of *ca.* $8-12$ qubits, and various other practical difficulties (i.e. limited frequency resolution, and the intrinsic decoherence in these systems) may make it difficult to go even that far. These limitations are not intrinsic to NMR *per se*, however, in that for example polarizations approaching unity can be obtained in crystalline solids at temperatures of 4K, while at the same time increasing the intrinsic decoherence times of the spins to hours or more. Additionally, it is in principle possible to use gradient methods to *spatially* label the spins, thereby circumventing frequency resolution problems. By these means coherences among up to *ca.* 10^{11} spins have been created, *and* refocused, in the laboratory [61]. Regardless of the technology by which a large-scale quantum computer is ultimately implemented, it is certain that both NMR and geometric algebra will remain essential tools for its development.

Chapter 15

Geometric Feedforward Neural Networks and Support Multivector Machines

Eduardo Bayro Corrochano and Refugio Vallejo

15.1 Introduction

The representation of the external world in biological creatures appears to be definable in terms of geometry. We can formalize the relationships between the physical signals of external objects and the internal signals of a biological creature by using extrinsic vectors coming from the world and intrinsic vectors representing the world internaly. We can also assume that the external world and the internal world have different reference coordinate systems. If we consider the acquisition and coding of knowledge as a distributed and differentiated process, it is imaginable that there should exist various domains of knowledge representation obeying different metrics which can be modelled using different vectorial basis. How it is possible that nature could have acquired through evolution such tremendous representation power for dealing with complicated geometric signal processing [13]? Pellionisz and Llinàs [15, 16] claim in a stimulating series of articles that the formalization of the geometrical representation seems to be the dual expression of extrinsic physical cues performed by the intrinsic central nervous system vectors. These vectorial representations, related to reference frames intrinsic to the creature, are covariant for perception analysis and contravariant for action synthesis. The authors explain that the geometric mapping between these two vectorial spaces can be implemented by a neural network which performs as a metric tensor [16].

In view of this line of thought, the geometric interpretation of Clifford algebra by Hestenes [11], offers an alternative to tensor analysis that has been employed since 1980 by Pellionisz and Llinàs for the theory of the perception and action cycle (PAC). Tensor calculus is covariant, which means that it requires transformation laws for getting coordinate independent relations. Clifford algebra or geometric algebra is more attractive than tensor analysis because it is coordinate free, and because it includes spinors which the tensor theory does not. The computational efficiency of geometric algebra has also been confirmed in various challenging areas of mathematical physics [6]. The other mathematical system used in neural networks is matrix analysis. Geometric algebra better captures the geometric characteristics of the problem independently of a coordinate reference system and offers also other computational advantages that matrix algebra does not, e.g. the bivector representation of linear operators in the null cone, incidence relations (meet and join operation), and the conformal group in the horosphere.

Attempts at applying the geometric algebra to neural geometry have been already made in [8, 9, 4, 2]. In this paper, first we show the generalization of the standard feedforward networks in the geometric algebra and then we give an introduction of the use of the SV–Machines in the geometric algebra framework. Using SV–Machines we can generate straightforwardly two layer networks and RBF networks, which is an important issue. In this way we expand the sphere of applicability of the SV–Machines for the treatment of multivectors.

The paper is organized as follows. Section two reviews the computing principles of feedforward neural networks, giving their most important characteristics. Section three deals with the extension of the multilayer perceptron (MLP) to complex and quaternionic MLPs. Section four presents the generalization of the feedforward neural networks in geometric algebra. Section five describes the generalized learning rule across different geometric algebras, and it presents experiments which compare geometric neural networks with real valued MLPs. Section six introduces the Support Multivector Machines and gives experiments. The last section discusses the applicability of the geometric feedforward neural nets and the Support Multivector Machines.

15.2   Real Valued Neural Networks

The approximation of nonlinear mappings using neural networks is useful in various areas of signal processing, such as pattern classification, prediction, system modelling and identification. This section reviews the fundamentals of standard real valued feedforward architectures.

Cybenko [5] used for the approximation of a continuous function $g(\mathbf{x})$, the superposition of weighted functions

$$y(\mathbf{x}) = \sum_{j=1}^{N} w_j \sigma_j (\mathbf{w}_j^T \mathbf{x} + \theta_j), \tag{2.1}$$

where $\sigma(.)$ is a continuous discriminatory function like a sigmoid, $w_j \in \mathcal{R}$ and $\mathbf{x}, \theta_j, \mathbf{w}_j \in \mathcal{R}^n$. Finite sums of the form of Eq. (2.1) are dense in $C^0(I_n)$, if $|g_k(\mathbf{x}) - y_k(\mathbf{x})| < \varepsilon$ for a given $\varepsilon > 0$ and all $\mathbf{x} \in [0, 1]^n$. This is called a *density theorem* and is a fundamental concept in approximation theory and nonlinear system modelling [5, 10].

A structure with k outputs y_k, having several layers using logistic functions, is known as the *Multilayer Perceptron* (MLP) [20]. The output of any neuron of a hidden layer or of the output layer can be represented in similar way,

$$o_j = f_j(\sum_{i=1}^{N_i} w_{ji} x_{ji} + \theta_j)$$

$$y_k = f_k(\sum_{j=1}^{N_j} w_{kj} o_{kj} + \theta_k), \tag{2.2}$$

where $f_j(\cdot)$ is logistic and $f_k(\cdot)$ is logistic or linear. Linear functions at the outputs are often used for pattern classification. In some tasks of pattern classification, a hidden layer is necessary, whereas in some tasks of automatic control two hidden layers may be required. Hornik [10] showed that standard multilayer feedforward networks are able to approximate accurately any measurable function to a desired degree. Thus they can be seen as *universal approximators*. In case of a training failure, we should attribute an error to an inadequate learning, an incorrect number of hidden neurons or a poorly defined deterministic relation between the input and output patterns.

Poggio and Girosi [18] developed the Radial Basis Function (RBF) network, which consists of a superposition of weighted Gaussian functions,

$$y_j(\mathbf{x}) = \sum_{i=1}^{N} w_{ji} G_i(D_i(\mathbf{x} - \mathbf{t}_i)) \tag{2.3}$$

where y_j is the j-output, $w_{ji} \in \mathcal{R}$, G_i is a Gaussian function, D_i a $N \times N$ dilatation diagonal matrix and $\mathbf{x}, \mathbf{t}_i \in \mathcal{R}^n$. The vector $\mathbf{t}_i$ is a translation vector. This architecture is supported by the regularization theory.

15.3   Complex MLP and Quaternionic MLP

A MLP is defined in the complex domain when its weights, activation function and outputs are complex valued. The selection of the activation function is not a trivial matter. For example, the extension of the sigmoid function from $\mathcal{R}$ to $\mathcal{C}$,

$$f(z) = \frac{1}{(1 + e^{-z})} \tag{3.4}$$

where $z \in \mathcal{C}$, is not allowed because this function is analytic and unbounded [12]; similarly for the functions $\tanh(z)$ and e^{-z^2}. These kinds of activation functions exhibit problems with convergence in training due to their singularities. The necessary conditions that a complex activation $f(z) = a(x, y) + ib(x, y)$ has to fulfill are: $f(z)$ is nonlinear in x and y, the partial derivatives a_x, a_y, b_x and b_y exist $(a_x b_y \not\equiv b_x a_y)$ and $f(z)$ is not entire. Accordingly, Georgiou and Koutsougeras [12] proposed

$$f(z) = \frac{z}{c + \frac{1}{r}|z|} \tag{3.5}$$

where $c, r \in \mathcal{R}^+$. These authors extended the usual real back-propagation learning rule for the *Complex Multilayer Perceptron* (CMLP).

Arena et al. [1] introduced the Quaternionic MLP (QMLP) which is an extension of the CMLP. The weights, activation functions and outputs of this net are represented in terms of quaternions [22]. They choose the following non-analytic bounded function

$$
\begin{aligned}
f(q) &= f(q_0 + q_1 i + q_2 j + q_3 k) \tag{3.6}\\
&= (\frac{1}{1 + e^{-q_0}}) + (\frac{1}{1 + e^{-q_1}})i + (\frac{1}{1 + e^{-q_2}})j + (\frac{1}{1 + e^{-q_3}})k,
\end{aligned}
$$

where $f(\cdot)$ is now the function for quaternions. These authors proved that superpositions of such functions approximate accurately any continuous quaternionic function defined in the unit polydisc of $\mathcal{C}^n$. The extension of the training rule along the lines of the CMLP was done in [1].

15.4   Geometric Algebra Neural Networks

Real, complex and quaternionic neural networks can be further generalized in the geometric algebra framework. The weights, the activation functions and the outputs will be now represented using multivectors. In the real valued neural networks of section 15.2, the vectors are multiplied with the weights using the scalar product. For geometric neural networks the scalar product will be replaced by the geometric product.

15.4.1   The activation function

The activation function of Eq. (3.5), used for the CMLP, was extended by Pearson and Bisset [14] for a type of Clifford MLP by applying different Clifford algebras, including the quaternion algebra. We propose here an activation function which affects each multivector basis element. This function was introduced independently by the authors [4] and is in fact a generalization of the function of Arena et al [1]. The function for a n-dimensional multivector m is given by

$$f(m) = f(m_0 + m_i \sigma_i + m_j \sigma_j + m_k \sigma_k + \cdots + m_{ij}\sigma_i \wedge \sigma_j + \cdots +$$

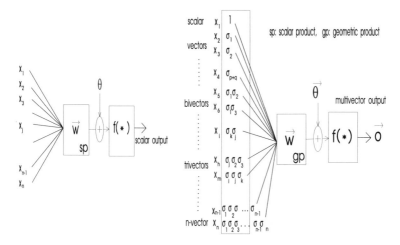

FIGURE 15.1. *McCulloch-Pitts Neuron and Geometric Neuron.*

$$+m_{ijk}\sigma_i \wedge \sigma_j \wedge \sigma_k + \cdots + m_n \sigma_1 \wedge \sigma_2 \wedge \cdots \wedge \sigma_n)$$
$$= (m_0) + f(m_i)\sigma_i + f(m_j)\sigma_j + f(m_k)\sigma_k + \cdots + f(m_{ij})\sigma_i \wedge \sigma_j +$$
$$+ \cdots + f(m_{ijk})\sigma_i \wedge \sigma_j \wedge \sigma_k + \cdots + f(m_n)\sigma_1 \wedge \sigma_2 \wedge \ldots \wedge \sigma_n, \quad (4.7)$$

where $\boldsymbol{f}(\cdot)$ is written in bold to be distinguished from the one used for a single argument $f(\cdot)$. The values of $f(\cdot)$ can be of the sigmoid or Gaussian type.

15.4.2   The geometric neuron

The *McCulloch-Pitts neuron* uses the scalar product of the input vector and its weight vector [20]. The extension of this model to the *geometric neuron* requires the substitution of the scalar product with the Clifford or geometric product, i.e.

$$\mathbf{w}^T \mathbf{x} + \theta \qquad \Rightarrow \qquad \boldsymbol{wx} + \boldsymbol{\theta} = \boldsymbol{w} \cdot \boldsymbol{x} + \boldsymbol{w} \wedge \boldsymbol{x} + \boldsymbol{\theta} \quad (4.8)$$

Figure 15.1 shows in detail the McCulloch-Pitts neuron and the geometric neuron. This figure also depicts how the input pattern is formated in a specific geometric algebra. The geometric neuron outputs a richer kind of pattern. Let us illustrate this with an example in $\mathcal{G}_{3,0,0}$

$$
\begin{aligned}
o &= \boldsymbol{f}(\boldsymbol{wx} + \boldsymbol{\theta}) &(4.9)\\
&= \boldsymbol{f}(s_0 + s_1\sigma_1 + s_2\sigma_2 + s_3 k + s_4\sigma_1\sigma_2 + s_5\sigma_1 k + s_6\sigma_2 k + s_7\sigma_1\sigma_2 k)\\
&= f(s_0) + f(s_1)\sigma_1 + f(s_2)\sigma_2 + f(s_3)k + f(s_4)\sigma_1\sigma_2 + \ldots +\\
&\quad + f(s_5)\sigma_1 k + f(s_6)\sigma_2 k + f(s_7)\sigma_1\sigma_2 k,
\end{aligned}
$$

where $\boldsymbol{f}$ is the activation function defined in Eq. (4.7) and $s_i \in \mathcal{R}$. If we use the McCulloch-Pitts neuron in the real valued neural networks the output is simply the scalar given by

$$o = f(\sum_i^N w_i x_i + \theta). \qquad (4.10)$$

The geometric neuron outputs a signal with more geometric information

$$o = f(\boldsymbol{wx} + \boldsymbol{\theta}) = f(\boldsymbol{w} \cdot \boldsymbol{x} + \boldsymbol{w} \wedge \boldsymbol{x} + \boldsymbol{\theta}) \qquad (4.11)$$

It has both a scalar product like the McCulloch-Pitts neuron,

$$\boldsymbol{f}(\boldsymbol{w} \cdot \boldsymbol{x} + \theta) = f(s_0) \equiv f(\sum_i^N w_i x_i + \theta) \qquad (4.12)$$

and also the outer product given by

$$\begin{aligned}
\boldsymbol{f}(\boldsymbol{w} \wedge \boldsymbol{x} + \boldsymbol{\theta} - \theta) \;=\; & f(s_1)\sigma_1 + f(s_2)\sigma_2 + f(s_3)\mathsf{k} + f(s_4)\sigma_1\sigma_2 + \dots + \\
& + f(s_5)\sigma_1\mathsf{k} + f(s_6)\sigma_2\mathsf{k} + f(s_7)\sigma_1\sigma_2\mathsf{k}. \qquad (4.13)
\end{aligned}$$

Note that the outer product gives the scalar cross-products between the individual components of the vector. This is nothing else than the multi-vector components of points or lines (vectors), planes (bivectors) and vo-lumes (trivectors). This characteristic can be used for the implementation of geometric preprocessing in the extended geometric neural network. To a certain extend, this kind of neural network resembles the higher order neural networks of [17]. However, an extended geometric neural network uses not only a scalar product of higher order, but also all the necessary scalar cross-products for carrying out a geometric cross-correlation. Figure (15.2) shows a geometric network with its extended first layer.

In conclusion, a geometric neuron can be seen as kind of *geometric co-rrelation operator* which, in contrast to the McCulloch-Pitts neuron, offers not only points but also higher grade multivectors like planes, volumes, ... , hyper-volumes for interpolation.

15.4.3 Feedforward geometric neural networks

Figure (15.3) depicts standard neural network structures for function a-pproximation in the geometric algebra framework. Here, the inner vector product has been extended to the geometric product and the activation functions are according to (4.7).

The equation (2.1) of the Cybenko's model in geometric algebra is

$$y(x) = \sum_{j=1}^N w_j f(\boldsymbol{w}_j \cdot \boldsymbol{x} + \boldsymbol{w}_j \wedge \boldsymbol{x} + \theta_j). \qquad (4.14)$$

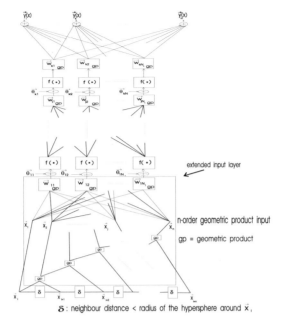

FIGURE 15.2. *Geometric neural network with extended input layer.*

The extension of the MLP is straightforward. The equations using the geometric product for the outputs of hidden and output layers are given by

$$o_j = f_j\left(\sum_{i=1}^{N_i} \boldsymbol{w}_{ji} \cdot \boldsymbol{x}_{ji} + \boldsymbol{w}_{ji} \wedge \boldsymbol{x}_{ji} + \theta_j\right)$$

$$y_k = f_k\left(\sum_{j=1}^{N_j} \boldsymbol{w}_{kj} \cdot \boldsymbol{o}_{kj} + \boldsymbol{w}_{kj} \wedge \boldsymbol{o}_{kj} + \theta_k\right) \tag{4.15}$$

In radial basis function networks, the dilatation operation, given by the diagonal matrix $\mathbf{D}_i$, can be implemented by means of the geometric product with a dilation $\boldsymbol{D}_i = e^{\alpha \frac{i\bar{i}}{2}}$ [11], i.e.

$$\mathbf{D}_i(\boldsymbol{x} - \boldsymbol{t}_i) \Rightarrow \boldsymbol{D}_i(\boldsymbol{x} - \boldsymbol{t}_i)\tilde{\boldsymbol{D}}_i \tag{4.16}$$

$$y_k(\boldsymbol{x}) = \sum_{j=1}^{N} \boldsymbol{w}_{kj} G_j(\boldsymbol{D}_j(\boldsymbol{x}_{ji} - \boldsymbol{t}_j)\tilde{\boldsymbol{D}}_j) \tag{4.17}$$

Note that in the case of the geometric RBF we are also using an activation function according to (4.7).

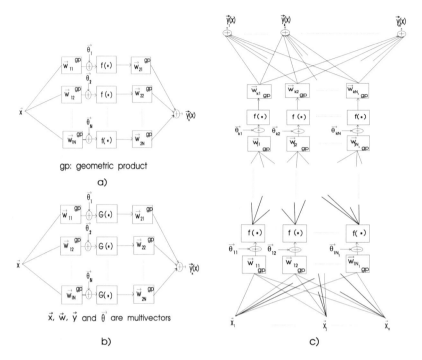

FIGURE 15.3. *Geometric Network Structures for Approximation: (a) Cybenko's (b) GRBF network (c) GMLP$_{p,q,r}$.*

15.5   Learning Rule

This section presents the *multidimensional generalization* of the gradient descent learning rule in geometric algebra. This rule can be used for the Geometric MLP (GMLP) and for tuning the weights of the Geometric RBF (GRBF). Previous learning rules for the real valued MLP, complex MLP [12] and the quaternionic MLP [1] are special cases of this extended rule.

15.5.1   *Multi-dimensional back-propagation training rule*

The norm of a multivector x for the learning rule is given by

$$|x| = (x|x)^{\frac{1}{2}} = \left(\sum_A [x]_A^2 \right)^{\frac{1}{2}}. \tag{5.18}$$

The geometric neural network with n inputs and m outputs approximates the target mapping function

$$\mathcal{Y}_t : (\mathcal{G}_{p,q,r})^n \to (\mathcal{G}_{p,q,r})^m, \tag{5.19}$$

where $(\mathcal{G}_{p,q,r})^n$ is the n-dimensional module over the geometric algebra $\mathcal{G}_{p,q,r}$ [14]. The error at the output of the net is measured according to the metric

$$E = \frac{1}{2} \int_{\boldsymbol{x} \in \boldsymbol{X}} ||\mathcal{Y}_w - \mathcal{Y}_t||^2, \tag{5.20}$$

where $\boldsymbol{X}$ is some compact subset of the Clifford module $(\mathcal{G}_{p,q,r})^n$ involving the product topology derived from equation (5.18) for the norm and where $\mathcal{Y}_w$ and $\mathcal{Y}_t$ are the learned and target mapping functions, respectively. The back-propagation algorithm [20] is a procedure for updating the weights and biases. This algorithm is a function of the negative derivative of the error function (Eq. (5.20)) with respect to the weights and biases themselves. The computing of this procedure is straightforward and here we will only give the main results. The updating equation for the multivector weights of any hidden $j-$layer is

$$\boldsymbol{w}_{ij}(t+1) = \eta \Big[\big(\sum_{k}^{N_k} \delta_{kj} \otimes \overline{\boldsymbol{w}_{kj}}\big) \odot \boldsymbol{F}'(\boldsymbol{net}_{ij}) \Big] \otimes \overline{\boldsymbol{o}_i} + \alpha \boldsymbol{w}_{ij}(t), \tag{5.21}$$

for any $k-$output with a non-linear activation function

$$\boldsymbol{w}_{jk}(t+1) = \eta \Big[(\boldsymbol{y}_{k_t} - \boldsymbol{y}_{k_a}) \odot \boldsymbol{F}'(\boldsymbol{net}_{jk}) \Big] \otimes \overline{\boldsymbol{o}_j} + \alpha \boldsymbol{w}_{jk}(t), \tag{5.22}$$

and for any $k-$output with a linear activation function,

$$\boldsymbol{w}_{jk}(t+1) = \eta (\boldsymbol{y}_{k_t} - \boldsymbol{y}_{k_a}) \otimes \overline{\boldsymbol{o}_j} + \alpha \boldsymbol{w}_{jk}(t). \tag{5.23}$$

In the above, $\boldsymbol{F}$ is the activation function defined in equation (4.7), t is the update step, η and α are the *learning rate* and the momentum respectively, $\otimes$ is the Clifford or geometric product, $\odot$ is the scalar product and $\overline{(\cdot)}$ is the multivector anti–involution (reversion or conjugation).

In the case of the non-Euclidean $\mathcal{G}_{0,3,0}$ $\overline{(\cdot)}$ corresponds to the simple conjugation. Each neuron consists now of p+q+r units, each for a multivector component. The biases are also multivectors and are absorbed as usual in the sum of the activation signal called here $\boldsymbol{net}_{ij}$. In the learning rules, Eqs. (5.21)- (5.23), the way how the geometric product and the anti–involution are computed varies depending on the geometric algebra being used [19]. As illustration we give the conjugation required in the learning rule for the quaternion algebra : $\bar{x} = x_0 - x_1\sigma_1 - x_2\sigma_2 - x_3\sigma_1\sigma_2$, where $\boldsymbol{x} \in \mathcal{G}_{0,2,0}$.

15.6 Experiments Using Geometric Feedforward Neural Networks

The power of the method of learning using bivectors is demonstrated by u-sing the XOR function. Figure 15.4 shows that the geometric nets GMLP$_{0,2,0}$

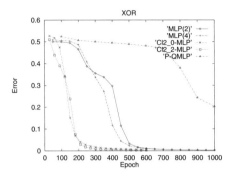

FIGURE 15.4. *Learning XOR using the MLP(2), MLP(4), $GMLP_{0,2,0}$, $GMLP_{2,0,0}$ and P-QMLP.*

and $GMLP_{2,0,0}$ have a faster convergence rate than the MLP and the P-QMLP, the quaternionic Multilayer Perceptron of Pearson [14], which uses the activation function given by the equation (3.5). Figure 15.4 shows the MLP with 2- and 4-dimensional input vectors. Since the MLP(4), working also in 4D, can not beat the GMLP, it can be claimed that the better performance of the geometric neural network is not only due to the higher dimensional quaternionic inputs, but rather to the algebraic advantages of the geometric neurons of the net.

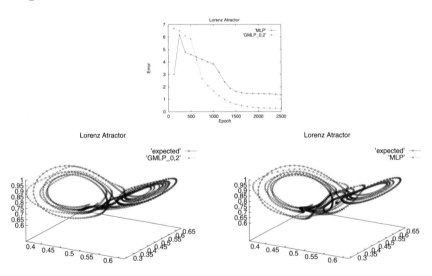

FIGURE 15.5. *a) Training error b) Prediction by $GMLP_{0,2,0}$ and expected trend c) Prediction by MLP and expected trend.*

Let us show another application of a geometric multi–layer perceptron which distinguishes geometric information in a chaotic process. For that we used the well known Lorenz attractor (σ=3, r=26.5 and b=1) with the initial conditions [0,1,0] and sample rate 0.02 sec. A 3-12-3 MLP and a 1-4-1 GMLP$_{0,2,0}$ were trained in a time interval of from 12 to 17 seconds to perform a 8 τ step ahead prediction. The next 750 samples, unseen during training, were used for the test. Figure 15.5.a shows the error during training. Note that the GMLP$_{0,2,0}$ converges faster than the MLP. Figures 15.5b-c show that the GMLP$_{0,2,0}$ predicts better than the MLP. Analyzing the covariance parameters of the MLP [0.96815, 0.67420,0.95675], and those of the GMLP$_{0,2,0}$ [0.9727, 0.93588, 0.95797], we can see that the MLP requires more time to process the geometry involved in the second variable, because the convergence is slower. As a result, the MLP loses the ability to predict well on the other side of the looping (see Figure 15.5.b) . In contrast, the geometric net captures at an early stage the geometric characteristics of the attractor, so it can not fail in its prediction on the other side of the looping.

15.7   Support Vector Machines in Geometric Algebra

The *Support Vector Machine* (SVM) approach of Vladimir N. Vapnik [21] applies optimization methods for learning. Using SV–Machines we can generate a type of two layer networks and RBF networks as well as networks with other kernels. Our idea is to generate neural networks using the SV–Machines in geometric algebra. In this way we are using SV–Machines for the processing of multivectors. We will call our approach *Support Multivector Machine* (SMVM). Let us review briefly the SVM and then explain the SMVM.

15.7.1   *Support vector machines*

The SVM maps the input space R^d into a high–dimensional *feature space* H, given by $\Phi : R^d \Rightarrow H$, satisfying a Kernel $K(\boldsymbol{x}_i, \boldsymbol{x}_j) = \Phi(x_i) \cdot \Phi(x_j)$ which fulfill the *Mercer's condition* [21]. The SMV constructs an optimal hyperplane in the feature space which divides the data into two clusters.

SV–Machines build the mapping

$$f(\boldsymbol{x}) = sign\left(\sum_{supportvectors} y_i\alpha_i K(\boldsymbol{x}_i, \boldsymbol{x}) - b \right). \qquad (7.24)$$

We find the coefficients α_i in the separable case (and analogously in the non–separable case), which maximizes the functional

$$W(\alpha) = \sum_{i=1}^{l} \alpha_i - \frac{1}{2} \sum_{i,j}^{l} \alpha_i \alpha_j y_i y_j K(x_i, x_j) \tag{7.25}$$

subject to the constraints $\sum_{i=1}^{l} \alpha_i y_i = 0$, where $\alpha_i \geq 0$, i=1,2,...,l. This functional coincides with the functional for finding the optimal hyperplane.
   Examples of SV Machines include:

- polynomial learning machines $K(\boldsymbol{x}, \boldsymbol{x}_i) = [(\boldsymbol{x} \cdot \boldsymbol{x}_i) + 1]^d$

- radial basis functions machines $K_\gamma(|x - x_i|) = exp\{-\gamma|\boldsymbol{x} - \boldsymbol{x}_i|^2\}$

- two layer neural networks $K(\boldsymbol{x}, \boldsymbol{x}_i) = S\Big(v(\boldsymbol{x} \cdot \boldsymbol{x}_i) + c\Big).$

15.7.2   Support multivector machines

A *Support Multivector Machine* (SMVM) maps the multivector input space R^{2n} of $\mathcal{G}_{p,q,r}$ $(n = p + q + r)$ into a high–dimensional feature space R^{n_f} $(n_f >= 2^n)$

$$\Phi : R^{2^n} \Rightarrow R^{n_f}. \tag{7.26}$$

The SMVM constructs an optimal separating hyperplane in the multivector feature space by using, in the nonlinear case, the kernels

$$\overset{N\,graded\,spaces}{\underset{m=1}{\sum}} K_m(\boldsymbol{x}_{mi}, \boldsymbol{x}_{mj}) = \overset{N\,graded\,spaces}{\underset{m=1}{\sum}} \Phi(x_{mi}) \cdot \Phi(x_{mj}), \tag{7.27}$$

which fulfill the Mercer's conditions. SMV–Machines for multivectors $\boldsymbol{x}$ of a geometric algebra $\mathcal{G}_{p,q,r}$ are implemented by the multivector mapping

$$\{y_1, y_2, ..., y_m\} = \overset{N\,graded\,spaces}{\underset{m=1}{\sum}} f_m(\boldsymbol{x}_{mi})$$

$$= \overset{N\,graded\,spaces}{\underset{m=1}{\sum}} sign\Big(\underset{support\,vectors}{\sum} y_{mi}\alpha_{mi} K(\boldsymbol{x}_{mi}, \boldsymbol{x}_m) - b_m\Big), \tag{7.28}$$

where m denotes the grade of the spaces. The coefficients α_{mi} in the the separable case (and analogously in the non–separable case), are found by maximizing the functional

$$W(\alpha_i^m) = W\Big(\overset{N\,graded\,spaces}{\underset{m=1}{\sum}} \sum_{i}^{l} \alpha_{mi}\Big) = \tag{7.29}$$

$$\sum_{m=1}^{\overset{N\ graded\ spaces}{}} \sum_{i=1}^{l} \alpha_{mi} - \frac{1}{2} \sum_{m=1}^{\overset{N\ graded\ spaces}{}} \sum_{i,j}^{l} \alpha_{mi}\alpha_{mj}y_{mi}y_{mj}K(x_{mi}, x_{mj})$$

subject to the constraint

$$\sum_{m=1}^{\overset{N\ graded\ spaces}{}} \sum_{i=1}^{l} \alpha_{mi}y_{mi} = 0, \qquad (7.30)$$

where $\alpha_{mi} \geq 0$, for m=1,...,N graded spaces and i=1,2,...,l. This functional coincides with the functional for finding the optimal separating hyperplane for the multivector feature space.

15.8 Experimental Analysis of Support Multivector Machines

This section presents the use of the SMVM using RBF kernels to find support multivectors; this illustrates the geometric role of the support multivectors. The second experiment applies SMVM for the task of robot grasping. In this experiment, the coding of the data using the motor algebra simplifies the complexity of the problem.

15.8.1 Finding support multivectors

This section shows the use of SVM for finding the support vectors in the 2D and 3D cases. The examples show how we can depict the feature space and how the SVM and the SMVM work. First, we take two clusters in R^2 and apply a nonlinear SVM and a nonlinear SMVM, using in both cases a RBF kernel. Figure 15.6.a shows the original 2D data, Figure 15.6.b the support vectors found by SVM and 15.6.c the support multivectors found by the SMVM in $\mathcal{G}_{2,0,0}$. The multivectors used were $m_i = x_i + \frac{1}{3}\sum_{j=i}^{i+2} x_j + (x_i - x_{i+1}) \wedge (x_i - x_{i+2})$. Note that the areas follow the borders of the clusters, whereas in the SVM case some of the support vectors are in the interior of the clusters.

In the following experiment we are interested in the separation of the two clouds of 3D. points as shown in Figure 15.7.a. Note that the points of the clouds lie on a non–linear surface. Figure 15.7.b shows individual support vectors found by the linear SVM when the curvature of the data manifold changes, and 15.7.c shows the individual support multivectors found by the linear SMVM in $\mathcal{G}_{3,0,0}$. The multivectors coding the data are of the form: $m_i = x_i + \frac{1}{4}\sum_{j=i}^{i+3} x_j + (x_i - x_{i+1}) \wedge (x_i - x_{i+2}) \wedge (x_i - x_{i+3})$, which represent pyramidal volumes. Note that the width of the volume basis follows the curvature change of the surface. This experiment shows how the support multivector captures the geometric properties of the manifold. It is interesting to note that only a linear SMVM is sufficient to find the optimal support multivector. In Figure 15.7.c, a multivector lies inside of the lower cluster which was not considered by the SMVM, whereas in the SVM case (see Figure 15.7.b) it was unnecessarily used as a support vector.

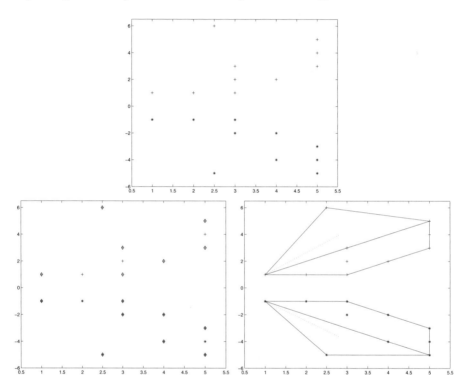

FIGURE 15.6. *2D case using RBF Kernel: a) two 2D clusters data b) SVM (support vectors indicated with diamonds) c) SMVM using $\mathcal{G}_{2,0,0}$ (support multivectors by triangles).*

15.8.2   *Estimation of 3D rigid motion*

In this experiment, we show the importance of the input data coding and the use of the SMVM for estimation. The problem is to estimate the Euclidean motion necessary to move an object to a certain point along a 3–D nonlinear curve. The task might be to move the grasper of a robot manipulator to a specific curve point. In order to linearize the model of the motion of a point, we used the geometric algebra $\mathcal{G}_{3,0,1}^{+}$ or *motor algebra* [3]. Working in a 4D space, we simplify the complexity of the motor estimation

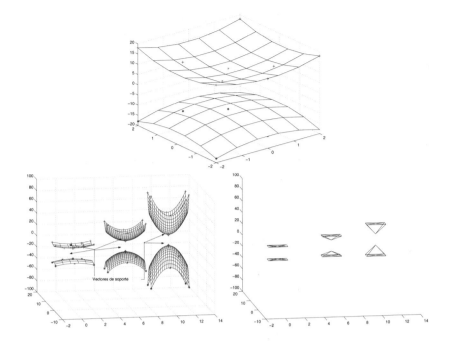

FIGURE 15.7. *3D case: a) two 3D clusters. b) A linear SVM by changing the curvature of the data surface (one support vector). c) A linear SMVM using $\mathcal{G}_{3,0,0}$ by changing the curvature of the data surface.*

necessary to carry out the 3D motions along the curve. We assumed that the trajectory is known and prepared the training data as couples of 3D position points $\boldsymbol{x}_i$ and the 3D rigid motions, coded in the motor algebra $\mathcal{G}_{3,0,1}^{+}$ as follows

$$\boldsymbol{P}_i = 1 + I\boldsymbol{x}_i \tag{8.31}$$

$$\boldsymbol{M}_i = \boldsymbol{T}_i \boldsymbol{R}s_i, \tag{8.32}$$

where the rotor $\boldsymbol{R} = \cos(\frac{\theta_i}{2}) + \sin(\frac{\theta_i}{2})$ rotates about the screw axis line $\boldsymbol{L}$, and the translator $\boldsymbol{T}_i = 1 + I\frac{d_i \boldsymbol{n}_i}{2}$ corresponds to translating the distance $d_i \in R$ along $\boldsymbol{L}s_i$. Considering motion along a non–linear path, we take the lines connecting to sampled points of the curve as the screw lines $\boldsymbol{L}s_i$. With the estimated motor $\boldsymbol{M}_j^e$ for the position $\boldsymbol{P}_j$, obtained using the SMVM, we can move the grasper to the new position j

$$\boldsymbol{X}_j = \boldsymbol{M}_j^e \boldsymbol{X}_l \overline{\widetilde{\boldsymbol{M}}}_j^e = \boldsymbol{T}_j^e \boldsymbol{R}_j^e \boldsymbol{X}_l \widetilde{\boldsymbol{R}}_j^e \boldsymbol{T}_j^e, \tag{8.33}$$

where $\boldsymbol{X}_l$ stands for any point on the grasper in its final position.

The training data for the SMVM consists of the couples $\boldsymbol{P}_i$ as input data and the $\boldsymbol{M}_i$ as output data. The training was done so that the SMVM could estimate the motors $\hat{\boldsymbol{M}}_j^e$ for the unseen points $\boldsymbol{P}_j$. Since the points $\boldsymbol{P}_i$ are given by the three first bivector components, and the outputs by the eight motor components of $\boldsymbol{M}_i$, we used an SMVM architecture with three inputs and eight outputs. We used three inputs because the points are of the form

$$\boldsymbol{P}_j = 1 + I\boldsymbol{x}_j \equiv [1, 0, 0, 0, x_{j1}, x_{j2}, x_{j3}, 0]. \qquad (8.34)$$

Since the first component of all the $\boldsymbol{P}_j$ is "1", we can ignore it. After the training we tested whether the SMVM can or can not estimate the right 3D rigid motion. The Figure 15.8.a (left top) shows a nonlinear motion path; we took a bunch of points of this curve and their correspondent motors for training the SMVM architecture. For recall, we selected three arbitrary points which were not used during the training of the SMVM. The estimated motors for these points were applied to the object in order to move it to these particular positions of the curve. We can see in Figure 15.8.a that the estimated motors are very good. Thereafter we trained a real valued MLP with 3 input nodes, 10 hidden nodes and eight output nodes trained with the same training data and with a convergence error of 0.001. Figure 15.8.b (right top) shows the motion approximated by the net. However, we see that the SMVM in Figure 15.8.a has a bit better performance. Then we carried out similar experiments using noisy data. For that we add 1% of uniform distributed noise to the three components of the position vectors and to the eight components of the motors multivectors as well. We trained the SMVM and MLP with this noisy data. The latter used again a convergence error of 0.001. The results for the case of noisy data show that the SMVM behaves better than the MLP.

15.9   Conclusions

In the literature there are basically two mathematical systems used in neural computing: the tensor algebra and the matrix algebra. In contrast, here the authors use the coordinate-free geometric algebra for the analysis and design of feedforward neural networks. The paper shows that real-, complex- and quaternion–valued neural networks are just particular cases of the more general geometric algebra multidimensional neural networks, and that some of them can be generated by using Support Multivector Machines. In particular, the generation of RBF for neurocomputing in geometric algebra is easier than using the SMVM, which allows us to find the optimal parameters automatically. The use geometric algebra in SVM offers both new tools and new understanding of SV–Machines for multi–dimensional learning. The experiments demonstrate the advantages of using geometric neural networks. In particular, the experiment of grasping shows

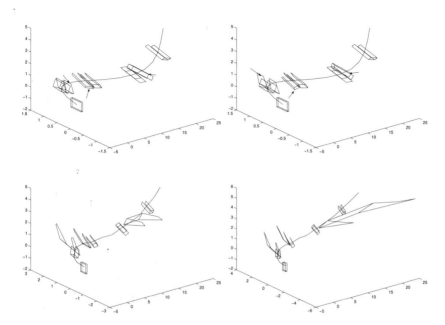

FIGURE 15.8. *From top to bottom: estimated three grasper positions (indicated with an arrow) using noise free data and noisy data 1% a) left column: estimation using a SMVM in $\mathcal{G}_{3,0,1}$. b) right column: approximation using a 3–10–8 MLP.*

the importance of choosing a geometric algebra suitable for the preparing of the input and output data.

Chapter 16

Image Analysis Using Quaternion Wavelets

Leonardo Traversoni

16.1 Introduction

The idea for Quaternion Wavelets comes from the need to represent evolving objects without the use of sequential pictures of the object in different positions. We present a short introduction to ordinary wavelets, emphazising those concepts that are needed to translate the ideas to a quaternion framework. We also set up the quaternionic framework for the theory, because even when it is known, it has been used in so many different ways that a coherent picture becomes very difficult.

Step by step, we first represent a static object, the complex human head. From this very first step, the advantages of our approach become evident.

There are antecedents to our work. M. Mitrea [1] published a paper about Clifford Wavelets, a the more exhaustive aproach to the subject, years before I published [3]. My old paper treats quaternion wavelets from a completely different point of view. In this paper, I try to develop a little more the ideas of Mitrea in terms of a "static" use of quaternion wavelets, and also some of my own ideas related to its "dynamic" use.

There are some tools normally used in 2D or 3D for wavelets or B-splines that must be first defined in the quaternionic environment, so that we can use them later.

16.1.1 Wavelets

We will consider wavelets of one real variable in order to later derive the wavelets of one quaternion variable. Consider the space $L^2(\mathbb{R})$ of all measurable functions f defined on the real line $\mathbb{R}$ that satisfy:

$$\int_{-\infty}^{\infty} |f(x)|^2 < \infty \tag{1.1}$$

These functions decay very rapidly to zero at $\pm\infty$. We also need the definitions of internal product and norm in $L^2(\mathbb{R})$:

$$< f, g >= \int_{-\infty}^{\infty} f(x)\overline{g(x)}dx \tag{1.2}$$

$$||f||_2 =< f, f >^{1/2} \tag{1.3}$$

Now let $\mathbb{Z}$ be the set of integers, $\mathbb{Z} = \{..., -1, 0, 1, ...\}$. In order to cover all of $\mathbb{R}$ we define the set of functions

$$\psi(x - k), \quad k \in \mathbb{Z} \tag{1.4}$$

If we don't want single frequency waves, we must consider

$$\psi_{j,k} = \psi(2^j x - k), \quad j, k \in \mathbb{Z} \tag{1.5}$$

A function $\psi \in L^2(\mathbb{R})$ is called an orthogonal wavelet if the family $\psi_{j,k}$ is an orthonormal basis of $L^2(\mathbb{R})$

The simplest example of an orthogonal wavelet is the Haar function:

$$\psi_H = \begin{cases} 1 & \text{for} & 0 \le x < 1/2 \\ -1 & \text{for} & 1/2 \le x < 1 \\ 0 & \text{otherwise} & . \end{cases} \tag{1.6}$$

16.1.2 Multirresolution analysis

Any wavelet generates a direct sum decomposition of $L^2(\mathbb{R})$. Let us consider for each $j \in \mathbb{Z}$, the closed subspaces:

$$V_j = ... + W_{j-2} + W_{j-1} \ j \in \mathbb{Z} \tag{1.7}$$

which have the following properties:

$$1) \quad ... \subset V_{-1} \subset V_0 \subset V_1 \subset ... \tag{1.8}$$

$$2) \quad clos_{L^2}\left(\bigcup_{j \in Z} V_j\right) = L^2(\mathbb{R}) \tag{1.9}$$

$$3) \quad \bigcap_{j \in Z} V_j = \{0\} \tag{1.10}$$

$$4) \quad V_{j+1} = V_j + W_j, \quad j \in \mathbb{Z} \tag{1.11}$$

$$5) \quad f(x) \in V_j \leftrightarrow f(2x) \in V_{j+1}, \quad j \in \mathbb{Z} \tag{1.12}$$

If the reference subspace V_0 is generated by a single function $\phi \in L^2(\mathbb{R})$,

$$V_0 = clos_{L^2(\mathbb{R})}(\phi_{0,k} : k \in \mathbb{Z}) \tag{1.13}$$

where

$$\phi_{j,k} = 2^{j/2}\phi(2^j x - k), \tag{1.14}$$

then all the subspaces V_j are also generated by ϕ.

A function ϕ is called a "scaling function" and generates a Multiresollution Analysis(MRA), if it generates a nested sequence of closed subspaces V_j that satisfy conditions 1), 2), 3) and 5). A typical example of scaling functions are the mth order cardinal splines for $m \in \mathbb{Z}$.

16.1.3   Cardinal splines

The cardinal splines are no other than polynomial spline functions with equally spaced simple knots. Let $\mathbb{Z}$ be the set of all integers taken as knots, π_m the collection of algebraic polynomials with degrees of at most m, and let $F^m = F(\mathbb{R})$ the collection of all functions $f, f', ..., f^{(m)}$ which are everywhere continuous.

<u>Definition</u> ([2]). For each positive integer m the space S_m of cardinal splines of order m, and with the knot sequence $\mathbb{Z}$, is the collection of all functions $f \in F^{m-2}$ such that the restrictions of f to any interval $[k, k + 1), k \in \mathbb{Z}$ are also in π_{m-1}. That is

$$f|_{[k,k+1)} \in \pi_{m-1}, \quad k \in \mathbb{Z} \tag{1.15}$$

For example, the first order cardinal B-Spline N_1 is the characteristic function of the unit interval $[0, 1)$. For $m \geq 2$, N_m may be defined by integral convolution

$$N_m(x) = \int_{-\infty}^{\infty} N_{m-1}(x - t)N_1(t)dt = \int_0^1 N_{m-1}(x - t)dt \tag{1.16}$$

From another point of view, the subspace V_0 generated by N_m is exactly the definition we gave earlier for the cardinal B-spline. All the other subspaces V_j may now be specified by

$$V_j = \{f \in F^{m-2} \bigcap L^2(\mathbb{R}) : f|_{[\frac{k}{2^j}, \frac{k+1}{2^j})} \in \pi_{m-1}, \quad k \in \mathbb{Z}\} \tag{1.17}$$

16.1.4   Decomposition and reconstruction

It follows that $\{V_j\}$ is generated by the scalar function $\phi \in L^2(\mathbb{R})$ and $\{W_j\}$ is generated by some wavelet $\psi \in L^2(\mathbb{R})$. Every function $f \in L^2(\mathbb{R})$ can be approximated uniquely, given $N \in \mathbb{Z}$, as closely as desired by some $f_N \in V_N$,

$$f_N = f_{N-1} + g_{N-1} \tag{1.18}$$

where $f_{N-1} \in V_{N-1}$ and $g_{N-1} \in W_{N-1}$. Repeating this process, we get

$$f_N = g_{N-1} + g_{N-2} + \ldots + g_{N-M} + f_{N-M} \tag{1.19}$$

This is called the "wavelet decomposition". By suitable choosing M, we can obtain a f_{N-M} sufficiently "blurred". For all $x \in \mathbb{R}$, we have that

$$\phi(x) = \sum_k p_k \phi(2x - k) \tag{1.20}$$

$$\psi(x) = \sum_k q_k \phi(2x - k) \tag{1.21}$$

The above functions are called the scale relations of the scaling function and wavelet. Since both $\phi(2x)$ and $\phi(2x - 1)$ are in V_1 and $V_1 = V_0 + W_0$, there are four l^2 sequences $\{a_{-2k}\}$, $\{b_{-2k}\}$, $\{a_{1-2k}$ and $\{b_{1-2k}\}$ such that

$$\phi(2x) = \sum_k [a_{-2k}\phi(x - k) + b_{-2k}\psi(x - k)] \tag{1.22}$$

$$\phi(2x - 1) = \sum_k [a_{-2k}\phi(x - k) + b_{-2k}\psi(x - k)] \tag{1.23}$$

$$\tag{1.24}$$

Combining both of these formulas, we get

$$\phi(2x - l) = \sum_k [a_{l-2k}\phi(x - k) + b_{l-2k}\psi(x - k)], \tag{1.25}$$

which is the decomposition formula.

Using the decomposition formula, we get the decomposition algorithm

$$c_k^{j-1} = \sum_l a_{l-2k} c_l^j \tag{1.26}$$

$$d_k^{j-1} = \sum_l b_{l-2k} c_l^j, \tag{1.27}$$

and the composition algorithm

$$c_k^j = \sum_l [p_{k-2l} c_l^{j-1} + q_{k-2l} d_l^{j-1}] \tag{1.28}$$

16.1.5 Hilbert spaces of quaternionic valued functions

Let $V_{4,s}$ be a 4-dimensional real linear vector space with the basis (e_1, e_2, e_3, e_4), and where s is an integer such that $0 \le s \le 4$. Let a bilinear form

$$(v|w), \qquad v, w \in V_{4,s} \tag{1.29}$$

be given on $V_{4,s}$ such that

$$(e_i|e_j) = 0, \quad i \neq j \tag{1.30}$$
$$(e_i|e_i) = 1, \quad i = 1, ..., s \tag{1.31}$$
$$(e_i|e_i) = -1, \quad i = s+1, ..., 4 \tag{1.32}$$

We call $V_{4,s}$ a quadratic linear space. If $v = \sum_{i=1}^{4} v_i e_i \in V_{4,s}$ then the associated quadratic form is:

$$(v|v) = \sum_{i=1}^{s} v_i^2 - \sum_{i=s+1}^{4} v_i^2 \tag{1.33}$$

Consider now the 2^4 dimensional real linear space $C(V_{4,s})$ defined by the basis

$$\{e_A = e_{h_1...h_r} : A = (h_1, ..., h_r) \in PN, \ 1 \leq h_1 < .... < h_r \leq 4\} \tag{1.34}$$

Let N be the set $\{1, 2, 3, 4\}$. A product is defined on $C(V_{4,s})$ by

$$e_A e_B = (-1)^{4((A \cap B) \ S)}(-1)^{p(A,B)} e_{A \Delta B} \tag{1.35}$$

Here $S = \{1, .., s\}$,

$$p(A, B) = \sum_{j \in B} p(A, j), \ p(A, j) = \#\{i \in A : i > j\},$$

e_0 is the identity element,

$$e_i^2 = 1, \quad i = 1, ..., s$$

$$e_i^2 = -1, \quad i = s+1, ..., 4,$$

and where for $1 \leq h_1 < .... < h_r \leq 4$,

$$e_{h_1} \cdot e_{h_2} ..... e_{h_r} = e_{e_{h_1}...e_{h_r}}.$$

For any vectors $v, w \in V_{4,s}$, we have

$$vw + wv = 2(v|w) \tag{1.36}$$

In this way $C(V_{4,s})$ becomes a real, linear, associative, but non conmutative algebra called the Universal Clifford algebraover $V_{4,s}$.

The linear subspace of $C(V_{4,s})$, spanned by the $\binom{n}{p}$ elements e_A, where $\#A = p$, is denoted by C_p and is called the space of p-vectors. C_0 is the 1-dimensional subspace of 0-vectors or scalars, C_1 is the 4-dimensional subspace of 1-vectors or vecotrs of $V_{4,s}$ with the basis $\{e_1, e_2, e_3, e_4\}$, C_4 is the 1-dimensional subspace of 4-vectors or pseudo-scalars with basis $\{e_{1..4}\}$.

Geometrically each p-vector is a piece of an oriented p-dimensional subspace of the underlying space $V_{4,s}$ with a certain magnitude.

The product of two 1-vectors has a symmetric and an antisymmetric part,

$$vw = \frac{1}{2}(vw + wv) + \frac{1}{2}(vw - wv) \tag{1.37}$$

or

$$vw = (v|w) + v \wedge w, \tag{1.38}$$

the inner product $(v|w)$ coincides with the bilinear form on $V_{4,s}$. The outer product is an antisymmetric 2-vector which vanishes when the vectors are collinear.

16.1.6 Modules over quaternions

There are right and left quaternion modules denoted with the subscripts (r) and (l).

Let $X_{(l)}$ be a unitary left quaternionic module. Thus, $X_{(l)}+$ is an abelian group with an operation $(\lambda, f) \rightarrow \lambda f$ from $\mathbb{H} \times X$ into $X_{(l)}$, defined such that for all $\lambda, \mu \in \mathbb{H}$ and $f, g \in X_{(l)}$:

1) $(\lambda + \mu)f = \lambda f + \mu f$

2) $(\lambda\mu)f = \lambda(\mu f)$

3) $\lambda(f + g) = \lambda f + \lambda g$

4) $e_0 f = f$.

If $X_{(l)}$ and $Y_{(l)}$ are unitary left quaternionic modules, then a function $T : X_{(l)} \rightarrow Y_{(l)}$ is a left quaternionic linear operator if for all $f, g \in X_{(l)}$ and $\lambda \in \mathbb{H}$:

$$T(\lambda f + g) = \lambda T(f) + T(g) \tag{1.39}$$

Let $X_{(l)}$ be a unitary left quaternionic module. Then a family P of functions $p : X_{(l)} \rightarrow R$ is said to be a proper system of semi-norms on $X_{(l)}$ if the following conditions are fullfilled:

1) There exists a constant $C_0 \geq 1$ such that for all $p \in P$, $\lambda \in \mathcal{H}$ and $f, g \in X_{(l)}$:

    1.1) $p(f + g) \leq p(f) + p(g)$

    1.2) $p(\lambda f) \leq C_0|\lambda|_0 p(f)$ and $p(\lambda f) = |\lambda| p(f)$ if $\lambda \in \mathbb{R}$

2) For any finite number of functions $p_1, p_2, ..., p_k \in P$, there exist $p \in P$ and $C > 0$ such that for all $f \in X_{(l)}$:

$$\sup_{j=1,...,k} p_j(f) \le C_p(f) \tag{1.40}$$

3) If $p(f) = 0$ for all $p \in P$ then $f = 0$.

16.1.7 Hilbert quaternion modules

Let $H_{(r)}$ be a unitary right quaternion module. Then a function

$$(,) : H_{(r)} \times H_{(r)} \to \mathbb{IH} \tag{1.41}$$

is said to be an inner producton $H_{(r)}$ if for all $f, g, h \in H_{(r)}$ and $\lambda \in \mathbb{IH}$

1) $(f, g + h) = (f, g) + (f, h)$
2) $(f, g\lambda) = (f, g)\lambda$
3) $(f, g) = \overline{(g, f)}$
4) $< \tau_{e_0}, (f, f) >\ge 0$ and $< \tau_{e_0}, (f, f) >= 0$ iff $f = 0$
5) $< \tau_{e_0}, (f\lambda, f\lambda) >\le |\lambda|_0^2 < \tau_{e_0}, (f, f) >$ it follows that:
$(f, 0) = (0, f) = 0$ and that $(f\lambda, g) = \overline{\lambda}(f, g)$
    Putting

$$||f||^2 =< \tau_{e_0}, (f, f) > \tag{1.42}$$

for each $f \in H_{(r)}$, $|| \cdot ||$ becomes a proper norm on $H_{(r)}$ which is a right quaternion normed module.

16.1.8 Hilbert modules with reproducing kernel

Let $H_{(r)}$ be the unitary Hilbert quaternion module consisting of quaternion–valued functions defined on some set F. Then a function $K : F \times F \to \mathbb{IH}$ is called reproducing kernelof $H_{(r)}$ if for any fixed $t \in F$,

1) $K(., t) \in H_{(r)}$
2) $f(t) = (K(., t), f)$ for all $f \in H_{(r)}$.

In this case, $H_{(r)}$ is said to be a unitary right Hilbert quaternion module with reproducing kernel.

16.1.9 Kernel

Let R^{m+1} with $m \ge 1$ be an Euclidean space.

Points in $\mathbb{IR}^{m+1}$ are denoted by $x = (x_0, x_1, ..., x_m)$, or by $(x_0, \vec{x})$. We can see that $\vec{x} = (x_1, ..., x_m)$ lies in the hyperplane $x_0 = 0$, which is identified with $\mathbb{IR}^m$.

On the other hand, the vector subspace H_1 of the quaternion algebra $\mathbb{H}$ has dimension 3. Assuming that $m \leq 3$, for $x \in \mathbb{R}^{m+1}$, and $\vec{x} \in \mathbb{R}^m$, we write

$$x = \sum_{i=0}^{m} e_i x_i \quad \vec{x} = \sum_{i=1}^{m} e_i x_i \quad \overline{x} = \sum_{i=0}^{m} \overline{e_i} x_i. \qquad (1.43)$$

In the particular case when $m = 3$,

$$x = \sum_{i=0}^{3} e_i x_i \quad \vec{x} = \sum_{i=1}^{3} e_i x_i \quad \overline{x} = \sum_{i=0}^{3} \overline{e_i} x_i \qquad (1.44)$$

, and the Euclidean norm and the quaternion norm of x differ by the constant $|x|_0 = 2^{3/2}|x|$.

Let Ω be an open subset of R^{3+1}. Functions f defined in Ω with values in $\mathbb{H}$ are of the form

$$f(x) = \sum_{H} e_H f_H(x) \qquad (1.45)$$

where the functions f_H are real valued. Whenever f has a property such as continuity, or differentiability, it is clear that all the components f_H also possess the property. The conjugate of the function f, $\overline{f}$ is defined by

$$\overline{f} = \sum_{H} \overline{e_H} f_H. \qquad (1.46)$$

Let D denote the differential operator

$$D = \sum_{i=0}^{3} e_i \partial_{x_i} \qquad (1.47)$$

Then

$$Df = \sum_{i,H} e_i e_H \partial_{x_i} f_H \quad and \quad fD = \sum_{i,H} e_H e_i \partial_{x_i} f_H. \qquad (1.48)$$

The conjugate operator is given by

$$\overline{D} = \sum_{i=0}^{3} \overline{e_i} \partial_{x_i} \qquad (1.49)$$

Definition A function $f \in C^1(\Omega, H)$ is said to be left or right monogenic in Ω if $Df = 0$ or $fD = 0$ in Ω.

The components of a left monogenic fuction f are

$$\sum_{i,H} e_i e_H \partial_{x_i} f_H, \qquad (1.50)$$

so the condition of monogenicity is equivalent to a sistem of 2^3 linear homogeneous first order partial differential equations with constant coefficients. If $f(\vec{x}) = \sum_{i=1}^{3} e_i f(\vec{x}_i)$ the sistem takes the form

$$div\ f = 0 \tag{1.51}$$

$$rot\ f = 0 \tag{1.52}$$

The set of left (right) monogenic functions in Ω is denoted by $M_{(r)}(\Omega; H)$ and $M_{(l)}(\omega; H)$. Under the ordinary laws of addition and multiplication, it forms a right or left module.

Note that $D\overline{D} = \overline{D}D = \Delta_{3+1}e_0$, where Δ_{m+1} denotes the Laplacian in $\mathbb{R}^{3+1}$. Any monogenic function in Ω is harmonic, and hence infinitely differentiable and $\mathbb{H}$ -analytic in Ω. Also, each of its components is real analytic in Ω.

Denoting by $a(\Omega; H)$ the bi-$\mathbb{H}$-module of $\mathbb{H}$-valued analytic functions in Ω, we thus have

$$M_{(r)}(\Omega; H) \subset a_{(r)}(\Omega; H) \subset E_{(r)}(\Omega; H), \tag{1.53}$$

where $E(\Omega; H)$ denotes the unitary bimodule of all $\mathbb{H}$-valued functions in Ω

Proposition If Ω is open and star shaped with respect to the origin, and $u : \Omega \to \mathbb{R}$ is harmonic in Ω, then the function

$$f(x) = u(x)e_0 + \int_0^1 t^2 \overline{D}u(tx)\, x\, dt - \left[\int_0^1 t^2 \overline{D}u(tx)\, x\, dt\right]_0 \tag{1.54}$$

is left monogenic in Ω and its scalar part is exactly the function u.

Introducing spherical coordinates

$$x_0 = r\cos\theta_1 \tag{1.55}$$
$$x_1 = r\sin\theta_1\cos\theta_2 \tag{1.56}$$
$$x_2 = r\sin\theta_1\sin\theta_2\cos\theta_3 \tag{1.57}$$
$$x_3 = r\sin\theta_1\sin\theta_2\sin\theta_3\sin\theta_4 \tag{1.58}$$

where $0 < r < +\infty,\ 0 < \theta_1, \theta_2 \leq \pi,\ 0 < \theta_3 < 2\pi$ for a point $x \in R^4$, we can write

$$x = r\omega \qquad \overline{x} = r\overline{\omega} \tag{1.59}$$

where

$$\omega = \sum_{i=0}^{3} e_i\omega_i\ , \qquad \overline{\omega} = \sum_{i=0}^{3} \overline{e_i}\omega_i \tag{1.60}$$

for

$$\omega_i = \frac{x_i}{|x|} = \frac{x_i}{r}, \quad i = 0, 1, 2, 3 \tag{1.61}$$

Notice that ω represents a point on the unit sphere S^3. For the operators D and $\overline{D}$ the following spherical form, valid in R^4, is obtained

$$D = \omega \partial_r + \frac{1}{r} \partial_\omega, \quad \overline{D} = \overline{\omega} \partial_r + \frac{1}{r} \overline{\partial}_\omega \tag{1.62}$$

where

$$\partial_\omega = \sum_{i=1}^{3} \frac{1}{\left| \frac{\partial \omega}{\partial \theta_i} \right|^2} \cdot \frac{\partial \omega}{\partial \theta_i} \cdot \partial_{\theta_i} \tag{1.63}$$

and

$$\left| \frac{\partial \omega}{\partial \theta_i} \right|^2 = \sin^2 \theta_1 \dots \sin^2 \theta_{i-1} \tag{1.64}$$

<u>Definition</u> The operators

$$\Gamma = \overline{\omega} \partial_\omega \ \text{and} \ \Gamma^* = \omega \overline{\partial}_\omega \tag{1.65}$$

are called the spherical Cauchy Riemann operators, and have the respective adjoint operators

$$\Gamma^@ = \overline{\partial}_\omega \omega \ \text{and} \ \Gamma^{@*} = \partial_\omega \overline{\omega} \tag{1.66}$$

Note that we can write the generalized Cauchy Riemann operator D and its conjugate $\overline{D}$ in the form

$$D = \omega \left(\partial_r + \frac{1}{r} \Gamma \right) = \left(\partial_r + \frac{1}{r} \Gamma^{@*} \right) \omega \tag{1.67}$$

$$\overline{D} = \overline{\omega} \left(\partial_r + \frac{1}{r} \Gamma^* \right) = \left(\partial_r + \frac{1}{r} \Gamma^@ \right) \overline{\omega} \tag{1.68}$$

The spherical Cauchy Riemann operators can be used to decompose the Laplace Beltrami operator Δ_4^* appearing in the spherical expression for the Laplacian

$$\Delta_4 = \partial_{r^2}^2 + \frac{4}{r} \partial_r + \Delta_4^* \tag{1.69}$$

We thus find that

$$\Delta_4 = \overline{D} D = \overline{D} \omega \overline{\omega} D \tag{1.70}$$

$$= \left(\partial_r + \frac{1}{r} \Gamma^@ \right) \left(\partial_r + \frac{1}{r} \Gamma \right) \tag{1.71}$$

$$= \partial_{r^2}^2 + \frac{1}{r} \left(\Gamma + \Gamma^@ \right) \partial_r + \frac{1}{r^2} \tag{1.72}$$

$$\left(\Gamma^{@}\Gamma - \Gamma\right) \tag{1.73}$$

The so called Cauchy kernel E in R^4 is given by

$$E(x) = \frac{1}{\omega_4} \frac{\overline{x}}{|x|^4}, \quad x \neq 0 \tag{1.74}$$

where:

$$\omega_4 = 2\pi^{4/2} \frac{1}{\Gamma(4/2)} \tag{1.75}$$

is the area of the unit sphere S^3

The function E enjoys the following properties

1) $E \in M(\mathbb{R}^4; \mathbb{H})$ since $E \in a(\mathbb{R}^4; \mathbb{H})$, $DE = ED = 0$ in $\mathbb{R}^4$, and $\lim_{x \to \infty} E(x) = 0$

2)The Cauchy kernel E also defines a left and a right $\mathbb{H}$-distribution, since

$$DE = ED = \delta \tag{1.76}$$

We write $E \in D^*_{(r)}(\mathbb{R}^4; H)$ as for $E \in D^*_{(l)}(\mathbb{R}^4; \mathbb{H})$ It folllows that the function E is a left and a right fundamental solution of the operator D

We now have all of the definitions that are required in order to construct quaternion wavelets and quaternionic splines

16.2   The Static Approach

First of all we must explain why we use the word "static". Quaternions, in fact, may be used to represent motions, so any quaternionic function can also represent a movement or a sucession of movements. In our approach, we do not use such properties, but rather simply manipulate a tridimensional image by using Haar quaternion wavelets.

The first use of this approach is due to Mitrea [1], where he introduces Clifford WaveletsIn what follows, Clifford algebras will be denoted by $\mathbb{R}_{(n)}$ and complex Clifford algebras by $\mathbb{C}_{(n)}$.

We wish to construct systems of Clifford valued wavelet-like bases adapted to a Clifford valued measure $b(x)dx$ in $\mathbb{R}^m$ where $b : \mathbb{R}^m \to \mathbb{R}^{n+1} \subset \mathbb{C}_{(n)}$. Since Clifford algebras are not commutative, we will use a pair of Clifford algebra valued functions, denoted by $\{lO^L_{j,k}\}_{j,k}$ and $\{lO^R_{j,k}\}_{j,k}$. We will call them **Clifford wavelets**, if they satisfy the following properties.

Properties

1) Cancellation properties: $< lO^L_{j,k}, lO^R_{j',k'} >_b = \delta_{j,j'} \delta_{k,k'}$

2) They form a Riesz frame for L^2:

$$f = \sum < f, lO^R_{j,k} >_b lO^L_{j,k} = \sum lO^R_{j,k} < lO^L_{j,k}, f >_b \tag{2.77}$$

$$||f||^2 = \sum |<f, \mathsf{IO}_{j,k}^R >_b|^2 = \sum |<\mathsf{IO}_{j,k}^L, f >_b|^2 \qquad (2.78)$$

for any L^2-integrable Clifford algebra valued function f. More generally,

$$< f_1, f_2 >_b = \int_{R^m} f_1(x)b(x)f_2(x)dx \qquad (2.79)$$

Definition Let $\mathcal{H}$ an arbitrary complex separable Hilbert space, and V a closed submodule (left or right) of $\mathcal{H}_{(n)}$. A continuous endomorphism T of V is called δ-accretive on V if

$$Re[Tx, x] \geq \delta ||x||_{(2)}^2 \quad x \in V \qquad (2.80)$$

A form $B : \mathcal{H}_{(2)} \times \mathcal{H}_{(n)} \to \mathbb{C}_{(n)}$ is called δ-accrettive if
1)$B(\cdot, \cdot)$ is bilinear, that is $B(\alpha x + \beta y, z) = \alpha B(x, z) + \beta B(y, z)$ and
$B(x, t\alpha + z\beta) = B(x, y)\alpha + B(x, z)\beta$ for all $\alpha, \beta \in \mathsf{IC}_{(2)}$ and $x, y, z \in \mathcal{H}_{(n)}$
2) $B(\cdot, \cdot)$ is continuous on V
3)$Re\, B(x, \overline{x}) \geq \delta ||x||_{(2)}^2$ for any $x \in \mathcal{H}_{(n)}$

16.3   Clifford Multiresolution Analyses of $L^2(\mathbb{R}^m) \otimes \mathbb{C}_{(n)}$

We will abreviate Clifford multiresollution analysis by CMRA
We will consider $\mathcal{H} = L^2(\mathbb{R}^m)$ with the involution given by the conjugation of complex valued functions, and

$$B(f, g) = \int_{\mathbb{R}^m} f(x)b(x)g(x)dx \quad f, g \in L^2(\mathbb{R}^m)_{(n)} \qquad (3.81)$$

Here $b : \mathbb{R}^m \to \mathbb{R}^{n+1} \subset \mathbb{C}_{(n)}$ is a L^∞ function with $Re\, b(x) \geq \delta > 0$, and B is a δ-accretive form on $L^2(\mathbb{R}^m)_{(n)}$. Consider now $\{V'_k\}_k$ a multiresolution analysis of $L^2(\mathbb{R}^m)$ that is a family of closed subspaces for which
1)$\bigcap_{-\infty}^{+\infty} V'_k = \{0\}$ and $\bigcup_{-\infty}^{+\infty} V'_k$ is dense in $L^2(\mathbb{R}^m)$
2)For any $k \in \mathbb{Z}$, $f(x) \in V'_k \Longleftrightarrow f(2x) \in V'_{k+1}$
3)For any $j \in \mathbb{Z}$, $f(x) \in V'_k \Rightarrow f(x - j) \in V'_k$
4)There exists $\phi(x) \in V'_0$ such that $\{\phi(x - j)\}$ is an orthonormal basis for V'_0
The functions

$$\phi_{j,k} = 2^{km/2}\phi(2^k x - j), \quad k \in Z, j \in Z^m \qquad (3.82)$$

form an orthonormal basis for V'_k, and there exist $2^m - 1$ functions$\{\psi_\epsilon\}$ in V'_1 with the same regularity and decay, as the functions ϕ, which form an orthonormal basis of the wavelet space $W'_0 = V'_1 \ominus V'_0$.

Note that $\{\psi_{\epsilon,j,k}\}_{\epsilon,j}$, defined by

$$\psi_{\epsilon,j,k}(x) = 2^{km/2}\psi_\epsilon(2^k x - j)\,,\quad k \in Z\,,\; j \in Z^m \tag{3.83}$$

is an orthonormal basis for $W'_k = V'_{k+1} \ominus V'_k$

It is important to note that for the above CMRA of $L^2(\mathbb{R}^m)_{(2)}$, it can be proved that there exists a dual pair of wavelet bases $\{lO^L_{\epsilon,j,k}\}_{\epsilon,j,k}$ and $\{lO^R_{\epsilon,j,k}\}_{\epsilon,j,k}$ which are r-regular.

16.4   Haar Quaternionic Wavelets

Having the definition above, given by Mitrea, we can adapt it to quaternions to obtain a Quaternion Multiresolution Analyses (QMRA) which is a particular case of the CMRA defined above.

We will first build a quaternion Haar wavelet. Then we will show that we can use quaternion wavelets to do in 3D what is usually done in 2D with ordinary wavelets in a pyramid algorithm.

For any $k \in \mathbb{Z}$, let Υ_k denotes the collection of all dyadic cubes

$$Q = Q_{k,v} = \{x \in \mathbb{R}^3\,;\; 2^{-k}v_i \le x_i \le 2^{-k}(v_i + 1)\} \tag{4.84}$$

$$i = 1,2,3\}\; v \in \mathbb{Z}^3 \tag{4.85}$$

having the side length $l(Q) = 2^{-k}$, and let $\Upsilon = \bigcup_{k \in \mathbb{Z}} \Upsilon_k$ Each dyadic cube $Q \in \Upsilon$ has 2^3 "children". For example, for $m = 3$ each cube has the 8 children

$$\{Q^j\}_{j=1}^{2^3} = \{Q' \in \Upsilon_{k+1}\,;\; Q' \subset Q\} \tag{4.86}$$

The QMRA of $L^2(\mathbb{R}^3)_{(2)}$ uses

$$V_k = \{f \in L^2(\mathbb{R}^3)_{(2)}\} \tag{4.87}$$

where f is piecewise constant on the cubes of $\Upsilon\}$

For each $Q \in \Upsilon$, we now build a family of $2^3 - 1$ functions in V_{k+1}, (L. Traversoni A. Gonzalez [4] (1998)), denoted by $\{\theta_{Q,i}\}_{i=1}^{2^3-1}$ such that

1) $\displaystyle\int_{\mathbb{R}^3} lo_{Q,i}(x)b(x)dx = \int_{\mathbb{R}^3} b(x)lo_{Q,i}(x)dx = 0,$ $\quad$ (4.88)

$$i = 1,2,...,2^3 - 1.$$

2) $\displaystyle\int_{\mathbb{R}^3} lo_{Q,i}(x)b(x)lo_{Q,j}(x)dx = \delta_{i,j}\; \forall\, i,j \tag{4.89}$

We can take

$$lo_{Q,i} = a_i\left(\sum_{j=1}^{i}\chi_{Q^j}\right) - b_{i+1}\chi_{Q^{i+1}} \tag{4.90}$$

where

$$a_i = \left(\sum_{j=1}^{i} 3(Q^j) \right)^{-1} \tag{4.91}$$

$$b_{i+1} = 3(Q^{i+1})^{-1} \tag{4.92}$$

The Haarquaternion wavelets are defined by

$$1O_{Q,i}^L = M(Q,i)^{-1/2}1o_{Q,i} \quad i = 1, 2, ..., 2^3 - 1 \tag{4.93}$$

$$1O_{Q,i}^R = 1o_{Q,i} M(Q,i)^{-1/2} \quad i = 1, 2, ..., 2^3 - 1 \tag{4.94}$$

where

$$M(Q,i) = \left(\sum_{j=1}^{i} 3(Q^j) \right)^{-1} \left(\sum_{j=1}^{i} 3(Q^j) \right) 3(Q^{i+1})^{-1} \tag{4.95}$$

$$|M(Q,i)| \simeq |Q|^{-1} \tag{4.96}$$

and where $|Q|$ is the euclidean volume of the cube.

It is very important to note that the wavelets either vanish or are quaternion–valued in $\mathbb{R}_{(2)}$.

16.4.1   Decomposition and reconstruction for quaternion wavelets

Now, just as in the real case, we can write analogously $\{V_j\}$ generated by a scale function $\phi \in L^2(\mathbb{R}^3_{(2)})$. In this case the functions are the cubes, and the $\{W_j\}$ are generated by some wavelet $\psi \in L^2(\mathbb{R}^3_{(2)})$. Every function $f \in L^2(\mathbb{R}^3_{(2)})$ can be aproximated uniquely, given $N \in \mathbb{Z}$, as closely as desired by some $f_N \in V_N$,

$$f_N = f_{N-1} + g_{N-1} \tag{4.97}$$

where $f_{N-1} \in V_{N-1}$ and $g_{N-1} \in W_{N-1}$. Repeating this process,

$$f_N = g_{N-1} + g_{N-2} + ... + g_{N-M} + f_{N-M}. \tag{4.98}$$

16.4.2   A biomedical application

Suppose now that we have a set of 3D data, for example, tomography data for the interior of the brain. This 3D data is given in the form of a collection of 2D pictures of slices, and the device which makes the image gives a color to each pixel in order to diferenciate softer than harder tissues. The position

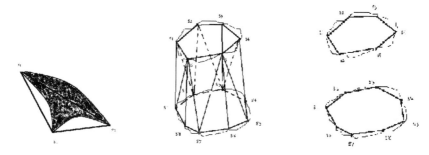

FIGURE 16.1. a) A B-spline adjusted to the contour. b) Slices are united by triangles, in each slice polygons are created as limits of each distinguishable tissue. c) A surface is adjusted in each triangle.

of each slice is known to be 1.2 mm. The horizontal width of each pixel is about 0.1mm, so if we transform each pixel in a "boxel", a cube of 0.1mm of side of the same colour, we still have voids between slices of about 1.1mm. This means that 11 boxels are needed to give continuity to an image of a tumor. Now, we want to represent each boxel by a quaternion of the form $q_i = \{\cos\alpha, x\sin\alpha, y\sin\alpha, z\sin\alpha\}$, where x, y, z are the spacial coordinates of each boxel with respect to the baricenter of the head, and we let α be the color.

We can now apply the quaternion Haar wavelets to this data, but first we must interpolate the data in order to fill in the gaps. The problem is how to interpolate in such a way as to take advantage of the properties of quaternions. There are many possible strategies. We can cite for example the classical work by Schumaker [9], where pattern recognition is used to distinguish between the different tissues. The idea used in his approach has the 3 steps

In the Figure 16.1.a, the border is defined by a curve given by pixels, and a B-spline is adjusted to such data in order to reduce the number of data points, giving a polygonal P_{s_i} that represents the border for the ith slice. Repeating the procedure, we get polygonals for the adjacent slices $P_{s_{i-1}}$ and $P_{s_{i+1}}$. In Figure 16.1.b, a triangulation is constructed between polygonals and in Figure 16.1.c, surfaces are adjusted to each triangle in order to have C_1 continuity between the triangles.

At the end, continuity between the contiguous slices is ensured by performing one interpolation for each different tissue. Instead, we choose a local interpolation scheme more amenable to our means. The function to be interpolated is the color, represented by the angle of rotation in the scalar part of the quaternion. Interpolation is performed between the "natural neighbors" of the point considered in 3D (there are several techniques of determining such neighbors, we can cite, Traversoni[7] and Sukumar [6]).

To minimize the computations necessary, we first reduce the data by

FIGURE 16.2. The local cloud of points to be interpolated and the zones in the two layers where are the points.

filtering, then we create the interpolated points, and finally we apply the reconstructing procedure to the new set. (see figure 1 and 2)

The formula we used to interpolate is a combination of the known formulas for the plane

$$I(p_i) = \sum_{i=1}^{n} \frac{p_i \sin(w_i \theta)}{\sin(\theta)} \tag{4.99}$$

Here the weights w_i are a function of the position.

Once the interpolation of the gaps between the slices has been completed, we can use quaternion wavelets, and in particular, the Haar quaternion wavelet on the resulting data.

The first experiment is to compress and decompress the data set. To do this, we use the compression-decompression algorithm

$$c_k^{j-1} = \sum_l a_{l-2k} c_l^j \tag{4.100}$$

$$d_k^{j-1} = \sum_l b_{l-2k} c_l^j \tag{4.101}$$

or the composition algorithm

$$c_k^j = \sum_l [p_{k-2l} c_l^{j-1} + q_{k-2l} d_l^{j-1}] \tag{4.102}$$

We present in figures 3 and 4 a detail (a piece of the nose) and the whole image of the head, as can be seen in image 4 we have some oscillation problems that produce artifacts in some places.

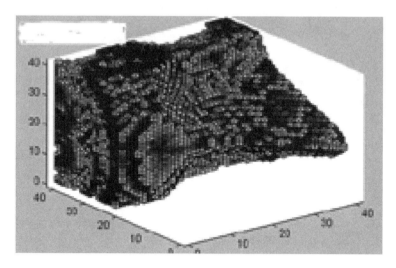

FIGURE 16.3. A detail of the figure (the nose) showing the boxels.

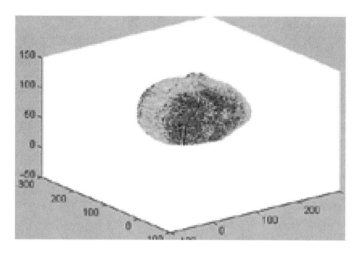

FIGURE 16.4. The full image of the head, showing the same colours provided by the magnetic resonance device.

16.5   A Dynamic Interpretation

Quaternions can represent rotations and movements in general. For example, in a dynamic environment, a Haarwavelet may be a full turn to the left or to the right, instead of 1 or -1 when we deal with real numbers.

Reconstructions of movements given by several positions of a rigid body have been done by Jüttler and others using dual quaternions and quaternionic B spline. More complex objects may be represented, such as moving surfaces, by using quaternionic wavelets.

Consider, for example, the case of a sphere and a surface moving over it. Let S be a surface over the unit sphere and let $P = \{p_1, p_2...p_n\}$ be a set of given perturbations, each with its given trajectory. Let $t = \{t_1, t_2, ..., t_n\}$ be the times of each of the perturbations, and $l = \{l_1, l_2, ..., l_n\}$ the positions of the surface S.

We are going to express the set t by a set of unit quaternions and the perturbations by another set of quaternions. Multiplying them together, gives the expression of a moving perturbation. Let $MP_i = q_t * q_l$ where $*$ stands for the quaternionic multiplication. What we want to know is the state of the perturbation, first for a static point at any time, as well as the same state of perturbation of the point moving in a given trajectory.

Let $u = \{u_0, u_1, u_2, u_3\}$ be a quaternionic function represented by

$$\{\cos \phi, x \sin phi, y \sin \phi, z \sin \phi\} \qquad (5.103)$$

where $\phi = f(t)$ and $f(t)$ is a function of a real variable t. As ϕ is the angle of the rotation represented by the quaternion, the rotation is around a fixed (x, y, z) axis. Of course, ϕ can a wavelet, $\phi \in L^2(\mathbb{R})$.

We can reconstruct the movement of a rotating object by using wavelets on its angular velocity at several instants in time. We will use the wavelet of a spline, following Chui [2]:

$$\psi_m(x) = \sum_n q_n N_m(2x - n) \qquad (5.104)$$

$$Q_m(z) = \frac{1}{2} \sum_n q_n z^n = \left(\frac{1-z}{2}\right)^m \sum_{k=0}^{2m-2} N_{2m}(k+1)(-z)^k \qquad (5.105)$$

where, since Q_m is the product of two polynomials from the sequence $\{q_n\}$, the convolution of these two polynomials is

$$q_n = \frac{(-1)^n}{2^{m-1}} \sum_{l=0}^m \binom{m}{l} N_{2m}(n+1-l) \quad n = 0, ..., 3m - 2 \qquad (5.106)$$

The same may be done with the coefficients (x, y, z) of the vectorial part, which will represent changes of the direction of the axis, or even dilations of the axis.

16.6   Global Interpolation

We will start by using radial basis functionsfor global interpolation on the skewfield of the quaternions.

We first introduce the Teodorescu transform

$$(T_G u)(x) = -\int_G e(x-y)u(y)dy \qquad (6.107)$$

where

$$e(x) = \frac{1}{\sigma_n} \frac{\overline{w}(x)}{|x|^{n-1}} \qquad (6.108)$$

$$w(x) = \frac{x}{|x|} \qquad (6.109)$$

$$x = \underline{x} = \sum_{i=1}^{3} x_i e_i. \qquad (6.110)$$

We also define, following Guerlebeck [5], a IH-regular quaternionic spline,

$$S_p(x) = \int_\Gamma e(y-x)\alpha(y)sp(y)d\Gamma_y \qquad (6.111)$$

Note that such splines are monogenic functions, monogenic functions. Let $P_{\mathbb{IH}}$ be the space of monogenic polynomials on IH, Delanghe [8].

A quaternionic regular polynomial has the form:

$$P_m(x) = \frac{1}{m!} \sum_{\mu_1,\dots,\mu_m} z_{\mu_1} \cdot \dots \cdot z_{\mu_m} \qquad (6.112)$$

where

$$z_k = x_0 e_k - x_k e_0 \quad (k = 1,2,3) \qquad (6.113)$$

and where $\{e_0, e_1, e_2, e_3\}$ is a basis of IH and $x = \sum_{i=0}^{3} x_i e_i$ and $e_0^2 = e_0$ and $e_i e_j + e_j e_i = -2\delta_{ij}$

This leads to the idea of a "moving thin plate" spline. We can change it into a quaternionic wavelet, where the scaling function is a quaternionic spline. Movement is obtained by the variation of each source or sink (at the data points), and interpolation between them may be done using a quaternionic polynomial b-spline or cardinal quaternionic wavelet splines.

16.7   Dealing with Trajectories

One of the most widely used applications of quaternions has been the reconstruction of trajectories, of points, and of even planes and surfaces. Generally trajetories are interpolated using B splines and wavelets. The usual interpolation for trajectories is made using dual quaternions, see Jüttler [11] and Bayro [10]:

Consider the translational

$$Q_{tras}^{(i)} = 2 + \epsilon \sum_{j=0}^{3} d_j^3 \left(\frac{t - t_i}{t_{i+1} - t_i} \right) \vec{p}_j^{(i)} \tag{7.114}$$

and the rotational

$$Q_{rot}^{(i)} = 2 + \epsilon \sum_{j=0}^{3} d_j^3 \left(\frac{t - t_i}{t_{i+1} - t_i} \right) C_j^{(i)} \tag{7.115}$$

Multiplication of the translational and rotational gives a quaternion describing any possible movement. Chui [2] has shown that B splines can in general be used as scaling functions, as we have already seen in the introduction.

We can then make a multiresollution analysis by using the quaternion B-splines described above. This might seem trivial and not significantly different than the more usual treatments. The difference is in the application that allows us to use wavelet procedures in the trajectory reconstruction of blurry objects (for example, the blurryness due to bad reception of radar data which hides the dimensions, orientation, and position of the object).

16.8   Conclusions

Combining Computer Aided Geometric Design (CAGD) techniques and wavelet techniques offers the most promising way of developing quaternion wavelets of quaternion splines. Using quaternion wavelets allows us to add movement and further develop Computer Aided Geometric Mechanics. For example, from several images of a beating heart, the movement can be reproduced not as in a film, but by approximating the equations of the movement. Unfortunately, our computational skills don't allow us at the moment to make commercially competitive programs. But I think that the potential of the technique is important enough to justify trying to improve the computational techniques.

All of the programs used here were implemented in MATLAB, with a lot of constraints in memory and in the speed of the computations needed to do real time animations. At present the programs are very slow. Much work is still necessary to realize the full potential of the software for CAGD and the full animation of images based on quaternion wavelet techniques.

Part VI

Applications to
Engineering and Physics

Chapter 17

Objects in Contact: Boundary Collisions as Geometric Wave Propagation

Leo Dorst

17.1 Introduction

17.1.1 Towards a 'systems theory' of collision

The motivation behind this work is to make the computation of collision-free motions of robots efficiently computable. For translational motions, the boundary of permissible translations of a reference point is obtained from the obstacles and the robot by a kind of dilation, 'thickening' the obstacle (see below for details) to produce the forbidden states in the configuration space of translations. The intuitive similarity of this operation to convolution suggests that we might be able to find a kind of Fourier transformation, in the sense that we might separate the shapes into independent 'spectral components' and combine those simply; after which the collision boundary would be obtained by the inverse transformation. This would enable the development of a 'systems theory' for collisions.

This was indeed done for two-dimensional boundaries [1], using a Legendre transformation and its coordinate-free counterpart (which is related to the polar curves of projective geometry). However, using classical differential geometry the generalization to the m-dimensional case was not straightforward. In this Chapter we do it using geometric algebra, which easily captures the geometric intuition in simple, computable expressions and allows compact derivations of fairly advanced results.

17.1.2 Collision is like wave propagation

This paper actually treats the mathematics of geometric wave propagation according to Huygens' principle, in which points on a wave front become secondary sources (also called propagators), of which the forward caustic generates the new wave front. The reason is that for arbitrary shapes of propagators, the generation of the secondary wave front is mathematically very closely related to the collision problem which we really want to treat (but the wave propagation formulation has some advantages).

This can be seen as follows, using Figure 17.1.

- *Huygens wave propagation*
  When we perform a Huygens wave propagation for a finite time interval, we place copies of a 'propagator' $\mathcal{A}$ at each position on a wave front $\mathcal{B}$ (see Figure 17.1a). By Huygens' principle, the forward caustic of these secondary wave fronts then forms the resulting propagated front, which we denote $\mathcal{A} \check{\oplus} \mathcal{B}$ (see Figure 17.1b). Each point P of $\mathcal{B}$ in this way locally 'causes' a point in the result (easiest to see when $\mathcal{A}$ is convex and $\mathcal{B}$ is differentiable). By performing a linear approximation to $\mathcal{B}$ at P, it is clear that at every point of the resulting caustic, the tangent is equal to that of the point which caused it, and equal to the tangent at the corresponding point of the 'propagator' $\mathcal{A}$.

- *Collision detection*
  Now consider Figure 17.1c, which depicts the collision of a movable object (the robot) $\mathcal{A}'$ with a fixed obstacle $\mathcal{B}$. The position of $\mathcal{A}'$ is indicated by that of its *reference point*, which is some fixed point that moves with it (of we were to allow rotations, we would need to specify a reference frame, but for now a reference point is enough). This point is prevented from moving freely due to the collision at P. Computing such local contacts for all translational motions of $\mathcal{A}'$ generates the boundary of the 'free space' for the reference point of $\mathcal{A}'$. A linear approximation shows that the tangents at P of $\mathcal{B}$, $\mathcal{A}'$ and the reference point at the resulting boundary are proportional (with the outwardly directed normal vector of $\mathcal{A}'$ at P having an opposite sense).

This shows that the two operations of wave propagation and collision are mathematically closely related. Indeed, the boundary of free space in the collision problem is precisely $(-\mathcal{A}') \check{\oplus} \mathcal{B}$, i.e. the propagation of $\mathcal{B}$ with the propagator $-\mathcal{A}' \equiv \{-a \mid a \in \mathcal{A}'\}$. Thus an analysis of either is applicable to the other.

In this Chapter, we will use the wave propagation terminology, since it has the simplest relationship between directed tangents (no opposite orientations). We first come up with a representation for proper boundaries which unifies the 'hypersurface' aspects of a boundary with the assignment of a local 'inside'. Then we analyze the wave propagation in terms of this representation; it will turn out to be the 'Fourier-like' representation we were looking for, in which wave propagation becomes separable into a simple operation on the 'spectral components' (these are the tangent hypersurfaces of the objects involved).

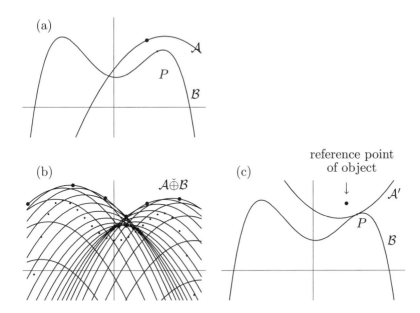

FIGURE 17.1. Wave propagation and collision are mathematically similar (see text).

17.1.3 Related problems

In fact, operations similar to wave propagation occur in many places in science and engineering. In *computer graphics*, 'growing' objects by thickening some specified skeleton shape involves a thickening by spheres (or other shapes), which is clearly equivalent to one step of Huygens wave propagation. In *image analysis*, the field of 'mathematical morphology' for object selection, originally designed to mimic the selective filtering of grains by sieves, involves extensive use of the 'dilation' operation – which is essentially 'growing' the object geometrically. The 'distance transform' and the related techniques of 'skeletonization', which involve computing the shortest distance to the nearest object, are also example of operations with a wave-propagation-like structure. In *robotics*, some investigators have approached the 'path planning problem' in this manner, with the waves computing the distance function for the shortest path in the state space of the robot between initial state and goal state. In *milling* (carving away excess material to end up with a desired shape), the carving bit performs a destructive collision with the object to produce the result; the relationship between those three is again essentially Figure 17.1c. In *scanning tunneling microscopy*, an atomic probe (of unknown shape) is moved over an unknown atomic surface such that voltage between them is constant; this implies a contact operation on the equi-potential surfaces, which happens according to Figure 17.1c, in good approximation. Again, this gives the same relationship between the surface, the probe, and the measured surface of positions.

Our analysis of the mathematics of geometric wave propagation is in principle applicable to all these fields. However, most of these applications contain an essential element which we will *not* treat here: in sharply concavely curved sections of $\mathcal{B}$, a locally tangent contact of $\mathcal{A}'$ can be precluded by an intersection of $\mathcal{A}'$ elsewhere along $\mathcal{B}$. This happens in the central section of Figure 17.1b, and part of the caustic then becomes excluded (e.g. you would not send your milling bit there, it would carve out unwanted sections). Mathematically, it is easier to treat these parts on a par with the rest, and in a genuine wave propagation they would be observed. For the robotics collision application which motivated this work, these parts must be treated: if those concave parts of the obstacle $\mathcal{B}$ have been observed somehow, a path planning algorithm should already exclude the corresponding parts of $\mathcal{A} \bar{\oplus} \mathcal{B}$ from consideration, even though later observations of other sections of $\mathcal{B}$ may exclude them eventually as impossible collisions.

17.2   Boundary Geometry

17.2.1   The oriented tangent space

In an m-dimensional *Euclidean* vector space $\mathcal{G}^1(\mathbf{I}_m)$, with pseudoscalar $\mathbf{I}_m$, we consider an object, noting specifically its boundary. This boundary is an $(m-1)$-dimensional hypersurface, with locally two 'sides': an *inside* and an *outside*. Assume the boundary to be smooth (we will not treat edges in this Chapter); then at every point $\mathbf{p}$ of the boundary, the boundary surface has a local tangent space with pseudoscalar $\mathbf{I}[\mathbf{p}]$ of grade $m-1$ (which we will mostly denote by $\mathbf{I}$, with $\mathbf{p}$ understood; throughout this Chapter we will use square brackets for non-linear arguments, round brackets for linear arguments). We may represent this tangent space by a dual vector $\mathbf{n}[\mathbf{p}]$ (with again $\mathbf{p}$ mostly understood as the parameter), defined by the geometric product with the inverse pseudoscalar:

$$\mathbf{n}[\mathbf{p}] = -\mathbf{I}[\mathbf{p}] \, \mathbf{I}_m^{-1}. \tag{2.1}$$

The '$-$' is introduced here to avoid awkward signs later on. We will call $\mathbf{n}[\mathbf{p}]$ the *normal vector* to $\mathbf{I}[\mathbf{p}]$. Since we are only interested in rigid body transformations, we can restrict our treatment to Euclidean geometry, in which the normal vector is well-behaved (the normal vector of a transformed object is the transformation of the normal vector), so using duals

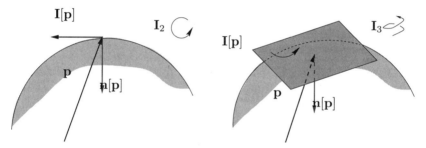

FIGURE 17.2. The inward pointing normal n[p] related to the orientation of the local tangent space I[p], in two-dimensional $\mathcal{G}(\mathbf{I}_2)$ (left) and 3-dimensional $\mathcal{G}(\mathbf{I}_3)$ (right).

is permitted. We wish to denote the notion of 'inside' geometrically in the boundary representation, to make it more than merely the representation of the boundary surface. This involves orienting $\mathbf{n}[\mathbf{p}]$ (and hence $\mathbf{I}[\mathbf{p}]$). The usual convention for a circular blob in 2D (with the usual right-handed pseudoscalar $\mathbf{I}_2$) is 'when following the contour, the object is at the left-hand side'. So with a tangent $\mathbf{I}[\mathbf{p}]$ in the direction of motion, we have $\mathbf{I}[\mathbf{p}]\,\mathbf{I}_2$ as *inward* pointing direction. Therefore $\mathbf{n}[\mathbf{p}] = -\mathbf{I}[\mathbf{p}]\,\mathbf{I}_2^{-1} = \mathbf{I}[\mathbf{p}]\,\mathbf{I}_2$ is the *inward pointing normal vector*. We generalize this to m-dimensional space, deriving the sign of $\mathbf{I}[\mathbf{p}]$ using Equation (2.1) from the desire to have $\mathbf{n}[\mathbf{p}]$ be the locally *inward* pointing normal vector. Thus the tangent spaces have been oriented properly, whether represented by $\mathbf{I}[\mathbf{p}]$ or by $\mathbf{n}[\mathbf{p}]$.

17.2.2   Differential geometry of the boundary

The boundary surface at position $\mathbf{p}$ in $\mathcal{G}^1(\mathbf{I}_m)$ has directed tangent $\mathbf{I}[\mathbf{p}]$; when we move along the boundary surface, the tangent will change. The description of these changes can be found in [3](Chapters 4 & 5), and we briefly repeat and extend the elements relevant to our analysis.

So let $\mathbf{n}$ denote the inside pointing local unit normal vector, as a differentiable function of the position $\mathbf{p}$ on the boundary. Its direction can be derived from the first order differential structure of $\mathbf{p}$ (e.g., if the surface of the boundary is implicitly given by a scalar function ϕ as $\phi(\mathbf{p}) = 0$, then $\mathbf{n}[\mathbf{p}]$ is proportional to $\partial_{\mathbf{p}}\phi(\mathbf{p})$), but its sign must be explicitly determined by our notion of 'inside'. The second order differential structure of the boundary is obtained by differentiating such a properly oriented $\mathbf{n}$ using a vector derivative in some direction $\mathbf{a}$. We denote the resulting position-dependent linear function by $\underline{\mathbf{n}}(\cdot)[\mathbf{p}]$. So $\underline{\mathbf{n}}[\mathbf{p}] : \mathbf{I}_m[\mathbf{p}] \to \mathbf{I}[\mathbf{p}]$ is defined as the differential of $\mathbf{n}$:

$$\underline{\mathbf{n}}(\mathbf{a})[\mathbf{p}] \equiv (\mathbf{a} \cdot \partial)\mathbf{n}[\mathbf{p}], \qquad (2.2)$$

and in our notation we will mostly let the dependence on $\mathbf{p}$ be understood implicitly. Since $\underline{\mathbf{n}}(\mathbf{a})$ is a linear function of the vector argument $\mathbf{a}$, we may extend it to arbitrary multivector arguments as an outermorphism (i.e. a $\wedge$-preserving linear operator). Specifically, we can form $\underline{\mathbf{n}}(\mathbf{a}_{\wedge}\mathbf{a}_2 \wedge \cdots \mathbf{a}_{m-1})$, with the $\mathbf{a}_i$ forming a basis for the tangent space $\mathcal{G}^1(\mathbf{I}[\mathbf{p}])$ at $\mathbf{p}$. The quantity $\underline{\mathbf{n}}(\mathbf{I})$ thus denotes the volumetric change of $\mathbf{n}$ as we move over a local tangent volume the size of the tangent pseudoscalar $\mathbf{I}$; the ratio with $\mathbf{I}$ (which is the determinant of the mapping $\underline{\mathbf{n}}$) is related to the *directed Gaussian curvature* κ of the boundary at $\mathbf{p}$ by:

$$\kappa \equiv (-1)^{m-1}\underline{\mathbf{n}}(\mathbf{I})\mathbf{I}^{-1} = \underline{\mathbf{n}}(\widehat{\mathbf{I}})\mathbf{I}^{-1} \qquad (2.3)$$

(the $\widehat{}$ denotes grade inversion), where we obtain an extra sign factor relative to the standard convention for the definition of κ which uses the *outward pointing normal vector*, due to the fact that $\mathbf{n} \to -\mathbf{n}$ gives $\underline{\mathbf{n}}(\mathbf{I}) \to \underline{\mathbf{n}}(\widehat{\mathbf{I}})$ by Equation (2.2).

The vector $\underline{\mathbf{n}}(\mathbf{a})$ gives the change in the unit normal vector when moving in the $\mathbf{a}$-direction; this value is unique for the regular surfaces we treat. It is one of the properties of differentials that $\underline{\mathbf{n}}(\mathbf{P}(\mathbf{a})) = \underline{\mathbf{n}}(\mathbf{a})$ for all $\mathbf{a}$ (where $\mathbf{P}(\mathbf{a}) \equiv (\mathbf{a} \cdot \partial)\mathbf{p}$ is the projection onto the local tangent space) – and so a unique inverse to $\underline{\mathbf{n}}(\cdot)$ does not exist. Any $\mathbf{a}$ that projects to the same $\mathbf{P}(\mathbf{a})$ gives the same value for $\mathbf{n}$; but even when we limit the inverse to have values in the tangent space $\mathcal{G}^1(\mathbf{I}[\mathbf{p}])$, there may not be a unique solution. (An example is a cylinder with axis $\mathbf{z}$, where only the component of $\mathbf{P}(\mathbf{a})$ perpendicular to $\mathbf{z}$ determines the value of $\underline{\mathbf{n}}(\mathbf{a})$.) Therefore the inverse of $\underline{\mathbf{n}}(\cdot)$ usually produces a *set* of vectors in the space with pseudoscalar $\mathbf{I}_m$ based at $\mathbf{p}$ (we will denote this space by $\mathcal{G}^1(\mathbf{I}_m[\mathbf{p}])$). We prefer to limit the values to the local tangent space $\mathcal{G}^1(\mathbf{I}[\mathbf{p}])$, and so define:

$$\underline{\mathbf{n}}^{-1}[\mathbf{p}] : \mathcal{G}^1(\mathbf{I}[\mathbf{p}]) \to \mathcal{G}^1(\mathbf{I}[\mathbf{p}]) : \quad \underline{\mathbf{n}}^{-1}(\mathbf{m})[\mathbf{p}] \equiv \{\mathbf{a} \mid \underline{\mathbf{n}}(\mathbf{a})[\mathbf{p}] = \mathbf{m}\}. \quad (2.4)$$

(In words, $\underline{\mathbf{n}}^{-1}$ gives the set of 'tangent velocities' at the point $\mathbf{p}$ required to produce the change $\mathbf{m}$ in $\mathbf{n}$.) Such set-valued functions can be added, using the *Minkowski sum* (denoted by $\oplus$) as set addition:

$$A \oplus B = \{a + b \mid a \in A, b \in B\}, \qquad (2.5)$$

where the '+' denotes is the vector addition. Note that if one of the arguments is $\emptyset$, then so is the result.

17.3   The Boundary as a Geometric Object

In the representation of the boundary so far, we required a description of the position $\mathbf{p}$ (of which the differential structure gives us the direction of the local tangent space, characterizable by $\mathbf{I}[\mathbf{p}]$ or $\mathbf{n}[\mathbf{p}]$) and an orientation sign to specify 'inside' (which then gives the proper sign to $\mathbf{I}[\mathbf{p}]$ or $\mathbf{n}[\mathbf{p}]$). Thus the boundary is not yet a *single geometric object* in an embedding space.

In [1], a single representation was found in a homogeneous embedding in projective $(m+1)$-space (for $m = 2$ only); it is actually structurally more clear to embed in the null-cone of the Minkowski space with Clifford algebra $\mathcal{Cl}_{m+1,1}$, and we do so now.

17.3.1   Embedding in $\mathcal{Cl}_{m+1,1}$

We embed the boundaries of the Euclidean space $\mathcal{G}^1(\mathbf{I}_m)$ through a conformal split in the higher dimensional space $\mathcal{G}^1(I_{m+2})$ with pseudoscalar $I_{m+2} \equiv E\mathbf{I}_m$. Here E is a bivector defined as

$$E \equiv e_0 \wedge e^0, \tag{3.6}$$

where e_0 and e^0 span the two extra dimensions. We choose them to be two reciprocal null vectors perpendicular to $\mathbf{I}_m$, so they satisfy

$$(e_0)^2 = (e^0)^2 = 0, \quad e_0 \cdot e^0 = 1, \quad e_0 \cdot \mathbf{I}_m = e^0 \cdot \mathbf{I}_m = 0. \tag{3.7}$$

We are therefore in the Clifford algebra $\mathcal{Cl}_{m+1,1}$ of a Minkowski space (with e_0 and e^0 spanning the null cone), and $E^2 = 1$. As Chapters 1 and 3 have shown, this representation can be used to extend the more commonly used projective split (with e_0 as splitting vector) to an *isometric* embedding of Euclidean m-space onto the *horosphere* in $\mathcal{G}^1(I_{m+2})$. (The horosphere is the intersection of the null cone with the plane $x \cdot e^0 = 1$.) This powerful embedding was introduced into geometric algebra in [2] and [4]; Chapter 1 introduces two splits to produce this representation; we prefer the *additive split* as in the original formulation of [4] (using e^0 for their $-e$ since we want proper reciprocity between e_0 and e^0). In that split, a point $\mathbf{p}$ is represented as the null vector: $p' = e_0 + \mathbf{p} - e^0\mathbf{p}^2/2$. We will use **bold** for elements of the algebra $\mathcal{G}(\mathbf{I}_m)$, and the usual *math* font for elements of the larger algebra $\mathcal{G}(I_{m+2})$.

Intuitively, e_0 is the representation of a point at the origin, and $-e^0$ is the representation of (the direction of) the point at infinity. In this Chapter, we will only need to embed *flats*, i.e. offset linear subspaces such as 0-dimensional points, 1-dimensional lines, etc., since those are the tangent spaces used to describe the boundaries. A flat with tangent $\mathbf{I}$ at the position $\mathbf{p}$ is represented in the additive split as (see [4]):

$$e^0 \wedge p' \wedge \mathbf{I} = e^0 \wedge (e_0 + \mathbf{p}) \wedge \mathbf{I} \tag{3.8}$$

(leave off '$e^0\wedge$' to retrieve the usual homogeneous representation). For brevity, we will denote $e_0 + \mathbf{p}$, the homogeneous representation of the point at $\mathbf{p}$, by p. Note that $e^0 \wedge p' = e^0 \wedge p$, so that we may substitute p for p' in this equation.

Instead of with Equation (3.8), it is somewhat more convenient to work with its dual in our embedding space $\mathcal{G}(E\mathbf{I}_m)$, using $(E\mathbf{I}_m)^{-1} = \mathbf{I}_m^{-1}E$:

$$(e^0 \wedge p \wedge \mathbf{I})\mathbf{I}_m^{-1}E \quad = \quad -(e^0 \wedge p) \cdot \mathbf{n}E = p \cdot \left(e^0 \cdot (E\mathbf{n})\right)$$

$$= \; p \cdot (e^0 \mathbf{n}) = \mathbf{n} - e^0 (\mathbf{p} \cdot \mathbf{n})$$

We define this as the representation $B(\mathbf{n})[\mathbf{p}]$ of the boundary at $\mathbf{p}$, so:

$$B(\mathbf{n})[\mathbf{p}] \equiv p \cdot (e^0 \mathbf{n}) = \mathbf{n} - e^0 (\mathbf{p} \cdot \mathbf{n}). \tag{3.9}$$

This represents the flat by its normal vector and its scalar support $\mathbf{p} \cdot \mathbf{n}$ as an object in $\mathcal{Cl}_{m+1,1}$. This representation is indexed by $\mathbf{n}$, and has an extra component in the e^0-direction; this scalar is the *support* of the flat. Over all $\mathbf{n}$ occurring in the boundary, this is therefore essentially the extended Gauss map: the sphere of directions, augmented by a scalar function specifying the directed support, see [5]. We will soon see that $(B(\mathbf{n})[\mathbf{p}])^2 = 1$, so that the extended Gauss-sphere is geometrically embedded as an actual sphere in Minkowski space (rather than as a scalar-valued function on the tangent space sphere, which is the usual description).

17.3.2 Boundaries represented in $\mathcal{G}(E\mathbf{I}_m)$

We view a boundary as a collection of tangent flats, and assume throughout this Chapter that this collection is differentiable; so we limit ourselves to 'regular boundaries' in this sense. (This does not preclude the treatment of swallowtail catastrophes in the propagation result, as we will show in Section 17.4.5 – despite the characterization of such a curve in classical differential geometry as non-regular.)

Since the representation Equation (3.9) contains $\mathbf{n}$ explicitly, we will view the boundary as 'indexed by $\mathbf{n}$'. We are then required to view the positions $\mathbf{p}$ as a function of $\mathbf{n}$, so we write $\mathbf{p}[\mathbf{n}]$ – we use square brackets since $\mathbf{p}$ is generally non-linear in $\mathbf{n}$. Also, this is not a single-valued function, since the same tangent may occur at different locations if the object is not convex. Once we have done this, the representation becomes a set-valued function of $\mathbf{n}$ only, which we denote by $B(\mathbf{n}_{\mathcal{A}})$ for a boundary $\mathcal{A}$ – though we mostly omit the subscript if the context is clear.

The representation $B(\mathbf{n})$ has some very nice properties:

- *representation commutes with differentiation*
  We have to be careful here, since it depends whether we differentiate $B(\mathbf{n})[\mathbf{p}]$ relative to variations in $\mathbf{p}$ (defining $\mathbf{n}$ as a function of $\mathbf{p}$) or in $\mathbf{n}$ (with $\mathbf{p}$ as a function of $\mathbf{n}$). Differentiating relative to $\mathbf{n}$, we obtain:

$$
\begin{aligned}
(\mathbf{m} \cdot \boldsymbol{\partial}_\mathbf{n}) B(\mathbf{n})[\mathbf{p}[\mathbf{n}]] &= (\mathbf{m} \cdot \boldsymbol{\partial}_\mathbf{n}) \left(p[\mathbf{n}] \cdot (e^0 \mathbf{n}) \right) \\
&= \mathbf{P}(\underline{\mathbf{n}}(\mathbf{m})) \cdot (e^0 \mathbf{n}) + p[\mathbf{n}] \cdot (e^0 \mathbf{m}) \\
&= p[\mathbf{n}] \cdot (e^0 \mathbf{m}) = B(\mathbf{m})[\mathbf{p}[\mathbf{n}]], \tag{3.10}
\end{aligned}
$$

since $\mathbf{P}(\underline{\mathbf{n}}(\mathbf{m}))$ is an element of the tangent space at $\mathbf{p}[\mathbf{n}]$, and therefore perpendicular to both e^0 and $\mathbf{n}$. This thus gives the commutative

relationship between differentiation and representation:
$(\mathbf{m} \cdot \boldsymbol{\partial_n})B(\mathbf{n}) = B((\mathbf{m} \cdot \boldsymbol{\partial_n})\mathbf{n})$. It is convenient to use a shorthand for the differential, as in [3]: define $\underline{B}(\mathbf{a}) \equiv (\mathbf{a} \cdot \boldsymbol{\partial_n})B(\mathbf{n})$, then

$$\underline{B}(\mathbf{m})[\mathbf{p}] = B(\mathbf{m})[\mathbf{p}].$$

(Be careful to use both expressions at the same value of $\mathbf{p}[\mathbf{n}]$, *not* at $\mathbf{p}[\mathbf{m}]$!) Since we need to preserve the norm of $\mathbf{n}$ for the representation $(\mathbf{n}^2 = 1)$, we will only use differentials for which $0 = (\mathbf{m} \cdot \boldsymbol{\partial_n})\mathbf{n}^2 = 2\mathbf{m} \cdot \mathbf{n}$, i.e. $\mathbf{m}$ is perpendicular to $\mathbf{n}$.

On the other hand, taking the derivative of $B(\mathbf{n})[\mathbf{p}]$ to $\mathbf{p}$ yields

$$\begin{aligned}
(\mathbf{a} \cdot \boldsymbol{\partial_p})B(\mathbf{n})[\mathbf{p}] &= (\mathbf{a} \cdot \boldsymbol{\partial_p})\left(p \cdot (e^0\mathbf{n}[\mathbf{p}])\right) \\
&= \mathbf{P}(\mathbf{a}) \cdot (e^0\mathbf{n}) + p \cdot \left(e^0\underline{\mathbf{n}}(\mathbf{a})\right) \\
&= p \cdot \left(e^0\underline{\mathbf{n}}(\mathbf{a})\right) = B(\underline{\mathbf{n}}(\mathbf{a}))[\mathbf{p}],
\end{aligned}$$

so that now $B(\cdot)$ needs to be evaluated at $\underline{\mathbf{n}}(\mathbf{a})$ rather than at $\mathbf{a}$.

- *any tangent multivector u based at* $\mathbf{p}$ *is represented as* $B(\underline{\mathbf{n}}(u))[\mathbf{p}]$
  The derivation above shows that a tangent vector $\mathbf{a}$ at $\mathbf{p}$ is represented as $B(\underline{\mathbf{n}}(\mathbf{a}))[\mathbf{p}]$; since this is linear in $\mathbf{a}$ we can extend it as an outermorphism to any tangent blade at $\mathbf{p}$, and then by linearity to any tangent multivector. For a scalar α, this gives $B(\underline{\mathbf{n}}(\alpha))[\mathbf{p}] = B(\alpha)[\mathbf{p}] = \alpha$, as it should.

- *any multivector u from the differential space based at* $\mathbf{p}$ *is represented as* $B(u)[\mathbf{p}]$
  This result is very useful, but a bit hard to formulate. By the *differential tangent space* at $\mathbf{p}$ we mean the space $\mathcal{G}^1(\underline{\mathbf{n}}(\mathbf{I}))$; the *differential space* at $\mathbf{p}$ is then the space spanned by $\mathbf{n}$ and the differential tangent space. Equation (3.10) shows that the vectors $\mathbf{m}$ from the differential tangent space are represented as $B(\mathbf{m})[\mathbf{p}]$. This is a linear map, and can be extended by outermorphism to all of the differential tangent space. But since $\mathbf{n}$ is represented by $B(\mathbf{n})[\mathbf{p}]$, which is of the same form, we can even extend the representation to *any* multivector of $\mathcal{G}(\mathbf{n} \wedge \underline{\mathbf{n}}(\mathbf{I}))$ based at $\mathbf{p}$, i.e. any multivector of the differential space at $\mathbf{p}$. So the representation of such a u at $\mathbf{p}$ is $p \cdot (e^0 u)$. For a scalar α, this gives $B(\alpha)[\mathbf{p}] = \alpha$, as it should.

- *the representation at* $\mathbf{p}$ *commutes with the geometric product*
  For scalars, this holds by linearity. For vectors $\mathbf{m}_1$ and $\mathbf{m}_2$ in the differential space at $\mathbf{p}$:

$$\begin{aligned}
B(\mathbf{m}_1)\,B(\mathbf{m}_2) &= \left(p \cdot (e^0\mathbf{m}_1)\right)\left(p \cdot (e^0\mathbf{m}_2)\right) \\
&= \left(\mathbf{m}_1 - e^0 p \cdot \mathbf{m}_1\right)\left(\mathbf{m}_2 - e^0 p \cdot \mathbf{m}_2\right)
\end{aligned}$$

$$
\begin{aligned}
&= \quad \mathbf{m}_1\mathbf{m}_2 - e^0\left((\mathbf{p}\cdot\mathbf{m}_1)\mathbf{m}_2 - \mathbf{m}_1(\mathbf{p}\cdot\mathbf{m}_2)\right) + 0 \\
&= \quad (\mathbf{m}_1\mathbf{m}_2) - e^0\left(\mathbf{p}\cdot(\mathbf{m}_1\mathbf{m}_2)\right) = p\cdot\left(e^0(\mathbf{m}_1\mathbf{m}_2)\right) \\
&= \quad B(\mathbf{m}_1\mathbf{m}_2).
\end{aligned}
$$

This result for vectors extends naturally to the whole geometric algebra of the differential space at $\mathbf{p}$. Note how this uses $e^0 e^0 = 0$; this is why we like the embedding in Minkowski space. This specifically means that the representation is isometric; for instance $(B(\mathbf{n}))^2 = B(\mathbf{n}^2) = B(1) = 1$, so all $B(\mathbf{n})$ reside on a sphere in Minkowski space; this is the embedding of the extended Gaussian sphere of directions we referred to earlier.

We can derive a similar result for the tangent space at $\mathbf{p}$: the representation of elements of the tangent algebra also commutes with the geometric product; however, we will not need that in this Chapter.

- *the representation is invertible if $\kappa \neq 0$*
  Observe that $p = e_0 + \mathbf{p}$ is perpendicular to the representation of any of the elements of the differential space at $\mathbf{p}$, since
  $p\cdot B(u) = p\cdot\left(p\cdot(e^0 u)\right) = (p\wedge p)\cdot(e^0 u) = 0$. This gives m independent conditions, and thus determines p, if and only if the differential tangent space $\mathbf{n}(\mathbf{I})$ is $(m-1)$-dimensional (perpendicularity to $\mathbf{n}$ provides the one extra condition required). Since $\underline{\mathbf{n}}(\mathbf{I}) = \kappa\widehat{\mathbf{I}}$ by Equation (2.3), and $\mathbf{I}$ is known to be $(m-1)$-dimensional for the regular surfaces we study, this requires that $\kappa \neq 0$.

So when $\kappa \neq 0$, we have a proportional image of the full-rank tangent space at $\mathbf{p}$ present as the tangent space to our representation at $B(\mathbf{n})[\mathbf{p}]$. The perpendicularity of p to this and to $\mathbf{n}$ gives m constraints, and is therefore sufficient to determine p by duality in the embedding space of $B(\mathbf{n}) \wedge B(\mathbf{I}) = B(\mathbf{n}\wedge\mathbf{I}) = B(-\mathbf{I}_m) = -B(\mathbf{I}_m)$. And indeed:

$$e^0 \wedge p = -B(\mathbf{I}_m)\mathbf{I}_m^{-1}E. \tag{3.11}$$

This is the flat of the point at $\mathbf{p}$, so that $\mathbf{p}$ is retrievable as:
$\mathbf{p} = e_0 \cdot \left(B(\mathbf{I}_m)\mathbf{I}_m^{-1}E\right) - e_0$, the usual formula in the additive split. The computation of Equation (3.11) is straightforward:

$$
\begin{aligned}
-B(\mathbf{I}_m)(E\mathbf{I}_m)^{-1} &= \quad -\left(p\cdot(e^0\mathbf{I}_m)\right)\mathbf{I}_m^{-1}E = -p\wedge\left(e^0\mathbf{I}_m\mathbf{I}_m^{-1}E\right) \\
&= \quad -p\wedge\left(e^0 E\right) = -(p\wedge e^0) = e^0 \wedge p.
\end{aligned}
$$

(When $\kappa = 0$, the dual of the largest possible tangent space of the representation produces a flat of equivalent positions with the same local shape as at $\mathbf{p}$; for instance, for a cylinder we obtain a line parallel to its axis. This need not be a disadvantage, since we will see that all such points behave similarly under wave propagation; this thus allows for effective lumping in propagation algorithms. We plan to investigate that.)

We thus have (in those non-flat cases) in $B(\mathbf{n}) = \mathbf{n} - e^0(\mathbf{p}[\mathbf{n}] \cdot \mathbf{n})$ an invertible representation of a boundary in terms of a '*spectrum*' of tangent directions indexed by $\mathbf{n}$ (or dually by $\mathbf{I}$), with '*amplitude*' $\mathbf{p}[\mathbf{n}] \cdot \mathbf{n}$ marked off in the e^0-direction. The advantage of the embedding space is that the original boundary and its dual representation reside in the same space $\mathcal{G}(E\mathbf{I}_m)$, so that the transition between them is purely algebraic and geometric (through perpendicularity).

17.3.3 Example: a spherical boundary

Let us do an example: the representation of m-dimensional boundaries of which the surface is a sphere – we are interested in both spherical blobs and in spherical holes.

The position of points on a sphere with center $\mathbf{c}$ and radius ρ can be defined by the scalar function equation $\phi(\mathbf{p}) = 0$, with $\phi(\mathbf{p}) = (\mathbf{p}-\mathbf{c})^2 - \rho^2$. Differentiation to $\mathbf{p}$ yields for the unit normal vector $\mathbf{n}$:

$$\mathbf{n} = \pm \frac{\partial_{\mathbf{p}}\phi(\mathbf{p})}{|\partial_{\mathbf{p}}\phi(\mathbf{p})|} = \pm \frac{\mathbf{p} - \mathbf{c}}{|\rho|}.$$

This needs to be oriented properly to point inwards. For a spherical blob, the inward pointing normal is positively proportional to $\mathbf{c}-\mathbf{p}$, for a spherical hole to $\mathbf{p} - \mathbf{c}$. We can therefore use the radius ρ of the sphere to indicate which it is; we prefer to denote blobs by a positive radius, holes by a negative radius, so we obtain

$$\mathbf{n} = \frac{\mathbf{c} - \mathbf{p}}{\rho} \tag{3.12}$$

as the inward pointing normal for a spherical boundary, whether hole or blob. Then to make $B(\mathbf{n})$, we need to express $\mathbf{p}$ in terms of $\mathbf{n}$, which is simply $\mathbf{p} = \mathbf{c} - \rho\mathbf{n}$. This gives:

$$B(\mathbf{n}) = \mathbf{n} - e^0(\mathbf{p} \cdot \mathbf{n}) = \mathbf{n} - e^0(\mathbf{c} \cdot \mathbf{n} - \rho)$$

as the representation of the spherical boundary. This representation satisfies $B(\mathbf{n})^2 = 1$ and $c \cdot B(\mathbf{n}) = \rho$ (with $c = e_0 + \mathbf{c}$); so in the embedding space it is on the intersection of the Gaussian sphere in Minkowski space with the plane with normal vector c, at distance ρ. This is illustrated in Figure 17.3a for the 2-dimensional case of the circular blob. The Minkowski sphere looks like a cylinder to our Euclidean eyes; the intersection with the plane $c \cdot B(\mathbf{n}) = \rho$ gives the tilted ellipse depicted. Figure 17.3b depicts the circular hole and its representation, which simply has the opposite sign.

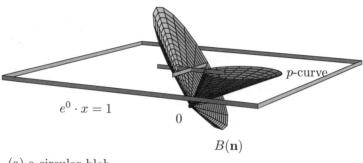

$e^0 \cdot x = 1$

p-curve

0

$B(\mathbf{n})$

(a) a circular blob

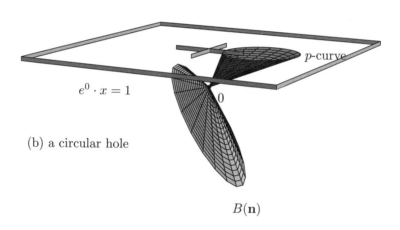

$e^0 \cdot x = 1$

p-curve

0

$B(\mathbf{n})$

(b) a circular hole

$B(\mathbf{n})$

FIGURE 17.3. The representation of a circular hole and a blob in $\mathcal{G}(\mathbf{I}_2)$. For the p-curve the vertical axis denotes e_0; since $p = e_0 + \mathbf{p}$, this curve resides in the plane $e^0 \cdot x = 1$ which is indicated. For the $B(\mathbf{n})$ curves, the vertical axis is e^0, and the curve resides on the extended Gaussian sphere $B(\mathbf{n})^2 = 1$, which due to the Minkowski geometry of $\mathcal{G}_{m+1,1}$ looks like a Euclidean cylinder in this projection. The curves have been made into cones to better indicate their spatial nature, and to help show that $p = e_0 + \mathbf{p}$ is everywhere perpendicular to $B(\mathbf{I}_m) = B(\mathbf{n}) \wedge B(\mathbf{I})$. Since $e_0 \cdot e^0 = 1$, the axes e_0 and e^0 are parallel under the duality involved in the representation, which is why we can draw them this way.

The representation of the Gauss sphere in a space shared with the original curve makes the dual (i.e. perpendicular) nature of the representation geometrically explicit.

We may check the differential relationships:

$$
\begin{aligned}
\underline{B}(\mathbf{m}) &\equiv (\mathbf{m} \cdot \boldsymbol{\partial_n}) \left(\mathbf{n} - e^0(\mathbf{c} \cdot \mathbf{n} - \rho)\right) \\
&= \mathbf{m} - e^0(\mathbf{c} \cdot \mathbf{m}) = (e_0 + \mathbf{c} - \rho\,\mathbf{n}) \cdot (e^0\mathbf{m}) = B(\mathbf{m}),
\end{aligned}
$$

since $\mathbf{m} \cdot \mathbf{n} = 0$. (Note that according to Equation (3.10), $B(\mathbf{m})$ should *not* be interpreted as $\mathbf{m} - e^0(\mathbf{c} \cdot \mathbf{m} - \rho)$, simply substituting $\mathbf{m}$ for $\mathbf{n}$ in the expression for $B(\mathbf{n})$; the latter is shorthand for $B(\mathbf{n})[\mathbf{p}[\mathbf{n}]]$ and we need $B(\mathbf{m})[\mathbf{p}[\mathbf{n}]]$, not $B(\mathbf{m})[\mathbf{p}[\mathbf{m}]]$!)

To retrieve position and curvature from the representation, we take the derivative in the embedding space. With the above, we obtain through outermorphism:

$$
B(\mathbf{I}_m) = B(\mathbf{n}) \wedge B(-\mathbf{I}) = \mathbf{I}_m - e^0(\mathbf{c} \cdot \mathbf{I}_m + \rho\mathbf{I})
$$

and then for the dual of this:

$$
\begin{aligned}
-B(\mathbf{I}_m)\mathbf{I}_m^{-1}E &= -E + e^0(\mathbf{c} \cdot \mathbf{I}_m)\mathbf{I}_m^{-1}E + \rho\, e^0\mathbf{II}_m^{-1}E \\
&= -E + e^0(\mathbf{c} \wedge 1)E - \rho\, e^0\mathbf{n} \\
&= -E + e^0(\mathbf{c} - \rho\,\mathbf{n}) = e^0 \wedge (e_0 + \mathbf{c} - \rho\,\mathbf{n}).
\end{aligned}
$$

We retrieve $\mathbf{p}$ in terms of $\mathbf{n}$ from this by:

$$
\mathbf{p} = e_0 \cdot \left(e^0 \wedge (e_0 + \mathbf{c} - \rho\,\mathbf{n})\right) - e_0 = \mathbf{c} - \rho\,\mathbf{n},
$$

which is indeed the set of positions on the spherical boundary with inward pointing normal $\mathbf{n}$.

By the way, note that $\underline{\mathbf{n}}(\mathbf{a}) \equiv (\mathbf{a} \cdot \boldsymbol{\partial_p})\mathbf{n}[\mathbf{p}] = -\mathbf{a}/\rho$, by Equation (3.12). Therefore the Gaussian curvature is, by Equation (2.3), given by

$$
\kappa = \underline{\mathbf{n}}(\widehat{\mathbf{I}})\,\mathbf{I}^{-1} = 1/\rho^{m-1},
$$

as it should be.

17.3.4 Boundaries as direction-dependent rotors

Equation (3.9) for $B(\mathbf{n})[\mathbf{p}]$ can be written in an interesting alternative form:

$$
\begin{aligned}
B(\mathbf{n})[\mathbf{p}] &= \mathbf{n} - e^0(\mathbf{p} \cdot \mathbf{n}) = (1 - e^0\mathbf{p}/2)\,\mathbf{n}\,(1 + e^0\mathbf{p}/2) \\
&= \exp(-e^0\mathbf{p}/2)\,\mathbf{n}\,\exp(e^0\mathbf{p}/2).
\end{aligned}
$$

Thus the $\mathbf{n}$-representation can be constructed from a normal vector $\mathbf{n}$ via the general versor equation $\underline{U}(\mathbf{x}) = U\mathbf{x}\widehat{U}^{-1}$, using the $\mathbf{n}$-dependent versor

$$
U = T_{\mathbf{p}} \equiv \exp(-e^0\mathbf{p}[\mathbf{n}]/2) = 1 - e^0\mathbf{p}[\mathbf{n}]/2. \tag{3.13}
$$

This is the versor $T_{\mathbf{p}}$ of a translation over $\mathbf{p}[\mathbf{n}]$ in the standard homogeneous model of a Euclidean space $\mathcal{G}^1(\mathbf{I}_m)$ in the Minkowski space $\mathcal{G}^1(E\mathbf{I}_m)$, see [4]. It is even a *rotor*, since $U\tilde{U} = (1 - e^0\mathbf{p}[\mathbf{n}]/2)(1 + e^0\mathbf{p}[\mathbf{n}]/2) = 1$. In this view, we can see an object boundary (as represented by $B(\mathbf{n})$) as an **n**-dependent translation $\mathbf{p}[\mathbf{n}]$ applied to the unit normal vector **n**. Since the latter is the representation of a point blob at the origin as a (trivial) function of its orientation, this provides the view:

> *Any object boundary can be represented as a deformation by orientation-dependent translation of a point blob at the origin.*

Non-convex objects may have a particular inward pointing normal vector **n** at different points **p**, so for those the function $\mathbf{p}[\mathbf{n}]$ should be considered set-valued.

17.3.5   The effect of Euclidean transformations

When a boundary is subjected to a transformation, its representation must change. If we represent the boundary as a rotor, the transformed rotors under common Euclidean operations transform in a straightforward manner, according to the rules in Table 17.1 (ignore the entry of wave propagation for now). These are easily proved by keeping track of what happens to the position and its differential (which gives **n**). As an example we treat the rotation of a boundary.

When the boundary rotates around **c** over a bivector angle characterized by a rotor **R** (we use boldface since this rotor is in $\mathcal{G}(\mathbf{I}_m)$), then **p** becomes $(\mathbf{R}(\mathbf{p} - \mathbf{c})\mathbf{R}^{-1} + \mathbf{c})$. This is achieved on its versor $T_{\mathbf{p}}$ by: $T_{\mathbf{p}'} = T_{\mathbf{c}}(\mathbf{R}(T_{-\mathbf{c}}T_{\mathbf{p}})\mathbf{R}^{-1})$, as is easily verified. Differentiating yields that **n** is rotated as well, to $\mathbf{n}' = \mathbf{R}\mathbf{n}\mathbf{R}^{-1}$. The $B(\mathbf{n})$-representation of the rotated boundary is achieved by applying $T_{\mathbf{p}'}$ to $\mathbf{n}'$ as a versor, which yields:

$$
\begin{aligned}
\underline{T}_{\mathbf{p}'}(\mathbf{n}') &= T_{\mathbf{p}'}\mathbf{n}'T_{\mathbf{p}'}^{-1} \\
&= \left(T_{\mathbf{c}}\mathbf{R}T_{-\mathbf{c}}T_{\mathbf{p}}\mathbf{R}^{-1}\right)\mathbf{R}\mathbf{n}\mathbf{R}^{-1}\left(T_{\mathbf{c}}\mathbf{R}T_{-\mathbf{c}}T_{\mathbf{p}}\mathbf{R}^{-1}\right)^{-1} \\
&= (T_{\mathbf{c}}\mathbf{R}T_{-\mathbf{c}})T_{\mathbf{p}}\,\mathbf{n}\,T_{\mathbf{p}}^{-1}(T_{\mathbf{c}}\mathbf{R}T_{-\mathbf{c}})^{-1}.
\end{aligned}
$$

Therefore the total result can be achieved by the application of a new versor to the original **n**. This new versor is $T_{\mathbf{p}}$ left-multiplied by:

$$
T_{\mathbf{c}}\mathbf{R}T_{-\mathbf{c}} = (1 - e^0\mathbf{c}/2)\mathbf{R}(1 + e^0\mathbf{c}/2) = \mathbf{R} - e^0(\mathbf{c}\cdot\mathbf{R}) \equiv R_{\mathbf{c},\mathbf{R}}. \quad (3.14)
$$

This is the entry in Table 17.1; other entries are derived similarly.

Equation (3.14) gives the rotated boundary in a representation which is 'parametrized' in terms of $\mathbf{R}\mathbf{n}\mathbf{R}^{-1}$, so it can not immediately be used in combination with other, **n**-based, boundaries without reparametrization. Such a reparametrization cannot in general be done globally, since $\mathbf{p}[\mathbf{n}]$

boundary operation	boundary rotor (spectrum)
null-boundary (a point blob)	1
arbitrary boundary creation	$T_{\mathbf{p}[\mathbf{n}]} = \exp(-e^0 \mathbf{p}[\mathbf{n}]/2)$
translation over $\mathbf{t}$	$T_{\mathbf{t}} T_{\mathbf{p}[\mathbf{n}]} = \exp(-e^0 \mathbf{t}/2) T_{\mathbf{p}[\mathbf{n}]}$
rotation (center $\mathbf{c}$, rotor $\mathbf{R}$)	$R_{\mathbf{c},\mathbf{R}} T_{\mathbf{p}[\mathbf{n}]} = (\mathbf{R} - e^0(\mathbf{c}\cdot\mathbf{R})) T_{\mathbf{p}[\mathbf{n}]}$
mirroring in hyperplane, support $\mathbf{d}$	$M_{\mathbf{d}} T_{\mathbf{p}[\mathbf{n}]} = (\mathbf{d} - e^0 \mathbf{d}^2/2) T_{\mathbf{p}[\mathbf{n}]}$
wave propagation by boundary $T_{\mathbf{q}}$	$T_{\mathbf{q}[\mathbf{n}]} T_{\mathbf{p}[\mathbf{n}]}$

TABLE 17.1. The boundary rotor $T_{\mathbf{p}[\mathbf{n}]}$ associated with common operations on the boundary.

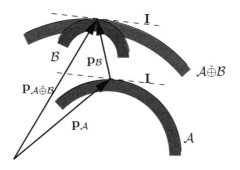

FIGURE 17.4. The definition of wave propagation.

may be quite arbitrary. However, it can be shown that for small central rotations, in the first-order approximation $\mathbf{R} = 1 - \mathbf{i}\phi/2$, the represented boundary becomes $B(\mathbf{n}) - e^0(\mathbf{n} \wedge \mathbf{p}) \cdot (\mathbf{i}\phi)$. Perhaps surprisingly, this does not involve derivatives of $\mathbf{n}$.

17.4 Wave Propagation of Boundaries

17.4.1 Definition of propagation

Propagation combines two boundaries $\mathcal{A}$ and $\mathcal{B}$ to produce a boundary $\mathcal{A}\check\oplus\mathcal{B}$ according to the following rules (which can be taken as the definition of propagation, or alternatively derived from a formulaic definition as in [1]):

Propagation definition:

- The resulting position vector after combining a point $\mathbf{p}_A$ on A and a point $\mathbf{p}_B$ on B is the position $\mathbf{p}_A + \mathbf{p}_B$:

$$\mathbf{p}_{A \tilde{\oplus} B} = \mathbf{p}_A + \mathbf{p}_B \qquad (4.15)$$

- The points $\mathbf{p}_A$ and $\mathbf{p}_B$ *must* have the same inward pointing normal vector (to A and B, respectively), and this is also the inward pointing normal vector at the resulting position in the resulting boundary. Symbolically:

$$\mathbf{n}_{A \tilde{\oplus} B}[\mathbf{p}_{A \tilde{\oplus} B}] = \mathbf{n}_A[\mathbf{p}_A] = \mathbf{n}_B[\mathbf{p}_B]. \qquad (4.16)$$

These conditions together fully determine the propagation result and the dependence of its geometrical properties on the geometrical properties of A and B.

17.4.2 Propagation in the embedded representations

Since $B(\mathbf{n})$ is an invertible description of a boundary, if we can construct the representation of the wave propagation result then we know what the resulting boundary is. But this is extremely simple, since the representation lends itself to direct implementation of the definition of propagation of Equation (4.15) and Equation (4.16).

Let $\mathbf{p}_A[\mathbf{n}]$ be defined as: $\mathbf{p}_A[\mathbf{n}] \equiv \{\mathbf{x} \in A \mid \mathbf{n}_A[\mathbf{x}] = \mathbf{n}\}$, so as the set of all positions of the boundary where the inward normal vector is $\mathbf{n}$; and similarly for $\mathbf{p}_B[\cdot]$. Then the propagation result of $B(\mathbf{n}_A) = \mathbf{n}_A - e^0(\mathbf{p}_A[\mathbf{n}_A] \cdot \mathbf{n}_A)$ and $B(\mathbf{n}_B) = \mathbf{n}_B - e^0(\mathbf{p}_B[\mathbf{n}_B] \cdot \mathbf{n}_B)$ is by Equation (4.16) indexed by the same normal vector $\mathbf{n}$, and

$$\begin{aligned} B(\mathbf{n}_{A \tilde{\oplus} B}) &= \mathbf{n} - e^0\left(\mathbf{p}_{A \tilde{\oplus} B}[\mathbf{n}] \cdot \mathbf{n}\right) \\ &= \mathbf{n} - e^0\left((\mathbf{p}_A[\mathbf{n}] \oplus \mathbf{p}_B[\mathbf{n}]) \cdot \mathbf{n}\right) \\ &= \mathbf{n} - e^0\left((\mathbf{p}_A[\mathbf{n}] \cdot \mathbf{n}) \oplus (\mathbf{p}_B[\mathbf{n}] \cdot \mathbf{n})\right). \qquad (4.17) \end{aligned}$$

So basically, *the e^0 components add up at the same $\mathbf{n}$* (we must use $\oplus$ since $\mathbf{p}[\mathbf{n}]$ is a set-valued function).

In the direction-dependent translator representation of boundaries, we get:

$$\begin{aligned} T_{\mathbf{p}_{A \tilde{\oplus} B}[\mathbf{n}]} &= 1 - e^0(\mathbf{p}_A[\mathbf{n}] \oplus \mathbf{p}_B[\mathbf{n}])/2 \\ &= \bigcup \left(1 - e^0 \mathbf{p}_A[\mathbf{n}]/2 - e^0 \mathbf{p}_B[\mathbf{n}]/2\right) \\ &= \bigcup \left(1 - e^0 \mathbf{p}_A[\mathbf{n}]/2\right)\left(1 - e^0 \mathbf{p}_B[\mathbf{n}]/2\right) \\ &= T_{\mathbf{p}_A[\mathbf{n}]} \, T_{\mathbf{p}_B[\mathbf{n}]}, \end{aligned}$$

where we used that e^0 is a null vector perpendicular to $\mathbf{p}_A$ and $\mathbf{p}_B$, and wrote the Minkowski sum as a union of ordinary additions. So *wave propagation is represented as the product of boundary rotors* (at least if we overload the geometric product to work on sets of rotors, a straightforward extension).

17.4.3   A systems theory of wave propagation and collision

The above is analogous to what the Fourier transformation does for convolution: the convolution of two signals becomes a multiplication of their frequency spectra (a complex number $A(\omega)\exp(i\phi(\omega))$ for each frequency ω); propagation of two boundaries has become multiplication of their 'direction spectra' $T_{\mathbf{p[n]}}$ (a rotor $\exp(-e^0\mathbf{p[n]}/2)$ for each direction $\mathbf{n}$).

This algebraic analogy permits a transfer of ideas from linear systems theory to the treatment of wave propagation, collision detection, and the other related problems of Section 17.1.3. For instance, for linear systems the delta-function is the function with which convolution reproduces the convolution kernel; it is the input function which allows measurement of the system's response function [6] (for instance, the image of a point source gives the optical transfer function of a camera). In wave propagation, the point at the origin plays the same role: propagation from it provides the shape of the propagator; in robotics, you could measure the shape of a robot vehicle by tracking the position of a reference point as the vehicle collides with an infinitely thin pole at the origin. This is because the versor representation of a point equals 1, or equivalently $B(\mathbf{n}) = \mathbf{n}$); so this is the 'delta-boundary' for propagation.

In linear systems theory, if a delta-function is not available as a probe one may hope to reconstruct the system's response function by determining the common multiplicative factor in the Fourier transformation of a sufficiently rich set of responses (through a Wiener filter [6]); this can then be used to sharpen that data by deconvolution. Similarly, given sufficiently rich collision data, one could determine a common multiplicative versor (or common additive $\mathbf{n}$-dependent function in the $B(\mathbf{n})$-representation) by a similar procedure. This would enable determination of the shape of the atomic probe in the STM example of Section 17.1.3, and then of the actual unknown atomic surface by a 'de-dilation' of the measured surface. (But to do this fully, one would need to treat the unobservable parts of the collision boundary around the swallowtail catastrophes of Figure 17.1b properly – which we have not done yet.)

Such direct transfer of techniques from linear systems theory is possible because we have found a characteristic 'spectral' representation, which enables us to replace the involved effects of the collision operation as local operations in the spectral domain.

17.4.4 Matching tangents

The rotor result shows the algebraic analogy with the Fourier approach to convolution; yet the equivalent 'addition of e^0-components' is actually simpler to implement. However, we should not that either result is somewhat deceptively simple, since the demand that both $\mathbf{p}_A$ and $\mathbf{p}_B$ be written in terms of the *same* $\mathbf{n}$ may require an inversion and a reparametrization of either or both to obtain $\mathbf{p}$ as a function of $\mathbf{n}$ valid over a finite domain (if the boundaries were originally given in terms of parametrized position). Yet this can be done, if necessary numerically; and then the result is useful to construct the resulting boundary (in its $\mathbf{n}$-based representation form), and to derive its properties. Collision detection and wave propagation (and the other, equivalent operations mentioned in the introduction) can be done fully in this representation. (In some applications such as radar observations, the representation $B(\mathbf{n}) = \mathbf{n} - e^0(\mathbf{p} \cdot \mathbf{n})$ is even measured *directly*: a radar yields the distance $(\mathbf{p} \cdot \mathbf{n})$ of a tangent plane perpendicular to the direction $\mathbf{n}$ of the outgoing beam.) It is only when one desires the result to be drawn as a positional surface that the rather involved inversion formula Equation (3.11) needs to be invoked.

17.4.5 Examples of propagation

For a sphere at $\mathbf{c}$ with radius ρ, we found in Section 17.3.3 the representation $B(\mathbf{n}) = \mathbf{n} - e^0(\mathbf{c} \cdot \mathbf{n} - \rho)$. Therefore the dilation of two spheres (index 1 and 2) yields:

$$
\begin{aligned}
B(\mathbf{n}) &= \mathbf{n} - e^0(\mathbf{c}_1 \cdot \mathbf{n} - \rho_1) - e^0(\mathbf{c}_2 \cdot \mathbf{n} - \rho_2) \\
&= \mathbf{n} - e^0\left((\mathbf{c}_1 + \mathbf{c}_2) \cdot \mathbf{n} - (\rho_1 + \rho_2)\right),
\end{aligned}
$$

which is immediately recognizable as the sphere with center $(\mathbf{c}_1 + \mathbf{c}_2)$ and radius $(\rho_1 + \rho_2)$. This is what we would expect as the result; but note that it is also valid for spherical holes (ρ negative). The propagation of a hole with radius ρ by a blob with radius ρ is a point (with radius zero). Or, in terms of robot collision avoidance, a spherical robot of radius ρ in a hole with radius ρ cannot move, its only permissible position is $(\mathbf{c}_1 + \mathbf{c}_2)$. Figure 17.5a depicts the wave propagation on a parabolic concave boundary, and it was generated by adding the representations of a parabola $(\mathbf{x} \cdot \mathbf{e}_2) = \frac{1}{2}(\mathbf{x} \cdot \mathbf{e}_1)^2$ (with 'inside' in the $(-\mathbf{e}_2)$-direction) and a circular blob, and then 'inverting' the result to a positional boundary. Note the occurrence of swallowtail catastrophes in some of the wave fronts. Classically, these have been considered hard to treat, and even non-differentiable; however they are fully differentiable in our *directed* representation of the tangent space. In fact, these *spatial cusps correspond to inflection points* in the $\mathbf{n}$-representation, and are thus nothing more unusual than a sign change in the curvature of $B(\mathbf{n})$. This is illustrated in Figure 17.5b, which shows how the boundaries and their representations coexist in $\mathcal{Cl}_{m+1,1}$. The depiction is similar

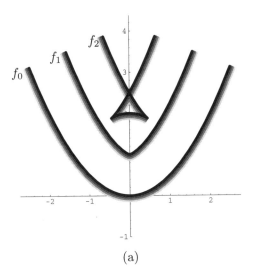

(a)

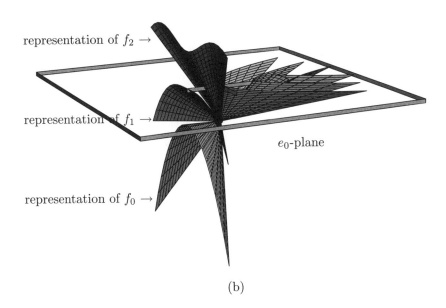

(b)

FIGURE 17.5. Circular wave propagation of the parabola-shaped boundary f_0 (see text for explanation).

to Figure 17.3, but we have rescaled $B(\mathbf{n})$ and drawn its intersection with the plane $\mathbf{e}_2 = -1$, where it is in fact the Legendre transformation of the boundary, see [1]. (In this example, we obtain $\mathbf{n}' - \frac{1}{2}e^0(\mathbf{n}'^2 - 1)$ where $\mathbf{n}' = \mathbf{n}/(\mathbf{n} \cdot \mathbf{e}_2)$.) The shift in e^0 (by the radius ρ of the circle, since the centered circular blob has $B(\mathbf{n}) = \mathbf{n} + e^0\rho$) results in the development of inflection points in the representation, which correspond to the cusps. The locations where the boundary surface self-intersects (important for the analysis of the 'millability' of surfaces) correspond to non-local properties of the representation; the intersection point at $\frac{1}{2}(1 + \rho^2)\mathbf{e}_2$ for $\rho > 1$ corresponds to a straight part of the convex hull of $B(\mathbf{n})$ (for some more details see [1]).

17.4.6   Analysis of propagation

Now that we have a convenient representation of wave propagation, we can derive many properties of the geometry of the result, for instance:

> *The propagated boundary* $\mathcal{C} = \mathcal{A} \check{\oplus} \mathcal{B}$ *obeys the 'velocity law'*
> *which relates velocities on the propagated boundary to those on*
> *the propagators:*
>
> $$\mathbf{n}_{\mathcal{A}\check{\oplus}\mathcal{B}}^{-1}(\mathbf{m})[\mathbf{p}_{\mathcal{A}} \oplus \mathbf{p}_{\mathcal{B}}] = \mathbf{n}_{\mathcal{A}}^{-1}(\mathbf{m})[\mathbf{p}_{\mathcal{A}}] \oplus \mathbf{n}_{\mathcal{B}}^{-1}(\mathbf{m})[\mathbf{p}_{\mathcal{B}}]. \quad (4.18)$$
>
> *The result is $\emptyset$ for $\mathbf{m}$ not in the common range of $\mathbf{n}_{\mathcal{A}}(\cdot)[\mathbf{p}_{\mathcal{A}}]$*
> *and $\mathbf{n}_{\mathcal{B}}(\cdot)[\mathbf{p}_{\mathcal{B}}]$.*

Proof: Introduce three tangent vectors $\mathbf{a}$, $\mathbf{b}$ and $\mathbf{c}$, to measure the derivative on each of the surfaces, and use the chain rule of [3] to rewrite them in terms of derivatives of $\mathbf{p}[\mathbf{n}]$: $\mathbf{a} = \mathbf{P}_{\mathcal{A}}(\mathbf{a}) = (\mathbf{a} \cdot \partial_\mathbf{p})\mathbf{p}_{\mathcal{A}} = (\mathbf{n}_{\mathcal{A}}(\mathbf{a}) \cdot \partial_\mathbf{n})\,\mathbf{p}_{\mathcal{A}}[\mathbf{n}]$ and similarly $\mathbf{b} = \mathbf{P}_{\mathcal{B}}(\mathbf{b}) = (\mathbf{b} \cdot \partial_\mathbf{p})\mathbf{p}_{\mathcal{B}} = (\mathbf{n}_{\mathcal{B}}(\mathbf{b}) \cdot \partial_\mathbf{n})\,\mathbf{p}_{\mathcal{B}}[\mathbf{n}]$ and $\mathbf{c} = \mathbf{P}_{\mathcal{C}}(\mathbf{c}) = (\mathbf{c} \cdot \partial_\mathbf{p})\mathbf{p}_{\mathcal{C}} = (\mathbf{n}_{\mathcal{C}}(\mathbf{c}) \cdot \partial_\mathbf{n})\,\mathbf{p}_{\mathcal{C}}[\mathbf{n}]$. Now select these such that $\mathbf{n}_{\mathcal{A}}(\mathbf{a}) = \mathbf{n}_{\mathcal{B}}(\mathbf{b}) = \mathbf{n}_{\mathcal{C}}(\mathbf{c}) = \mathbf{m}$. We then find from the above that these tangents add as position vectors: $\mathbf{c} = (\mathbf{m} \cdot \partial_\mathbf{p})\mathbf{p}_{\mathcal{C}}[\mathbf{n}] = (\mathbf{m} \cdot \partial_\mathbf{p})\,(\mathbf{p}_{\mathcal{A}}[\mathbf{n}] + \mathbf{p}_{\mathcal{B}}[\mathbf{n}]) = \mathbf{a} + \mathbf{b}$. Our selection of $\mathbf{m}$ implies that $\mathbf{a} \in \mathbf{n}_{\mathcal{A}}^{-1}(\mathbf{m})$, $\mathbf{b} \in \mathbf{n}_{\mathcal{B}}^{-1}(\mathbf{m})$ and $\mathbf{c} \in \mathbf{n}_{\mathcal{C}}^{-1}(\mathbf{m})$; therefore, over all possibilities of choosing $\mathbf{a}$ and $\mathbf{b}$ given $\mathbf{c}$, we obtain $\mathbf{n}_{\mathcal{C}}^{-1}(\mathbf{m}) = \mathbf{n}_{\mathcal{A}}^{-1}(\mathbf{m}) \oplus \mathbf{n}_{\mathcal{B}}^{-1}(\mathbf{m})$. The right hand side produces $\emptyset$ for any element not common to both sets contributing to the Minkowski sum; hence only elements in both ranges contribute – which implies that $\mathbf{m}$ must be in the common range of $\mathbf{n}_{\mathcal{A}}(\cdot)$ and $\mathbf{n}_{\mathcal{B}}(\cdot)$ at $\mathbf{p}_{\mathcal{A}}$ and $\mathbf{p}_{\mathcal{B}}$, respectively. $\square$

This interaction of the local differential geometries can produce involved results, especially for surfaces with torsion. However, there is an interestingly simple property when we 'lump' over all tangent directions at $\mathbf{p}$:

In m-dimensional wave propagation, Gaussian curvatures add reciprocally:

$$\kappa_{\mathcal{C}}^{-1} = \kappa_{\mathcal{A}}^{-1} + \kappa_{\mathcal{B}}^{-1} \tag{4.19}$$

(locally, at every triple of corresponding points).

Proof: Extending Equation (4.18) as a linear outermorphism to all of $\mathcal{G}(\mathbf{I})$ at the appropriate points we get: $\underline{\mathbf{n}}_{\mathcal{A} \dot{\oplus} \mathcal{B}}^{-1}(\mathbf{I}) = \underline{\mathbf{n}}_{\mathcal{A}}^{-1}(\mathbf{I}) + \underline{\mathbf{n}}_{\mathcal{B}}^{-1}(\mathbf{I})$ for each triple of corresponding points. The Gaussian curvature is related to $\underline{\mathbf{n}}(\mathbf{I})$ by Equation (2.3): $\underline{\mathbf{n}}(\widehat{\mathbf{I}}) = \kappa \mathbf{I}^{-1}$, so that $\underline{\mathbf{n}}^{-1}(\mathbf{I}) = \widehat{\mathbf{I}}/\kappa$, at each point, and Equation (4.19) follows after division by $\widehat{\mathbf{I}}$. $\square$

One of the obvious consequences is that locally flat parts (where $\kappa = 0$) stay locally flat after propagation.

17.5   Conclusions

This Chapter demonstrates that the rather involved operation of wave front propagation in m-dimensional space can be represented as a geometric product of direction-dependent rotors. These rotors represent boundaries in Euclidean m-space, within a Minkowski space of dimension $(m + 1, 1)$, as *direction-dependent translations* of the point object at the origin. This representation combines well with Euclidean operations on the boundaries, as Table 17.1 showed. We plan to use it to analyze differential properties of the propagation operation (some first results were shown).

Alternatively, and computationally somewhat more convenient, propagation can be represented as an addition of scalar support functions on the Gauss sphere of directions; in our representation this is a sphere in Minkowski space, with the support function geometrically represented as the e^0-component. Such representation have been used before (e.g. [5]); but their relevance for the propagation-type interactions of boundaries appears to be new; and we now have them for arbitrary dimensionality.

We hope to apply this spectral representation of wave propagation to some of the practical problems of Section 17.1.3 which have, in essence, the same mathematical structure; notably to the prevention of robot collisions which was our original motivation. This will require the development of efficient algorithms based on our representation.

Acknowledgments

This work was performed while on a sabbatical with David Hestenes' group at Arizona State University, Tempe, AZ, USA.

Chapter 18

Modern Geometric Calculations in Crystallography

G. Aragon, J.L. Aragon, F. Davila, A. Gomez and M.A. Rodriguez

18.1 Introduction

The aim of mathematical crystallography is the classification of periodic structures by means of different equivalence relationships, yielding the well known crystallographic classes and Bravais lattices [1]. Periodicity (crystallinity) has been the paradigm of classical crystallography . Recently, more systematical attention has been paid to structures which are not orthodox crystals. For example, some generalizations, involving curved spaces with non-Euclidean metrics, were developed for the understanding of random and liquid crystalline structures [2]. However, the first step away from orthodox crystalline order, represented by the 230 crystallographic space groups , was motivated by the appearance of quasicrystals in 1984 [3]. Since then, crystallography has been the subject of deep revisions. Quasicrystals are metallic alloys whose diffraction patterns exhibit sharp spots (like a crystal) but non-crystallographic symmetry. This means that the lattice underlying the atomic structure cannot be periodic. So, crystallography faces a non-crystalline but perfectly ordered structure. There are also many other directions in which classical crystallography can be generalized, by relaxing or altering various requirements, to include structures which are ordered but do not follow the exact paradigm of crystallization [4].

Quasiperiodic structures can be obtained by an appropriate projection from an N-dimensional periodic lattice ($N > 3$). This embedding allows to recover, in a higher dimensional space, crystallographic concepts such as lattice translational symmetry, reciprocal lattice and unit cell . Extending crystallography to N-dimensions was then an obvious step (4-dimensional crystallography was completely worked out in 1978 in [5]).

The purpose of this work is to show that geometric algebra is a powerful language in which to carry out various crystallographic calculations in spaces of arbitrary dimensions. With the application of concepts from geometric algebra, it is shown how to find the densest planes in 6-dimensional lattices and how to analyze the problem of coincidences between two lattices in arbitrary dimensions. The first problem is directly related with the possible external shapes (facets) of real quasicrystals and the second one has provided helpful answers to the complex problems that arise in the description of grain and twin boundaries. We hope that our results encourage more research in the formulation of crystallography in a concise geometrical language which is valid in any dimension.

18.2   Quasicrystals

Quasicrystals are metallic alloys whose diffraction patterns exhibit sharp spots but non-crystallographic symmetry. The sharp spots in the diffraction pattern mean that the structure has long-range order but the forbidden symmetry implies that the lattice underlying the atomic structure cannot be periodic. The lattice is quasiperiodic, so it is referred to as a *quasilattice* [6] .

After the discovery of the first quasicrystalline $Al_{0.86}Mn_{0.14}$ alloy, with icosahedral symmetry [3], a large variety of quasicrystals were obtained; most of them are quasiperiodic in the plane and present periodicity in the perpendicular direction. Symmetries found so far correspond to 10-fold (decagonal), 8-fold (octagonal) and 12-fold (dodecagonal) rotation [7]. Even a 1-dimensional quasicrystal has been discovered; the diffraction pattern shows a periodic plane, and a perpendicular quasiperiodic row of spots [8].

The first example of a non-periodic tiling of the plane with 5-fold symmetry was the "Penrose Tiling" . In 1974, Roger Penrose discovered an aperiodic, highly ordered, pentagonal tiling which consists of a pair of tiles (rhombi) and a set of matching rules that forces non-periodicity [9]. It was latter demonstrated that the Penrose Tiling have an optical diffraction pattern with numerous sharp peaks [10]. The set of vertices of the Penrose pattern was then the first example of a quasilattice.

The use of matching rules to generate quasiperiodic tilings is very restrictive and it was necessary to develop more efficient methods to generate infinite quasilattices of any desired symmetry and dimension. In what follows, we shall discuss the most popular and elegant: the so-called Cut and Projection Method [11, 12] .

The Cut and Projection Method regards n-dimensional quasilattices as projections of a subset of a lattice in an N-dimensional space $\mathbf{R}^N$, $(N > n)$. The simplest example is the generation of a 1-dimensional quasicrystal, consisting of a tile of two types of intervals, by projecting points from a 2-dimensional square lattice (Figure 1.1). Let us denote by $E^{\parallel}$ the line to be tiled and consider the strip obtained by shifting the unit square Γ_2 of the lattice along $E^{\parallel}$. The points inside of the strip are projected orthogonally onto $E^{\parallel}$ producing a sequence of projected points. If the slope of $E^{\parallel}$, with respect to the canonical basis of $\mathbf{R}^2$, is rational, the sequence is periodic, otherwise a quasiperiodic sequence is obtained. In particular, if the slope is given by $1/\tau$, τ the golden mean , the distribution of segments on the line obeys a Fibonacci sequence . The generalization to obtain 2- and 3-dimensional quasilattices is straightforward [12]. For example, we shall consider the most useful case of quasilattices with icosahedral symmetry which are obtained by projecting a cubic lattice in the 6-dimensional space.

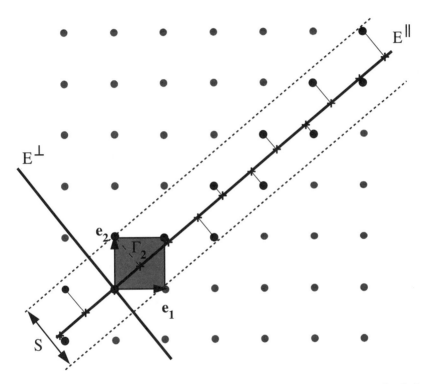

FIGURE 18.1. Illustration of the cut and projection method in two-dimensions. If the slope of $E^{\parallel}$ is irrational, points inside the strip S project on $E^{\parallel}$ in a ordered non-periodic structure.

Let us consider the cubic lattice $\Lambda_P = \mathbf{Z}^6 \subset \mathbf{R}^6$ equipped with an orthonormal basis $\{e_1, e_2, e_3, e_4, e_5, e_6\}$ and let us denote by Γ_6 the unit hypercube. Now, let $E^{\parallel}$ be the 3-dimensional "physical space" of the quasilattice and assume that it does not contain any points of the lattice Λ_P. A "strip" S in Λ_P is generated by shifting Γ_6 along $E^{\parallel}$, i.e., $S = \Gamma_6 \oplus E^{\parallel}$. If $P^{\parallel} : \Lambda_P \to E^{\parallel}$ is the projection from the lattice Λ_P onto the physical space, a quasilattice Q_P is generated in $E^{\parallel}$ by projecting all the points of Λ_P that fall inside the strip S. That is

$$Q_P = P^{\parallel}\left(\Lambda_P \cap S\right).$$

$\mathbf{R}^6$ can be decomposed as $\mathbf{R}^6 = E^{\parallel} \oplus E^{\perp}$ where $E^{\perp}$ is the orthogonal complement to $E^{\parallel}$ that is often called "phason space" . If $P^{\perp} : \Lambda_P \to E^{\perp}$ is the projection onto $E^{\perp}$, then the projection of all points inside the strip onto $E^{\perp}$ defines an "acceptance domain" $K = P^{\perp}\left(\Lambda_P \cap S\right)$. The name comes from the fact that a point $x \in \Lambda_P$ is inside the strip S if $P^{\perp}\left(x\right) \in K$. In this case, K is a rhombic triacontahedron.

The physical space $E^{\parallel}$ is fixed by the condition that

$$P^{\parallel}\left(e_i\right) = \varepsilon_i, \quad i = 1, \ldots, 6, \tag{2.1}$$

where $\{\varepsilon_1, \varepsilon_2, \varepsilon_3, \varepsilon_4, \varepsilon_5, \varepsilon_6\}$ are vectors pointing to the upper vertices of an icosahedron (they line up with the six 5-fold symmetry axes, as shown in Figure 1.2) . Relative to the orthonormal basis, the projection matrix is [12]:

$$P^{\parallel} = \begin{bmatrix} b & a & -a & -a & a & a \\ a & b & a & -a & -a & a \\ -a & a & b & a & -a & a \\ -a & -a & a & b & a & a \\ a & -a & -a & a & b & a \\ a & a & a & a & a & b \end{bmatrix}, \tag{2.2}$$

where $a = 1/\sqrt{20}$ and $b = 1/2$. The projector onto $E^{\perp}$ is given by $P^{\perp} = \mathbf{1} - P^{\parallel}$ where $\mathbf{1}$ is the 6×6 identity matrix.

In six dimensions, there exists three Bravais lattices invariant under the icosahedral group [13]: simple cubic (P), face centered (F) and body centered (I). Defined as in [14]

$$\Lambda_P = \left\{ \sum_{i=1}^{6} n_i e_i \mid n_i \in \mathbf{Z} \right\}, \tag{2.3}$$

$$\Lambda_F = \left\{ \sum_{i=1}^{6} n_i e_i \mid \sum_{i=1}^{6} n_i = 0 \ (\mathrm{mod}\ 2) \right\}, \tag{2.4}$$

$$\Lambda_I = \left\{ \sum_{i=1}^{6} \frac{n_i}{2} e_i \mid n_i = n_j \ (\mathrm{mod}\ 2) \right\}. \tag{2.5}$$

Most of the experimentally observed icosahedral quasicrystals are of the P type, however F icosahedral quasicrystals have been obtained in some alloys such as $Al_{65}Cu_{20}Fe_{15}$ [15], $Al_{65}Cu_{20}Ru_{15}$ [16], $Al_{70}Pd_{20}Mn_{10}$ and $Al_{70}Pd_{20}Re_{10}$ [17]. The last case (I) has not yet been observed experimentally .

18.3   The Morphology of Icosahedral Quasicrystals

It is well known that crystals frequently grow with polyhedral external shapes (habits) . This fact can be understood on the basis that crystals consist of identical units that are repeated according to a lattice. The lattice periodicity also explains why the various facets (flat faces of the polyhedra) of crystals form always the same angles among themselves (*law of constancy of angles*) and why the directions of the normals to the facets can be indexed (Miller indices) with small integers (*law of rational indexes*).

The name "morphologically important planes" comes from the fact that the observed facets tend to correspond to the densest lattice planes (*Bravais rule*) , which at the same time are those with the largest distances among themselves to keep constant lattice volume. The densest lattices planes have small values of the Miller indices h, k, l [18].

One of the facts that verified the icosahedral nature of some alloys was the macroscopic solidification morphology, which was first reported within the Al_6CuLi_3 phase [19]. The macroscopic external shape of this phase clearly consist of stacked triacontahedra (the triacontahedron is a non-regular polyhedron with 30 rhombus-shaped faces, 32 vertices, 60 edges and icosahedral symmetry). More evidence of icosahedral morphology was later found in stable icosahedral alloys; a pentagonal dodecahedron morphology was observed in $Al_{65}Cu_{20}Fe_{15}$ [15] and $Ga_{16}Mn_{32}Zn_{52}$ [20]; rapidly quenched $AlMn$ quasicrystals also exhibit small crystallites with a dodecahedral morphology. From the experimental observations one can conclude that, as a rule, quasicrystals belonging to the P 6-dimensional lattice [Equation (2.3)] show the polyhedral shape of a rhombic triacontahedron while the more likely shape of F quasicrystals [Equation (2.4)] is a pentagonal dodecahedron. Quasicrystals of type I have no yet been observed experimentally.

Bravais' Law of reticular density states that the important faces of a crystal should be those parallel to the densest lattice planes (which at the same time are those with the largest distances among themselves). It has been surprisingly successful in predicting the morphology of crystals of many compositions [18]. This idea can be extended to quasicrystals in a direct manner using the Cut and Projection Method described in Section 18.2. The assumption is made that the important quasilattice planes in quasicrystals are also those with the highest density and largest separation. Thus, a hypothesis can be formulated that relevant planes in $\mathbf{R}^3$ are those that correspond (via projection) to the densest lattice hyperplanes in $\mathbf{R}^6$. In what follows, we solve this problem for the 6-dimensional lattices P, F and I. We obtain that considering only the densest quasilattice planes, the most likely external form of the ideal icosahedral quasicrystals are the triacontahedron and the pentagonal dodecahedron, for the P and F cases, respectively, in agreement with experimental observations. The I phase, yet unobserved, should prefer the shape, according our results, of a

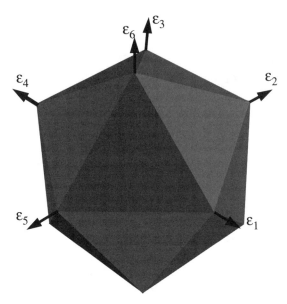

FIGURE 18.2. The projected basis vectors ε_i of Λ_P are six vertices of an icosahedron.

triacontahedron.

Our procedure can be summarized this way:

1. Consider a plane p of the quasilattice characterized by the 2-blade B_p.

2. Find the sublattice $\Lambda_W \subset \Lambda_i$, $i = P, F, I$, whose points project onto the plane p. The density ρ of points in p is directly related to the density of Λ_W.

3. Find the average density $\bar{\rho}$ of the family of planes parallel to p. The more dense this family of planes is, the more likely it is that a facet will appear and develop.

We present in detail only the case of planes invariant under the 5-fold rotations (the 5-fold planes) in a P quasilattice. A similar approach may be applied to the other cases including quasilattices of types F and I.

Consider a plane (facet) in a quasilattice whose normal points in the direction of one of the 5-fold axes. Using the numbering shown in Fig. 1.2 for the vectors pointing to the vertices of an icosahedron, it can be readily seen that two vectors in this plane are

$$\varepsilon_3 - \varepsilon_2 = P^{\parallel}(e_3) - P^{\parallel}(e_2),$$

and

$$\varepsilon_1 - \varepsilon_2 = P^{\parallel}(e_1) - P^{\parallel}(e_2),$$

so a 5-fold plane p is characterized by a 2-blade

$$B_p = (\varepsilon_3 - \varepsilon_2) \wedge (\varepsilon_1 - \varepsilon_2). \qquad (3.6)$$

Now consider the subspace $W \subset \mathbf{R}^6$ defined as

$$W = Span\{\varepsilon_3 - \varepsilon_2, \varepsilon_1 - \varepsilon_2, \varepsilon_3^{\perp} - \varepsilon_2^{\perp}, \varepsilon_1^{\perp} - \varepsilon_2^{\perp}, \varepsilon_6^{\perp}\},$$

where ε_1, ε_2 and ε_3 are as before and $\varepsilon_i^{\perp} = P^{\perp}(e_i)$, $i = 1, 2, 3, 6$. Since $P^{\parallel}(\varepsilon_i) = \varepsilon_i$ and $P^{\parallel}(\varepsilon_i^{\perp}) = 0$, all vectors in W project onto the plane characterized by (3.6). Then the 5-dimensional subspace W projects onto the 5-fold plane of the quasilattice. The sublattice $\Lambda_W \subset \Lambda_p$ is therefore

$$\Lambda_W = W \cap \Lambda_P.$$

A basis for Λ_W can be obtained by noticing that a vector $x = \sum_{i=1}^{6} x_i e_i$ lies in Λ_W if and only if x_i are integers and if

$$B_W \wedge x = 0, \qquad (3.7)$$

where $B_W = (\varepsilon_3 - \varepsilon_2) \wedge (\varepsilon_1 - \varepsilon_2) \wedge (\varepsilon_3^{\perp} - \varepsilon_2^{\perp}) \wedge (\varepsilon_1^{\perp} - \varepsilon_2^{\perp}) \wedge \varepsilon_6^{\perp}$.

With respect to the orthonormal basis $\{e_i\}_{i=1}^{6}$, the components of vectors ε_i and $\varepsilon_i^{\perp}$ form the i-th column of the projection matrix $P^{\parallel}$ and $P^{\perp}$, respectively [see Equation (2.2)]. Using these values, we obtain that (3.7) is fulfilled if $x_1 + x_2 + x_3 + x_4 + x_5 + x_6/\sqrt{5} = 0$, i.e., $x_6 = 0$ and $x_1 + x_2 + x_3 + x_4 + x_5 = 0$. So, a basis for Λ_W is given by

$$\{(1,0,0,0,-1,0), (0,1,0,0,-1,0), (0,0,1,0,-1,0), (0,0,0,1,-1,0)\}$$

or

$$\{e_1 - e_5, e_2 - e_5, e_3 - e_5, e_4 - e_5\}. \qquad (3.8)$$

As we can see, the lattice Λ_W that projects onto the 5-fold plane characterized by B_p is 4-dimensional. One can obtain the pattern of projected points onto this 5-fold plane by the Cut and Projection Method in Λ_W, with the strip $S_W = (S \cap \Lambda_W)$.

There is a 2-dimensional lattice $\Lambda_{W^{\perp}}$, orthogonal to Λ_W, which projects onto an 1-dimensional space orthogonal to the quasilattice plane, i.e., projects onto a space spanned by the normal to the 5-fold plane. By considering the 4-blade

$$B_W = (e_1 - e_5) \wedge (e_2 - e_5) \wedge (e_3 - e_5) \wedge (e_4 - e_5),$$

TABLE 18.1. Generators of the primitive cells of sublattices Λ_W associated with three P-type quasilattice planes in three dimensions.

Direction	Generators in 3D	Generators of Λ_W
2-fold	$\{\varepsilon_1, \varepsilon_2\}$	$\{e_1, e_2, e_3 - e_5, e_4 - e_6\}$
3-fold	$\{\varepsilon_1 + \varepsilon_2, \varepsilon_2 + \varepsilon_3\}$	$\{e_1 - e_3, e_2 + e_3, e_4 - e_6, e_5 - e_6\}$
5-fold	$\{\varepsilon_3 - \varepsilon_2, \varepsilon_1 - \varepsilon_2\}$	$\{e_1 - e_5, e_2 - e_5, e_3 - e_5, e_4 - e_5\}$

a 2-blade $B_{W^\perp}$ that characterizes a 2-dimensional space containing $\Lambda_{W^\perp}$ can be found. It must satisfy the condition $B_W B_{W^\perp} = \lambda e_{12\ldots6}$, where $\lambda \in \mathbf{R}$. Therefore

$$B_{W^\perp} = \frac{\lambda e_{12\ldots6}}{B_W},$$

which gives

$$B_{W^\perp} = \frac{\lambda}{5}(e_1 + e_2 + e_3 + e_4 + e_5)e_6 = \frac{\lambda}{5}(e_1 + e_2 + e_3 + e_4 + e_5) \wedge e_6,$$

so a basis for $\Lambda_{W^\perp}$ is

$$\{e_1 + e_2 + e_3 + e_4 + e_5, e_6\}.$$

The pattern of vertices along the normal to the 5-fold plane, that gives the sequence of separations between planes, is derived similarly by the cut and projection method but in the 2-dimensional lattice

$$\Lambda_{W^\perp} = P_{W^\perp}(\Lambda_P),$$

with a strip $S_{W^\perp} = P_{W^\perp}(S)$, where $P_{W^\perp}$ denotes the orthogonal projection onto the subspace $W^\perp$. For our purposes, all that matters is the lattice Λ_W generated by (3.8), whose points project onto the 5-fold plane and which contains information about the density of vertices in this plane.

The procedure described above was applied also to the 6-dimensional lattices of types F and I. Calculations were made to determine the morphological importance of high symmetry planes, i.e., 5-fold, 3-fold and 2-fold facets. Results for the three type of the 6-dimensional lattices are summarized in Tables 18.1-18.3 which give generators of three different quasilattice planes in three dimensions and a basis for the 4-dimensional lattices Λ_W associated with each plane in terms of the orthonormal basis.

For a given type of lattice (Λ_P, Λ_F or Λ_I) and for a given type of facet (2-fold, 3-fold, 5-fold or any other), the average occupation density (number of vertices per unit area) of the family of planes parallel to the facet is

TABLE 18.2. Generators of the primitive cells of sublattices Λ_W associated with three F-type quasilattice planes in three dimensions.

Direction	Generators in 3D	Generators of Λ_W
2-fold	$\{-\varepsilon_1 - \varepsilon_2, \varepsilon_1 - \varepsilon_2\}$	$\{-e_1 - e_2, e_1 - e_2,$
		$e_3 - e_4 - e_5 + e_6, e_4 - e_6\}$
3-fold	$\{-\varepsilon_1 - \varepsilon_2, \varepsilon_5 - \varepsilon_6\}$	$\{-e_1 - e_2, e_1 - e_3, e_4 - e_5, e_5 - e_6\}$
5-fold	$\{\varepsilon_1 - \varepsilon_2, \varepsilon_2 - \varepsilon_3\}$	$\{e_1 - e_2, e_2 - e_3, e_3 - e_4, e_4 - e_5\}$

TABLE 18.3. Generators of the primitive cells of sublattices Λ_W associated with three I-type quasilattice planes in three dimensions.

Direction	Generators in 3D	Generators of Λ_W
2-fold	$\{\varepsilon_1, \varepsilon_2\}$	$\{e_1, e_2, e_3 - e_5,$
		$(-e_1 - e_2 - e_3 + e_5 + e_5 - e_6)/2\}$
3-fold	$\{\varepsilon_1 + \varepsilon_2, \varepsilon_2 + \varepsilon_3\}$	$\{-e_2 - e_3 - e_4 + 2e_5 - e_6,$
		$e_1 + 2e_2 + e_3 + e_4 - 2e_5 + e_6,$
		$-e_1 - e_2 - e_4 + 2e_5 - e_6, e_4 - e_5\}$
5-fold	$\{\varepsilon_3 - \varepsilon_2, \varepsilon_1 - \varepsilon_2\}$	$\{e_1 - e_5, e_2 - e_5, e_3 - e_5, e_4 - e_5\}$

TABLE 18.4. Volumes of the primitive cells of lattices Λ_W and average density of planes for each case.

Lattice	Direction	$V(\Lambda_W)$	$\bar{\rho}$
P	2-fold	2	0.500
	3-fold	3	0.333
	5-fold	$\sqrt{5}$	0.447
F	2-fold	4	0.250
	3-fold	3	0.333
	5-fold	$\sqrt{5}$	0.447
I	2-fold	1	1.000
	3-fold	3	0.333
	5-fold	$\sqrt{5}$	0.447

inversely proportional to the volume $V(\Lambda_W)$ of the primitive cell of the 4-dimensional lattice Λ_W that projects onto one plane of the family:

$$\bar{\rho} \propto \frac{1}{V(\Lambda_W)}. \tag{3.9}$$

From the generators of each Λ_W in Tables 18.1-18.3, the volume of the unit cell in each case can be calculated. Table 18.4 shows volumes of the unit cells calculated for the three families of planes of the three quasilattices Λ_P, Λ_F and Λ_I. The average density $\bar{\rho}$ has been calculated using the approximation (3.9) and is given in the last column.

From Table 18.4, the following predictions about the ideal shape of the icosahedral quasilattices can be made:

- In the case P, the largest occupation density is carried by the planes with a normal along the 2-fold axis of the icosahedron. The polyhedral shape should have big 2-fold facets resembling a rhombic triacontahedron.

- In the case F, the densest planes are those of the 5-fold family. In this case we have a polyhedron with big 5-fold facets. The ideal shape is therefore a pentagonal dodecahedron.

- In the yet unobserved case I we have the same result as in the P case but with larger 2-fold facets. This resembles also a rhombic triacontahedron.

All these prediction are in agreement with experimental observations

18.4   Coincidence Site Lattice Theory

Coincidence Site Lattice (CSL) theory has provided partial answers to the complex problems that arise in the description of grain and twin boundaries [21] . Mathematically, the problem can be stated as follows:

Let Λ be a lattice in $\mathbf{R}^n$ and let $R \in O(n)$ be an orthogonal transformation . R is called a *coincidence isometry* if $\Lambda \cap R\Lambda$ is a sublattice of Λ. The problem is therefore to identify and characterize coincidence isometries of a given lattice Λ.

This problem has been worked out from several points of view. M.A. Fortes [22] developed a matrix theory of CSL in arbitrary dimensions, including a method to calculate a basis for the coincidence lattice through a particular factorization of the matrix defining the relative orientation. Also Duneau *et al.* [23] developed a matrix theory where the parameters of the coincidence lattice are evaluated by means of a method based on the Smith normal form for integer matrices. M. Baake [24] used complex numbers and quaternions to solve the coincidence problem in dimensions ≤ 4. Finally, in Aragón *et al.* [25] a weak coincidence criterion is proposed and 4-dimensional lattices are used to characterize CSL in the plane.

Here, a different point of view is adopted. We analyze the problem in terms of reflections by finding conditions under which a given reflection is of coincidence . The use of reflections instead of rotations suggest the use of geometric algebra as a natural tool for this problem. It is found that any arbitrary coincidence isometry can be decomposed as a product of coincidence reflections in vectors of the lattice Λ.

18.4.1   Basics

Here we present a brief summary of basic concepts related to lattices and coincidence lattices. For this purpose we adopt definitions and notation given by Baake [24], who has formulated the CSL in more mathematical terms.

Definition 18.1.   *A discrete subset $\Lambda \subset \mathbf{R}^n$ is called a lattice, of dimension n, if it is spanned as $\Lambda = \bigoplus\limits_{i=1}^{n} \mathbf{Z}a_i$, with $\{a_1, a_2, \ldots, a_n\}$ a set of linearly independent vectors of $\mathbf{R}^n$. These vectors form a basis of the lattice.*

The lattice Λ is isomorphic to the free Abelian group of order n. It lead us to define the concept of sublattice:

Definition 18.2.   *Let Λ be a lattice in $\mathbf{R}^n$. A subset $\Lambda' \subset \Lambda$ is called a sublattice of Λ if it is a subgroup of finite index, i.e., $[\Lambda : \Lambda'] < \infty$ (the number of right lateral classes is finite). It is also said that Λ is a superlattice of Λ'.*

The following two definitions are central for the coincidence problem:

Definition 18.3.   *Two lattices Λ_1 and Λ_2 are called commensurate, denoted by $\Lambda_1 \sim \Lambda_2$, if and only if $\Lambda_1 \cap \Lambda_2$ is a sublattice of both Λ_1 and Λ_2.*

Definition 18.4.   *Let Λ be a lattice in $\mathbf{R}^n$. An orthogonal transformation $R \in O(n)$ is called a coincidence isometry of Λ if and only if $R\Lambda \sim \Lambda$. The integer $\Sigma(R) := [\Lambda : \Lambda \cap R\Lambda]$ is called the coincidence index of R with respect to Λ. If R is not a coincidence isometry then $\Sigma(R) := \infty$. Two useful sets are also defined:*

$$
\begin{aligned}
OC(\Lambda) &:= \{R \in O(n) \mid \Sigma(R) < \infty\}, \\
SOC(\Lambda) &:= \{R \in OC(\Lambda) \mid \det(R) = 1\}.
\end{aligned}
$$

Coincidence transformations are usually worked out using matrices. In what follows, we recall this approach in order to state a result that will be used later.

Definition 18.5.   *Let Λ be a lattice in $\mathbf{R}^n$ with basis $\{a_1, a_2, \ldots, a_n\}$; the structure matrix N, of Λ, is defined through the relation $a_i = Ne_i = \sum_{j=1}^{n} N_{ji} e_j$, where $\{e_1, e_2, \ldots, e_n\}$ is the canonical basis of $\mathbf{R}^n$.*

Structure matrices can be useful to formulate the problem of coincidence between two lattices (for a proof see Ref. [22]):

Theorem 18.1.   *[Grimmer] Let Λ_1 and Λ_2 two lattices in $\mathbf{R}^n$ with structure matrices N_1 and N_2, respectively. Then $\Lambda_1 \sim \Lambda_2$ if and only if $N_1^{-1} N_2$ has rational entries.*

Notice that if $\Lambda_2 = R\Lambda_1$, where R is a coincidence isometry, then $N_2 = RN_1$ and, in this case, $N_1^{-1} RN_1$ is a rational matrix. In case of a hypercubic lattice, N_1 is a diagonal matrix and then we can state the following

Theorem 18.2.   *Let Λ be an hypercubic lattice in $\mathbf{R}^n$ and let R be an orthogonal transformation, then, $R \in OC(\Lambda)$ if an only if their matrix, with respect to the canonical basis, has rational entries.*

In the next section we formulate the problem of coincidences between lattices, using reflections as the primitive isometry, in the language of geometric algebra. We restrict ourselves to the case of hypercubic lattices $\Lambda = \mathbf{Z}^n$.

18.4.2 Geometric algebra approach

Let $\mathbf{Z}^n = \bigoplus_{i=1}^{n} \mathbf{Z}e_i$ a n-dimensional hypercubic lattice with the canonical basis $\{e_1, e_2, \ldots, e_n\}$. Any reflection R of a vector x, in the hyperplane $H_s = s^\perp$, is written as $R(x) = -sxs^{-1} = -\lambda sx \,(\lambda s)^{-1}$, for $\lambda \neq 0$. We now shall find conditions under which the reflection $R(x) = -sxs^{-1}$ is a coincidence reflection.

We have that

$$R(x) = -sxs^{-1} = \frac{-sxs}{s^2},$$

where $s = \sum_{k=1}^{n} \alpha_k e_k$, $\alpha_k \in \mathbf{Z}$. In terms of the canonical basis, R reads:

$$se_i s = \left(\sum_{k=1}^{n} \alpha_k e_k \right) e_i \left(\sum_{j=1}^{n} \alpha_j e_j \right)$$

$$= \left(\alpha_i^2 - \sum_{k \neq i} \alpha_k^2 \right) e_i + \sum_{j=1(j \neq i)}^{n} 2\alpha_i \alpha_j e_j.$$

Therefore

$$R(e_i) = \frac{\left(\sum_{k \neq i}^{n} \alpha_k^2 - \alpha_i^2 \right) e_i - \sum_{j=1(j \neq i)}^{n} 2\alpha_i \alpha_j e_j}{\sum_{l=1}^{n} \alpha_l^2}, \qquad (4.10)$$

from which we have that, in terms of the canonical basis, the entries R_{ij} of the matrix R are:

$$R_{ij} = \begin{cases} \dfrac{\sum_{k \neq i}^{n} \alpha_k^2 - \alpha_i^2}{\sum_{l=1}^{n} \alpha_l^2} & \text{for } i = j, \\[4mm] -\dfrac{2\alpha_i \alpha_j}{\sum_{l=1}^{n} \alpha_l^2} & \text{for } i \neq j. \end{cases}$$

This proves that if the α_i ($i = 1, 2..n$) are rational then the matrix R has rational entries and, due to Grimmer's theorem, represents a coincidence.

Conversely, now let us prove that if R is a coincidence reflection, then there exists t with rational entries such that $R(x) = -sxs^{-1} = -txt^{-1}$.

Since R is a coincidence reflection, $R_{i,j} \in \mathbf{Q}$ $\forall i, j$. Let $s = \sum_{k=1}^{n} \alpha_k e_k$, be a vector such that $R(x) = -sxs^{-1}$. Since $s \neq 0$, not all the α_k can be zero, in particular (renumbering the basis if necessary), let $\alpha_1 \neq 0$. Then, since

$$R_{i,j} = \delta_{i,j} - \frac{2\alpha_i \alpha_j}{\sum_h \alpha_h^2}$$

it follows that $(\alpha_i \alpha_j)/\left(\sum_h \alpha_h^2\right) \in \mathbf{Q}$ and, in particular

$$(\alpha_1 \alpha_j)/\left(\sum_h \alpha_h^2\right) \in \mathbf{Q}.$$

Call $Q_j = (\alpha_1 \alpha_j)/\left(\sum_h \alpha_h^2\right)$, $Q_j \in \mathbf{Q}$. Then define $t = \sum_j Q_j e_j$ (t has then rational coordinates). But

$$t = \sum_j Q_j e_j = \frac{\alpha_1}{\sum_h \alpha_h} \sum \alpha_i e_j = \frac{\alpha_1}{\sum_h \alpha_h} s,$$

and since $R(x) = -sxs^{-1}$

$$R(x) = -\left(\frac{\sum_h \alpha_h^2}{\alpha_1} t\right) x \left(\frac{\alpha_1}{\sum_h \alpha_h^2} t^{-1}\right) = -txt^{-1},$$

which completes the proof.

From the above results, we arrive to the following

Proposition 18.1.   *Let* $\mathbf{Z}^n = \bigoplus_{i=1}^{n} \mathbf{Z}e_i$, *then the transformation* $R(x) = -sxs^{-1}$ *is a coincidence reflection, i.e.* $R(x) \in OC(\mathbf{Z}^n)$, *if and only if there exists a vector* t, *with rational components, such that* $R(x) = -sxs^{-1} = -txt^{-1}$.

Immediate consequences of the above propositions are the following corollaries:

Corollary 18.1.   *Let* $\mathbf{Z}^n = \bigoplus_{i=1}^{n} \mathbf{Z}e_i$, *then the transformation* $R(x) = -sxs^{-1}$ *is a coincidence reflection, i.e.* $R(x) \in OC(\mathbf{Z}^n)$, *if and only if there exists a vector* t, *with integer components, such that* $R(x) = -sxs^{-1} = -txt^{-1}$.

Proof: Let $t' = \sum_{j=1}^{n} \frac{\alpha_j}{\beta_j} e_j$ which fulfills the conditions of proposition 18.1, with $\alpha_j, \beta_j \in \mathbf{Z}$. Consider $\gamma = \mathrm{lcm}\{\beta_j, \ j = 1, 2, \ldots, n\}$. Therefore $t = \gamma t'$ has integer entries. $\square$

Corollary 18.2.   *Let* $\mathbf{Z}^n = \bigoplus_{i=1}^{n} \mathbf{Z}e_i$, *and consider the following orthogonal transformation*

$$R(x) = (-1)^m \left(\prod_{i=1}^{m} s_i\right) x \left(\prod_{i=1}^{m} s_i\right)^{-1}.$$

If each vector s_i has rational or integer entries, then $R(x) \in OC(\mathbf{Z}^n)$.

At this point, we have that in a n-dimensional hypercubic lattice $\mathbf{Z}^n = \bigoplus_{i=1}^{n} \mathbf{Z}e_i$, transformations given by

$$R(x) = (-1)^m \left(\prod_{i=1}^{m} s_i\right) x \left(\prod_{i=1}^{m} s_i\right) = R_{s_1} R_{s_2} \ldots R_{s_k}(x),$$

are of coincidence if $s_i \in \mathbf{Z}^n$ (we use R_{s_i} to denote the reflection by a vector s_i.) Note that the treatment given above is also valid for the case of orthogonal transformations, since by Cartan-Dieudonné theorem any orthogonal transformation can be decomposed as the product of a finite number of R_{s_i}, $i = 1, 2 \ldots k$ ($k \leq 2n$) [26]. Thus in general, since $OC(\Lambda)$ is a group under composition, for an arbitrary number of reflections one has

Theorem 18.3.   *Let $\Lambda = \bigoplus_{i=1}^{n} \mathbf{Z}a_i$, be a lattice and let R be an orthogonal transformation. If $R_{s_i} \in OC(\Lambda)$, for $i = 1, 2, \ldots, k$, then $R \in OC(\Lambda)$.*

This is also a consequence of Cartan-Dieudonné's theorem.

The reciprocal is not always true; it is possible to have two reflections R_{s_1}, $R_{s_2} \notin OC(\Lambda)$ and such that $R = R_{s_1} R_{s_2} \in OC(\Lambda)$. Although the reciprocal of the theorem does not hold, we can see that if $R = R_{s_1} R_{s_2}$, $s_i \in \mathbf{R}^2$ and $s_i^2 = 1$ ($i = 1, 2$), in terms of the canonical basis R reads

$$R(e_1) = \left((s_1 \cdot s_2)^2 - |s_1 \wedge s_2|^2\right) e_1 - 2(s_1 \cdot s_2)|s_1 \wedge s_2|e_2$$

$$R(e_2) = -2(s_1 \cdot s_2)|s_1 \wedge s_2|e_1 + \left((s_1 \cdot s_2)^2 - |s_1 \wedge s_2|^2\right)$$

Then $R \in OC(\Lambda)$ if and only if the geometric product $s_1 s_2$ has rational coordinates. Thus, in general, we have

Proposition 18.2.   *Let $\Lambda = \bigoplus_{i=1}^{n} \mathbf{Z}a_i$, be a lattice and let R_{s_1} and R_{s_2}, $s_i \in \mathbf{R}^n$, $(i = 1, 2)$ be two reflections. Then $R_{s_1} R_{s_2} \in OC(\Lambda)$ if and only if the geometric product $s_1 s_2$ has rational components.*

It can be proved iterating the Equation (4.10) and evaluating the geometric product in the canonical basis.

From this proposition it follows that

Theorem 18.4.   Let $\Lambda = \bigoplus_{i=1}^{n} \mathbf{Z}a_i$, be a lattice and let R be an orthogonal transformation. Then $R \in OC(\Lambda)$ if and only if there are exist R_{s_i}, $i = 1, 2, \ldots, k$ ($k \leq 2n$) such that the geometric product $s_i s_{i+1}$, $(i = 1, 2, \ldots, k - 1)$ has rational components for k even. If k odd, it is also required that s_k has rational components.

The proof is based on induction, using the Cartan-Dieudonné's theorem, the proposition 18.2 and the fact that $OC(\Lambda)$ is a group under composition.

These results are valid for any hypercubic lattice since the $OC-$group is isomorphic to $OC(\mathbf{Z}^n)$. Then for the purpose of studying coincidences, one may well restrict the attention to the reflections R_{s_j}, the vectors s_j and the rotations $R_{s_i} R_{s_{i+1}}$, $i = 1, 2, \ldots, k$, $(k \leq 2n)$, such that both s_j and $s_i s_{i+1}$ has rational components.

18.5   Conclusions

In this work we show that geometric algebra is the natural algebraic setting for performing modern crystallographic calculations. The algebra incorporates in a single system all the tools required for doing geometry and for dealing with spaces of arbitrary dimensions.

In the first part we show how geometric algebra can be used to study the fundamental problem of finding the densest planes in quasicrystals and of predicting their external shape.

In the second half of this contribution we present a preliminary examination of the possible role of geometric algebra in the study of grain boundaries. The mathematical problem of finding coincidence lattices is addressed in this language which, we believe, is the natural one for most of modern crystallography.

Acknowledgments

This work was supported by CONACYT (grants 25237-E and 25125-A), DGAPA-UNAM (grants IN-119698 and IN-107296), and to Program for the Professional Development in Automation (grant from the UAM-A and Parker-Haniffin México.)

Chapter 19

Quaternion Optimization Problems in Engineering

Ljudmila Meister

19.1   Introduction

The interconnection between algebraic and geometric descriptions of space-time properties has attracted and ravished mathematicians since the time of Euclid. The last two centuries have been marked by several great contributions to this subject; among them Clifford and Grassmann algebras and Hamilton's quaternions. Quaternions were invented by Hamilton to simplify mathematical modelling of rigid body motion in three dimensions. A fascinating history of quaternions is presented in many books, see, for instance, [1]. The joint work of many mathematicians revealed the fundamental connections of Clifford's, Grassmann's, and Hamilton's approaches. The result of all this work is Geometric Algebra (see [13]).

The demands of different engineering areas, such as space and terrestrial navigation, photogrammetry, computer vision, robotics, and others, revived a Worldwide interest in the problems of orientation, and in particular, in the orientation of a rigid body in a 3-dim space.

The main problem of orientation in space and terrestrial navigation is determining the location and attitude of a vehicle. (Attitude means the orientation of a vehicle in space with respect to an inertial reference frame. Although widely adopted by engineers, this term is sometimes confusing to mathematicians.) Conventionally, the control systems of vehicles are of two main types [7]: control systems based on gyrostabilized platforms (usually called inertial navigation systems), and control systems without gyrostabilized platforms (usually called strap-down navigation systems). In the first case, a vehicle has an on-board inertial reference frame, and all measurements and succeeding calculations are realized with respect to this reference frame. In the second case, there is no on-board inertial reference frame: measurement equipment is rigidly fixed and is rotating together with the vehicle. It is in the second case that quaternions provide an extremely convenient mathematical apparatus, and this is the reason of the growing interest to quaternions in space technology. Application of quaternions leads to a significant reduction of computer time and memory requirements without any loss in accuracy (see [9]). The analysis of the problem of attitude determination, including a quaternion algebra approach and a vast list of references, are given in [22].

The main object of the present work is to show a general background of applications of quaternions to various problems of orientation in different engineering areas. In each case the initial questions are reduced to quaternion optimization problems. To illustrate the theory we use examples from photogrammetry and navigation.

Since our main interest concerns applications of quaternions to engineering, we will try to avoid using special knowledge in mathematics which may be beyond the scope of engineers. We collect, for the readers' sake, the most important properties of quaternions, even though some of them have been presented and discussed by other authors (see, for instance, [11, 16]). A vast list of references on applications of quaternions in photogrammetry, geodesy, and navigation was given by the author in [19].

19.2   Properties of Quaternions

19.2.1   Notations and definitions

We recall briefly the main notations, definitions, and properties of quaternions. Quaternions are generalized complex numbers of the form

$$A = a_0 + a_1\mathbf{i} + a_2\mathbf{j} + a_3\mathbf{k} \tag{2.1}$$

where the coefficients a_0, a_1, a_2, a_3 are real numbers (or functions) and $\mathbf{i}$, $\mathbf{j}$, $\mathbf{k}$ are imaginary units. The real part of A is denoted by Re $A = a_0$. The sum of all the other terms makes up the imaginary part Im A of A. If Re $A = 0$ the quaternion A is called imaginary.

Arithmetic operations with quaternions are similar to those for ordinary complex numbers. Products of the imaginary units are defined as follows

$$\begin{aligned} \mathbf{ii} = \mathbf{jj} = \mathbf{kk} = -1 \quad & \mathbf{ki} = -\mathbf{ik} = \mathbf{j} \\ \mathbf{ij} = -\mathbf{ji} = \mathbf{k} \quad & \mathbf{jk} = -\mathbf{kj} = \mathbf{i}. \end{aligned} \tag{2.2}$$

Multiplying two quaternions according to these rules, one gets the coefficients of their product. Quaternion multiplication is associative, distributive, but, in general, not commutative. However, for arbitrary quaternions we have

$$\mathrm{Re}(AB) = \mathrm{Re}(BA). \tag{2.3}$$

The following important assertion is easily proved:

Lemma 1 *If $Re(AB) = 0$ for an arbitrary quaternion B, then $A = 0$.*

Quaternion multiplication and matrices.

In engineering applications it is sometimes useful to rewrite quaternion equations in a matrix form. There is a one-to-one correspondence between quaternions and 4-vectors

$$A = a_0 + a_1\mathbf{i} + a_2\mathbf{j} + a_3\mathbf{k} \longrightarrow \mathbf{v}_A = (a_0, a_1, a_2, a_3)^T. \qquad (2.4)$$

(We consider column vectors; the superscript "T" means transposition.)

Let $\mathbf{v}_A$ and $\mathbf{v}_B$ be the 4-vectors which correspond to the quaternions A and B, respectively. The matrix form of the product AB is

$$\mathbf{v}_{AB} = G_1(A)\mathbf{v}_B = G_2(B)\mathbf{v}_A \qquad (2.5)$$

where the matrices $G_1(A)$ and $G_2(B)$ are given by

$$G_1(A) = \begin{pmatrix} a_0 & -a_1 & -a_2 & -a_3 \\ a_1 & a_0 & -a_3 & a_2 \\ a_2 & a_3 & a_0 & -a_1 \\ a_3 & -a_2 & a_1 & a_0 \end{pmatrix}; \quad G_2(B) = \begin{pmatrix} b_0 & -b_1 & -b_2 & -b_3 \\ b_1 & b_0 & b_3 & -b_2 \\ b_2 & -b_3 & b_0 & b_1 \\ b_3 & b_2 & -b_1 & b_0 \end{pmatrix}$$

Using mathematical induction, one can prove that for an arbitrary number of factors $A_1 \dots A_n$ one has

$$\mathbf{v}_{A_1 \dots A_n} = G_1(A_1) \dots G_1(A_{m-1})G_2(A_n) \dots G_2(A_{m+1})\mathbf{v}_{A_m}, \quad m = 1, \dots, n \quad (2.6)$$

This assertion allows us to avoid difficulties that are connected with the noncommutativity of quaternion multiplication.

Example 1 *Given $A = BCD, B \neq 0, D \neq 0$, what are the components of C? Using (2.6) one gets $\mathbf{v}_A = G_1(B)G_2(D)\mathbf{v}_C$. It then follows that $\mathbf{v}_C = G_2^{-1}(D)G_1^{-1}(B)\mathbf{v}_A$.*

More properties of the matrices G_1 and G_2 are discussed in [19].

19.2.2 Quaternions and vectors

In engineering problems we usually operate with vectors in 3-dim space. Let $\mathbf{e}_1, \mathbf{e}_2, \mathbf{e}_3$ be the basis vectors of a 3-dim right-handed orthogonal reference frame S. Each vector $\mathbf{r}$ can be represented in S as follows

$$\mathbf{r}_S = r_1\mathbf{e}_1 + r_2\mathbf{e}_2 + r_3\mathbf{e}_3$$

with the components r_1, r_2, r_3 relative to S. One can identify the imaginary units $\mathbf{i}, \mathbf{j}, \mathbf{k}$ with the basis vectors and consider the components of $\mathbf{r}$ as coefficients of the imaginary quaternion

$$R_S = r_1\mathbf{i} + r_2\mathbf{j} + r_3\mathbf{k}.$$

We say that R_S is the associated quaternion to $\mathbf{r}_S$. On the other hand, coefficients of an arbitrary imaginary quaternion can be considered as coordinates of a certain vector in 3-dim space. From now on we will use the word "vector" when speaking of the associated imaginary quaternion and vice versa.

Using the correspondence between vectors and imaginary quaternions, one finds that

$$A = a_0 + \mathbf{a} \tag{2.7}$$
$$A + B = a_0 + b_0 + \mathbf{a} + \mathbf{b}, \quad cA = ca_0 + c\mathbf{a} \tag{2.8}$$
$$AB = a_0 b_0 - \mathbf{a} \cdot \mathbf{b} + a_0 \mathbf{b} + b_0 \mathbf{a} + \mathbf{a} \times \mathbf{b} \tag{2.9}$$

where $\mathbf{a} \cdot \mathbf{b}$ is the scalar product and $\mathbf{a} \times \mathbf{b}$ is the vector product of the vectors $\mathbf{a}$ and $\mathbf{b}$. The collinearity, orthogonality, and coplanarity of vectors can now be directly expressed in terms of the algebraic operations between their associated quaternions,

$$orthogonality \ (\mathbf{a} \perp \mathbf{b}) : AB + BA = 0 \tag{2.10}$$
$$collinearity \ (\mathbf{a} \parallel \mathbf{b}) : AB - BA = 0 \tag{2.11}$$
$$coplanarity \ (\mathbf{a}, \mathbf{b}, \mathbf{c} \in the \ same \ plane) :$$
$$C(AB - BA) + (AB - BA)C = 0. \tag{2.12}$$

When dealing with quaternions as abstract mathematical objects, their imaginary parts are treated as vectors relative to the same reference frame. Things are different when one works with quaternions associated with vectors which have been measured during engineering experiments. We have to take into consideration what reference frames were used in making the measurements. We cannot draw any conclusions about the relative position of the vector parts of quaternions if they are relative to different reference frames. Fortunately, it is not necessary to transfer all vectors to the same reference frame before going to the associated quaternions. It is possible to transform quaternions directly from one reference frame into another. Thus, when dealing with quaternions, the situation is similar to the one when dealing with vectors: one has to distinguish particular reference frames to which the quaternions (or the vectors) are related. We will use a subscript to indicate a particular reference frame.

A trigonometric form of quaternions.

Every quaternion $A = a_0 + \mathbf{a}$ can be represented in the trigonometric form:

$$A = \|A\|(\cos\alpha + \mathbf{e}\sin\alpha) \tag{2.13}$$

where $\|A\| = (a_0{}^2 + a_1{}^2 + a_2{}^2 + a_3{}^2)^{1/2}$ is the norm of A, where the unit vector $\mathbf{e}$, the axis of A, is given by $\mathbf{e} = \mathbf{a}/|\mathbf{a}|$, $|\mathbf{a}|^2 = a_1{}^2 + a_2{}^2 + a_3{}^2$, and where

α, the angle of A, is given by $\cos \alpha = a_0/\|A\|$, $\sin \alpha = |\mathbf{a}|/\|A\|$, $0 \leq \alpha < \pi$. If A is a normalized quaternion, i.e. $\|A\| = 1$, then

$$A = \cos \alpha + \mathbf{e} \sin \alpha. \tag{2.14}$$

A quaternion given by (2.14) is also called a "spinor".

19.2.3   *Quaternion description of rotations*

Rotations of vectors.

Quaternions give a very clear algebraic description of rotations in 3-dim space. Let S be an arbitrary 3-dim right-handed orthogonal inertial reference frame with the basis vectors identified with $\mathbf{i}, \mathbf{j}, \mathbf{k}$ and let $\mathbf{r}_1$ be an arbitrary vector in S with the associated quaternion R_1. The result $\mathbf{r}_2$ of the counterclockwise rotation of $\mathbf{r}_1$ about the axis $\mathbf{e}$ by the angle 2α is given by

$$R_2 = A R_1 A^* \tag{2.15}$$

where the quaternion A is as in (2.14) and A^* is the conjugate quaternion $A^* = a_0 - \mathbf{a}$. (Note, that $\mathbf{r}_1, \mathbf{r}_2$ and the axis $\mathbf{e}$ of A are relative to the same reference frame.) If a rotation is described by the quaternion A we call it an A-rotation.

Rotations of reference frames.

Let the reference frame S_1 be rotated as a rigid body to a new position S_2 by the quaternion A, i.e. $A : S_1 \longrightarrow S_2$. This means that every basis vector of S is transformed by the A-rotation. We assume that the axis of A is given in the frame S_1. The new components of a fixed vector $\mathbf{r}$ relative to the rotated reference frame S_2 are the components of the quaternion

$$R_{S_2} = A^* R_{S_1} A. \tag{2.16}$$

Composition of rotations.

Let A and B be two normalized quaternions such that $A : S \longrightarrow S'$ and $B : S' \longrightarrow S''$. If the axes of A and B are both given in the same reference frame S, then the resulting quaternion $C : S \longrightarrow S''$ is given by

$$C = B_S A_S. \tag{2.17}$$

If the axis of A is given in S while the axis of B is given in S', then the order of the factors changes:

$$C = A_S B_{S'}. \tag{2.18}$$

Indeed, since $B_{S'} = A_S^* B_S A_S$, we have $B_S = A_S B_{S'} A_S^*$. Substituting this into (2.17), we get (2.18).

Example 2 *Let S be a 3-dim right-handed orthogonal reference frame and let it be rotated by the quaternions A, B, C, D as follows*

$$S \xrightarrow{A_S} S_1 \xrightarrow{B_S} S_2 \xrightarrow{C_{S_2}} S_3 \xrightarrow{D_{S_2}} S_4.$$

The subscripts indicate the reference frame in which the axes of the quaternions are given. The resulting quaternion $T_S : S \longrightarrow S_4$ is

$$T_S = B_S A_S D_{S_2} C_{S_2}.$$

The consideration above shows that we can choose the reference frame where quaternions have the simplest form for a particular calculation.

Rotations and parallel translations.

Now assume that a given A-rotation $S_1 \longrightarrow S_2$ is supplemented with a parallel displacement. Let **b** be a position vector of the origin of S_2 with respect to S_1 and B be the associated quaternion. A parallel displacement has no effect when operating on free vectors.

If **r** is a position vector with the endpoint M relative to S_1 and **p** is the position vector of M relative to S_2, then we have

$$P_{S_2} = A^*(R_{S_1} - B_{S_1})A = A^* R_{S_1} A - B_{S_2}. \tag{2.19}$$

See Figure 19.1.

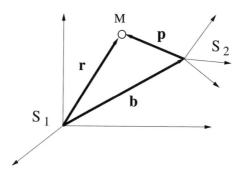

FIGURE 19.1. Rotation and parallel translation of the position vector.

Rotations which depend on time.

Let S_1 and S_2 be two 3-dim right-handed orthogonal reference frames with the same origin. Let S_1 be a fixed frame and S_2 be rotating about S_1 with a rotational velocity $\mathbf{w}(t)$ which depends on time. The associated quaternions are $W_{S_1}(t)$ relative to S_1, and $W_{S_2}(t)$ relative to S_2. The quaternion $T_{12}(t)$ corresponding to this rotation is a solution to the differential equations [12]:

$$\frac{d}{dt}T_{12} = \frac{1}{2}W_{S_1}T_{12} \quad or \quad \frac{d}{dt}T_{12} = \frac{1}{2}T_{12}W_{S_2} \tag{2.20}$$

where $\|T_{12}(t)\| = 1$ for every t.

19.3   Extremal Problems for Quaternions

19.3.1   Differentiation with respect to quaternions

As we shall see later, various applied problems can be formulated as extremal problems (minimum or maximum) of a function of a quaternion argument. To find the points of extrema we have to use differential calculus. Derivatives of functions can be defined in two ways: "strong"(or Fréchet's) derivative and "weak" (or Gâteaux's) derivatives . To find points of extrema we have to employ the strong derivative, which is not suitable for calculations (see [18]). Since quaternions form a 4-dim linear normed space, the strong and the weak derivatives are equivalent, if they both exist and are continuous. Thus, we can use the weak derivative, which is defined as follows:

Definition 19.1.   *Let $f(X)$ be a scalar function of the quaternion variable X and let H be an arbitrary quaternion. The derivative of $f(X)$ with respect to X in the direction of H is*

$$f'_{X,H}(X) = \lim_{\varepsilon \to 0} \frac{d}{d\varepsilon} f(X + \varepsilon H)$$

where ε is a scalar.

One can use the standard techniques for finding these derivatives.

Example 3 *Find the derivative of the function $f(X) = \|X\|^2$. Since $\|X\|^2 = XX^*$ we have*

$$f'_{X,H}(X) = \lim_{\varepsilon \to 0} \frac{d}{d\varepsilon}\|X + \varepsilon H\|^2 = \lim_{\varepsilon \to 0} \frac{d}{d\varepsilon}(X + \varepsilon H)(X + \varepsilon H)^* =$$

This must not be confused with the weak derivative in distribution theory.

$$\lim_{\varepsilon \to 0} \frac{d}{d\varepsilon}(XX^* + \varepsilon HX^* + \varepsilon XH^* + \varepsilon^2 HH^*) = HX^* + XH^* = 2Re(HX^*).$$

Generally, derivatives with respect to quaternion variables depend on the quaternion variable. It is similar to the directional derivative of a function of a vector variable.

Definition 19.1 can be naturally extended to the case of several quaternion variables.

19.3.2 Minimization of loss functions

In engineering applications, particularly in navigation and optimal control theory, minimizing functions are usually called **loss functions**. We shall use the same terminology.

Let $f(X)$ be a scalar loss function of a quaternion argument. At the points of extrema,

$$f'_{X,H}(X) = 0 \tag{3.21}$$

for every H. The condition (3.21) for an extrema depends on an arbitrary quaternion H. Fortunately, for many applied problems the loss functions has a special structure that makes it possible to eliminate this arbitrary quaternion. The following theorem holds.

Theorem 1 *If a scalar loss function $f(X)$ of the quaternion variable X has the form*

$$f(X) = \sum_{n=1}^{N} \|F_n(X)\|^2 \tag{3.22}$$

where $F_n(X)$ are polynomial quaternion functions with quaternion coefficients

$$F_n(X) = A_{n1} + A_{n2}X + XA_{n3} + X^2A_{n4} + \dots \tag{3.23}$$

then it is possible to eliminate the arbitrary quaternion H from the equation of extremum (3.21).

The proof of this theorem is based on the properties of quaternions, and especially on Lemma 1.

19.3.3 Conditional extremum problems

When we have to find the points of extrema of a loss function under certain conditions, for example for normalized quaternions, we get the so-called conditional extremum problems. The condition $\|X\| = 1$ leads to the following problem:

Problem 1 *Let $f(X)$ be a scalar loss function of a quaternion variable. Find the point of extremum under the condition $\|X\|^2 = 1$.*

To solve such a problem, we employ the Lagrange multiplier method which consists of looking for extremum of the auxiliary Lagrange function

$$F(X, \gamma) = f(X) + \gamma(\|X\|^2 - 1)$$

where γ is an unknown scalar, called the Lagrange multiplier. Differentiating $F(X, \gamma)$ with respect to X and equating the derivative to zero, we get

$$F'_{X,H}(X, \gamma) = 0. \tag{3.24}$$

Equation (3.24) and the condition $\|X\|^2 = 1$ give a set of two equations in the unknowns X and γ.

19.3.4   The least-squares method for quaternions

Another example of a quaternion conditional extremum problem is the problem of quaternion interpolation. We formulate this problem as follows:

Problem 2 *Let $X_1, X_2, \ldots, X_N$ be the measured values of a normalized quaternion. We wish to find the optimal value of the quaternion which minimizes the least-squares loss function*

$$f(X) = \sum_{p=1}^{N} \|X - X_p\|^2$$

under the condition $\|X\|^2 = 1$.

The auxiliary Lagrange function is

$$F(X, \gamma) = \sum_{p=1}^{N} \|X - X_p\|^2 + \gamma(\|X\|^2 - 1).$$

Differentiating by X and equating the derivative to zero gives

$$
\begin{aligned}
F'_{X,H}(X, \gamma) &= \lim_{\varepsilon \to 0} \frac{d}{d\varepsilon} \left(\sum_{p=1}^{N} \|X + \varepsilon H - X_p\|^2 + \gamma(\|X + \varepsilon H\|^2 - 1) \right) \\
&= 2 \sum_{p=1}^{N} \operatorname{Re}(XH^* - X_p H^*) + 2\gamma \operatorname{Re}(XH^*) = 0
\end{aligned}
$$

where H is an arbitrary quaternion. Using Lemma 1, we get the algebraic equation

$$\sum_{p=1}^{N}(X - X_p) + \gamma X = 0,$$

which does not contain H. Hence,

$$X = \frac{\sum_{p=1}^{N} X_p}{N + \gamma}. \tag{3.25}$$

The Lagrange multiplier γ is determined by the condition that $\|X\|^2 = 1$:

$$\|X\|^2 = XX^* = \frac{(\sum_{p=1}^{N} X_p)(\sum_{p=1}^{N} X_p)^*}{(N + \gamma)^2} = 1.$$

Then we get

$$|N + \gamma| = \sqrt{(\sum_{p=1}^{N} X_p)(\sum_{p=1}^{N} X_p)^*}. \tag{3.26}$$

The right-hand side in (3.26) is not a quaternion but a real number since

$$(\sum_{p=1}^{N} X_p)(\sum_{p=1}^{N} X_p)^* = \|\sum_{p=1}^{N} X_p\|^2.$$

Thus,

$$|N + \gamma| = \|\sum_{p=1}^{N} X_p\|$$

and we get the two possible solutions

$$\gamma_1 = \|\sum_{p=1}^{N} X_p\| - N \;\; or \;\; \gamma_2 = -\|\sum_{p=1}^{N} X_p\| - N.$$

Finally, we select as the optimal value of X

$$X = \frac{\sum_{p=1}^{N} X_p}{\|\sum_{p=1}^{N} X_p\|}.$$

19.4   Determination of Rotations

19.4.1   Unknown rotation of a vector

Let us consider now the "inverse problem":

Problem 3 *Let* **r** *and* **p** *be two measurements of the same unit vector* **before** *and* **after** *a rotation in the reference frame* S. *The associated quaternions are* R *and* P *correspondingly,* $\|R\| = \|P\| = 1$. *We wish to find the unknown rotation, i.e. the axis and the angle of a normalized quaternion* X, *such as*

$$P = XRX^* \tag{4.27}$$

with the condition that $\|X\| = 1$.

Problems of this kind are quite common, for example in robotics or when calibrating measurement equipment. We consider three methods for solving this problem: a method based on the algebraic properties of quaternions, a method based on geometric considerations, and the method of conditional extrema.

Algebraic method.

Multiplying (4.27) by XR^* on the right gives

$$PXR^* = X. \tag{4.28}$$

Now, using the matrix form of quaternion multiplication (2.6) gives

$$G_1(P)G_2(R^*)\mathbf{v}_X = \mathbf{v}_X. \tag{4.29}$$

Thus, we have reduced the initial problem to the well-known eigenvector problem of matrices, which is discussed thoroughly in many books on Linear Algebra (see, for example, [14]). From the properties of the matrices G_1 and G_2, it follows that the (4×4) matrix $G_1(P)G_2(R^*)$ is symmetric and orthogonal. Hence, it has only real eigenvalues, which are equal to 1 or -1. The sum of the eigenvalues of a matrix is equal to the trace of this matrix, i.e. to the sum of elements on the main diagonal. It is easy to verify that the trace of $G_1(P)G_2(R^*)$ is 0, therefore, this matrix has only two different eigenvalues, $\lambda_1 = 1$ and $\lambda_2 = -1$, each of them being of the second order. This means that Equation (4.29) has a solution, which is the eigenvector corresponding to the eigenvalue $\lambda = 1$. This solution can be found in the usual way (see [10]). Since every eigenvector is determined up to an arbitrary constant, we can choose this constant such that $\|X\| = 1$.

Since the eigenvalue $\lambda = 1$ is of the second order, the solution is not unique but depends on one free parameter. This will be more transparent from the geometric considerations given below.

Geometric method.

First, note that the linear quaternion equation (4.27) is invariant under rotations of the reference frame. In fact, let S' be a new reference frame which is obtained by the T-rotation of S, i.e. $S' = TST^*$. Multiplying (4.27) by T from the right and by T^* from the left leads to

$$T^*PT = T^*XRX^*T. \tag{4.30}$$

Since $\|T\|^2 = T^*T = 1$, we can write

$$T^*PT = T^*XTT^*RTT^*X^*T. \tag{4.31}$$

Let $A = T^*PT$, $B = T^*RT$, $Y = T^*XT$, then Equation (4.31) becomes

$$A = YBY^* \tag{4.32}$$

where all quaternions have imaginary parts relative to the new reference frame S'. This means that the Equation (4.27) is invariant under the rotation. This fact allows us to simplify the linear quaternion equations by choosing an appropriate reference frame.

To solve Equation (4.27) we assume without loss of generality that the axis $\mathbf{i}$ coincides with the bisector of the angle α between the vectors $\mathbf{r}$ and $\mathbf{p}$, and that these vectors are located in the plane spanned by $\mathbf{i}, \mathbf{j}$. (See Figure 19.2.)

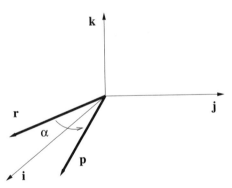

FIGURE 19.2. Relative position of the vectors r and p.

Clearly, the axis $\mathbf{x}$ of the quaternion X (the axis of rotation) is an arbitrary straight line lying in the plane spanned by the axes $\mathbf{i}, \mathbf{k}$. Hence, the unknown quaternion X must be of the form $X = x_0 + x_1\mathbf{i} + x_3\mathbf{k}$, where $x_0^2 + x_1^2 + x_3^2 = 1$.

The angle of the rotation δ can be found from simple geometric considerations. Let D be the point of the orthogonal projections of the end points M and N of $\mathbf{r}$ and $\mathbf{p}$ on the vector $\mathbf{x}$ (see Figure 19.3).

Due to symmetry, these projections coincide, and we have the following relation

$$\tan \frac{\delta}{2} = \frac{\sin \frac{\alpha}{2}}{\cos \frac{\alpha}{2} \sin \beta} \tag{4.33}$$

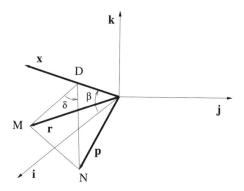

FIGURE 19.3. Connection between the angles (MD and ND are orthogonal to x).

where β is the angle between **x** and the axis **i**. We apply the trigonometric formulas

$$\sin\alpha = 2\sin\frac{\alpha}{2}\cos\frac{\alpha}{2}; \quad \cos\alpha = 2\cos^2\frac{\alpha}{2} - 1$$

and rewrite (4.33) as

$$\tan\frac{\delta}{2} = \frac{\sin\alpha}{(\cos\alpha + 1)\sin\beta} \tag{4.34}$$

The components of **x** are $(\cos\beta, 0, \sin\beta)$, so the trigonometric form of X is

$$X = \cos\frac{\delta}{2} + (\mathbf{i}\cos\beta + \mathbf{k}\sin\beta)\sin\frac{\delta}{2}.$$

Therefore, the coefficients of X are

$$x_0 = \cos\frac{\delta}{2}; \quad x_1 = \cos\beta\sin\frac{\delta}{2}; \quad x_3 = \sin\beta\sin\frac{\delta}{2}.$$

The geometric method clearly shows that the solution is not unique, but rather there is a set of solutions depending on a single parameter (δ and β are connected by (4.34)). To get a unique solution, we must impose an additional condition. This could be the condition of choosing the minimal angle of rotation. This condition implies that

$$\beta = \pi/2, \quad \tan\frac{\delta}{2} = \frac{\sin\alpha}{(\cos\alpha + 1)}$$

and $X = \cos(\delta/2) + \mathbf{k}\sin(\delta/2)$. It is the particular solution which corresponds to a rotation about the axis **k** by the angle $\delta = \alpha$.

The assumption that the axis of rotation lies in the common plane with **r** and **p** leads to $\beta = 0$ and $\delta = \pi$, so that $X = \mathbf{i}$. This is the solution which corresponds to the rotation about the axis **i** by the angle π.

Another example of an additional condition could be a restriction on the location of the axis of rotation, used in robotics (see, for instance, [21]).

Method of conditional extrema.

This method is used when the vectors **r** and **p** are measured with measurement errors. Then we have a nonzero difference,

$$Q = XRX^* - P$$

where Q is the quaternion associated with the vector of errors, which leads to the following quaternion conditional extremum problem:

Problem 4 *Find the quaternion X which minimizes the loss function*

$$f(X) = \|XRX^* - P\|^2 \tag{4.35}$$

under the condition $\|X\| = 1$.

The auxiliary Lagrange function is

$$F(X) = f(X) + \gamma(\|X\|^2 - 1)$$

where γ is the unknown scalar Lagrange multiplier. Differentiating $F(X)$, one gets

$$
\begin{aligned}
F'_{X,H}(X) &= \lim_{\varepsilon \to 0} \frac{d}{d\varepsilon} F(X + \varepsilon H) \\
&= 2\mathrm{Re}(XRX^* - P)(HRX^* + XRH^*)^* + 2\gamma \mathrm{Re}XH^* \\
&= 2\mathrm{Re}(XRX^* - P)XR^*H^* + 2\mathrm{Re}(XRX^* - P)HR^*X^* + 2\gamma \mathrm{Re}XH^*
\end{aligned}
\tag{4.36}
$$

Let us transform the second term on the right of (4.36) separately. Using Equation (2.3) and identities $R^* = -R$ (which hold for imaginary quaternions), and $\mathrm{Re}A = \mathrm{Re}A^*$ (for arbitrary quaternions), we get

$$
\begin{aligned}
2\mathrm{Re}(XRX^* - P)HR^*X^* &= 2\mathrm{Re}HR^*X^*(XRX^* - P) \\
&= 2\mathrm{Re}(XRX^* - P)^*XRH^* = 2\mathrm{Re}(XRX^* - P)XR^*H^*.
\end{aligned}
\tag{4.37}
$$

Substituting (4.37) in (4.36), we get

$$F'_{X,H}(X) = 4\mathrm{Re}(XRX^* - P)XR^*H^* + 2\gamma \mathrm{Re}XH^* = 0. \tag{4.38}$$

Finally, applying Lemma 1 to Equation (4.38), we obtain an algebraic quaternion equation which does not contain the arbitrary quaternion H

$$2(XRX^* - P)XR^* + \gamma X = 0. \tag{4.39}$$

After a simple transformation, (4.39) becomes

$$PXR^* = \frac{2 + \gamma}{2} X. \tag{4.40}$$

This equation is similar to Equation (4.28), so it has solutions if

$$\frac{2+\gamma}{2} = 1 \quad or \quad \frac{2+\gamma}{2} = -1. \tag{4.41}$$

We must choose the maximal eigenvalue. In fact, we have

$$f(X) = \|XRX^* - P\|^2 = (XRX^* - P)(XRX^* - P)^*$$
$$= 1 - PXR^*X^* - XRX^*P^* + 1 = 2 - 2\mathrm{Re}(PXR^*X^*) \tag{4.42}$$

Hence, $f(X)$ is minimal when $\mathrm{Re}(PXR^*X^*)$ is maximal. But from (4.40), it follows that $\mathrm{Re}(PXR^*X^*) = (2+\gamma)/2$, therefore $(2+\gamma)/2$ must be maximal.

Thus, again we get an eigenvector problem whose solution can be obtained in the standard way.

19.4.2 Rotation of several vectors

When we have N unit vectors $\mathbf{r}_1, \ldots, \mathbf{r}_N$ which are measured before the rotation, and $\mathbf{p}_1, \ldots, \mathbf{p}_N$ which are measured after it, then we have to consider the loss function

$$f(X) = \sum_{n=1}^{N} \|XR_nX^* - P_n\|^2 \tag{4.43}$$

with $\|X\| = 1$. The same considerations as before lead to the equation

$$\sum_{n=1}^{N} P_n X R_n^* = \frac{2N+\gamma}{2} X. \tag{4.44}$$

Transferring (4.44) to the matrix form gives

$$\sum_{n=1}^{N} G_1(P_n)G_2(R_n^*)\mathbf{v}_X = \frac{2N+\gamma}{2}\mathbf{v}_X. \tag{4.45}$$

The matrix $G = \sum_{n=1}^{N} G_1(P_n)G_2(R_n^*)$ is symmetric since it is a sum of symmetric matrices, but, in general, it is not orthogonal. Therefore, G has exactly 4 real eigenvalues, and, as before, we choose the maximal one. The corresponding eigenvector gives the solution to the problem.

19.5 The Main Problem of Orientation

The main problem of orientation of a rigid body can be formulated as follows. Let S_1 and S_2 be two 3-dim right-handed orthogonal reference frames. Let S_1 be an inertial reference frame and S_2 be a body-fixed reference frame with the origin at the center of mass of the rigid body. Given a set of measurements, we wish to find the attitude and the position of the body with respect to S_1 at a certain moment t. This means that we have to find a normalized quaternion X of the X-rotation and the imaginary quaternion B associated with the position vector $\mathbf{b}$ of the origin of S_2.

According to the available measurements, we have different variants of the problem. We consider here the most important cases.

19.5.1   Orientation based on free vectors

Some free vectors are measured with respect to both reference frames: let $\mathbf{r}_n$, $n = 1, \ldots, N$, be results of measurement with respect to S_1 and $\mathbf{p}_n$, $n = 1, \ldots, N$, be results of measurement with respect to S_2. The associated quaternions are R_{nS_1} and P_{nS_2}, respectively.

Such problem arises, for example, in space navigation, when unit vectors of a stars' directions are measured with respect to an inertial frame (S_1) and a body-fixed reference frame (S_2). The distances between every star and the origins of S_1 and S_2 are infinitely large in comparison with the distance between the origins. Therefore, the vectors $\mathbf{r}_n$ and $\mathbf{p}_n$ have to be equal. In space navigation this problem is often quoted as "Wahba problem" [22]. The geometry of the problem is given in Figure 19.4.

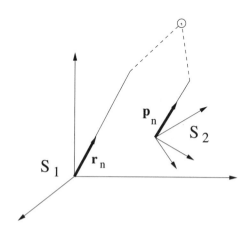

FIGURE 19.4. Geometry of orientation based on free vectors.

Mathematical model of the problem.

The reference frames S_1 and S_2 are connected by the X-rotation, hence $S_2 = X S_1 X^*$ and we have

$$P_{nS_2} = X^* R_{nS_1} X, \ n = 1, \dots, N. \tag{5.46}$$

According to the least-squares method, in the presence of measurement errors, we have to minimize the scalar loss function

$$f(X) = \sum_{n=1}^{N} \|X^* R_{nS_1} X - P_{nS_2}\|^2 \tag{5.47}$$

under the condition $\|X\| = 1$. Denoting $Y = X^*$ we get the same loss function as in (4.43). Note, that only attitude can be determined in this case.

This problem has been discussed by many authors (see, the references given in [19, 22]).

19.5.2   *Roto-translation problem*

Now, let the vectors $\mathbf{r}_n$ be the measured position vectors of the objects M_n relative to S_1 and the vectors $\mathbf{p}_n$ be the measured position vectors of these objects relative to S_2. In general, the reference frames will have different unit lengths with an unknown scale factor. One has to find the scale factor, the attitude of S_2, and the location of its origin. The geometry is similar to that given in Figure 19.1.

Mathematical model of the problem.

Let ρ be an unknown scale factor. According to (2.19), we get the loss function

$$f(X, B, \rho) = \sum_{n=1}^{N} \|X^* R_{nS_1} X - \rho P_{nS_2} - B_{S_2}\|^2 \tag{5.48}$$

where $\|X\| = 1$.

The complete solution of the problem in quaternions was given by F. Sanso [20]. He also observed that if one chooses the origins of S_1 and S_2 at the centers of gravity of the points $\mathbf{r}$ and $\mathbf{p}$, respectively, then the translation vector $\mathbf{b}$ will vanish and (5.48) reduces to a form similar to (5.47).

19.5.3   *Photo exterior orientation problem*

Let $\mathbf{r}_n$ be the measured position vectors of the objects M_n relative to S_1, and $\mathbf{p}_n^0$ be the measured position vectors of their images M_n^0 on a photo relative to the photographic reference frame S_2. These reference frames also have different unit lengths with an unknown scale factor. One has to find the scale factor, the location of the perspective center of the camera objective lens at the moment of exposure (the origin O_2 of S_2) and the attitude of the camera in space (the attitude of S_2). Geometry of this case is shown in Figure 19.5. This is the well-known problem of photo exterior orientation in photogrammetry (see, for instance, [24]).

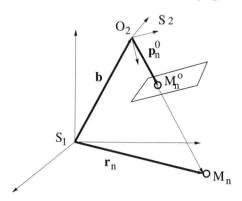

FIGURE 19.5. Geometry of photo exterior orientation.

Mathematical model of the problem.

This case is different from the preceding one because the position vectors $\mathbf{p}_n$ of the objects M_n relative to S_2 are not known. If we introduce the scale factors μ_n such as $\mathbf{p}_n = \mu_n \mathbf{p}_n^0$, then we can use the same loss function as before. Denoting by $P_{nS_2}^0$ the quaternion associated with the vector $\mathbf{p}_n^0$, we get

$$f(X, B, \mu_n) = \sum_{n=1}^{N} \|X^* R_{nS_1} X - \mu_n \rho P_{nS_2}^0 - B_{S_2}\|^2 \qquad (5.49)$$

with $\|X\| = 1$. However, this unnessesarily increases the number of the unknowns.

To avoid the additional scale factors, we can use another loss function based on the condition of collinearity (2.11). Since the object and its image lie on the same line, the vectors $\mathbf{r}_n - \mathbf{b}$ and $\mathbf{p}_n^0$ are collinear, and using the quaternion condition of collinearity (2.11), we get the loss function

$$f(X, B) = \sum_{n=1}^{N} \|P_{nS_2}^0 X^* (R_{nS_1} - B_{S_1}) X - X^* (R_{nS_1} - B_{S_1}) X P_{nS_2}^0\|^2 (5.50)$$

with $\|X\| = 1$. This problem was considered by the author in [4].

19.5.4   Orientation based on coplanar vectors

Assume now that only the unit vectors $\mathbf{r}_n^0$ and $\mathbf{p}_n^0$ of the directions to several objects M_n are measured with respect to S_1 and S_2, respectively, and that the distances between the objects and origins of the reference frames are comparable. The geometry of this case is given in Figure 19.6.

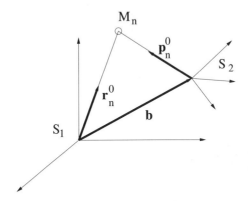

FIGURE 19.6. Geometry of unit coplanar vectors.

Mathematical model of the problem.

Introducing the unknown scale factors λ_n and μ_n, such as $R_{nS_1} = \lambda_n R_{nS_1}^0$ and $P_{nS_2} = \mu_n P_{nS_2}^0$, one can use the same loss function as in the roto-translation problem, but now the loss function depends on the additional unknowns λ_n and μ_n

$$f(X, B, \lambda_n, \mu_n) = \sum_{n=1}^{N} \| \lambda_n X^* R_{nS_1}^0 X - \mu_n P_{nS_2}^0 - B_{S_2} \|^2 \qquad (5.51)$$

with $\|X\| = 1$. A quaternion approach to this problem in computer vision was thoroughly treated by J.Lasenby [17].

19.5.5   Relative orientation of a stereo-pair

Let $\mathbf{r}_n^0$ be position vectors of the images M_{n1}^0 of the objects M_n on the first photo measured in the photographic reference frame S_1, and let $\mathbf{p}_n^0$ be position vectors of the images M_{n2}^0 of the same objects on the second photo measured in the photographic reference frame S_2. We wish to find the position vector $\mathbf{b}$ of the origin of S_2 and the attitude of S_2 with respect to S_1. In photogrammetry this problem is known as the problem of relative orientation of a stereo-pair. It has a geometric interpretation similar to the previous case (see Figure 19.7).

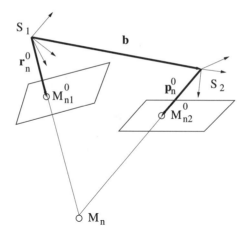

FIGURE 19.7. Geometry of relative orientation of a stereo-pair.

Mathematical model of the problem.

Again the position vectors of the objects M_n are not known. We could introduce the unknown scale factors as before, but in order to avoid additional unknowns we use another form of the loss function. Since every three vectors $\mathbf{r}_n^0, \mathbf{p}_n^0, \mathbf{b}$ have to lie in the same plane, their associated quaternions have to satisfy the condition of coplanarity (2.12), and we get the loss function

$$
\begin{aligned}
f(X, B) &= \sum_{n=1}^{N} \| (R_{nS_2}^0 P_{nS_2}^0 - P_{nS_2}^0 R_{nS_2}^0) B_{S_2} \\
&+ B_{S_2} (R_{nS_2}^0 P_{nS_2}^0 - P_{nS_2}^0 R_{nS_2}^0) \|^2
\end{aligned}
\tag{5.52}
$$

where $R_{nS_2}^0 = X^* R_{nS_1}^0 X$ and $\|X\| = 1$ as before.

It should be pointed out that for this loss function we have to use an additional condition of the form $\|B\| = b$. Otherwise, we would get the trivial minimum $B = 0$. The quaternion approach to this problem was considered by the author in [2], and the reduction of the problem for space navigation was given in [6].

We have presented here only the mathematical models of the above problems using quaternions. This overview shows that various orientation problems in different engineering areas can be considered as quaternion conditional extremum problems. The quaternion language enables short algebraic formulations and clear geometric interpretations which reveal the similarities of the problems.

19.6   Optimal Filtering and Prediction

In this section, we study the orientation of a system when not only measurements of some vectors are given, but when the equation of motion of the system is also known. We start with some definitions.

19.6.1   Random quaternions

First, we give very briefly some notations and definitions of random quaternions. A random quaternion $A = a_0 + a_1\mathbf{i} + a_2\mathbf{j} + a_3\mathbf{k}$ is a quaternion with random variable coefficients a_0, a_1, a_2, a_3, where $\mathbf{i}$, $\mathbf{j}$, $\mathbf{k}$ are imaginary units as before. Let $E[a_l]$ and $\sigma[a_l]$ be the mean value and the standard deviation of the coefficients $a_l, l = 0, 1, 2, 3$, respectively.

Due to the one-to-one correspondence between quaternions and 4-vectors, it is possible to use the same methods as in the case of random vectors.

The mean value $E[A]$ (the mathematical expectation) of A is

$$E[A] = E[a_0] + E[a_1]\mathbf{i} + E[a_2]\mathbf{j} + E[a_3]\mathbf{k}. \qquad (6.53)$$

The covariance (4×4) matrix $C[A]$ of A is given by

$$C[A] = \begin{pmatrix} c_{11}[A] & c_{12}[A] & c_{13}[A] & c_{14}[A] \\ c_{21}[A] & c_{22}[A] & c_{23}[A] & c_{24}[A] \\ c_{31}[A] & c_{32}[A] & c_{33}[A] & c_{34}[A] \\ c_{41}[A] & c_{42}[A] & c_{43}[A] & c_{44}[A] \end{pmatrix} \qquad (6.54)$$

The terms on the diagonal of the matrix $C[A]$ are the variances of the coefficients of A, i.e. $c_{ll}[A] = E[a_{l-1}^2] - E^2[a_{l-1}] = \sigma^2[a_{l-1}], l = 1, 2, 3, 4$. The off-diagonal elements are the covariances of the coefficients $c_{lp}[A] = E[(a_{l-1} - E[a_{l-1}])(a_{p-1} - E[a_{p-1}])]; \; l, p = 1, 2, 3, 4, \; l \neq p$. Using the 4-vector $\mathbf{v}_A$ corresponding to the random quaternion A, equality (6.54) can be expressed in the vector form

$$C[A] = E[(\mathbf{v}_A - E[\mathbf{v}_A])(\mathbf{v}_A - E[\mathbf{v}_A])^T].$$

Let us consider the quaternion

$$\sigma[A] = \sigma[a_0] + \sigma[a_1]\mathbf{i} + \sigma[a_2]\mathbf{j} + \sigma[a_3]\mathbf{k}$$

The following identity is easy to prove

$$\|\sigma[A]\|^2 = E[\|A - E[A]\|^2] \qquad (6.55)$$

Indeed, since $E^*[A] = E[A^*]$, we have

$$E[\|A - E[A]\|^2] = E[(A - E[A])(A - E[A])^*]$$

$$= E[AA^*] - E[E[A]A^*] - E[AE^*[A]] + E[E[A]E^*[A]]$$

$$= E[AA^*] - E[A]E[A^*] = \sum_{l=0}^{3} E[a_l^2] - E^2[a_l] = \sum_{l=1}^{4} c_{ll} = \|\sigma[A]\|^2.$$

Hence, $\|\sigma[A]\|^2 = E[\|A - E[A]\|^2]$ is the variance of A and $\|\sigma[A]\|$ is the standard deviation of A. It follows from the consideration above that the trace of the covariance matrix $C[A]$ is equal to the variance of A. More details on random quaternions can be found in [23].

19.6.2   Optimal filtering and prediction for single-stage rotations

Let us consider a system moving with respect to a reference frame S_1. The system with the 3-dim position vector $\mathbf{x}$ makes a discrete rotation about the unit axis $\mathbf{a}_0$ through the angle $2\alpha_0$ from the state $\mathbf{x}_0$ to the state $\mathbf{x}_1$. The relation between these two states of the system is

$$X_1 = A_0 X_0 A_0^* + W_0 \tag{6.56}$$

where X_0 and X_1 are the quaternions associated with the position vectors. The known normalized quaternion $A_0 = \cos\alpha_0 + \mathbf{a}_0 \sin\alpha_0$ corresponds to the rotation, and W_0 is a random quaternion with the mean $E(W_0) = 0$ and known covariance. We introduced the quaternion W_0 because the system may be subjected to random disturbances, for example, the axis and the angle of rotation could be known only up to measurement errors.

We also assume that X_0 is a random quaternion, with the known mean $\hat{X}_0$ and the known covariance. As usually, we consider W_0 and X_0 to be independent.

Equation (6.56) is called the "equation of dynamics" of the system.

We wish to estimate the system's position vector in the state 1 by using a 3-dim measured vector $\mathbf{z}_1$, connected with $\mathbf{x}_1$ by

$$Z_1 = B_1^* X_1 B_1 + V_1 \tag{6.57}$$

where Z_1 is the quaternion associated with $\mathbf{z}_1$, B_1 is a known quaternion, and V_1 is a random quaternion with zero mean and known covariance. The random quaternions Z_1 and V_1 are assumed to be independent of the state quaternion X. Equation (6.57) is called the "equation of observation".

The form of Equation (6.57) means that, in general, measurements could be relative to another reference frame, say S_2, which is obtained from S_1 by $S_2 = B_1 S_1 B_1^*$. Without loss of generality we can assume that $\|B_1\| = 1$.

Using the matrix form of quaternion multiplication (2.6), we can rewrite Equations (6.56) and (6.57). Then, we get the ordinary problem of optimal filtering for a single-stage linear transitions (see [8] p.359). The solution of the problem is known as the Kalman filter [15]. The matrix form of the solution is quite burdensome. We shall present the quaternion approach to the problem.

It is clear that X_1 is also a random quaternion. From Equation (6.56), we get the mean value

$$\bar{X}_1 = A_0 \hat{X}_0 A_0^*. \tag{6.58}$$

It is natural to consider this mean value as an *a priori* prediction of X_1, before taking measurements of state 1.

A reasonable estimate of X_1, which takes into account the prediction $\bar{X}_1$ and the measurements Z_1, is the least-squares estimate, which is optimal and which is denoted by $\hat{X}_1$. This optimal estimate has the value of X which minimizes the scalar loss function

$$f(X) = \|X - \bar{X}_1\|^2 + \|Z_1 - B_1^* X B_1\|^2. \tag{6.59}$$

The mean value of the first term on the right-hand side is the variance of the error of the prediction, and the mean value of the second term is the variance of the error of the measurement. We have to minimize both terms, and this leads to the loss function (6.59).

In control theory, the "weighted-least-squares estimate" is usually used. We present only the least-squares method to make all calculations clearer. The case of the weighted-least-squares estimate was considered by the author in [3].

Using the same technique of differentiation as before, we derive

$$f'_{X,H} = 2\mathrm{Re}\left((X - \bar{X}_1)H^* - (Z_1 - B_1^* X B_1)B_1^* H^* B_1\right)$$
$$= 2\mathrm{Re}\left((X - \bar{X}_1)H^* - B_1(Z_1 - B_1^* X B_1)B_1^* H^*\right) = 0. \tag{6.60}$$

Hence, by applying Lemma 1 to Equation (6.60), we get

$$X - \bar{X}_1 - B_1(Z_1 - B_1^* X B_1)B_1^* = 0.$$

The optimal estimate we are looking for is the following

$$\hat{X}_1 = \frac{1}{2}\left(\bar{X}_1 + B_1 Z_1 B_1^*\right). \tag{6.61}$$

The value $\hat{X}_1$ is the optimal estimate *after* the measurement has been made. It depends on the prior expected value $\bar{X}_1$, which is the estimate of X_1 *before* measurements are made.

The coefficient $1/2$ in (6.61) appeares because B_1 is assumed to be normalized. Otherwise there would be the coefficient $1/(1 + \|B_1\|^4)$.

19.6.3   Optimal filtering and prediction for multi-stage rotations

Now, assume that the system considered above undergoes several discrete rotations, one after another, about the axes $\mathbf{a}_n$ by the angles $2\alpha_n$, where $n = 0, 1, \ldots, N-1$. See Figure 19.8. After every rotation we make measurements of the system's state. We would like to have the optimal estimate of the position vector of the state N.

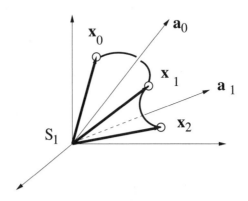

FIGURE 19.8. Motion of the system.

In this case, Equation (6.56) becomes

$$X_{n+1} = A_n X_n A_n^* + W_n, \quad n = 0, 1, \ldots, N-1 \tag{6.62}$$

where $A_n = \cos\alpha_n + \mathbf{a}_n \sin\alpha_n$, the mean values and covariance matrices of X_0 and W_n are known for every n, and the vector of the system's state does not depend on W_n for every n.

We now have several equations of observation of the form (6.57):

$$Z_n = B_n^* X_n B_n + V_n, \quad n = 1, \ldots, N, \tag{6.63}$$

where $\|B_n\| = 1$ for every n. The mean value and the covariance matrix of V_n are known, and V_n does not depend on X_0 and W_n for every n.

As before, the optimal prediction of the state X_N based on previous the measurements is $\bar{X}_N = A_{N-1}\hat{X}_{N-1}A_{N-1}^*$, and we get the optimal estimate

$$\hat{X}_n = \frac{1}{2}\left(\bar{X}_n + B_n Z_n B_n^*\right). \tag{6.64}$$

Estimation of the error covariance matrix before and after the measurements are made can be obtained in the same way as in ordinary control theory.

Another case of the system's motion based on the equation of dynamics (2.20) was considered by the author in [5].

19.6.4 The law of large numbers

An interesting interpretation of the law of large numbers for quaternions follows. Let $A_n = B_n = 1$ for every n, and suppose that the equation of dynamics does not include the random quaternion W. We then have

$$X_{n+1} = X_n \tag{6.65}$$
$$Z_n = X_n + V_n. \tag{6.66}$$

Using equations (6.65) - (6.66), we wish to find a constant quaternion X due to making repeated measurements containing errors. From (6.64) we have the following optimal estimate of X_N:

$$\hat{X}_N = \frac{1}{2}\left(\bar{X}_N + Z_N\right) \tag{6.67}$$

where the prediction $\bar{X}_n$ is given by

$$\bar{X}_N = \hat{X}_{N-1} = \frac{1}{2}\left(\bar{X}_{N-1} + Z_{N-1}\right). \tag{6.68}$$

Substituting (6.68) into (6.67), and repeating the process for all subscripts, we finally arrive at

$$\hat{X}_N = \frac{1}{2^N}\left(\bar{X}_0 + Z_1 + 2Z_2 + 4Z_3 + \cdots + 2^{N-1}Z_N\right). \tag{6.69}$$

This result means that the later measurements have a larger weight.

We have presented the theoretical background of the application of quaternions to different problems in engineering. Quaternion algebra gives a fresh approach to old problems as well, allowing us to focus on essential points, and to use similar treatments for problems from different fields. Using quaternions, we have rather simple and elegant analytic methods which clarify the essence of the problems. Developing the corresponding numerical algorithms and mathematical software has great practical value and is a subject of for special consideration.

19.7 Summary

We presented theoretical background of applications of quaternions to problems from different engineering areas. We used examples mainly from photogrammetry and space navigation, but specialists working in such areas as robotics, computer vision and flight simulators, etc. can recognize their problems immediately and employ the same or similar methods. Many engineers and designers have already discovered the wonderful world of quaternions and have started to use them extensively. Quaternion algebra gives a fresh approach to old problems as well, allowing us to focus on essential points, and to use similar treatments for problems from different fields. Using quaternions we have rather simple and elegant analytic methods which clarify the essence of problems. Developing the corresponding numerical algorithms and mathematical software has a great practical importance and is a subject of a special consideration.

Chapter 20

Clifford Algebras in Electrical Engineering

William E. Baylis

20.1 Introduction

The Maxwell-Lorentz equations that underpin all of electrical engineering are intrinsically relativistic. Even in problems confined to nonrelativistic velocities, the fundamental relativistic symmetries of the underlying theory imply important relations that often hold the keys to solutions of electromagnetic problems and may suggest significant insights into the underlying phenomena.

Electrical engineers usually employ noncovariant notation with vectors in three-dimensional physical space. Unfortunately, such notation is ill-suited to reveal basic relativistic symmetries of the phenomena under investigation. The reason for the engineers' choice is clear: the geometric and physical significance of vectors is much more apparent than that of the indexed tensor elements traditionally employed in relativistic formalism. On the other hand, the notation of differential forms, which has been promoted [12] over the last three decades for use in general relativity, is well adapted for distinguishing topologically invariant properties from those that depend on the metric. But while it certainly offers a covariant notation for problems in special relativity, its very flexibility in distinguishing between forms and vectors requires a level of abstraction that is unnecessary for most problems in metric spaces. Abstractions to metric-free formulations are not generally appreciated by engineers who have little need to distinguish, say, between electromagnetic fields that are 2-forms from those that are 2-vectors.

This paper explores an alternative approach based on Clifford's geometric algebras. Geometric algebras of various sizes have been applied to a variety of engineering problems.[3] Of course the complex numbers themselves are elements of the geometric algebra $Cl_{0,1}$ on the vector space of metric signature -1. We write $\mathbb{C} \simeq Cl_{0,1} \simeq Cl_{0,2}^+$, which also states the equivalence of complex numbers and the subalgebra of even elements of $Cl_{0,2}$. The Clifford algebra $Cl_{0,2} \simeq \mathbb{H} \simeq \mathbb{C}^2$ is the algebra of Hamilton's quaternions, which can be expressed as noncommuting products of pairs of complex numbers. The quaternion algebra $\mathbb{H}$ is useful for describing and calculating rotations in 3-

dimensional space.[1] It and the dual quaternion algebra $\mathbb{H} \oplus \mathbb{H}$, comprising sums $q + \varepsilon q'$ where $q, q' \in \mathbb{H}$ and ε is a commuting nilpotent, are being used for robotics and imaging, as described elsewhere in this volume. While the utility of quaternions in computing three-dimensional orientations was well appreciated in the space program, their usage has mushroomed since their discovery by software writers of video games, for which they allow efficient computations with minimal storage for smooth rotations of solid objects.

One may argue that the algebra isomorphism $\mathbb{H} \simeq C\ell_{0,2}$ does not naturally model the symmetry of physical space $\mathbb{R}^3$, because in $C\ell_{0,2}$ two of the unit quaternions, say $\mathbf{i}, \mathbf{j}$, are selected as basis vectors whereas the third is a bivector: $\mathbf{k} = \mathbf{ij}$. The geometric algebra based on $\mathbb{R}^3$ with signature -3 may be a more natural choice for the quaternions, but because its volume element $\mathbf{ijk} = \mathbf{k}^2 = -1$ is just the negative of the unit scalar, it is a non-universal algebra. Like the universal Clifford algebra $C\ell_{0,2}$ to which it is isomorphic, its vectors all square to negative numbers. This "disease" (as detractors of quaternions referred to it)[1] is avoided in the approach that recognizes the quaternion algebra as isomorphic to the even subalgebra of $C\ell_3$, the geometric algebra of vectors in physical space with signature $+3$: $\mathbb{H} \simeq C\ell_3^+$. Pure quaternions are all bivectors of $C\ell_3$ rather than vectors. The full algebra $C\ell_3$ has been used in the form of the Pauli matrix algebra,[2, 4] in which the Pauli spin matrices represent orthonormal basis vectors of the underlying space, and it is isomorphic to the quaternions over the complex numbers, $C\ell_3 \simeq \mathbb{H} \times \mathbb{C}$. In contrast to $\mathbb{H}$, $C\ell_3$ allows one to distinguish vectors and bivectors, and compared to the complex quaternions, its geometric application and interpretation is usually more intuitive. The complex numbers are just the center (commuting part) of $C\ell_3$.

The algebra $C\ell_3$ is in turn isomorphic to the even subalgebra of $C\ell_{1,3}$, which together with $C\ell_{3,1}$ are geometric algebras based on Minkowski spacetime. Although these algebras are best known to physicists in the matrix forms of the Dirac algebra of relativistic quantum theory, they have also been applied to relativistic problems in classical electrodynamics [8]. Still larger geometric algebras have been used for projective treatments of vision problems, as described in other chapters of this volume, and applications of Clifford analysis and extensions of wavelets appear promising in potential theory and signal processing.

Geometric algebras are algebras of vectors, and in particular, $C\ell_3$ is the algebra of the vectors of physical space and thus the natural extension of the vector analysis most familiar to engineers. The purpose of this chapter is to show that $C\ell_3$ provides a framework that seemlessly extends this familiar vector analysis to an efficient covariant description of relativistic electromagnetic phenomena, thereby providing a powerful but easily accessible tool to electrical engineers. The extension uses paravectors of $C\ell_3$, that is, sums of scalars and vectors, that constitute a four-dimensional linear space with a natural Minkowski spacetime structure. Several explicit applications to electrical engineering problems are given.

20.2   Structure of $C\ell_3$

Geometric algebras assume an associative product of vectors in a metric space that is distributive over addition and satisfies the fundamental axiom for any vector $\mathbf{v}$,

$$\mathbf{vv} = \mathbf{v} \cdot \mathbf{v}, \tag{2.1}$$

that is, the square of the vector is just its square length. Elements of $C\ell_3$ include all products of 3-dimensional vectors and their real linear combinations.

If an orthonormal basis of physical space is written $\{\mathbf{e}_1, \mathbf{e}_2, \mathbf{e}_3\}$, axiom (2.1) implies that

$$\mathbf{e}_j\mathbf{e}_k + \mathbf{e}_k\mathbf{e}_j = 2\delta_{jk}, \tag{2.2}$$

where the Kronecker delta δ_{jk} is the metric tensor of the space. It follows that elements of $C\ell_3$ form a linear space of dimension $2^3 = 8$ spanned by the basis

$$\{1, \mathbf{e}_1, \mathbf{e}_2, \mathbf{e}_3, \mathbf{e}_2\mathbf{e}_3, \mathbf{e}_3\mathbf{e}_1, \mathbf{e}_1\mathbf{e}_2, \mathbf{e}_1\mathbf{e}_2\mathbf{e}_3\} \text{ over } \mathbb{R}$$

taken over the reals. Subspaces include the real scalars with the basis $\{1\}$, the real vectors of physical space with the basis $\{\mathbf{e}_1, \mathbf{e}_2, \mathbf{e}_3\}$, the bivectors with the basis $\{\mathbf{e}_2\mathbf{e}_3, \mathbf{e}_3\mathbf{e}_1, \mathbf{e}_1\mathbf{e}_2\}$, the trivectors with the basis $\{\mathbf{e}_1\mathbf{e}_2\mathbf{e}_3\}$, the quaternions (even elements) spanned by $\{1, \mathbf{e}_2\mathbf{e}_3, \mathbf{e}_3\mathbf{e}_1, \mathbf{e}_1\mathbf{e}_2\}$, and the *paravectors*,[11] spanned by $\{1, \mathbf{e}_1, \mathbf{e}_2, \mathbf{e}_3\}$.

The center of $C\ell_3$ is spanned by $\{1, \mathbf{e}_1\mathbf{e}_2\mathbf{e}_3\}$, and since $(\mathbf{e}_1\mathbf{e}_2\mathbf{e}_3)^2 = -1$, the volume element itself can be identified with the unit imaginary i. With this identification, bivectors are imaginary vectors, for example

$$\mathbf{e}_1\mathbf{e}_2 = \mathbf{e}_1\mathbf{e}_2\mathbf{e}_3\mathbf{e}_3 = i\mathbf{e}_3. \tag{2.3}$$

Every element of $C\ell_3$ is thus a complex paravector, and the 8-dimensional real linear space of $C\ell_3$ is reduced to the 4-dimensional space spanned by

$$\{\mathbf{e}_0, \mathbf{e}_1, \mathbf{e}_2, \mathbf{e}_3\} \text{ over } \mathbb{C}, \tag{2.4}$$

where to emphasize the role of paravectors as vectors in a space of four dimensions, I have defined $\mathbf{e}_0 \equiv 1$. This definition lets us express paravectors with the summation convention

$$x = x^\mu \mathbf{e}_\mu = x^0 + x^k \mathbf{e}_k, \tag{2.5}$$

where repeated Greek indices (one upper, one lower) are summed over values $0, 1, 2, 3$, and repeated Latin indices are summed over values $1, 2, 3$.

20.2.1   Clifford dual and conjugations

The Clifford dual of any element $x \in C\ell_3$ is conveniently defined by

$$^*x = x\,(\mathbf{e}_1\mathbf{e}_2\mathbf{e}_3)^{-1} = -ix. \tag{2.6}$$

In particular, the dual of a bivector is the orthogonal vector, for example

$$^*(\mathbf{e}_1\mathbf{e}_2) = \mathbf{e}_3 \,. \tag{2.7}$$

The antiautomorphic conjugations (involutions) of elements $x = x^\mu\mathbf{e}_\mu$ of $C\!\ell_3$ are defined by

$$\text{Clifford conjugation } x \to \bar{x} = x^0\mathbf{e}_0 - x^k\mathbf{e}_k \tag{2.8}$$

$$\text{Hermitian conjugation } x \to x^\dagger = x^{\mu*}\mathbf{e}_\mu \,. \tag{2.9}$$

They can be combined into the *grade automorphism* $x \to \bar{x}^\dagger$ and used to isolate different parts of an element:

$$\text{scalar part } \langle x \rangle_S = \frac{1}{2}\left(x + \bar{x}\right) \tag{2.10}$$

$$\text{vector part } \langle x \rangle_V = \frac{1}{2}\left(x - \bar{x}\right) \tag{2.11}$$

$$\text{real part } \langle x \rangle_{I\!R} = \frac{1}{2}\left(x + x^\dagger\right) \tag{2.12}$$

$$\text{imaginary part } \langle x \rangle_\Im = \frac{1}{2}\left(x - x^\dagger\right) \tag{2.13}$$

$$\text{even part } \langle x \rangle_+ = \frac{1}{2}\left(x + \bar{x}^\dagger\right) \tag{2.14}$$

$$\text{odd part } \langle x \rangle_- = \frac{1}{2}\left(x - \bar{x}^\dagger\right) \,. \tag{2.15}$$

20.3   Paravector Model of Spacetime

The metric of 4-dimensional paravector space is determined by the Euclidean metric of physical space. We note that although x^2 is not generally a scalar, the product $x\bar{x}$ always is and can be used as the quadratic form to specify the inverse of any invertible element:

$$x^{-1} = \frac{\bar{x}}{x\bar{x}} \,, \tag{3.16}$$

If (and only if) $x\bar{x} = 0$, the element x is not invertible. The associated symmetric bilinear form for elements $x = x^\mu\mathbf{e}_\mu$ and $y = y^\nu\mathbf{e}_\nu$ is found from the quadratic forms of $x + y$, x, and y :

$$\langle x\bar{y} \rangle_S = \frac{1}{2}\left(x\bar{y} + y\bar{x}\right) \tag{3.17}$$

$$\equiv x^\mu y^\nu \eta_{\mu\nu} \,, \tag{3.18}$$

where the metric-tensor elements

$$\eta_{\mu\nu} = \langle \mathbf{e}_\mu\bar{\mathbf{e}}_\nu \rangle_S = \begin{cases} 1, & \mu = \nu = 0 \\ -1, & \mu = \nu = 1,2,3 \\ 0, & \mu \neq \nu \end{cases} \tag{3.19}$$

are seen to take the Minkowski spacetime form.

Paravectors of physical space thus have a natural spacetime metric and can serve to model vectors in spacetime. Their scalar parts are their time components. The paravectors x, y are said to be *orthogonal* iff $\langle x\bar{y} \rangle_S = 0$, and real paravectors can be classified as timelike, spacelike, and null (or lightlike) as usual. Some examples are

- spacetime momentum : $p = E/c + \mathbf{p} = mc u$ is timelike: $p\bar{p} = m^2 c^2 > 0$.

- photon wave vector: $k = \frac{\omega}{c}\left(1 + \hat{\mathbf{k}}\right)$ is lightlike (null): $k\bar{k} = 0$. It is orthogonal to itself!

- charge current $j = \rho c + \mathbf{j}$ is timelike.

- paravector potential $A = \phi/c + \mathbf{A}$ (can be lightlike, spacelike, or time-like; it depends on the gauge choice).

- Gradient operator $\partial = \mathbf{e}_\mu \partial^\mu = c^{-1}\partial_t - \nabla$.

20.3.1 Spacetime planes: biparavectors

Linear combinations of paravectors p and q lie in the spacetime plane given by the *biparavector* $\langle p\bar{q} \rangle_V = p^\mu q^\nu \langle \mathbf{e}_\mu \bar{\mathbf{e}}_\nu \rangle_V$. The biparavector space is the span of $\{\mathbf{e}_1\bar{\mathbf{e}}_0, \mathbf{e}_2\bar{\mathbf{e}}_0, \mathbf{e}_3\bar{\mathbf{e}}_0, \mathbf{e}_2\bar{\mathbf{e}}_3, \mathbf{e}_3\bar{\mathbf{e}}_1, \mathbf{e}_1\bar{\mathbf{e}}_2\}$. It is seen to be the direct sum of vector and bivector spaces of $\mathcal{C}\ell_3$. Unit biparavectors are operators that rotate paravectors to orthogonal directions in a spacetime plane:

- $\mathbf{e}_1\bar{\mathbf{e}}_2 \left(v^1\mathbf{e}_1 + v^2\mathbf{e}_2\right) = v^1\mathbf{e}_2 - v^2\mathbf{e}_1$ elliptic (spacelike) plane

- $\mathbf{e}_1\bar{\mathbf{e}}_0 \left(v^1\mathbf{e}_1 + v^0\mathbf{e}_0\right) = v^1\mathbf{e}_0 + v^0\mathbf{e}_1$ hyperbolic (timelike) plane.

Imaginary biparavectors such as $\mathbf{e}_1\bar{\mathbf{e}}_2$ are bivectors of $\mathcal{C}\ell_3$ and generate rotations, whereas real biparavectors such as $\mathbf{e}_1\bar{\mathbf{e}}_0$ are vectors and genera-te boosts (see the subsection "Lorentz Transformations and Covariance" below).

Clifford duals in $\mathcal{C}\ell_3$ can be given the spacetime form ${}^*x \equiv x \left(\mathbf{e}_1\bar{\mathbf{e}}_2\mathbf{e}_3\bar{\mathbf{e}}_0\right)$. They generalize Hodge duals to arbitrary (not necessarily homogeneous) elements. Note that dual spacetime planes are orthogonal in the sense that every paravector in one is orthogonal to every paravector in the other. For example, the spacetime planes $\mathbf{e}_1\bar{\mathbf{e}}_2$ and $\mathbf{e}_0\bar{\mathbf{e}}_3$ are dual and orthogonal to each other:

$$
\begin{aligned}
{}^*\left(\mathbf{e}_1\bar{\mathbf{e}}_2\right) &= \left(\mathbf{e}_1\bar{\mathbf{e}}_2\right)\left(\mathbf{e}_1\bar{\mathbf{e}}_2\mathbf{e}_3\bar{\mathbf{e}}_0\right) \\
&= \mathbf{e}_0\bar{\mathbf{e}}_3 .
\end{aligned}
\tag{3.20}
$$

A linear combination of dual biparavectors is also a biparavector, but it represents a pair of orthogonal spacetime planes rather than a single plane. Biparavectors that represent single spacetime planes are called *simple* and are distinguished by the fact that they square to real scalars.

20.3.2 Lorentz transformations and covariance

Lorentz transformations of a spacetime vector p are defined to be linear transformations that leave p real and $p\bar{p}$ invariant. Restricted (proper, orthochronous) transformations are spacetime rotations. They take the spinorial form

$$p \to LpL^\dagger, \ L\bar{L} = 1. \tag{3.21}$$

The transformation elements $L \in SL(2, \mathbb{C})$ can be expressed in several equivalent forms: $L = BR = \pm \exp(\mathbf{W}/2)$, $\mathbf{W} = \frac{1}{2}W^{\mu\nu}\langle \mathbf{e}_\mu \bar{\mathbf{e}}_\nu \rangle_V$, where $B = B^\dagger$ is a boost and $R = \bar{R}^\dagger$ is a spatial rotation. In particular, spatial rotations, which form the subgroup $SU(2) \subset SL(2, \mathbb{C})$, are realized by transformation elements of the form

$$R = \exp(\mathbf{\Theta}/2), \tag{3.22}$$

where $\mathbf{\Theta}$ is a bivector whose magnitude is the angle of rotation and whose "direction" gives the rotation plane. The elements of the basis tetrad $\{\mathbf{e}_\mu\}$ are paravectors and therefore transform according to (3.21). The transformed basis tetrad comprises $\mathbf{u}_\mu = L\mathbf{e}_\mu L^\dagger$, which includes the proper velocity $u = \mathbf{u}_0 = LL^\dagger$ of the transformed tetrad.

From the form (3.21) one can derive the transformation law for biparavectors. Thus, if p and q are spacetime vectors,

$$p\bar{q} \to \left(LpL^\dagger\right)\overline{\left(LqL^\dagger\right)} = Lp\bar{q}\bar{L}. \tag{3.23}$$

It may be noted that the transformation biparavector $\mathbf{W}$ can be expressed in the biparavector basis of either the original or transformed tetrad; the coefficients $W^{\mu\nu}$ are the same for both cases. This is easily proved: $L = LL\bar{L} = \exp(L\mathbf{W}\bar{L}) = \exp\left(\frac{1}{2}W^{\mu\nu}\langle \mathbf{u}_\mu \bar{\mathbf{u}}_\nu \rangle_V\right)$ and works for boosts, rotations, or a combination of the two.

The Lorentz transformations (3.21) and (3.23) can be passive, active, or a combination. Only the relative orientation and motion of the observed frame with respect to the observer is significant. Generally, a tetrad $\{\mathbf{u}^\mu\}$

Lounesto[10] has pointed out the need for the minus sign in the expression $L = \pm\exp(\mathbf{W}/2)$, since expressions of the form $-\exp(\mathbf{W}'/2)$ cannot be put in the form $+\exp(\mathbf{W}/2)$ when $(\mathbf{W}')^2 = 0$, $\mathbf{W}' \neq 0$. We note (i) that these are the only transformations for which the minus sign is required, (ii) that they are a set of measure zero, and (iii) even they can be reached as a limit of the transformations $+\exp(\mathbf{W}/2)$.

represents a frame that is moving with proper velocity $\mathbf{u}_0$ with respect to the observer. Usually any observer will reduce expressions to his or her own frame and determine coefficients on the rest tetrad $\{\mathbf{e}_\mu\}$ with $\mathbf{e}_0 = 1$. It is important to distinguish absolute frames from tetrads, which indicate frames relative to the observer. Two observers in relative motion can use the rest tetrad $\{\mathbf{e}_\mu\}$ to describe phenomena in their distinct rest frames. (See [5] for more discussion.)

Any split of a paravector into scalar and vector parts, or of a biparavector into real and imaginary parts, is not covariant, since different observers will determine different components in their rest frames. Nevertheless, in $\mathcal{C}\ell_3$, relativistic relations can be expressed without splits and without components, relating whole paravectors and their products so that all terms transform in the same way. Such relations are covariant. The fundamental equations of physics are all expected to be covariant. A number of examples are given below.

20.4   Using Relativity at Low Speeds

20.4.1   Maxwell and continuity equations

The paravector model of spacetime in $\mathcal{C}\ell_3$ offers insights and simplifications at low speeds as well as high. Covariance restricts equation forms and provides for compact notation and derivations. For example, Maxwell's (microscopic) equation

$$\bar{\partial}\mathbf{F} = Z_0 \bar{j} \tag{4.24}$$

relates the gradient of the biparavector electromagnetic field (*Faraday*[12]) $\mathbf{F}$

$$\mathbf{F} = c \left\langle \partial \bar{A} \right\rangle_V = \mathbf{E} + ic\mathbf{B} \tag{4.25}$$

to the paravector current j, where the impedance of free space is $Z_0 = (\varepsilon_0 c)^{-1} = 4\pi \times \dot{3}0$ Ohm. The relations (4.24) and (4.25) are covariant: both sides of each transform in the same way, and both are valid in every inertial frame. Maxwell's equation is easily split into real and imaginary scalar and vector parts to give the usual four vector equations, and the relation (4.25) for the field is readily expanded in unit biparavectors $\langle \mathbf{e}_\mu \bar{\mathbf{e}}_\nu \rangle_V$ to relate tensor components $F^{\mu\nu}$ to derivatives of the paravector potential as well as to the electric and magnetic fields in any frame. Maxwell's equation (4.24) leads in one step to the continuity equation when one realizes that $\partial\bar{\partial}$ is a scalar operator and $\mathbf{F}$ has no scalar part:

$$\left\langle \partial\bar{\partial}\mathbf{F} \right\rangle_S = 0 = \left\langle \partial\bar{j} \right\rangle_S . \tag{4.26}$$

The number $\dot{3} \equiv 2.99792458$ arises from the defined value of the speed of light.

20.4.2 Conducting screen

The covariant form (4.24) of Maxwell's equation immediately gives current densities j for any field configuration $\mathbf{F}$. For example, for an ideal plane capacitor, the field is $\mathbf{F} = \mathbf{e}_1 E_0 \left[\theta\left(x^1\right) - \theta\left(x^1 - d\right) \right]$, where $\theta\left(x^1\right)$ is the Heaviside step function and we have assumed a uniform field $\mathbf{e}_1 E_0$ in the region $0 < x^1 < d$. Differentiation of the step functions gives Dirac delta functions for the charge density,

$$\bar{j} = c\varepsilon_0 \bar{\partial}\mathbf{F} = c\varepsilon_0 E_0 \left[\delta\left(x^1\right) - \delta\left(x^1 - d\right) \right]$$
$$= \rho c\,, \tag{4.27}$$

which implies opposite surface charges $\varepsilon_0 E_0$ at the two plates. Similarly, the uniform interior magnetic field $\mathbf{F} = icB_0\mathbf{e}_3\theta\left(a - r\right)$ of a solenoid of cylindrical radius a requires the current density

$$\bar{j} = -i\hat{\mathbf{r}}\mathbf{e}_3 c^2 \varepsilon_0 B_0 \delta\left(r - a\right) \tag{4.28}$$

$$\mathbf{j} = \hat{\phi}\frac{B_0}{\mu_0}\delta\left(r - a\right)\,, \tag{4.29}$$

where $\hat{\phi} = \mathbf{e}_3 \times \hat{\mathbf{r}}$, and thus an azimuthal current of B_0/μ_0 per unit length in the surface of the solenoid.

Consider next an external field incident on a thin conducting screen. Let $\mathbf{n}$ be a unit vector normal to the screen (see Fig. 20.1) and let $\mathbf{F}_-\left(x\right)$ and $\mathbf{F}_+\left(x\right)$ be the electromagnetic field on the two sides as a function of the spacetime position x.

EndExpansion

The total field can be written

$$\mathbf{F}(x) = \mathbf{F}_-(x) + \theta\left(\langle x\mathbf{n}\rangle_S\right)\left[\mathbf{F}_+(x) - \mathbf{F}_-(x)\right]\,, \tag{4.30}$$

where

$$\bar{\partial}\mathbf{F}_\pm(x) = 0\,. \tag{4.31}$$

Maxwell's equation gives the current in the screen

$$\bar{j}(x) = \varepsilon_0 c\bar{\partial}\mathbf{F}(x) = \varepsilon_0 c\delta\left(\langle x\mathbf{n}\rangle_S\right)\mathbf{n}\left[\mathbf{F}_+(x) - \mathbf{F}_-(x)\right]\,. \tag{4.32}$$

Since j is real, the real components of $\mathbf{F}_+$ and $\mathbf{F}_-$ perpendicular to $\mathbf{n}$, as well as the imaginary components parallel to $\mathbf{n}$, must be equal at $\langle x\mathbf{n}\rangle_S = 0$:

$$(\mathbf{B}_+ - \mathbf{B}_-) \cdot \mathbf{n} = 0 = (\mathbf{E}_+ - \mathbf{E}_-) \times \mathbf{n}\,. \tag{4.33}$$

In particular, if $\mathbf{F}_+$ vanishes, then the components $\mathbf{B}_- \cdot \mathbf{n}$ and $\mathbf{E}_- \times \mathbf{n}$ of $\mathbf{F}_-$ must vanish in the screen.

The current j in the screen can be expressed as

$$j(x) \equiv K(x)\,\delta(\mathbf{x} \cdot \mathbf{n})\,, \tag{4.34}$$

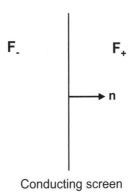

Conducting screen

FIGURE 20.1. The screen lies in the plane at $\langle xn \rangle_S \equiv \mathbf{x} \cdot \mathbf{n} = 0$; n is the vector dual and hence normal to the screen.

where $K(x) = \Sigma(x)\, c + \mathbf{K}(x)$ is the *surface current* in the screen

$$K(x) = \varepsilon_0 c \left[\mathbf{F}_+(x) - \mathbf{F}_-(x) \right] \mathbf{n}, \ \langle xn \rangle_S = 0, \tag{4.35}$$

comprising both a vector current $\mathbf{K}(x)$ and c times the scalar *surface charge* $\Sigma(x)$:

$$\Sigma(x) = \varepsilon_0 \left[\mathbf{E}_+(x) - \mathbf{E}_-(x) \right] \cdot \mathbf{n}$$
$$\mathbf{K}(x) = \mathbf{n} \times \left[\mathbf{B}_+(x) - \mathbf{B}_-(x) \right] / \mu. \tag{4.36}$$

Let $\mathbf{F}_0$ be the radiation field incident on the screen from the left, $\langle xn \rangle_S < 0$, and assume the screen fully blocks the radiation: $\mathbf{F}_+ = 0$. The currents in the screen must radiate the field $-\mathbf{F}_0$ toward the right in order to cancel $\mathbf{F}_0$. By reflection symmetry, whatever field the current in the screen radiates to the right, it also radiates to the left. That is, the part $\mathbf{F}_- - \mathbf{F}_0$ of the field on the left of the screen that is not the incident field $\mathbf{F}_0$ must be the reflection in the screen of $-\mathbf{F}_0$ on the right. Remembering that $\mathbf{n}\bar{x}^\dagger\mathbf{n}$ is the reflection of any element x in the *in* plane, we find the reflected field at the real spacetime position x to be

$$\mathbf{F}_1(x) = \mathbf{F}_-(x) - \mathbf{F}_0(x)$$
$$= -\mathbf{n}\bar{\mathbf{F}}_0^\dagger(\mathbf{n}\bar{x}\mathbf{n})\,\mathbf{n} = \mathbf{n}\mathbf{F}_0^\dagger(\mathbf{n}\bar{x}\mathbf{n})\,\mathbf{n}. \tag{4.37}$$

In the screen, $\langle x\mathbf{n}\rangle_S = \mathbf{x}\cdot\mathbf{n} = 0$ and $x = \mathbf{n}\bar{x}\mathbf{n}$. Therefore,

$$K(x) = -\varepsilon_0 c\mathbf{F}_-(x)\,\mathbf{n} = -\varepsilon_0 c\left[\mathbf{F}_0(x)\,\mathbf{n} + \mathbf{n}\mathbf{F}_0^\dagger(x)\right]$$
$$= -2\varepsilon_0 c\,\langle\mathbf{F}_0(x)\,\mathbf{n}\rangle_{I\!R}\;. \tag{4.38}$$

A plane wave $\mathbf{F}_0 = \left(1 + \hat{\mathbf{k}}_0\right)\mathbf{E}_0(s)$, where $s = \langle k_0\bar{x}\rangle_S$, thus induces a current
$$K(x) = 2\varepsilon_0 c\left[\left(\hat{\mathbf{k}}_0\times\mathbf{E}_0\right)\times\mathbf{n} - \mathbf{E}_0\cdot\mathbf{n}\right]\;.$$

The reflected wave is

$$\mathbf{F}_1(x) = -\mathbf{n}\bar{\mathbf{F}}_0^\dagger(\mathbf{n}\bar{x}\mathbf{n})\,\mathbf{n} = \left(1 + \hat{\mathbf{k}}_1\right)\mathbf{E}_1(s_1)\;, \tag{4.39}$$

where $\mathbf{k}_1 = -\mathbf{n}k_0\mathbf{n}$, $\mathbf{E}_1 = \mathbf{n}\mathbf{E}_0\mathbf{n}$, and $s_1 = \langle\bar{k}_0\mathbf{n}\bar{x}\mathbf{n}\rangle_S = \langle k_1\bar{x}\rangle_S$ with $k_1 = \omega/c + \mathbf{k}_1 = \mathbf{n}\bar{k}_0\mathbf{n}$. The simplest configuration to handle is normal incidence: let $\hat{\mathbf{k}}_0 = \mathbf{n}$. Then $\mathbf{k}_1 = -\mathbf{k}_0$, $\mathbf{E}_1 = -\mathbf{E}_0$, $s_1 = \omega t - \mathbf{k}_1\cdot\mathbf{x}$, and

$$\mathbf{F}_1 = -\left(1 - \hat{\mathbf{k}}_0\right)\mathbf{E}_0(s_1)$$
$$K(x) = 2\varepsilon_0 c\mathbf{E}_0(\omega t)\;. \tag{4.40}$$

More interesting is oblique incidence. Let $\mathbf{n} = \mathbf{e}_3$, $\hat{\mathbf{k}}_0 = \mathbf{e}_3\cos\theta - \mathbf{e}_2\sin\theta$, and $\mathbf{E}_0 = E_0(s)\left(\mathbf{e}_2\cos\theta + \mathbf{e}_3\sin\theta\right)$. Then

$$\mathbf{F}_0(x) = \left(1 + \hat{\mathbf{k}}_0\right)\mathbf{E}_0(s)$$
$$= E_0(s)\left(\mathbf{e}_2\cos\theta + \mathbf{e}_3\sin\theta - \mathbf{e}_2\mathbf{e}_3\right)\;, \tag{4.41}$$

where

$$s = \langle k_0\bar{x}\rangle_S = \omega t - \frac{\omega}{c}\left(x^3\cos\theta - x^2\sin\theta\right)\;. \tag{4.42}$$

The surface current in the screen is

$$K(x) = -2\varepsilon_0 c\,\langle\mathbf{F}_0(x)\,\mathbf{e}_3\rangle_{I\!R} = 2\varepsilon_0 cE_0(s)\left(\mathbf{e}_2 - \sin\theta\right)\;, \tag{4.43}$$

with $x^3 = 0$, and the reflected wave is

$$\mathbf{F}_1(x) = -\mathbf{e}_3\bar{\mathbf{F}}_0^\dagger\mathbf{e}_3$$
$$= -E_0(s_1)\left(\mathbf{e}_2\cos\theta - \mathbf{e}_3\sin\theta + \mathbf{e}_2\mathbf{e}_3\right)\;, \tag{4.44}$$

where $s_1 = \omega t - \mathbf{k}_1\cdot\mathbf{x} = \omega t + \left(x^3\cos\theta + x^2\sin\theta\right)\omega/c$. For a monochromatic wave with, for example, $E_0(s) = E_0(0)\cos s$, we thus find a ripple of current and a fluctuating charge density in the screen.

The same result is obtained by boosting the current $j(x) = K(x)\,\delta(\langle x\mathbf{n}\rangle_S)$ and $\mathbf{F}$ for the case of normal incidence in the $-\mathbf{e}_2$ direction by an appropriate amount. When the incident radiation has the orthogonal polarization, namely $\mathbf{E} = E_0(s)\,\mathbf{e}_1$, but the same propagation vector as in the oblique case above, then $K = 2\varepsilon_0 c E_0(s)\,\mathbf{e}_1\cos\theta$.

The paravector model is also useful for finding the general solution of Maxwell's equation (4.24) for the electromagnetic field in terms of the current (Jeffimenko), which is easily derived in the compact form (see [5] for details):

$$\mathbf{F}(x) = \frac{Z_0}{4\pi}\int d^4y\,\frac{\partial\bar{j}(y)}{R^0}\delta\!\left(R^0 - |\mathbf{R}|\right)$$

$$= \frac{Z_0}{4\pi}\int d^3\mathbf{y}\left\langle\frac{\hat{\mathbf{R}}\bar{j}(y) + R\partial_0\bar{j}(y)}{R^2}\right\rangle_{ret,V}. \tag{4.45}$$

The real and imaginary parts of this result are known as the generalized Coulomb-Faraday and generalized Biot-Savart laws, respectively, and were first derived by O. Jeffimenko.[7]

20.4.3 Lorentz force

The ability to easily relate inertial frames provides insight into many relations and shows, for example, how the Lorentz-force equation follows directly from covariance and the definition of the electric field. The electric field at a point x is defined to be the force per unit charge on a test charge at rest at x. This corresponds to the Lorentz force $\dot{p}_{\text{rest}} = e\mathbf{E}_{\text{rest}}$ in the instantaneous rest frame of the charge e, where $\mathbf{E}_{\text{rest}} = \langle\mathbf{F}_{\text{rest}}\mathbf{e}_0\rangle_{I\!R}$ is the electric field at the position of the charge and $\mathbf{F}$ is the covariant electromagnetic field.

Let Λ be the particular Lorentz transformation that boosts the charge from rest to its motion in the lab, where the field is $\mathbf{F} = \Lambda\mathbf{F}_{\text{rest}}\bar{\Lambda}$. By letting the dot continue to refer to differentiation with respect to the rest-frame time (i.e., the Lorentz-invariant proper time), we find

$$\dot{p} = \Lambda\dot{p}_{\text{rest}}\Lambda^\dagger = e\Lambda\mathbf{E}_{\text{rest}}\Lambda^\dagger$$

$$= e\Lambda\,\langle\mathbf{F}_{\text{rest}}\rangle_{I\!R}\,\Lambda^\dagger = e\,\langle\Lambda\mathbf{F}_{\text{rest}}\Lambda^\dagger\rangle_{I\!R}$$

$$= e\,\langle\mathbf{F}\Lambda\Lambda^\dagger\rangle_{I\!R} = e\,\langle\mathbf{F}u\rangle_{I\!R}. \tag{4.46}$$

This is the paravector form of the usual Lorentz-force equation, from which it is easily read that $\dot{p}$ lies in the spacetime plane of $\mathbf{F}$ (assuming the field $\mathbf{F}$ is simple) in the direction orthogonal to the projection of u in that plane. The usual component form $\dot{p}^\mu = eF^{\mu\nu}u_\nu$ is recovered by expanding p and u in the paravector basis elements $\mathbf{e}_\lambda$, expanding $\mathbf{F}$ in unit biparavectors $\langle\mathbf{e}_\mu\bar{\mathbf{e}}_\nu\rangle_V$, using the asymmetry of its indices, and noting

$$\langle\langle\mathbf{e}_\mu\bar{\mathbf{e}}_\nu\rangle_V\,\mathbf{e}_\lambda\rangle_{I\!R} = \mathbf{e}_\mu\,\langle\bar{\mathbf{e}}_\nu\mathbf{e}_\lambda\rangle_S - \mathbf{e}_\nu\,\langle\bar{\mathbf{e}}_\mu\mathbf{e}_\lambda\rangle_S$$

$$= \eta_{\nu\lambda}\mathbf{e}_\mu - \eta_{\mu\lambda}\mathbf{e}_\nu. \tag{4.47}$$

20.4.4   Wave guides

Lorentz transformations are useful not only for appreciating basic symmetries of electromagnetic theory, but also for deriving and understanding practical results such as wave-guide modes. For example, we can start with a standing wave between two parallel plane conductors at $y = 0$ and $y = \pi c/\omega_0$:

$$\mathbf{F} = cB_0\mathbf{e}_1\mathbf{e}_2 \cos\left[\omega_0\left(t - \mathbf{e}_2 y/c\right)\right] \tag{4.48}$$

$$= icB_0\mathbf{e}_3 \cos\omega_0 t \cos\omega_0 y/c + cB_0\mathbf{e}_1 \sin\omega_0 t \sin\omega_0 y/c. \tag{4.49}$$

A boost by $L = u^{1/2}$ in the $\mathbf{e}_3$ direction, with $u = \gamma\left(1 + v\mathbf{e}_3/c\right)$, yields a field in the TE_{01} mode:

$$\mathbf{F} \to LF\bar{L} \tag{4.50}$$

$$= icB_0\mathbf{e}_3 \cos\omega_0 t \cos\omega_0 y/c + ucB_0\mathbf{e}_1 \sin\omega_0 t \sin\omega_0 y/c$$

$$= iB_0\left(c\mathbf{e}_3 \cos\omega_0 t \cos\omega_0 y/c + \gamma v\mathbf{e}_2 \sin\omega_0 t \sin\omega_0 y/c\right)$$

$$+ \gamma B_0\mathbf{e}_1 \sin\omega_0 t \sin\omega_0 y/c, \tag{4.51}$$

where in terms of the transformed (primed) coordinates,

$$\omega_0 t = \omega t' - k_g z',\, y = y' \tag{4.52}$$

$$\omega = \gamma\omega_0 = \sqrt{\omega_0^2 + k_g^2 c^2},\, k_g = \gamma v\omega_0/c^2. \tag{4.53}$$

One sees that the standing-wave frequency is just the cut-off frequency, and that the dispersion relation $\omega^2 = \omega_0^2 + k_g^2 c^2$ is a direct result of the Lorentz-factor relation, $\gamma^2 = 1 + \gamma^2 v^2/c^2$.

20.5   Relativity at High Speeds

Of course there are also applications in electrical engineering where relativistic treatments are not only convenient but necessary. These include the relativistic motion of charges in strong electromagnetic fields[9] and radiation from charges in relativistic motion.

20.5.1   Motion of charges in fields

The equation of motion (4.46) that we need to solve is simpler in its spinorial form, which governs the time evolution of the active Lorentz transformation $\Lambda(\tau) \in SL(2, \mathbb{C})$ that takes the charge from rest to its momentum p (and orientation) at proper time τ:

$$p = \Lambda mc\Lambda^\dagger. \tag{5.54}$$

The transformation element $\Lambda(\tau)$ is the spinorial form of the Lorentz transformation and is known as the eigenspinor of the charge. It is related to the charge's proper velocity by

$$u = \Lambda e_0 \Lambda^\dagger = \Lambda\Lambda^\dagger. \tag{5.55}$$

From the unimodular property of Λ, namely $\Lambda\bar{\Lambda} = 1$, the eigenspinor is seen to satisfy an equation of motion of the form

$$\dot{\Lambda} = \frac{1}{2}\Omega\Lambda \tag{5.56}$$

where the biparavector $\Omega \equiv 2\dot{\Lambda}\bar{\Lambda}$ gives the spacetime rotation rate of the charge. By taking the proper-time derivative of p (5.54) and substituting (5.56), we obtain

$$\dot{p} = \dot{\Lambda}mc\Lambda^\dagger + \Lambda mc\dot{\Lambda}^\dagger \tag{5.57}$$

$$= mc\left\langle \Omega\Lambda\Lambda^\dagger \right\rangle_{\mathbb{R}} = mc\left\langle \Omega u \right\rangle_{\mathbb{R}}. \tag{5.58}$$

This has the same form as the Lorentz-force equation (4.46), and a comparison identifies the spacetime rotation rate of the charge as proportional to the field $\mathbf{F}$ at the position x of the charge: $\Omega = e\mathbf{F}/mc$. Thus the Lorentz-force equation (4.46) can be replaced by the simpler spinorial form

$$\dot{\Lambda} = \frac{e\mathbf{F}}{2mc}\Lambda. \tag{5.59}$$

If the direction of the field $\mathbf{F}$ is constant at the position of the charge, the integral of (5.59) gives directly

$$\Lambda(\tau) = \exp\left(\frac{e}{2mc}\int_0^\tau d\tau' \mathbf{F}(\tau')\right)\Lambda(0). \tag{5.60}$$

Solutions can also be found[5] for charge motion in a plane-wave pulse, for which the field has the form of the null flag[13]

$$\mathbf{F} = \left(1 + \hat{\mathbf{k}}\right)\mathbf{E}(s), \tag{5.61}$$

where $\hat{\mathbf{k}}$ is the propagation direction and $\mathbf{E}(s)$ is the real electric field, taken perpendicular to $\hat{\mathbf{k}}$, as a function of the scalar parameter $s = \langle k\bar{x}\rangle_S = \omega t - \mathbf{k}\cdot\mathbf{x}$. The null paravector $k = \omega/c + \mathbf{k} = (\omega/c)\left(1 + \hat{\mathbf{k}}\right)$ is the nominal propagation paravector in the direction $\hat{\mathbf{k}}$, but since the wave is generally not monochromatic, the actual value of the scalar ω/c can be chosen for convenience. The functional form of $\mathbf{E}(s)$ is arbitrary, but for a circularly polarized Gaussian pulse it can have the form

$$\mathbf{E}(s) = \mathbf{E}(0)\exp\left(is\hat{\mathbf{k}} - as^2\right), \tag{5.62}$$

where α is a constant proportional to the inverse width of the wave packet. The field (5.61) is seen to be a null biparavector ($\mathbf{F}\bar{\mathbf{F}} = 0$) and hence simple. Its vector part is the electric field $\mathbf{E}\,(s)$ and its bivector part gives the plane whose normal is the magnetic field: $\hat{\mathbf{k}}\mathbf{E}\,(s) = ic\mathbf{B}$. The solution makes use of the important projector properties of the factor $1 + \hat{\mathbf{k}}$, namely

$$\left(1 - \hat{\mathbf{k}}\right)\mathbf{F} = \mathbf{F}\left(1 + \hat{\mathbf{k}}\right) = 0\,. \tag{5.63}$$

As simple as the spinorial equation (5.59) is, it appears at first to be impossible to solve, since $\mathbf{F}$ is to be evaluated at the spacetime position of the charge, and the position is known only after the equation is solved and u is integrated. However, a remarkable symmetry permits direct solution. The key to the solution is the surprising invariance of the wave paravector k not only in the lab, but also in the sequence of instantaneous rest frames of the accelerating charge. Because $\bar{k}c = \omega\left(1 - \hat{\mathbf{k}}\right)$, the spinorial Lorentz-force equation (5.59) in the field (5.61) gives

$$\bar{k}\dot{\Lambda} = \frac{e\bar{k}\mathbf{F}}{2mc}\Lambda = 0\,. \tag{5.64}$$

From conjugations of this result, $\overline{\dot{\Lambda}}k = 0 = k\overline{\dot{\Lambda}}^{\dagger}$, and it follows that k in the frame of the charge is constant:

$$k_{\text{rest}} = \frac{\omega_{\text{rest}}}{c}\left(1 + \hat{\mathbf{k}}\right) = \bar{\Lambda}k\bar{\Lambda}^{\dagger} = \text{const.} \tag{5.65}$$

Furthermore, the proper time-rate of change of the Lorentz scalar s is simply the fixed rest-frame frequency

$$\dot{s} = c\,\langle k\bar{u}\rangle_S = \omega_{\text{rest}}\,. \tag{5.66}$$

Consequently, the evolution (5.56) of the eigenspinor can be expressed

$$\frac{d\Lambda}{ds} = \frac{1}{\dot{s}}\dot{\Lambda} = \frac{e\mathbf{F}\Lambda}{2mc\omega_{\text{rest}}}$$
$$= \frac{e\mathbf{E}\,(s)\,\bar{k}\Lambda\,(s)}{2m\omega\omega_{\text{rest}}} = \frac{ek\mathbf{E}\,(s)\,\Lambda\,(0)}{2m\omega\omega_{\text{rest}}}\,, \tag{5.67}$$

since integration of $\bar{k}\dot{\Lambda} = 0$ implies that $\bar{k}\Lambda$ is constant, and $k\mathbf{E} = \mathbf{E}\bar{k}$. In terms of the paravector potential $A\,(s)$,

$$\mathbf{F} = c\,\langle\partial\bar{A}\,(s)\rangle_V = c\,\langle k\bar{A}'\,(s)\rangle_V\,, \tag{5.68}$$

where $A'\,(s) = dA\,(s)/ds$, and with the Lorenz-gauge condition $\langle\partial\bar{A}\rangle_S = 0$,

$$\mathbf{F} = ck\bar{A}'\,(s) = -cA'\,(s)\,\bar{k} \tag{5.69}$$

so that $d\Lambda/ds$ can be integrated to give $\Delta\Lambda \equiv \Lambda(s) - \Lambda(0)$ proportional to the change ΔA in the paravector potential:

$$\Delta\Lambda = \frac{ek\Delta\bar{A}}{2m\omega_{\text{rest}}}\Lambda(0). \tag{5.70}$$

This beautifully simple result can be used to find the proper velocity $u = \Lambda\Lambda^{\dagger}$ and spacetime position $x = c \int u d\tau$ of the charge at later times. The method has recently been extended by adding an axial magnetic or electric field to the plane-wave pulse. The extensions give exact solutions of the effective mass of the charge dressed by the pulse. They also provide analytical solutions for the autoresonant laser accelerator (ALA). By locking the proper cyclotron frequency of the charge onto the rest-frame frequency of circularly polarized radiation, the ALA can achieve impressive accelerations of the charge.[6]

20.5.2   Virtual photon sheets

The paravector potential of an accelerating point charge e following the world path $r(\tau)$ is the Liénard-Wiechert potential

$$A(x) = \frac{Z_0}{4\pi}\frac{eu}{\langle R\bar{u}\rangle_S}, \tag{5.71}$$

where $uc = dr/d\tau$ and it is understood that r and u are evaluated at the retarded time so that $R \equiv x - r(\tau)$ is a lightlike ray. Differentiation with allowance for the dependence of the retarded time τ on x gives the electromagnetic field (4.25):[5]

$$\mathbf{F}(x) = \frac{Z_0}{4\pi}\frac{e}{\langle R\bar{u}\rangle_S^3}\left(c\langle R\bar{u}\rangle_V + \frac{1}{2}R\bar{\dot{u}}u\bar{R}\right) \tag{5.72}$$

in which the first term is the transformed Coulomb field and the second term is a null flag that gives the radiation (acceleration) field.

If $r(\tau)$ is confined to a single timelike plane, the calculation of field lines is quite simple. One notes that the electromagnetic field propagates from the charge at the speed of light and that in the commoving frame of the charge, the field starts off as the Coulomb field. Consider streams of "virtual photons" that are emitted isotropically in the commoving frame. The emission angles are easily transformed to the lab frame. While each virtual photon moves in a fixed direction at the speed of light, the stream consists of photons emitted at different times from the moving charge. If the charge oscillates, the streams develop kinks, similar to the kinks in a stream of water from an oscillating hose even though the component droplets follow ballistic trajectories.

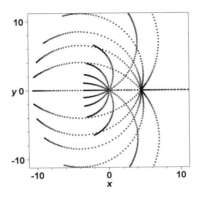

FIGURE 20.2. Virtual photons that form the electric field lines of a uniformly accelerated charge, as calculated by Maple V. The photons are shown both at the instant the charge turns around and later, after it has accelerated some distance along the x axis.

The virtual-photon streams sweep out sheets in spacetime. The tangent plane to the sheet at the observer position $x = r + R$ at retarded proper time $\tau = 0$ contains both the ray R and the proper-time derivative $d\left(r + R'\right)/d\tau$, where $R'\left(\tau\right)$ is defined to be equal to $R\left(0\right)$ in the commoving frame of the charge. In terms of the eigenspinor of the charge (see previous subsection),

$$R'\left(\tau\right) = \Lambda\left(\tau\right)\bar{\Lambda}\left(0\right)R\left(0\right)\bar{\Lambda}^{\dagger}\left(0\right)\Lambda^{\dagger}\left(\tau\right), \tag{5.73}$$

and the direction of $R'\left(\tau\right)$ will generally change if the charge is accelerating. From the evolution equation (5.56) for $\Lambda\left(\tau\right)$,

$$\frac{dR'}{d\tau} = \left\langle\mathbf{\Omega}R'\right\rangle_{\mathbb{R}} = \left\langle\mathbf{\Omega}R\right\rangle_{\mathbb{R}} \tag{5.74}$$

at $\tau = 0$. The tangent plane is thus given by the biparavector

$$\left\langle R\frac{d}{d\tau}\left(\bar{r} + \bar{R}'\right)\right\rangle_{V} = \left\langle R\bar{u}c + R\overline{\langle\mathbf{\Omega}R\rangle}_{\mathbb{R}}\right\rangle_{V}$$
$$= \left\langle R\bar{u}c + \frac{1}{2}R\bar{\mathbf{\Omega}}^{\dagger}\bar{R}\right\rangle_{V}. \tag{5.75}$$

Now both the position r of the charge and the direction of the ray R changes in time. If the acceleration is collinear with the velocity, $\mathbf{\Omega}$ is real and commutes with u so that

$$\dot{u} = \left\langle\mathbf{\Omega}u\right\rangle_{\mathbb{R}} = \mathbf{\Omega}u \tag{5.76}$$

$$\mathbf{\Omega}^{\dagger} = \bar{u}\dot{u} \tag{5.77}$$

and the tangent plane to the photon sheets is

$$\left\langle R\bar{u}c + \frac{1}{2}R\bar{u}u\bar{R} \right\rangle_V \qquad (5.78)$$

is exactly the plane of the electromagnetic field (5.72). Furthermore, it can be shown that the slice of such sheets at an instant in time give the electric-field lines. This is the basis for the virtual photon method, which allows the calculation of field lines with an extremely simple algorithm.[5] At each time step, virtual photons are launched at the Lorentz-transformed angles and continue to move in straight lines at the speed of light. The result of a Maple plot of the electric field lines for a uniformly accelerating charge is shown in Fig. 20.2.

20.6   Conclusions

The paravector model of $C\ell_3$ lets electrical engineers take advantage of the natural relativistic symmetries of the Maxwell-Lorentz theory while remaining close to their familiar vector notation. The powerful new tools offered by geometric algebra include projectors and eigenspinor methods.

Acknowledgments

The author thanks the Natural Sciences and Engineering Research Council of Canada for support of his research. Enlightening conversations with colleagues are also gratefully acknowledged, in particular with Dr. John Huschilt and with students Yuan Yao, Greg Trayling, Jacob Alexander, Shazia Hadi, and David Keselica.

Chapter 21

Applications of Geometric Algebra in Physics and Links With Engineering

Anthony Lasenby and Joan Lasenby

21.1 Introduction

While the early applications of geometric algebra (GA) were confined to physics, there has been significant progress over recent years in applying geometric algebra to areas of engineering and computer science. The beauty of using the same language for these applications is that both engineers and physicists should be able to understand the work done in each others fields. It is the aim of this paper to give brief outlines of the use of GA in the areas of relativity, quantum mechanics and gravitation – all using tools with which anyone working with GA should be familiar. Taking one particular area, multiparticle quantum mechanics, it is shown that the same mathematics may have some interesting applications in the fields of computer vision and robotics.

In this contribution, we review some of the physical applications in which a geometric algebra formulation is particularly helpful. This includes electromagnetism, quantum mechanics and gravitational theory. Then we show how some of these same techniques are of use and interest in engineering. More generally, we show how the availability of a unified mathematical language, able to span both disciplines, is an advantage in allowing professionals from each area to increase their understanding of previously inaccessible material, and make contributions outside their usual areas of expertise. As a case study involving new material, we examine a generalization to 4-d space, of the new conformal representation of 3-d Euclidean space being developed by Hestenes and collaborators (see e.g. chapter 1). This conformal representation is already finding application in robotics [3] and may also be important in interpolation of rigid body motion [4]. We show how the 4-d version has unexpected links with the mathematics of sophisticated objects called 'twistors', and perhaps even more surprisingly, with multiparticle quantum mechanics. These links then suggest a novel method for carrying out such interpolation, allowing consideration of *velocities* as well as positions in 3-d.

In order to start the discussion of applications of geometric algebra in physics, it is necessary to introduce the *spacetime algebra* or STA – the geometric algebra of relativistic 4-dimensional spacetime. This may seem overly complicated to someone who wishes to see examples of links between physics and engineering expressed in geometric algebra. It is true that relativity theory impinges hardly at all on most engineering practice and applications. However, we shall see below that setting up the STA at the start is useful in areas as diverse as computer vision, quantum computing and (as mentioned before) interpolation of rigid body motion. Also, it is what will allow us to consider applications in physics such as electromagnetism and gravitational theory. Thus this contribution begins with an introduction to the STA and shows briefly how a concept called the *projective split* allows an easy articulation between four dimensions and the concepts of ordinary 3-dimensional geometric algebra.

21.2   The Spacetime Algebra

The *spacetime algebra* or STA is the geometric algebra of Minkowski spacetime. We introduce an orthonormal frame of vectors $\{\gamma_\mu\}$, $\mu = 0 \ldots 3$, such that

$$\gamma_\mu \delta \gamma_\mu = \eta_{\mu\nu} = (+ - - -) \tag{2.1}$$

The STA has the basis

1	$\{\gamma_\mu\}$	$\{\gamma_\mu \wedge \gamma_\nu\}$	$\{i\gamma_\mu\}$	$i \equiv \gamma_0\gamma_1\gamma_2\gamma_3$
1 scalar	4 vectors	6 bivectors	4 trivectors	1 pseudoscalar

The pseudoscalar i *anti*-commutes with vectors.

At this point we note that generally, when working with single particle algebras, the standard has become to use I for the pseudoscalar; however, here, we will use i to denote the pseudoscalar to avoid later confusion when we discuss the multiparticle STA.

21.2.1   The spacetime split, special relativity, and electromagnetism

In special relativity (SR) we deal with a 4-dimensional space; the three dimensions of ordinary Euclidean space, and time. Suppose we have a stationary observer with whom we can associate coordinates of space and time; this observer will observe events from his *spacetime* position. Now suppose that we have another observer travelling at a velocity v – he too will observe events from his continuously changing spacetime position. Relative vectors for an observer moving with velocity v are modelled as bivectors, so $a \wedge v$ gives the vector a seen in the v frame. Usually we take $v = \gamma_0$ and define

$$\sigma_k \equiv \gamma_k \gamma_0 \qquad k = 1, 2, 3 \tag{2.2}$$

The even subalgebra of the STA is then the algebra of relative space, spanned by

$$1, \{\sigma_k\}, \{i\sigma_k\}, i \tag{2.3}$$

The distinction between relative vectors and relative bivectors is frame-dependent and the process of moving between a vector a in the 4-d STA and its representation $\boldsymbol{a}$ in the relative space is known as the *spacetime split*;

$$\boldsymbol{a} = a \wedge \gamma_0 \tag{2.4}$$

In practical geometric problems occuring in computer vision and computer graphics it is common to move up from our 3-d Euclidean space to work in a 4-d projective space, where non-linear transformations become linear and where intersections of lines, planes etc., are easy to compute. This extra dimension is analogous to the γ_0 in the STA (although in projective geometry one can have either a $(+, +, +, +)$ or a $(+, -, -, -)$ signature) and moving between projective space and Euclidean space can similarly be carried out using the *projective split*, given by

$$\boldsymbol{a} = \frac{a \wedge \gamma_0}{a \delta \gamma_0} \tag{2.5}$$

Here we are again relating the vectors in relative space (3-d) with bivectors in the higher (4-d) space. Alternatively, we can define the vector in 3-space, $\boldsymbol{a}$ as $a^j \gamma_j$, $j = 1, 2, 3$ and the associated vector in the higher space, a, by

$$a = a_0(a^j \gamma_j + \gamma_0), \quad j = 1, 2, 3 \tag{2.6}$$

so that we have

$$\boldsymbol{a} = \frac{(a \wedge \gamma_0)\gamma_0}{a \delta \gamma_0} \tag{2.7}$$

Both of the above interpretations have been used in the literature in discussions of projective and conformal geometry, [6, 7].

Returning to the STA, one can conventionally derive a coordinate transformation between the frames of two observers, and to move between these two frames one applies a matrix transformation known as a **Lorentz Boost**. Geometric algebra provides us with a beautifully simple way of dealing with special relativistic transformations using the simple formula for rotations that we will discuss below, namely $a' = Ra\tilde{R}$ ([8, 9]).

We have seen in other contributions in this volume that in any geometric algebra, rotations are achieved by quantities called *rotors*. A rotor R can be written as

$$R = \pm\exp(-B/2) \tag{2.8}$$

where B is a bivector representing the plane in which the rotation takes place. It is then easy to show that a rotation of a vector a to a vector a' is achieved by the equation

$$a' = Ra\tilde{R}$$

In 3-d R is made up of scalar and bivector parts while in 4-d it has scalar, bivector and pseudoscalar parts; in each case it has a double-sided action. We should stress here that a rotor is simply part of the algebra and need not have special operator status. We will see in the following sections that rotors are crucial quantities in much of physics, in particular, we will see that simple rotations in the STA will allow us to understand most of special relativity and will play an important role in quantum mechanics.

The Lorentz boost turns out to be simply a rotor R which takes the time axis to a different position in 4-d ; $R\gamma_0\tilde{R}$. So, in an elegant coordinate-free way we are able to give the transformations of SR an intuitive geometric meaning. All the usual results of SR follow very quickly from this starting point.

Moving now to electromagnetism, the electromagnetic field strength is given by the bivector

$$F \equiv \tfrac{1}{2}F^{\mu\nu}\gamma_\mu\wedge\gamma_\nu \tag{2.9}$$

where the Greek indices μ and ν run over $0, 1, 2, 3$. In the γ_0 frame this decomposes into bivectors of the form $\gamma_i\gamma_0$ and $\gamma_i\gamma_j$ $(i, j, = 1, 2, 3, i \neq j)$, so that we can write

$$F = \mathbf{E} + i\mathbf{B} \tag{2.10}$$

where $\mathbf{E}$ and $\mathbf{B}$ are the electric and magnetic fields and are given by $\mathbf{E} = E^k\sigma_k = \tfrac{1}{2}(F - \gamma_0F\gamma_0)$ and $\mathbf{B} = B^k\sigma_k = \tfrac{1}{2}(F + \gamma_0F\gamma_0)$. Here, sandwiching between γ_0 flips the sign of the $\gamma_i\gamma_0$ bivectors but leaves the $\gamma_i\gamma_j$ bivectors unaltered. This form of F explains the usefulness of complex numbers in electromagnetism. Now, let us define the 4-d gradient operator as

$$\nabla \equiv \gamma^\mu\frac{\partial}{\partial x^\mu} \tag{2.11}$$

It is not hard to show that the full Maxwell equations can then be written simply as

$$\nabla F = J \tag{2.12}$$

where J is the source current. The above formulation of electromagnetism is also being used in several engineering applications, e.g. surface scattering of EM waves from objects such as ships, antennae design etc.

As an example of the simplifications that this approach can afford, consider what the electric and magnetic fields look like under a Lorentz boost. The conventional complicated formulae for the transformation are now replaced by the result

$$\boldsymbol{E}' + i\boldsymbol{B}' = R(\boldsymbol{E} + i\boldsymbol{B})\tilde{R}$$

where dashes denote transformed quantities and R is the rotation in the STA representing the boost.

To illustrate this explicitly, and to make the link with the standard formulae, we consider a boost with velocity parameter u (so the actual velocity is $\tanh u$) in the x direction, where the original field is $\boldsymbol{E} = E\sigma_y$, i.e. an electric field in the y direction only with no magnetic field component. We have $R = e^{\frac{u}{2}\sigma_x}$ and so

$$
\begin{aligned}
\boldsymbol{E}' + i\boldsymbol{B}' &= E e^{\frac{u}{2}\sigma_x} \sigma_y e^{-\frac{u}{2}\sigma_x} \\
&= E e^{u\sigma_x} \sigma_y = E(\cosh u + \sigma_x \sinh u)\sigma_y \\
&= E(\cosh u\, \sigma_y + i\sigma_z \sinh u)
\end{aligned}
\tag{2.13}
$$

We can clearly see a $\boldsymbol{B}$ field is induced in the z-direction, with amplitude $E \sinh u$. Note the electromagnetic invariants arise immediately via the relation

$$F'^2 = RF\tilde{R}RF\tilde{R} = RF^2\tilde{R} = F^2 \tag{2.14}$$

since F^2 contains only scalar and pseudoscalar parts. Specifically we have

$$
\begin{aligned}
\text{Scalar part of } F^2 &= \boldsymbol{E}^2 - \boldsymbol{B}^2 \\
\text{Pseudoscalar part of } F^2 &= 2i\boldsymbol{E}\delta\boldsymbol{B}
\end{aligned}
\tag{2.15}
$$

and so $\boldsymbol{E}^2 - \boldsymbol{B}^2$ and $\boldsymbol{E}\delta\boldsymbol{B}$ are invariant under any Lorentz transformation (as may be checked for the example above).

The complicated tensor formula for the electromagnetic stress energy tensor becomes extremely simple in the STA. We find the flow of energy/momentum through a hypersurface normal to the vector n is given by

$$T(n) = \frac{1}{2}Fn\tilde{F} \tag{2.16}$$

That is, we just rotate n by F!

This then easily leads to the standard Heaviside and Poynting formulae for the energy density and momentum flow in a given frame (e.g. the γ_0 frame). (For further examples and details see e.g. [9, 10, 11] and [12].) The STA really does seem to capture the essence of electromagnetism in a very compact and useful formalism!

21.3 Quantum Mechanics

In non-relativistic quantum mechanics there are important quantities known as *Pauli spinors* – using these spinors we are able to write down the *Pauli equation* which governs the behaviour of a quantum mechanical state in some external field. The equation involves quantities called *spin operators* which are conventionally seen as completely different entities to the *states*. Using the 3-d geometric algebra we are able to write down the equivalent to the Pauli equation where the operators and states are all real-space multivectors – indeed the spinors become proportional to rotors of the type we have discussed earlier. The algebra of the $\{\sigma_i\}$ is isomorphic to the algebra of Pauli spin matrices.

To see how this works in a simple context, we consider the case of an electron in a magnetic field. A conventional quantum Pauli spinor $|\psi\rangle$, which is normally written as a two component complex column 'vector', is put into 1-1 correspondence with a GA spinor ψ (an even element of the geometric algebra of 3-d space) via:

$$|\psi\rangle = \begin{pmatrix} a^0 + ja^3 \\ -a^2 + ja^1 \end{pmatrix} \leftrightarrow \psi = a^0 + a^k i\sigma_k \qquad (3.17)$$

(Note the symbol j is used for the unit scalar imaginary of quantum mechanics). We are interested in how the electron spin behaves, and will ignore any spatial variation. It is then easy to show that the GA form of the Pauli equation for this setup is

$$\frac{d\psi}{dt} = \tfrac{1}{2}\gamma i\boldsymbol{B}\psi \qquad (3.18)$$

Here $\boldsymbol{B}$ is the magnetic field as described in the previous section and γ is the 'gyromagnetic ratio'. ($\gamma \approx e/m$ for an electron, where e and m are the electon charge and mass.) Any Pauli spinor can be decomposed as $\psi = \rho^{\frac{1}{2}}R$, where ρ is a scalar and R is a rotor. Substituting this form into (3.18), multiplying by $\tilde{\psi}$ and denoting time derivatives by an overdot, we obtain

$$\tfrac{1}{2}\dot{\rho} + \rho\dot{R}\tilde{R} = \tfrac{1}{2}\rho\gamma i\boldsymbol{B} \qquad (3.19)$$

It is straightforward to show that $R\tilde{R} = 1$ implies $\dot{R}\tilde{R}$ is a bivector. The right hand side of (3.19) is also a bivector, so we deduce $\dot{\rho} = 0$. The scale thus drops out of the problem and the dynamics reduces to the rotor equation

$$\dot{R} = \tfrac{1}{2}\gamma i \boldsymbol{B} R \tag{3.20}$$

The conventional approach is unable to work with this single rotor equation, but instead has to work with two coupled complex equations, one for each of the components of the quantum state. Although the underlying physics is the same, the rotor form is often significantly easier to solve (e.g., for a constant field $\boldsymbol{B} = B_0\sigma_3$ along the z-axis, we can immediately intergrate to find $\psi(t) = \exp(\gamma B_0 t i \sigma_3/2)\psi_0)$ and makes the analogue with the corresponding classical system much more transparent.

Relativistic quantum mechanics is conventionally described by the Dirac algebra, where the *Dirac equation* again tells us about the state of the particle in an external field. Here we use the 4-d spacetime geometric algebra with the algebra of the $\{\gamma_\mu\}$ isomorphic to that of the Dirac matrices. Again the *wavefunction* in conventional quantum mechanics becomes an instruction to rotate a basis set of axes and align them in certain directions – analogous to the theory of rigid body mechanics! We see therefore that there is a significant shift in interpretation; in GA, the states and operators no longer live in different spaces but are instead simply multivector elements of the geometric algebra.

Thus, with the STA, we can eliminate matrices and complex numbers from the Dirac theory. Suppose we start with the standard Dirac matrices:

$$\hat{\gamma}^0 = \begin{pmatrix} I & 0 \\ 0 & -I \end{pmatrix} \qquad \hat{\gamma}^i = \begin{pmatrix} 0 & \hat{\sigma}_i \\ -\hat{\sigma}_i & 0 \end{pmatrix} \tag{3.21}$$

where the $\{\hat{\sigma}_i\}$ are the usual Pauli spin matrices and I is the 2×2 identity matrix. A Dirac column spinor $|\psi\rangle$ maps onto an element of the 8-d even subalgebra (a spinor) of the STA via the following:

$$|\psi\rangle = \begin{pmatrix} a^0 + ja^3 \\ -a^2 + ja^1 \\ b^0 + jb^3 \\ -b^2 + jb^1 \end{pmatrix} \leftrightarrow \psi = \begin{array}{l} a^0 + a^k i\sigma_k \\ +(b^0 + b^k i\sigma_k)\mathsf{k} \end{array} \tag{3.22}$$

Dirac matrix operations are now replaced by:

$$\hat{\gamma}^\mu|\psi\rangle \leftrightarrow \gamma^\mu \psi \gamma_0 \qquad j|\psi\rangle \leftrightarrow \psi i\sigma_3 \tag{3.23}$$

This enables us to write the Dirac equation as

$$\nabla \psi i\sigma_3 - eA\psi = m\psi\gamma_0 \tag{3.24}$$

where $\nabla = \gamma^\mu \partial_\mu$ is the gradient operator defined in the previous section and A is the 4-potential of the external electromagnetic field. Note this equation — often referred to as the Hestenes form of the Dirac equation —

is independent of choice of matrix representation and is therefore the best form in which to expose the geometric content of Dirac theory.

In the conventional approach it is usual to define an additional operator $\hat{\gamma}_5 = -j\hat{\gamma}_0\hat{\gamma}_1\hat{\gamma}_2\hat{\gamma}_3$ — in our approach this is replaced by right multiplication by σ_3. Two of the main observables of Dirac theory (J and s, the so-called bilinear covariants) become:

$$J^\mu = \langle\bar{\psi}\hat{\gamma}^\mu\psi\rangle \quad \leftrightarrow \quad \langle\tilde{\psi}\gamma^\mu\psi\gamma_0\rangle = \gamma^\mu\delta J$$
$$s^\mu = \langle\bar{\psi}\hat{\gamma}_5\hat{\gamma}^\mu\psi\rangle \quad \leftrightarrow \quad \langle\tilde{\psi}\gamma^\mu\psi\gamma_3\rangle = \gamma^\mu\delta s$$

where $\bar{\psi}$ is the *Dirac adjoint*. The key quantities are the STA vectors J and s:

$$J = \psi\gamma_0\tilde{\psi} \qquad s = \psi\gamma_3\tilde{\psi} \qquad (3.25)$$

(Note that there exist a set of identities called the Fierz identities which, in the above formulation, reduce to simple vector manipulations.)

It is now possible to decompose the spinor ψ in a Lorentz invariant manner;

$$\psi\tilde{\psi} = \rho e^{i\beta} \quad = \quad \text{scalar + pseudoscalar} \qquad (3.26)$$

Using this decomposition we can write ψ as follows

$$\psi = (\rho e^{i\beta})^{1/2}R \qquad (3.27)$$

where R is a spacetime rotor. The observables now become

$$J = \rho R\gamma_0\tilde{R} \qquad s = \rho R\gamma_3\tilde{R} \qquad (3.28)$$

so the spinor reduces to an instruction to rotate the $\{\gamma_\mu\}$ frame onto the frame of observables. The STA framework for quantum mechanics has been applied in tunnelling theory [13] where it is capable of plotting streamlines representing the path of a particle inside a barrier. It is then easy to calculate *tunnelling times*, the time a particle spends within a barrier, – something which is much harder to do in conventional quantum mechanics where the concepts of imaginary time or momentum preclude straightforward calculations. Applications in electron scattering [14] have reformulated much of conventional theory allowing spin sums to be done straightforwardly and revealing rotor-structure at the heart of the formulation. For further details of applications to quantum theory, see [15]. The above once again illustrates that using geometric algebra one is able to deal with complex subjects such as relativistic quantum mechanics using those same tools used in current engineering applications of geometric algebra.

21.4   Gravity as a Gauge Theory

A gauge theory occurs if we stipulate that global symmetries must also be local symmetries – electromagnetism is a gauge theory where the symmetries are called *phase rotations*. Making these local, i.e. able to change arbitrarily from one spacetime position to the next, implies the introduction of *forces*. In geometric algebra, gravity can also be regarded as a gauge theory. If we require that the physics at all points of spacetime is invariant under arbitrary local displacements and rotations (recall that a 4-d rotation is a Lorentz boost), the *gauge field* that results is the gravitational field. Thus, the aim is to produce a gauge theory of gravity employing fields in a 'flat' background spacetime (defined by the STA); we then have no need for the complex notions of curved spacetime that are associated with Einstein's theory of general relativity. How can we construct such a theory without imposing some form of *absolute* Newtonian space? We start by ensuring that the following criteria are satisfied:

1. The physical content of a field equation must be unchanged under arbitrary local field displacements.

2. The physical content of a field equation must be unchanged under arbitrary local rotations of the fields.

In looking at how the resulting gauge theory differs from past gauge-theoretic approaches to gravity, we note the following points:

1. It is different from Poincaré gauge theory, which retains the ideas of a curved spacetime background.

2. There is no need to restrict to infinitesimal transformations; within GA we can work with finite rotations.

3. The need for principle 2 only emerges fully from a theory based on the Dirac equation.

To see mathematically what the symmetry constraints impose we first consider a relation of the type

$$a(x) = b(x) \tag{4.29}$$

which equates spacetime vectors at the same point. Now we introduce new fields

$$a'(x) \equiv a(x') \qquad b'(x) \equiv b(x') \tag{4.30}$$

where $x' = f(x)$ is some arbitrary (nonlinear) mapping between position vectors. The equation

$$a'(x) = b'(x) \tag{4.31}$$

has exactly the same content as the original equation, since the value of x is irrelevant provided it covers all of spacetime. This is true for *arbitrary* displacements.

In order to satisfy our previous conditions we require that this holds for all physical equations.

Next consider a relation of the type $a(x) = \nabla\phi(x)$. If we replace $\phi(x)$ with $\phi'(x) = \phi(x')$ we must now consider ∇ acting on the new scalar field $\nabla\phi(f(x)))$; by using the definition of the vector derivative it can be shown that [16]

$$\nabla\phi'(x) = \bar{\mathsf{f}}[\nabla_{x'}\phi(x')] \tag{4.32}$$

where

$$\mathsf{f}(a) = a\delta\nabla f(x) \tag{4.33}$$

Here, $\mathsf{f}(a) = \mathsf{f}(a,x)$ is a linear function of its vector argument, and a nonlinear function of position, $\bar{\mathsf{f}}(a)$ is its adjoint. The appearance of this function means that the equation does not have the required transformation property.

We repair this by replacing ∇ with a new derivative $\bar{\mathsf{h}}(\nabla)$, where $\mathsf{h}(a)$ is a linear function of a and has arbitrary position dependence; we call $\mathsf{h}(a) = \mathsf{h}(a,x)$ the *position gauge field*. The adjoint function is written $\bar{\mathsf{h}}(a)$. Under a local displacement, this is defined to transform as

$$\bar{\mathsf{h}}(a,x) \mapsto \bar{\mathsf{h}}'(a,x) = \bar{\mathsf{h}}[\bar{\mathsf{f}}^{-1}(a), x'] \tag{4.34}$$

This law ensures that the equation

$$a(x) = \bar{\mathsf{h}}[\nabla\phi(x)] \tag{4.35}$$

is now *covariant*, in the required manner, i.e. under a change of position the equation takes the same form but is evaluated at that new position.

Recovering General Relativity

Using the linear function h it is now possible to recover classical general relativity (GR). To do this we first introduce a set of local coordinates $x^\mu = x^\mu(x)$, with coordinate frames

$$e_\mu = \partial_\mu x \qquad e^\mu = \nabla x^\mu \tag{4.36}$$

we can then recover a metric as follows:

$$g_{\mu\nu} = \mathsf{h}^{-1}(e_\mu)\delta\mathsf{h}^{-1}(e_\nu) \tag{4.37}$$

This metric is then treated as a field in a flat background spacetime.

Rotations

As we have indicated in previous sections, rotations are often the key to the simplifications provided by GA. In this application it is again true that rotations are key to the novelty of this new approach, and also the key to *torsion*. Let us return to the equation $a(x) = b(x)$. Note that the physical content of this equation is unchanged if we replace a and b by

$$a'(x) = Ra(x)\tilde{R} \qquad b'(x) = Rb(x)\tilde{R} \qquad (4.38)$$

since $a = b \Rightarrow a' = b'$. The physics is unchanged, provided the absolute direction of the vector in the STA does not enter (the second of our two principles). Again, this argument holds for *arbitrary*, *local* rotations.

To ensure that relations of the type

$$a = \bar{\mathsf{h}}(\nabla\phi) \qquad (4.39)$$

remain unchanged, we are led to the transformation law

$$\bar{\mathsf{h}}(a) \mapsto \bar{\mathsf{h}}'(a) = R\bar{\mathsf{h}}(a)\tilde{R} \qquad (4.40)$$

for $\bar{\mathsf{h}}$ under local rotations.

What does general relativity have to say about this transformation? — surprisingly, *nothing!*

The metric $g_{\mu\nu}$ is unchanged by this transformation, as are the components of covariant quantities:

$$F_{\mu\nu} = F\delta[\mathsf{h}^{-1}(e_\mu) \wedge \mathsf{h}^{-1}(e_\nu)] \qquad (4.41)$$

Both F and h rotate to leave the components unchanged.

(Most of) classical general relativity can be formulated in the STA without mentioning the rotation gauge. But do we also need to consider the Poincaré group? In fact, it is already fully encompassed by allowing arbitrary displacements.

This then leads us to ask the question of whether we have to address the rotation group at all? The answer to this question is *Yes!*; it is indeed unavoidable in the Dirac theory. We can see this by recalling the fact that observables such as $J = \psi\gamma_0\tilde{\psi}$ imply the spinor transformation law

$$\psi \mapsto \psi' = R\psi \qquad (4.42)$$

Since this cannot be hidden, we are forced to introduce a new gauge field to make the Dirac theory invariant under local rotations.

Now let us look at the directional derivatives of $\nabla(R\psi)$:

$$\begin{aligned} a\delta\nabla(R\psi) &= a\delta\nabla R\psi + Ra\delta\nabla\psi \\ &= R[\tilde{R}a\delta\nabla R\psi + a\delta\nabla\psi] \end{aligned}$$

Note that the quantity $\tilde{R}a\delta\nabla R$ is a *bivector*. We now define the spinor covariant derivative as

$$D_a\psi \equiv a\delta\nabla\psi + \tfrac{1}{2}\Omega(a)\psi \qquad (4.43)$$

$\Omega(a)$ is a bivector-valued linear function of a, with nonlinear position dependence which has the transformation law

$$\Omega(a) \mapsto \Omega'(a) = R\Omega(a)\tilde{R} - 2a\delta\nabla R\,\tilde{R} \tag{4.44}$$

We are now able to write down the minimally-coupled Dirac equation:

$$\bar{\mathsf{h}}(\partial_a)D_a\psi i\sigma_3 = m\psi\gamma_0 \tag{4.45}$$

The $\{\partial_a, a\}$ construction is a frame-free way of writing a contraction (see [16] for further details).

Observables

We see that it is now possible to differentiate covariant vectors:

$$a\delta\nabla J = a\delta\nabla\psi\gamma_0\tilde{\psi} + \psi\gamma_0 a\delta\nabla\tilde{\psi} \tag{4.46}$$

which suggests that we define the derivative

$$\begin{aligned}
\mathcal{D}_a J &= (D_a\psi)\gamma_0\tilde{\psi} + \psi\gamma_0\widetilde{(D_a\psi)} \\
&= a\delta\nabla J + \Omega(a)\times J
\end{aligned}$$

This is the covariant derivative for multivectors, where $A\times B = \frac{1}{2}(AB - BA)$ represents the Hestenes commutator product [25].

From $\Omega(a)$ we define

$$\omega(a) = \Omega\mathsf{h}(a) \tag{4.47}$$

which is covariant under local displacements, and only sees the rotation group. When the rotation gauge is fixed, the quantities in $\omega(a)$ become physical observables (measurable). Classical general relativity has no analogue of these.

Note here that the full covariant derivative is

$$\mathcal{D} = \bar{\mathsf{h}}(\partial_a)\mathcal{D}_a. \tag{4.48}$$

The rest of the theory then proceeds by defining the following field strength tensor

$$[D_a, D_b]\psi = R(a\wedge b)\psi \tag{4.49}$$

By a double contraction we can get the Ricci scalar

$$\mathcal{R} = [\bar{\mathsf{h}}(\partial_b)\wedge\bar{\mathsf{h}}(\partial_a)]\delta R(a\wedge b) \tag{4.50}$$

This can then be used in an action principle requiring stationarity of

$$\int d^4x \det \mathsf{h}^{-1}(\frac{1}{2}\mathcal{R} - \kappa\mathcal{L}_m) \tag{4.51}$$

where the matter Lagrangian is $\mathcal{L}_m$ and with $\Omega(a)$ and $\bar{\mathsf{h}}(a)$ as the dynamical variables ($\kappa = 8\pi G$ is the gravitational coupling constant).

The result is a theory which locally reproduces the equations of the ECKS (Einstein, Cartan, Kibble, Sciama) extension of GR with the following notable differences:

- it sits in a topologically trivial flat spacetime

- has all the advantages of flat-space STA still available for calculations

- finite gauge rotations and displacements are allowed

- the torsion type is uniquely picked out ($\partial_{\Omega(a)}\mathcal{L}_m = \mathcal{S}(a) =$ torsion tensor. This must be of the Dirac type, i.e. $\partial_a \delta \mathcal{S}(a) = 0$ for minimal coupling)

- Physical observables and gauge covariant quantities of the theory are clearly picked out.

21.4.1   Some applications

In this section we briefly outline some of the applications of this gauge theory of gravity (GTG).

1. *Covariant and gauge-invariant calculation of cosmic microwave background (CMB) anisotropies.*
    The GTG approach provides a completely unified scheme for scalar, vector and tensor quantities. It has been applied very successfully to the gauge-invariant calculation of CMB anisotropies [17] and to the development of perturbations, where it recovers the covariant approach of Ellis and coworkers [5].

2. *Topological applications.*
    Despite sitting in a topologically trivial flat spacetime, the GTG can in fact be applied to some situations which would conventionally be thought of as involving topology. It is found that entities like *cosmic strings* are allowed and can be treated (similar to the Aharanov-Bohm effect in electromagnetism), but that *wormholes, kinks, Kruskal-Szekeres and all forms of double cover* are ruled out under this theory.

3. *Cosmic Stings*
    A new *spinning* cosmic string solution [18] has been found which corrected an earlier GR-based attempt.

4. *Singularities*
    The availability of integral theorems (Gauss etc.) means that we

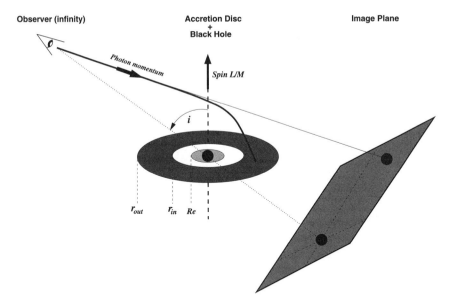

FIGURE 21.1. Setup for computing spectral lineshapes using the GTG approach.

can study the structure of singularities in new ways. For example, the singularity at the centre of the Kerr solution is revealed to be a ring of matter rotating at light-like velocity, but with a ring of *pure tension* stretched across it [19]. Such conclusions are gauge invariant.

5. Spectral lineshapes

A recent project [20] has concentrated on calculating spectral line-shapes from iron-line fluoresence in accretion discs around black holes in active galactic nuclei (AGN). Here the GTG provides an efficient calculational tool and gives a clear approach to the physical (gauge invariant) predictions, see figure 21.1. Results so far, for a particular active galactic nucleus, show that if a is the specific angular momentum of the black hole line and M is its mass, then $a/M > 0.9$ at 90 percent confidence, giving some of the first quantitative evidence for a *spinning* black hole. In this approach the 2nd order GR geodesic equations are replaced by first order equations for a rotor which describes the photon momentum. Integrating the rotor equations in such a setup has links with the procedures required when dealing with buckling beams and deforming elastic fibres (see below).

6. *Black holes*

In the GTG approach, black holes have a memory of the direction of time in which they were formed encoded in them. This means that the first order (in derivatives) nature of the GTG results in time-reversal properties which are slightly different than those predicted in GR based on metric (second order) theory. A full discussion of this may be found in [16].

7. *Spinning Black Holes*

The GTG has produced a new and very simple form of the Kerr solution for spinning black holes [21]. This is called the *Newtonian Kerr* and takes the form

$$\bar{\mathsf{h}}(a) = a - a\delta\hat{e}_u\sqrt{\frac{2M\sinh u}{L\cosh^2 u}}V \tag{4.52}$$

where we work in oblate spheroidal coordinates (t, u, ϕ, v), and the velocity vector V is given by

$$V = \frac{\cosh u\gamma_0 + \cos v\hat{\phi}}{\sqrt{\cosh^2 u - \cos^2 v}} \tag{4.53}$$

This provides a *global* solution which is not much more complicated than the Schwarzschild solution for stationary black holes.

21.4.2   Summary

This section has given an outline of how GA can be used to formulate a gauge theory of gravity and in the process reduces the tensor manipulations of general relativity to nothing more than linear algebra. The same tools are used throughout. Indeed it may be possible to use linear functions, which act in the same way as the h functions, to model elasticity. The concept of a frame of reference that varies in either space or time (or both) is also at the heart of much work that tries to understand deforming bodies. A very simple example is provided by a beam of uniform cross-section subject to some loading along its length. We can describe this deformation by splitting up the beam into very small segments and attaching a frame to the centre of mass of each segment. As the beam deforms and is subjected to torsional forces, we can describe its position at a given time by a series of translations and rotations specifying the positions and orientations of each element, see figure 21.2.

Current work [22] has focussed on rewriting conventional buckling equations in terms of GA which has the advantage of allowing us to deal with finite rotations and to interpolate resulting rotor fields. However, the methods outlined in this section present us with the possibility of employing more sophisticated techniques for such problems and for more general problems involving the deformation of long elastic fibres under given boundary conditions.

original configuration

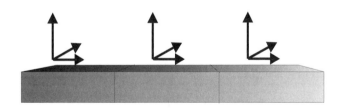

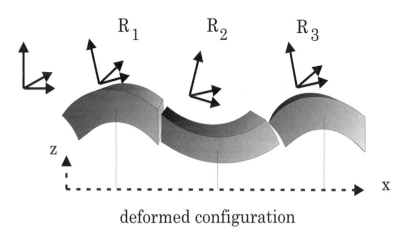

deformed configuration

FIGURE 21.2. Model of a beam split into very small segments – the deformation is described by the positon and orientation of each segment.

21.5 A New Representation of 6-d Conformal Space

A useful new application of geometric algebra to Euclidean geometry has been given by Hestenes *et al.* [2]. This uses a 5-d space to provide a conformal model of Euclidean geometry. Specifically two null vectors, e and e^* are adjoined to Euclidean space, which anticommute with the three basis vectors of Euclidean space and satisfy

$$e\delta e^* = 1. \tag{5.54}$$

Two key results are

1. If x and y are the 5-d vectors representing 3-d points **x** and **y**, then the inner product in 5-d gives a measure of the distance between the

points in 3-d:

$$x\delta y = -\frac{1}{2}\left(\mathbf{x} - \mathbf{y}\right)^2. \tag{5.55}$$

2. Secondly, both translations and rotations in 3-d are representable by *versor* multiplication in 5-d. We can write

$$x' = DxD^{-1}, \tag{5.56}$$

where $D = T_\mathbf{a}R$, R represents a rotation about a direction in 3-d and $T_\mathbf{a} = 1 + \frac{1}{2}ae$ is a translation in the direction $\mathbf{a}$. Note ae is a null bivector in the 5-d space, and so $T_\mathbf{a}$ can be written in the exponential form $\exp(\frac{1}{2}ae)$.

This new conformal model of Euclidean geometry appears to be rich in applications to computer vision and robotics (see chapter 13, where the authors use the conformal model for the algebra of incidence and for estimating Euclidean motion). One possible application is to the problem of the joint interpolation of rotational and translational motion of robot arms (see e.g. [4, 1]), where the ability to write the motion in versor form could be of great benefit. Here, we consider a similar model but applied to relativistic rather than Euclidean geometry. This can be achieved in two ways. Firstly, one could use a 6-d space in which two extra null vectors satisfying $e\delta e^* = 1$ have been added to a 4-d Lorentzian space. This is the obvious generalization of Hestenes' method to one dimension up, and should work very well as something to apply to relativistic problems (e.g. it may allow the problem of motion interpolation to be extended to include interpolation of *velocities* as well as positions – this is currently being investigated). However, as a novel method, one may instead use the 2-particle space of the 'multiparticle STA' (see [15, 12]), which is in fact 8-dimensional, and it is this we consider in detail here. The reason for wishing to stress this new method, is that it sheds wholly unexpected light on the links between such disparate concepts as multiparticle quantum mechanics, relativity, twistors, 2-spinors and the Hestenes conformal representation. Many of these are things which 'engineers' might never have expected to find out about, or to be related to things they wish to know, but here we show how they are in fact intimately related. In particular, we show that the re-expression of twistor theory in multiparticle GA, shows that the main results of the Hestenes conformal representation method are already-known aspects of twistor theory! The links between both of these and multiparticle quantum mechanics appear to be wholly new. (There is even an exciting hint in the work that it will allow a new and concrete expression of the particle physics concept of *supersymmetry*.) We give here just the bare outline of the method – a more detailed exposition is in preparation.

21.5.1 The multiparticle STA

To get started on this topic we need to understand aspects of the multiparticle theory within geometric algebra. The MSTA (Multiparticle SpaceTime Algebra) approach is capable of encoding multiparticle wavefunctions, and describing the correlations between them. The presentation here is hopefully complementary to the presentation given in the context of quantum computing by Havel (see chapter 14) and parallels that given in Chapter 11 of the Banff Lectures [12] by Lasenby, Gull and Doran and in the review paper by Doran *et al.* 'Spacetime Algebra and Electron Physics' [15].

The n-particle STA is created simply by taking n sets of basis vectors $\{\gamma^i_\mu\}$, where the superscript labels the particle space, and imposing the geometric algebra relations

$$\begin{array}{lll} \gamma^i_\mu \gamma^j_\nu + \gamma^i_\nu \gamma^j_\mu = 0, & 0.5 & i \neq j \\ \gamma^i_\mu \gamma^j_\nu + \gamma^i_\nu \gamma^j_\mu = 2\eta_{\mu\nu} & 0.5 & i = j. \end{array} \tag{5.57}$$

These relations are summarized in the single formula

$$\gamma^i_\mu \delta \gamma^j_\nu = \partial^{ij} \eta_{\mu\nu}. \tag{5.58}$$

The fact that the basis vectors from distinct particle spaces anticommute means that we have constructed a basis for the geometric algebra of a $4n$-dimensional configuration space. (Note the extra dimensions serve simply to label the properties of each individual particle, and should not be thought of as existing in anything other than a mathematical sense.)

Throughout, Roman superscripts are employed to label the particle space in which the object appears. So, for example, ψ^1 and ψ^2 refer to two copies of the same 1-particle object ψ, and not to separate, independent objects. Separate objects are given distinct symbols while the absence of superscripts denotes that all objects have been collapsed into a single copy of the STA.

21.5.2 2-Particle Pauli states and the
quantum correlator

As an introduction to the properties of the multiparticle STA, we first consider the 2-particle Pauli algebra and the spin states of pairs of spin-1/2 particles. As in the single-particle case, the 2-particle Pauli algebra is just a subset of the full 2-particle STA. A set of basis vectors is defined by

$$\begin{array}{lll} \sigma^1_i & = & \gamma^1_i \gamma^1_0 \end{array} \tag{5.59}$$
$$\begin{array}{lll} \sigma^2_i & = & \gamma^2_i \gamma^2_0 \end{array} \tag{5.60}$$

which satisfy

$$\sigma^1_i \sigma^2_j = \gamma^1_i \gamma^1_0 \gamma^2_j \gamma^2_0 = \gamma^1_i \gamma^2_j \gamma^2_0 \gamma^1_0 = \gamma^2_j \gamma^2_0 \gamma^1_i \gamma^1_0 = \sigma^2_j \sigma^1_i. \tag{5.61}$$

So, in constructing multiparticle Pauli states, the basis vectors from different particle spaces commute rather than anticommute. Using the elements $\{1, i\sigma_k^1, i\sigma_k^2, i\sigma_j^1 i\sigma_k^2\}$ as a basis, we can construct 2-particle states. Here we have introduced the abbreviation

$$i\sigma_i^1 \equiv i^1 \sigma_i^1 \qquad (5.62)$$

since, in most expressions, it is obvious which particle label should be attached to the i. In cases where there is potential for confusion, the particle label is put back on the i. The basis set $\{1, i\sigma_k^1, i\sigma_k^2, i\sigma_j^1 i\sigma_k^2\}$ spans a 16-dimensional space, which is twice the dimension of the direct product space of two 2-component complex spinors. For example, the outer-product space of two spin-1/2 states can be built from complex superpositions of the set

$$\begin{pmatrix} 1 \\ 0 \end{pmatrix} \otimes \begin{pmatrix} 1 \\ 0 \end{pmatrix}, \begin{pmatrix} 0 \\ 1 \end{pmatrix} \otimes \begin{pmatrix} 1 \\ 0 \end{pmatrix},$$
$$\begin{pmatrix} 1 \\ 0 \end{pmatrix} \otimes \begin{pmatrix} 0 \\ 1 \end{pmatrix}, \begin{pmatrix} 0 \\ 1 \end{pmatrix} \otimes \begin{pmatrix} 0 \\ 1 \end{pmatrix}, \qquad (5.63)$$

which forms a 4-dimensional complex space (8 real dimensions). Here the $(1,0)^T$ and $(0,1)^T$ symbols refer to the spin up and spin down states of conventional quantum mechanics, often written as $| \uparrow \rangle$ and $| \downarrow \rangle$ respectively. The dimensionality has doubled because we have not yet taken the complex structure of the spinors into account. While the role of j is played in the two single-particle spaces by right multiplication by $i\sigma_3^1$ and $i\sigma_3^2$ respectively, standard quantum mechanics does not distinguish between these operations. A projection operator must therefore be included to ensure that right multiplication by $i\sigma_3^1$ or $i\sigma_3^2$ reduces to the same operation. If a 2-particle spin state is represented by the multivector ψ, then ψ must satisfy

$$\psi i\sigma_3^1 = \psi i\sigma_3^2 \qquad (5.64)$$

from which we find that

$$\psi = -\psi i\sigma_3^1 i\sigma_3^2$$
$$\Rightarrow \quad \psi = \psi \tfrac{1}{2}(1 - i\sigma_3^1 i\sigma_3^2). \qquad (5.65)$$

On defining

$$E = \tfrac{1}{2}(1 - i\sigma_3^1 i\sigma_3^2), \qquad (5.66)$$

we find that

$$E^2 = E \qquad (5.67)$$

so right multiplication by E is a projection operation. (The relation $E^2 = E$ means that E is technically referred to as an 'idempotent' element.) It follows that the 2-particle state ψ must contain a factor of E on its right-hand side. We can further define

$$J = Ei\sigma_3^1 = Ei\sigma_3^2 = \tfrac{1}{2}(i\sigma_3^1 + i\sigma_3^2) \tag{5.68}$$

so that

$$J^2 = -E. \tag{5.69}$$

Right-sided multiplication by J takes on the role of j for multiparticle states.

The STA representation of a direct-product 2-particle Pauli spinor is now given by $\psi^1\phi^2 E$, where ψ^1 and ϕ^2 are spinors (even multivectors) in their own spaces. A complete basis for 2-particle spin states is provided by

$$
\begin{aligned}
\begin{pmatrix} 1 \\ 0 \end{pmatrix} \otimes \begin{pmatrix} 1 \\ 0 \end{pmatrix} &\quad\leftrightarrow\quad E \\
\begin{pmatrix} 0 \\ 1 \end{pmatrix} \otimes \begin{pmatrix} 1 \\ 0 \end{pmatrix} &\quad\leftrightarrow\quad -i\sigma_2^1 E \\
\begin{pmatrix} 1 \\ 0 \end{pmatrix} \otimes \begin{pmatrix} 0 \\ 1 \end{pmatrix} &\quad\leftrightarrow\quad -i\sigma_2^2 E \\
\begin{pmatrix} 0 \\ 1 \end{pmatrix} \otimes \begin{pmatrix} 0 \\ 1 \end{pmatrix} &\quad\leftrightarrow\quad i\sigma_2^1 i\sigma_2^2 E.
\end{aligned}
\tag{5.70}
$$

This procedure extends simply to higher multiplicities. All that is required is to find the 'quantum correlator' E_n satisfying

$$E_n i\sigma_3^j = E_n i\sigma_3^k = J_n \qquad \text{1for all } j,\ k. \tag{5.71}$$

E_n can be constructed by picking out the $j = 1$ space, say, and correlating all the other spaces to this, so that

$$E_n = \prod_{j=2}^n \tfrac{1}{2}(1 - i\sigma_3^1 i\sigma_3^j). \tag{5.72}$$

The value of E_n is independent of which of the n spaces is singled out and correlated to. The complex structure is defined by

$$J_n = E_n i\sigma_3^j, \tag{5.73}$$

where $i\sigma_3^j$ can be chosen from any of the n spaces. To illustrate this consider the case of $n = 3$, where

$$
\begin{aligned}
E_3 &= \tfrac{1}{4}(1 - i\sigma_3^1 i\sigma_3^2)(1 - i\sigma_3^1 i\sigma_3^3) \tag{5.74} \\
&= \tfrac{1}{4}(1 - i\sigma_3^1 i\sigma_3^2 - i\sigma_3^1 i\sigma_3^3 - i\sigma_3^2 i\sigma_3^3) \tag{5.75}
\end{aligned}
$$

and

$$J_3 = \tfrac{1}{4}(i\sigma_3^1 + i\sigma_3^2 + i\sigma_3^3 - i\sigma_3^1 i\sigma_3^2 i\sigma_3^3). \tag{5.76}$$

Both E_3 and J_3 are symmetric under permutations of their indices.

The above was framed for non-relativistic Pauli spinors, but in fact, the whole discussion also applies to Dirac spinors, since these are represented by even elements and multiplication by j is still right-sided multiplication by $i\sigma_3$. A significant feature of this approach is that all the operations defined for the single-particle STA extend naturally to the multiparticle algebra. The reversion operation, for example, still has precisely the same definition — it simply reverses the order of vectors in any given multivector. The spinor inner product also generalises immediately, to

$$(\psi, \phi)_S = \langle E_n \rangle^{-1} [\langle \tilde{\psi}\phi \rangle - \langle \tilde{\psi}\phi J_n \rangle i\sigma_3], \qquad (5.77)$$

where the right-hand side is projected onto a single copy of the STA. The factor of $\langle E_n \rangle^{-1}$ is included so that the state '1' always has unit norm, which matches with the inner product used in the matrix formulation.

21.5.3   A 6-d representation in the MSTA

Much more could be said about the properties and applications of the MSTA, but here we wish to use it in a novel linking-together of quantum mechanics, twistors and conformal geometry.

Let ϕ be a (single particle) Dirac spinor, and $r = t\gamma_0 + x\gamma_1 + y\gamma_2 + z\gamma_3$ be the position vector in 4-d space.

Consider the operator $\hat{r}$, mapping Dirac spinors to Dirac spinors, given by

$$\phi \mapsto \hat{r}(\phi) \equiv r\phi i\gamma_3 \tfrac{1}{2}(1 + \sigma_3). \qquad (5.78)$$

The operator $(1 + \hat{r})$ has the remarkable property of leaving the inner product between Dirac spinors invariant. Specifically, we have

$$\langle \tilde{\psi}'\phi' \rangle_S = \langle \tilde{\psi}\phi \rangle_S, \qquad (5.79)$$

where

$$\psi' = (1 + \hat{r})\psi \quad \text{and} \quad \phi' = (1 + \hat{r})\phi. \qquad (5.80)$$

(The subscript S applied in this single-particle case just means the scalar and $i\sigma_3$ parts only are taken.) This relation is true for any Dirac spinors ψ and ϕ. We note further $\hat{r}^2 = 0$, so we can write $(1 + \hat{r})$ in the 'rotor' form $e^{\hat{r}}$.

Now consider the following two-particle quantum state:

$$\epsilon = \left(i\sigma_2^1 - i\sigma_2^2 \right) \tfrac{1}{2}\left(1 - \sigma_3^1\right) \tfrac{1}{2}\left(1 - \sigma_3^2\right) \tfrac{1}{2}\left(1 - i\sigma_3^1 i\sigma_3^2\right), \qquad (5.81)$$

This is a relativistic generalisation of the non-relativistic Pauli singlet state (see Doran et al [15]). Specifically it can be shown that it obeys

$$R^1 R^2 \epsilon = R^1 \tilde{R}^1 \epsilon = \epsilon \qquad (5.82)$$

for any (Lorentz) rotor R, and is therefore relativististically invariant. We now use this to construct our first '6-d' point as follows:

$$\psi = e^{\hat{r}^1} e^{\hat{r}^2} \epsilon. \tag{5.83}$$

ψ here is a 2-particle wavefunction which provides a representation of the 4-d point r. We shall see shortly in what way it connects with 6 dimensions. Firstly, however, note that this ψ has vanishing norm viewed as a 2-particle wavefunction:

$$\langle \tilde{\psi}\psi \rangle_S = 0. \tag{5.84}$$

More generally, let

$$\phi = e^{\hat{s}^1} e^{\hat{s}^2} \epsilon, \tag{5.85}$$

correspond to some different 4-d position s. Then we find

$$\langle \tilde{\phi}\psi \rangle_S = -\frac{1}{4}(r - s)^2. \tag{5.86}$$

Just as with the 'horosphere' construction used by Hestenes, we see we have found a way of turning differences into products, except here it is taking place in a relativistic context.

The way the quantum state links with 6 dimensions is as follows. The state space for relativistic spinors describing two particles is 16 complex dimensional (as effectively the outer product of two Dirac spinors) and splits into a 10-d space symmetric under particle interchange (i.e. swapping of the 1 and 2 labels) and a 6-d space anti-symmetric under interchange. This 6-d space is 'complex' (i.e. with 12 real degrees of freedom), but we can define a 'real' subspace of it via taking the following as being the general point:

$$\psi_P = (V + W)\epsilon' + P^1 \epsilon i\gamma_3^1 + P^2 \epsilon i\gamma_3^2 + (V - W)\epsilon. \tag{5.87}$$

Here

$$\epsilon' = -\gamma_0^1 \gamma_0^2 \epsilon \gamma_0^1 \gamma_0^2 = \left(i\sigma_2^1 - i\sigma_2^2 \right) \tfrac{1}{2} \left(1 + \sigma_3^1 \right) \tfrac{1}{2} \left(1 + \sigma_3^2 \right) \tfrac{1}{2} \left(1 - i\sigma_3^1 i\sigma_3^2 \right), \tag{5.88}$$

and

$$P = T\gamma_0 + X\gamma_1 + Y\gamma_2 + Z\gamma_3. \tag{5.89}$$

V, W, T, X, Y and Z are the coordinates of a 6-d real space with metric

$$ds^2 = dT^2 + dV^2 - dW^2 - dX^2 - dY^2 - dZ^2. \tag{5.90}$$

The extra dimensions V and W allow the formation of the combinations $V + W$ and $V - W$, which correspond to null directions in the 6-d space. (These directions are the equivalent of the e and e^* introduced by Hestenes in the 5-d case.)

The representation of 4-d points proceeds via working with points on the 'null cone' in 6-d. For these points we relate the 6-d space to ordinary 4-d Lorentz space projectively via

$$t = \frac{T}{V - W}, \quad x = \frac{X}{V - W}, \quad y = \frac{Y}{V - W}, \quad z = \frac{Z}{V - W}. \tag{5.91}$$

The way this relates to our previous construction is as follows:

$$\psi_P = (V - W) \, e^{\hat{r}^1} e^{\hat{r}^2} \epsilon, \tag{5.92}$$

i.e. it is simply a scaled version of the state generated by the rotor construction. We can see this by taking the length of ψ_P, via the norm of the quantum state:

$$\langle \tilde{\psi}_P \psi_P \rangle = \tfrac{1}{2}(V^2 - W^2 + T^2 - X^2 - Y^2 - Z^2). \tag{5.93}$$

This being null implies

$$T^2 - X^2 - Y^2 - Z^2 = -(V^2 - W^2), \tag{5.94}$$

i.e.

$$t^2 - x^2 - y^2 - z^2 = -\left(\frac{V + W}{V - W}\right). \tag{5.95}$$

ψ_P is thus just

$$(V - W)\left(-|r|^2 \epsilon' + r^1 \epsilon \, i\gamma_3^1 + r^2 \epsilon \, i\gamma_3^2 + \epsilon\right), \tag{5.96}$$

which is $(V - W) \, e^{\hat{r}^1} e^{\hat{r}^2} \epsilon$ as claimed.

21.5.4   Link with twistors

The above has been framed as a mixture of 2-particle relativistic quantum mechanics (written in the MSTA) and conformal geometry. It also links directly with *twistor* theory (see e.g. Penrose & Rindler, Vol. 2 [23]). Twistors were introduced by Penrose as objects describing the geometry of spacetime at a 'pre-metric' level (partially in an attempt to allow an alternative route to quantum gravity). Instead of points and a metric, the idea is that twistors can represent incidence relations between null rays. Spacetime points and their metric relations then emerge as a secondary concept, corresponding to the points of intersection of null lines. As discussed in Lasenby, Doran & Gull [24], in geometric algebra twistors are translated as Dirac spinors with a particular position dependence. Specifically, a twistor, which is written in 2-spinor notation as

$$\mathsf{Z}^\alpha = (\omega^A, \pi_{A'}) \tag{5.97}$$

is translated as the Dirac spinor Z given by

$$Z = \phi - r\phi i\gamma_3 \tfrac{1}{2}(1 + \sigma_3), \tag{5.98}$$

where r is the 4-d position vector and ϕ is the (constant) Dirac spinor

$$\phi = \omega_0 \tfrac{1}{2}(1 + \sigma_3) - \pi i\sigma_2 \tfrac{1}{2}(1 - \sigma_3), \tag{5.99}$$

with ω_0 and π the geometric algebra Pauli spinors corresponding to the Penrose & Rindler 2-spinors ω_0^A and $\pi_{A'}$. (ω_0^A is ω^A evaluated at the origin.)

What we can observe now is that Z is none other than $e^{-\hat{r}}\phi$. This links twistors with the previous section. In twistor theory, given two twistors Z and X satisfying certain conditions, we can find a spacetime point corresponding to their intersection via forming the skew $\begin{bmatrix} 2 \\ 0 \end{bmatrix}$ twistor

$$\mathsf{R}^{\alpha\beta} = \mathsf{Z}^\alpha\mathsf{X}^\beta - \mathsf{X}^\alpha\mathsf{Z}^\beta \tag{5.100}$$

(see Penrose & Rindler, Vol. 2, [23], p. 65 and p. 305). Without going into the details, it turns out that what we have described in the previous section corresponds precisely to this construction, but instantiated in a concrete fashion in the MSTA. In particular, the twistor relation

$$\mathsf{S}^{\alpha\beta}\mathsf{R}_{\alpha\beta} = -(s^a - r^a)(s_a - r_a) \tag{5.101}$$

(Equation 6.2.30 of Penrose & Rindler, Vol. II) corresponds precisely to both our Equation 5.86 and the Hestenes 'horosphere' relation Equation 5.55. The latter is already prefigured therefore in twistor geometry.

It might be wondered why, if corresponding constructions exist in twistor theory, it is useful to have a version in geometric algebra. The advantages of the latter are twofold. Firstly, there is the economy of using a single algebraic system for all areas as different as quantum mechanics, conformal geometry, screw theory etc. Secondly, we can use the geometric algebra to do things which are not easily possible within twistor theory, but which extend its results in a very neat fashion. For example, in the next section we show how the full special conformal group of Lorentzian spacetime can be realized via very simple transformations in our two particle space. The corresponding operations would be much harder to display explicitly in twistor theory.

As a final remark in this area, we note that twistor theory encourages one to think about a *complexified* version of Lorentzian spacetime. The same occurs in our present constructions via the fact that the 2-particle antisymmetric space is actually 12-dimensional, allowing us to have a complex version of the 6-d conformal space. In order to understand some areas of practical computer vision, we apparently require a complex projective space; this is the case particularly for camera calibration using the concepts of the *absolute conic* and *absolute quadric*. A complex version of our 6-d conformal space may turn out to be very useful in allowing us to find a natural home for such entities in geometric algebra. This area is currently being explored.

21.5.5   The special conformal group

We now look briefly at how rotations, dilations, inversions, translations and special conformal motions in Lorentzian spacetime can be represented via simple transformations in our two particle space. This parallels the equivalent analysis in the 5-d case given by Hestenes for motions in Euclidean space, except that here they emerge in a (perhaps surprising) fashion as operations within relativistic quantum mechanics. In 3-d the importance of such motions is that they preserve the *angles* between vectors, and thus are next in generality as regards rigid body motion if we wish to go beyond the strictly Euclidean transformations of translation and rotation. In 4-d, they are of great interest in physics from the point of view of conformally invariant theories, such as electromagnetism and massless fields, and may be of interest in engineering for the description of rigid body motion where velocities and not just positions are specified, and also in projective spaces. We now describe in each case the required operation, and indicate why it works.

Translations:

Here we just need to note that the operators $e^{\hat{r}}$ for different r's are all mutually commutative. Thus if we have a point r in 4-d that we wish to move to $r + s$, where s is another 4-d position vector, we just need to carry out the transformation

$$\psi_P \mapsto \psi_P' = e^{(\hat{s}^1 + \hat{s}^2)}\psi_P. \tag{5.102}$$

Rotations

These are easily accomplished. Given a Lorentz rotor R, we rotate in the 2-particle space via

$$\psi_P \mapsto \psi_P' = R^1 R^2 \psi_P. \tag{5.103}$$

This works since e.g. the r^1 term in the expansion for ψ_P responds like

$$R^1 R^2 r^1 \epsilon\, i\gamma_3 = R^1 r^1 R^2 \epsilon\, i\gamma_3 = R^1 r^1 \tilde{R}^1 \epsilon\, i\gamma_3. \tag{5.104}$$

Inversions

The aim here, in the 4-d space, is to have $r \mapsto r/|r|^2$. Since the coefficient of ϵ' in the ψ_P expansion is $-|r|^2$, the way to achieve this in the 2-particle space would be to swap the roles of ϵ and ϵ'. We can achieve this by multiplying on the right by $i\sigma_2^1 i\sigma_2^2$, since this swaps both ideals. At the same time one finds

$$r^1 \epsilon i\gamma_3^1 i\sigma_2^1 i\sigma_2^2 = -r^2 \epsilon i\gamma_3^2, \qquad (5.105)$$

and *vice-versa*. Thus the required operation is

$$\psi_P \mapsto \psi_P' = \psi_P i\sigma_2^1 i\sigma_2^2. \qquad (5.106)$$

Dilations

Here in 4-d space we want r to transform to $e^\alpha r$, where α is a scalar. In the 2-particle space we need a rotor operation which can accomplish this. Like inversion, it is clear that we need to swap the roles of the $\frac{1}{2}(1 + \sigma_3)$ and $\frac{1}{2}(1 - \sigma_3)$ ideals, only this time it needs to happen in a gradual fashion. It is easy to show that the required operation is

$$\psi_P \mapsto \psi_P' = \psi_P e^{\alpha/2(\sigma_3^1 + \sigma_3^2)}. \qquad (5.107)$$

Special conformal motions

These motions are in fact composites of inversions and translations, so in a sense we have already done these. However, the resulting expression for the operation in the 2-particle space is quite neat, so we give the results explicitly. In 4-d space we want to achieve the motion

$$r \mapsto r \frac{1}{1 + sr}, \qquad (5.108)$$

where s is a constant vector. This can be generated via inverting r, translating by s and then inverting again (see Hestenes & Sobczyk [25], p. 218). In our case, the combination of two inversions amounts to changing the ideal used in $e^{\hat{r}}$ to its opposite, plus a change of sign for the vector. Thus if we define the new operator $\check{r}$ via

$$\phi \mapsto \check{r}(\phi) \equiv r\phi i\gamma_3 \tfrac{1}{2}(1 - \sigma_3), \qquad (5.109)$$

we see that the overall operation we want is

$$\psi_P \mapsto \psi_P' = e^{-(\check{s}^1 + \check{s}^2)}\psi_P. \qquad (5.110)$$

21.5.6   6-d space operations

Although above we have confined ourselves to setting up the basic correspondence between conformal operations and 'quantum' operations in the 2-particle space, it is of interest to relate these operations directly to the operations that would be carried out in a 6-d space generalising the 'horosphere' construction. The simplest version of such a space uses the representation discussed at the end of Hestenes & Sobczyk [25]. At the risk of causing great confusion, we shall stick with the original Hestenes & Sobczyk notation, which has e and $\bar{e}$, satisfying

$$e^2 = -\bar{e}^2 = 1, \tag{5.111}$$

as the new vectors which would be added to make up a $(1,3)$ space with vectors r say, up to a $(2,4)$ conformal space. The null vectors formed from e and $\bar{e}$ are defined by

$$n = e + \bar{e}, \quad \bar{n} = e - \bar{e}. \tag{5.112}$$

The crucial representation formula, relating (in this case) 4-d vectors r to their 6-d equivalents $F(r)$ is

$$F(r) = -(r - e)e(r - e) + (r - e)^2\bar{e}, \tag{5.113}$$

([25], eqn 3.14). Re-expressing this in terms of the null vectors, one finds

$$F(r) = r^2 n + 2r - \bar{n}. \tag{5.114}$$

We should compare this equation (5.114), with our quantum representation (5.96). It is clear that how they work is that (up to signs) the relativistic singlet state ϵ takes on the role of the null vector n, its version using the opposite ideals, ϵ', takes on the role of $\bar{n}$ and the middle term $r^1\epsilon\, i\gamma_3^1 + r^2\epsilon\, i\gamma_3^2$ is an expanded version (appropriate to the 2-particle space) of the vector $2r$.

It is now very interesting to compare some of the actions of the conformal group in the two approaches. Taking inversion as an example, this operation is not discussed in Hestenes & Sobczyk, but it is easy to see that we invert a 4-d point, $r \mapsto r/|r|^2$, via reflection in the unit vector e. Explicitly, we carry out

$$F(r) \mapsto eF(r)e. \tag{5.115}$$

This swaps the roles of n and $\bar{n}$. In the 2-particle case, we know inversion is accomplished by right multiplication by $i\sigma_2^1 i\sigma_2^2$, since this swaps the quantities ϵ and ϵ'. Thus the quantum operation of swapping the spin states (up $\longleftrightarrow$ down) of the 2-particles (which is what the $i\sigma_2^1 i\sigma_2^2$ multiplication achieves), parallels the operation of *reflection* in the 6-d space. This hints at a deep geometrical connection between the two spaces, which will be investigated further elsewhere.

21.6   Summary and Conclusions

In this contribution we have seen that geometric algebra is able to span an enormous range of physics and mathematical physics. From the rest of this volume it is clear that GA is useful in many areas of engineering also. Thus GA stands ready to be adopted as a useful and efficient tool by scientists and engineers in a wide variety of fields, with consequent benefit for mutual comprehensibility. Even areas considered as difficult as general relativity have been shown to be understandable within GA using just simple tools of linear function theory. The links between the new conformal representation of Euclidean geometry, twistors and multiparticle quantum theory have been shown to be both fascinating and unexpected. Much more work is possible along this direction, including the possible role of *complex* projective and conformal geometry, and of relativistic spaces in allowing representation of velocity as well as position transformations.

Part VII

Computational Methods in Clifford Algebras

Chapter 22

Clifford Algebras as Projections of Group Algebras

Vladimir M. Chernov

22.1   Introduction

Clifford algebras appeared as a result of the natural desire of mathematicians to extend a finite-dimensional vector space to an algebraic structure where the inner and outer products are defined in terms of a single geometric multiplication [14], [26], [1], [25]. This idea was most attractively developed by D.Hestenes [20], [21]), and was immediately accepted by some physicists.

The effective application of Clifford algebras to the Computer Sciences is based on two ideas. First, some problems in robotics and computer vision have a direct physical analog in higher dimensions [22], [16]. For example, the 8-dimensional geometric algebra of Euclidean space extends the 4-dimensional quaternion algebra for analysis and motion simulation in robotics [2], [23] computer vision [23], [15] and neurocomputing [3].

Second, Clifford algebras can be used for data representation and increasing the effectiveness of some signal processing algorithms. For example, data representation in the quaternion algebra or in the 2×2 matrix algebra permits effective multidimensional FFT algorithms "with multioverlapping" [7], [8].

In the first case, the elements of a Clifford algebra are easy to interpret geometrically. In the second case, the existence of a large Clifford algebra group of automorphisms greatly facilates calculations.

Signal (image) analysis and digital processing tasks have a clear physical origin. Their solution requires a significant amount of calculations and effective algorithmic support. For these tasks it is necessary to develop and use algebraic means which have both a clear physical interpretation and calculational efficiency.

Many effective methods of discrete signal processing are based on algorithms in which most arithmetic operations are multiplications by the constant numbers of some finite set (e.g. fast algorithms (FA) of discrete orthogonal transforms (DOT) [4]). Thus, the idea of increasing the dimension by the inclusion of "popular" constants seems to be very attractive.

This article introduces the corresponding formal structures, and gives e-xamples of their application to the synthesis of DOTs fast algorithms.

22.2   Group Algebras and Their Projection

22.2.1   Basic definitions and examples

Definition 1 *Let $\nu_j \in \mathbb{R}$, $\nu_1, \ldots, \nu_p > 0$, $\nu_{p+1}, \ldots, \nu_d < 0$, $q = d - p$. According to [1], the Clifford algebra $C(p, q)$ of a d-dimensional vector space with the basis $\{\mathbf{e}_1, \ldots, \mathbf{e}_d\}$ is the 2^d-dimensional associative algebra with the basis*

$$\{\mathbf{E}_A = \mathbf{e}_1^{\alpha_1} \ldots \mathbf{e}_d^{\alpha_d}; \quad A = (\alpha_1, \ldots, \alpha_d), \quad \alpha_j = 0, 1\}, \qquad (2.1)$$

together with the multiplication of the basis elements $\mathbf{E}_A$ induced by the multiplication of the elements $\mathbf{e}_j$,

$$\mathbf{e}_j^2 = \nu_j, \quad \mathbf{e}_i \mathbf{e}_j = -\mathbf{e}_j \mathbf{e}_i. \qquad (2.2)$$

Definition 2 *Let $\mathbf{G}$ be a $D-$element finite group. The group algebra $\mathbf{A}(\mathbf{F}, \mathbf{G})$ over the field $\mathbf{F}$ is a D-dimensional associative algebra*

$$\mathbf{A}(\mathbf{F}, \mathbf{G}) = \left\{ Q = \sum_{g_t \in \mathbf{G}} \gamma_t g_t; \quad \gamma_t \in \mathbf{F} \right\} \qquad (2.3)$$

with the multiplication rule

$$Q_1 Q_2 = \sum_{t=1}^{D} \left(\sum_{<p,q:\ g_p g_q = g_t>} \gamma_p^{(1)} \gamma_q^{(2)} \right) g_t. \qquad (2.4)$$

Example 1

The quaternion algebra $\mathbf{H}$ is usually understood as an associative $4-$dimensional algebra over $\mathbf{R}$ with the basis $\{1, \mathbf{i}, \mathbf{j}, \mathbf{k},\}$ and the multiplication rules:

$$\mathbf{i}^2 = \mathbf{j}^2 = \mathbf{k}^2, \qquad \mathbf{ij} = -\mathbf{ji} = \mathbf{k}. \qquad (2.5)$$

On the other hand, the finite quaternion group $\mathbf{H}_8$ contains 8 elements usually denoted [19] by

$$\mathbf{H}_8 = \{\pm 1, \pm \mathbf{i}, \pm \mathbf{j}, \pm \mathbf{k},\} \qquad (2.6)$$

with the multiplication rule (2.5) and the additional stipulation that

$$(-1)a = -a, (a \in \mathbf{H}_8).$$

The designation (2.6) is a consequence of the connection between the rule for multiplication in the group $\mathbf{H}_8$ and in the algebra $\mathbf{H}$ that has been found to be useful. Note, however, that the elements of the algebra $\mathbf{H}$ and the group $\mathbf{H}_8$ are of a fundamentally different kind. This difference becomes even more evident if for the elements of the group $\mathbf{H}_8$ we use "neutral" designations $a_1, ..., a_8$ (see [19], Ch.1., Example 5). We may employ another description of the group $\mathbf{H}_8$ in the form (2.6),

$$\mathbf{H}_8 = \{\varepsilon, \tau, I, J, IJ, \tau I, \tau J, \tau IJ\},$$

with the rules of group multiplication:

- *the element ε is the identity of the group* $\mathbf{H}_8$;

- *the element τ is an involution:* $\tau^2 = \varepsilon$;

- $IJ = \tau JI$.

The element τ is like the number (-1).
Let us consider an 8-dimensional group algebra $\mathbf{A}(\mathbf{R}, \mathbf{H}_8)$

$$\mathbf{A}(\mathbf{R}, \mathbf{H}_8) = \left\{ Q = \sum_{g_t \in \mathbf{H}_8} \gamma_t g_t; \quad \gamma_t \in \mathbb{R} ; \quad t = 1, \ldots, 8 \right\}.$$

with multiplication defined by

$$Q_1 Q_2 = \sum_{t=1}^{8} \left(\sum_{<p,q: \quad g_p g_q = g_t>} \gamma_p^{(1)} \gamma_q^{(2)} \right) g_t. \tag{2.7}$$

The mapping $\Psi : \mathbf{A}(\mathbf{R}, \mathbf{H}_8) \to \mathbf{H}$ defined by

$$\varepsilon \longmapsto 1, \quad \tau \longmapsto (-1), \quad I \longmapsto \mathbf{i}, \quad J \longmapsto \mathbf{j},$$

$$\Psi(Q_1 Q_2) = \Psi(Q_1)\Psi(Q_2)$$

is linearly extended to an homomorphism of the algebra $\mathbf{A}(\mathbf{R}, \mathbf{H}_8)$ into the algebra $\mathbf{H}$. In other words, only the homomorphism (projection) Ψ of the group algebra $\mathbf{A}(\mathbf{R}, \mathbf{H}_8)$ into $\mathbf{H}$ maps the involution τ into the number (-1). Thus, the quaternion algebra is a "projection" of the 8−dimensional group algebra $\mathbf{A}(\mathbf{R}, \mathbf{H}_8)$.

This process of generating Clifford algebras is not limited to the algebra $\mathbf{H} = C(0, 2)$. We can, for example, find another group algebra which projects into the Clifford algebra $G(1, 1)$ of the pseudo-Euclidean plane.

Example 2

Consider the 8-element dihedral group $\mathbf{D}_4$ (see [19], Ch.1., Example 1)

$$\mathbf{D}_4 = \{\varepsilon, \tau, \sigma^1, \sigma^2, \sigma^3, \tau\sigma^1, \tau\sigma^2, \tau\sigma^3\}, \tag{2.8}$$

where ε is identity element of the group $\mathbf{D}_4$, and group multiplication is defined by

$$\tau^2 = \varepsilon, \quad \sigma^4 = \varepsilon, \quad \tau\sigma = \sigma^3\tau, \tag{2.9}$$

and group algebra $\mathbf{A}(\mathbf{R}, \mathbf{D}_4)$ with multiplication given by (2.4) and (2.9). The Clifford algebra $G(1,1)$ is a 4-dimensional algebra with the basis $\{1, \mathbf{e}_1, \mathbf{e}_2, \mathbf{e}_1\mathbf{e}_2\}$ and the multiplication

$$\mathbf{e}_1^2 = 1, \quad \mathbf{e}_2^2 = -1 \quad \mathbf{e}_1\mathbf{e}_2 = -\mathbf{e}_2\mathbf{e}_1.$$

The mapping Ψ of the $8-$dimensional group algebra of the dihedral group into $G(1,1)$ is defined by the following rules

$$\varepsilon \longmapsto 1, \quad \tau \longmapsto \mathbf{e}_1, \quad \sigma^2 \longmapsto (-1), \quad \sigma \longmapsto \mathbf{e}_2,$$

$$\Psi(Q_1Q_2) = \Psi(Q_1)\Psi(Q_2), \quad Q_1, Q_2 \in \mathbf{A}(\mathbf{R}, \mathbf{D}_4)$$

and is linearly extended to an homomorphism of the algebra $\mathbf{A}(\mathbf{R}, \mathbf{D}_4)$ into the Clifford algebra $G(1,1)$.

Where can we go with these ideas?

22.2.2   Clifford algebras as projections of group algebras

The above examples lead us to the description of Clifford algebras as projections of group algebras.

Theorem 1 Let $C(p,q)$ the 2^d-dimensional Clifford algebra defined by the relations (2.1) and (2.2). Then there exists a $2^{d+1}-$ group $\mathbf{G}$ and an homomorphism $\Psi : \mathbf{A}(\mathbf{R}, \mathbf{G}) \longrightarrow C(p,q)$ such that

$$Im\Psi \cong C(p,q), \quad (q = d - p).$$

Proof Let τ be the central involution, and let the elements $g_1, \ldots, g_d, \tau$ satisfy the relations

$$\tau^2 = \varepsilon, \quad g_1^2 = \cdots = g_p^2 = \varepsilon, \quad g_{p+1}^2 = \cdots = g_d^2 = \tau;$$
$$\tau g_j = g_j\tau, \quad g_ig_j = \tau g_jg_i, \quad i,j = 1, \ldots, d.$$

The group $\mathbf{G}$ is generated by the elements $\tau, g_1, \ldots, g_d$ and contains the 2^{d+1} elements

$$\mathbf{G} = \{\tau^{\alpha_0}g_1^{\alpha_1}\cdots g_d^{\alpha_d}; \quad \alpha_k = 0,1; \quad k = 0, \ldots, d\}.$$

Now consider the group algebra $\mathbf{A}(\mathbf{R}, \mathbf{G})$ and the mapping $\Psi : \mathbf{A}(\mathbf{R}, \mathbf{G}) \longrightarrow C(p, q)$ induced by the mapping ψ,

$$\psi(\varepsilon) = 1, \quad \psi(\tau) = -1, \quad \psi(g_j) = \mathbf{e}_j.$$

The homomorphism Ψ maps $\mathbf{A}(\mathbf{R}, \mathbf{G})$ into the factor-algebra $\mathbf{A}(\mathbf{R},\mathbf{G})/\mathbf{J}$, where $\mathbf{J}$ is the main ideal, generated by the element $(\varepsilon + \tau)$. It is easy to see that the factor-algebra has dimension 2^d and that the multiplications rules are the same as (2.1) and (2.2). $\blacksquare$

22.3 Applications

The discrete orthogonal transforms (DOT)

$$\hat{x}(m) = \sum_{n=0}^{N-1} x(n) h_m(n), \quad m = 0, \ldots, N-1, \quad x(n) \in \mathbb{R}, \mathbb{C}, \quad (3.10)$$

where $x(n)$ is an input signal and $\{h_m(n)\}$ is the set of basis functions satisfying the orthogonality conditions

$$\sum_{n=0}^{N-1} h_p(n) h_q^*(n) = \delta_{pq},$$

where $*$ denotes complex conjugation. The DOT are one of the main means of digital signal processing and have been discussed in many publications (see [4], [24], [18], etc.).

22.3.1 Fast algorithms of the discrete Fourier transform

The well-known "overlapped" one-dimensional FFT is built on the possibility of obtaining computational advantage at the expense of the redundancy of the representation
of a real input signal $x(n)$. More exactly, the possibility of constructing an overlapped algorithm exists due to the presence in the field of complex number $\mathbb{C}$ of a non-trivial automorphism, complex conjugation, acting identically upon $\mathbb{R}$,

$$\hat{x}(m) = \sum_{n=0}^{N-1} x(n) \exp\{2\pi i \frac{mn}{N}\}, \quad m = 0, \ldots, N-1, \quad N = 2^r, \quad x(n) \in \mathbb{R}.$$

Let us define the auxiliary sequence

$$z(n) = x(2n) + ix(2n+1) = x_0(n) + ix_1(n).$$

$$\hat{z}(m) = \sum_{n=0}^{\frac{N}{2}-1} z(n) \exp\{2\pi i \frac{2mn}{N}\}, \quad m = 0, \ldots, \frac{N}{2} - 1.$$

Then, "partial spectra"

$$\hat{x}_0(m) = \sum_{n=0}^{\frac{N}{2}-1} x_0(n) \exp\{2\pi i \frac{2mn}{N}\}, \quad \hat{x}_1(m) = \sum_{n=0}^{\frac{N}{2}-1} x_1(n) \exp\{2\pi i \frac{2mn}{N}\}$$

can be found from

$$2\hat{x}_0(m) = \hat{z}(m) + \overline{\hat{z}(-m)}, \quad 2i\hat{x}_1(m) = \hat{z}(m) - \overline{\hat{z}(-m)}.$$

The full spectrum reconstruction is realized by the relation

$$\hat{x}(m) = \hat{x}_0(m) + \hat{x}_1(m) \exp\{2\pi i \frac{2m}{N}\}.$$

Note that the transition to the complex-conjugate does not require additional arithmetic operations.

It is known that the majority of fast algorithms (of the Cooley-Tuckey type) of the discrete Fourier transform (DFT) have the complexity

$$W(N) = \lambda N \log_2 N + O(N), \tag{3.11}$$

where the constant λ characterizes a particular scheme of the algorithm [4]. The complexity of the "ovelapped" FFT is

$$W(N) = \frac{1}{2}\lambda N \log_2 N + O(N).$$

If the DFT is multi-dimensional (in particular, a 2-D transform), the use of the techniques discussed above has the problem that the field $\mathbb{C}$ has "too few" automorphisms for the

separation of spectra. This leads to the necessity of embedding the field $\mathbb{C}$ into an algebraic structure possessing a sufficiently large number of trivially implemented automorphisms over $\mathbb{R}$. In the papers [7], [8], [6], [10], [9], I developed this approach to solve the problem of DOTs fast algorithms.

Tasks similar to those of DOTs FA complexity reduction

(the task for a multichannel signal) can be solved using similar methods. We

consider the following

examples of how the technique described above can be applied.

22.3.2   Fast algorithms for five Fourier spectra calculation in the algebra $\mathbf{A}(\mathbf{R}, \mathbf{S}_3)$

Let $\mathbf{S}_3 = \tau^2 = \varepsilon$, $\sigma^3 = \varepsilon$, $\tau\sigma = \sigma^2\tau$ be the six-element permutation group with the neutral element ε. The elements τ, σ satisfy the multiplication rules

$$\mathbf{S}_3 = \{\varepsilon, \sigma, \sigma^2, \tau, \tau\sigma, \tau\sigma^2\}.$$

Lemma 2 *Let* $\alpha, \beta, p, q \in \mathbf{F}$, *where* $\mathbf{F}$ *is a field. For the elements* X *and* Y *given by*

$$X = \alpha p + \beta q, \quad Y = \beta q + \alpha p,$$

only two multiplications in the field $\mathbf{F}$ *are required.*

Proof It is easy to show that

$$\begin{aligned} X &= (\alpha + \beta)\left(\frac{p+q}{2}\right) + (\alpha - \beta)\left(\frac{p-q}{2}\right), \\ Y &= (\alpha + \beta)\left(\frac{p+q}{2}\right) - (\alpha - \beta)\left(\frac{p-q}{2}\right). \quad \blacksquare \end{aligned}$$

Lemma 3 *The multiplication of* $z \in \mathbf{A}(\mathbf{F}, \mathbf{S}_3)$:

$$z = A\varepsilon + C\sigma + D\sigma^2 + B\tau + F\tau\sigma + H\tau\sigma^2 \tag{3.12}$$

by the element $w = x\sigma + y\sigma^2$, *requires 8 multiplications of elements in the field* $\mathbf{F}$.

Proof The direct calculation gives

$$\begin{aligned} zw &= \alpha_1\varepsilon + \alpha_2\tau + \alpha_3\sigma + \alpha_4\sigma^2 + \alpha_5\tau\sigma + \alpha_6\tau\sigma^2 \\ &= \left[(Dx + Cy)\varepsilon + (Hx + Fy)\tau\right] \\ &\quad + \left[(Ax + Dy)\sigma + (Cx + Ay)\sigma^2\right] \\ &\quad + \left[(Bx + Hy)\tau\sigma + (Fx + By)\tau\sigma^2\right]. \end{aligned}$$

As shown in Lemma 1, four multiplications are required to calculate

$$\begin{array}{ll} \beta_3 = Ax + Cy & \beta_5 = Bx + Hy \\ \beta_4 = Cx + Ay \quad \text{and} \quad & \beta_6 = Fx + By \end{array}.$$

Since the following relations are also true

$$\begin{aligned} \alpha_3 &= \beta_3 + y(D - C) = \beta_3 + \gamma_3, \quad \alpha_4 = \beta_4, \\ \alpha_5 &= \beta_5 - y(F - H) = \beta_5 - \gamma_5, \quad \alpha_6 = \beta_6, \end{aligned}$$

and

$$\alpha_1 = D(x + y) - \gamma_3, \quad \alpha_2 = H(x + y) + \gamma_5.$$

the Lemma is proved. $\blacksquare$

The following Lemma is easily established:

Lemma 4 *Let* $\omega = a + bi \in \mathbb{C}$, $\gamma = e^{\frac{2\pi}{3}}$. *Then there exist* $A, B \in \mathbb{C}$ *such that*

$$\omega = A\gamma + B\gamma^2. \tag{3.13}$$

Lemma 5 *Let* $g, h \in \mathbf{S}_3$, *and let the automorphism* Rev_g *be defined by*

$$Rev_g(h) = g^{-1}hg,$$

and

$$Rev_g(\mathbf{S}_3) = \{Rev_g(\varepsilon), Rev_g(\sigma), Rev_g(\sigma^2), Rev_g(\tau), Rev_g(\tau\sigma), Rev_g(\tau\sigma^2)\}.$$

Then,

$$\begin{aligned}
Rev_\varepsilon(\mathbf{S}_3) &= \{\varepsilon, \sigma, \sigma^2, \tau, \tau\sigma, \tau\sigma^2\}, \\
Rev_\sigma(\mathbf{S}_3) &= \{\varepsilon, \sigma, \sigma^2, \tau\sigma^2, \tau, \tau\sigma\}, \\
Rev_{\sigma^2}(\mathbf{S}_3) &= \{\varepsilon, \sigma, \sigma^2, \tau\sigma, \tau\sigma^2, \tau\}, \\
Rev_\tau(\mathbf{S}_3) &= \{\varepsilon, \sigma^2, \sigma, \tau, \tau\sigma^2, \tau\sigma\}, \\
Rev_{\tau\sigma}(\mathbf{S}_3) &= \{\varepsilon, \sigma^2, \sigma, \tau\sigma^2, \tau\sigma, \tau\}, \\
Rev_{\tau\sigma^2}(\mathbf{S}_3) &= \{\varepsilon, \sigma^2, \sigma, \tau\sigma, \tau, \tau\sigma^2\}.
\end{aligned}$$

Proof The proof is a direct verification. ∎
Note that automorphisms

$$\{Rev_\varepsilon(\mathbf{S}_3), Rev_\sigma(\mathbf{S}_3), Rev_{\sigma^2}(\mathbf{S}_3)\} = \mathcal{E}^+$$

do not permute the elements σ and σ^2, while automorphisms

$$\{Rev_\tau(\mathbf{S}_3), Rev_{\tau\sigma}(\mathbf{S}_3), Rev_{\tau\sigma^2}(\mathbf{S}_3)\} = \mathcal{E}^-$$

permute these elements.

Lemma 6 *Let* $z \in \mathbf{A}(\mathbf{R}, \mathbf{S}_3)$ *be given as in (3.12), where* $A = 0$. *Then the system of equations*

$$Rev_g(z) = \xi_g, \quad g \in \mathbf{S}_3 \setminus \{\varepsilon\} \tag{3.14}$$

has a unique solution for any $\xi_g \in \mathbf{A}(\mathbf{R}, \mathbf{S}_3)$.

Proof The proof is a direct verification. ∎
Let a five-channel input signal

$$X(n) = (x_1(n), \dots, x_5(n))$$

and the complex roots $\omega^q = \exp\{\frac{2\pi q}{N}\}$ be represented in the form (3.13):

$$\omega^q = \alpha_q\gamma + \beta_q\gamma^2, \quad q = 0, \dots, N-1.$$

Let us consider now the auxiliary transform (ADOT)

$$\hat{B}(m) = \sum_{n=0}^{N-1} B(n)(\alpha_{mn}\sigma + \beta_{mn}\sigma^2), \quad m = 0, \dots, N-1,$$

where

$$B(n) = x_1(n)\sigma + x_2(n)\sigma^2 + x_3(n)\tau + x_4(n)\tau\sigma + x_5(n)\tau\sigma^2.$$

The ADOT-spectrum is calculated by the canonical Cooley-Tuckey FFT scheme [4]:

$$\hat{B}\left(m + \delta\frac{N}{2}\right) = \sum_{n=0}^{\frac{N}{2}-1} B(2n)(\alpha_{2mn}\sigma + \beta_{2mn}\sigma^2) +$$

$$+ (-1)^\delta(\alpha_m\sigma + \beta_m\sigma^2) \sum_{n=0}^{\frac{N}{2}-1} B(2n+1)(\alpha_{2mn}\gamma + \beta_{2mn}\gamma^2),$$

where $\delta = 0, 1$; $m = 0, 1, \dots, \frac{N}{2} - 1$.

According to Lemma 2, the multiplicative complexity $M_B(N)$ of calculating the array $\hat{B}(m)$ satisfies the recurrent relation

$$M_B(N) = 2M_B\left(\frac{N}{2}\right) + 8 \times \frac{N}{2}.$$

Thus,

$$M_B(N) \le 4N \log_2 N.$$

The application of Lemma 5 to the system of equations

$$\begin{cases} Rev_g\left(\sum_{n=0}^{N-1} B(n)(\alpha_{mn}\sigma + \beta_{mn}\sigma^2)\right) = Rev_g(\hat{B}(m)), & \text{if} \quad Rev_g \in \mathcal{E}^+; \\ Rev_g\left(\sum_{n=0}^{N-1} B(n)(\alpha_{mn}\sigma + \beta_{mn}\sigma^2)\right) = Rev_g(\hat{B}(-m)), & \text{if} \quad Rev_g \in \mathcal{E}^-, \end{cases}$$

$$(3.15)$$

gives the arrays of the partial auxiliary spectra:

$$\hat{B}_j(m) = \sum_{n=0}^{N-1} x_j(n)(\alpha_{mn}\sigma + \beta_{mn}\sigma^2), \quad j = 1, \dots, 5.$$

The homomorphism $\Psi : \mathbf{A}(\mathbf{R}, \mathbf{S}_3) \to \mathbb{C}$ defined by the condition that

$$\sigma \longmapsto \gamma = e^{\frac{2\pi}{3}}$$

transforms the partial auxiliary spectra $\hat{B}_j(m)$ into the complex Fourier spectra

$$\hat{B}_j(m) \longmapsto \hat{x}_j(m) = \sum_{n=0}^{N-1} x_j(n) \exp\left\{\frac{2\pi i mn}{N}\right\}. \tag{3.16}$$

Since the solution of the system (3.15) and the calculations of the mapping (3.16) have a linear complexity, the multiplicative complexity $M(N)$ of calculating all five spectra is equal to

$$M(N) = 4N \log_2 N + O(N). \tag{3.17}$$

Note that the real multiplicative complexity of the separate calculation of the five spectra via the Cooley-Tuckey FFT is (see [4]) equal to

$$5M_{FFT}(N) = 5 \times \frac{3}{2} N \log_2 N + O(N).$$

The algorithm of the ADOT calculation, just considered, did not use the decomposition schemes similar to DFT which employ the trivial multiplications by $\pm i$ (e.g. Radix-4, Split-Radix FFT [17], etc.). Actually, we have the following equality,

$$i = (\gamma - \gamma^2) \frac{\sqrt{3}}{2}.$$

Thus, multiplications in the algebra $\mathbf{A}(\mathbf{R}, \mathbf{S}_3)$ by the inverse image of i are equivalent to the non-trivial multiplications by the element

$$I = (g - g^2) \frac{\sqrt{3}}{2}.$$

In the next subsection we give an example of an overlapped calculation of three complex Fourier spectra, which are free of the above-mentioned drawbacks.

22.3.3   Fast algorithm for three complex Fourier spectra with overlapping

Let $\mathbf{D}_4 = \mathbf{D}_4 = \{\varepsilon, \sigma, \sigma^2, \sigma^3, \tau, \tau\sigma, \tau\sigma^2, \tau\sigma^3\}$. be the 8-element dihedral group with the neutral element ε. The generating elements τ, σ satisfy the multiplication rules

$$\tau^2 = \varepsilon, \quad \sigma^4 = \varepsilon, \quad \tau\sigma = \sigma^3\tau.$$

Lemma 7 *The computation of the product zw, for*

$$z = a\tau + b\tau\sigma + c\tau\sigma^2 + d\tau\sigma^3, \quad w = x + y\sigma$$

requires no more than five real multiplications.

Proof. Since

$$zw = \alpha_0\tau + \alpha_1\tau\sigma + \alpha_2\tau\sigma^2 + \alpha_3\tau\sigma^3,$$

where

$$\alpha_3 = dx + ay - y(a - c), \quad \alpha_1 = bx + cy + y(a - c,)$$
$$\alpha_0 = ax + dy, \qquad\qquad \alpha_2 = cx + by,$$

the calculation of zw is reduced to the repeated application of Lemma 1 and to the calculation of product $y(a - c)$. ∎

Lemma 8 *The mapping Ψ of the group $\mathbf{D}_4$ into group $\mathbf{C}_4$ of complex 4th roots of unity, defined by*

$$\begin{aligned}
\Psi(\varepsilon) = \Psi(\tau\sigma^2) = 1, \quad & \Psi(\sigma) = \Psi(\tau\sigma) = i, \\
\Psi(\sigma^2) = \Psi(\tau) = -1, \quad & \Psi(\tau\sigma^3) = \Psi(\sigma^3) = -i
\end{aligned} \tag{3.18}$$

is a homomorphism.

Proof The kernel of the mapping (3.18) is the two-element subgroup $\mathbf{G}_0 = \{\varepsilon, \tau\sigma^2\}$, which is normal in $\mathbf{D}_4$ and $\mathbf{C}_4 \cong {}^{\mathbf{A}(\mathbf{R},\mathbf{D}_4)}/_{\mathbf{G}_0}$. The homomorphism Ψ can be $\mathbf{R}$−linearly extended to the algebras homomorphism $\Psi : \mathbf{A}(\mathbf{R}, \mathbf{D}_4) \longrightarrow \mathbb{C}$. ∎

Lemma 9 *Given any $\xi_1, \xi_2, \xi_3 \in \mathbf{A}(\mathbf{R}, \mathbf{D}_4)$, the elements $b, c, d \in \mathbf{A}(\mathbf{R}, \mathbf{D}_4)$ are uniquely determined by the relationships*

$$\begin{cases}
\mathrm{Rev}_\varepsilon(c\tau + b\tau\sigma + d\tau\sigma^3) = \xi_1 \\
\mathrm{Rev}_\tau(c\tau + b\tau\sigma + d\tau\sigma^3) = \xi_2 \\
\mathrm{Rev}_{\tau\sigma}(c\tau + b\tau\sigma + d\tau\sigma^3) = \xi_3
\end{cases} \tag{3.19}$$

Proof The system (3.19) can be rewritten in the form

$$\begin{cases}
c\tau + b\tau\sigma + 0 \cdot \tau\sigma^2 + d\tau\sigma^3 = \xi_1 \\
c\tau + d\tau\sigma + 0 \cdot \tau\sigma^2 + b\tau\sigma^3 = \xi_2 \\
0 \cdot \tau + b\tau\sigma + c\tau\sigma^2 + d\tau\sigma^3 = \xi_3
\end{cases}$$

which can easily be solved. ∎

Theorem 2 *There is an algorithm of overlapped calculations of three discrete Fourier spectra, which requires no more than*

$$M(N) = \frac{15}{8} N \log_2 N + O(N) \tag{3.20}$$

real multiplications.

Proof. Let $R(n), G(n), B(n)$ be three N-periodic real sequences. Let the auxiliary $\mathbf{A}(\mathbf{R}, \mathbf{D}_4)$-valued sequence be defined by

$$Z(n) = R(n)\tau + G(n)\tau\sigma + B(n)\tau\sigma^3,$$

and let

$$\alpha_{mn} = \cos\frac{2\pi mn}{N}, \quad \beta_{mn} = \sin\frac{2\pi mn}{N}, \quad W_{mn} = \alpha_{mn} + \beta_{mn}\sigma.$$

We will use the auxiliary transform defined by

$$\hat{Z}(m) = \sum_{n=0}^{N-1} Z(n)W_{mn}, \tag{3.21}$$

where

$$W_{mn} = \begin{cases} \sigma, & \text{for} \quad mn = \frac{N}{4}; \\ \sigma^3, & \text{for} \quad mn = \frac{3N}{4}. \end{cases}$$

The transform (3.21) can be realized by a scheme similar to the Radix-4 FFT [4]:

$$\hat{Z}(m) = \sum_{a=0}^{3} W^{am} \sum_{n=0}^{\frac{N}{4}-1} Z(4n+a)W_{4mn}. \tag{3.22}$$

This requires (see Lemma 6)

$$M_{ADFT}(N) = \frac{15}{8}N\log_2 N$$

real multiplications.

The $\mathbf{A}(\mathbf{R}, \mathbf{D}_4)$-valued spectra $\hat{R}(m), \hat{G}(m), \hat{B}(m)$ are calculated similarly to the algorithm described in the previous subsection, using $O(N)$ nontrivial real multiplications. The reconstruction of complex Fourier spectra is the substitution for $\tau, \tau\sigma, \tau\sigma^2, \tau\sigma^3$ of the complex numbers $(-1), i, 1, i,$ respectively, and does not require any non-trivial real multiplications. ∎

Corollary 1 *There exists an algorithm for calculating the three discrete 2-dimensional Fourier spectra for the RGB-image, which requires no more than*

$$M^*(N) = \frac{5}{4}N^2\log_2 N + O(N^2)$$

multiplications for each spectrum $\hat{R}(m_1, m_2), \hat{G}(m_1, m_2), \hat{B}(m_1, m_2)$.

In [27], a similar task was solved using the quaternionic DFT (QDFT):

$$\hat{Q}(m_1, m_2) = \sum_{n_1,n_2=0}^{N-1} e^{2\pi i\frac{n_1 m_1}{N}} Q(n_1, n_2) e^{2\pi j\frac{n_2 m_2}{N}}, \tag{3.23}$$

where

$$Q(n_1, n_2) = (iR(n_1, n_2) + jG(n_1, n_2) + kB(m_1, m_2)).$$

The transform (3.23) is introduced in [7] as an auxiliary transform, and is analyzed separately as an independent transform in [13], [12]. The applications of QDFT to "anisotropic" tasks is discussed in [5]. The RGB-spectra calculation method, proposed in [27] requires

$$M^*(N) = \frac{8}{3}N^2 \log_2 N + O(N^2)$$

real multiplications. This is two-times worse than the method considered above.

Remark: Corollary 1 assumes that the two-dimensional ADOT is realized by a row-column method. The use of the more complicated schemes of $2D - DFT$ reduction results in better estimates of the multiplicative complexity. In particular, the $2D - DFT$ with "multicovering" [11], in the version considered here, has the multiplicative complexity

$$M^*(N) = \frac{10}{27}N^2 \log_2 N + O(N^2).$$

22.3.4 Fast algorithms for discrete Fourier transforms with maximal overlapping

The examples considered above show that the increase in the size of the algebra **A** results in a greater complexity in the complex spectra reconstruction. For very large dimensions this complexity dominates. The "overlapped" algorithm of the DFT calculation considered below illustrates the limits of our method.

Detailed proofs of the following statements can be found in [10].

Let $N = 2^r$, $q = 2^s$, $T = 2^{r-s}$, and **G** be a q-element cyclic group with the generating element g, and $\mathbf{A}(\mathbf{R}, \mathbf{G}) = \mathbf{A}$ be a group $\mathbf{R}$-algebra.

Lemma 10 *There exists a set of* **A**-*valued functions*
$H(m.n)$ $(0 \le m, n < T - 1)$,

$$H(m + aTq^{-1}, n) = g^{an}H(m, n), \quad (a \in \mathbf{Z}) \tag{3.24}$$

and a homomorphism $\rho : \mathbf{A} \longrightarrow \mathbf{C}$ *over* $\mathbb{R}$, *such that*
$\rho(H(m, n)) = \omega^{mn}$, $\rho(g) = \gamma$, *where* ω, γ *are primitive* Tth *and* qth *roots of unity, respectively.*

In other words, the function $H(m, n)$ behaves like the exponential function, while not being such indeed.

Lemma 11 *Let* $\{b(n)\}$ *be the* T-*periodic* **A**-*valued sequence defined by*

$$b(n) = \sum_{\nu=o}^{q-1} b_\nu(n)g^\nu$$

and $b_\nu(n) = 0$ *for even* ν. *Let* $B(m)$ *be the auxiliary transform*

$$B(m) = \sum_{n=0}^{T-1} b(m)H(m,n), \quad (m = 0, 1, ..., T-1). \qquad (3.25)$$

Let $\{B_\nu(m)\}$ *be the set of "partial spectra" defined by*

$$B_\nu(m) = \sum_{n=0}^{T-1} b_\nu(m)H(m,n), \quad (\nu = 0, 1, \ldots, q-1)$$

of the auxiliary transform. Then there exists a set $\mathcal{E}$ *of automorphisms of the algebra* $\mathbf{A}$, *such that the calculation of the "complex partial spectra"*

$$\widehat{b}_\nu(m) = \sum_{n=0}^{T-1} b_\nu(n)\omega^{mn}, \quad (m = 0, 1, \ldots, T-1)$$

from the system

$$(\rho \circ \varepsilon_\tau)(B_\nu(m)) = (\rho \circ \varepsilon_\tau)\left(\sum_{n=0}^{T-1} b_\nu(n)H(m,n)\right), \quad \varepsilon_\tau \in \mathcal{E}, \qquad (3.26)$$

requires at least $O(2^s N)$ *real multiplications.*

The multiplication of the elements of the algebra $\mathbf{A}$ by the element g^k does not require real multiplications; it is reduced to a permutation of the components of the group algebra elements. It follows that the complexity of the transform calculation, by the "radix-q scheme"

$$B(m) = \sum_{n=0}^{T-1} b(n)H(m,n) = \sum_{a=0}^{q-1} g^{am} \sum_{n=0}^{\frac{T}{q}-1} b(qn+a)H(qm,n),$$

satisfies the following lemma.

Lemma 12 *For every positive integer* s *there exists an absolute constant* λ_1 *and an algorithm for computing the auxiliary transform (3.25) which requires*

$$M^*(T) \leq \lambda_1 s^{-1} T \log_2 T$$

real multiplications.

Let the input N−periodical sequence $x(n)$ be given as a union of the subsequences

$$\{x(n)\} = \bigcup_{\alpha=0}^{\frac{q}{2}-1} \{x(\frac{q}{2}n+\alpha)\} = \bigcup_\alpha \{x_\alpha(n)\}.$$

Introducing the new auxiliary **A**-valued sequence

$$b(n) = 0 \cdot g^0 + x_1(n)g^1 + 0 \cdot g^2 + \cdots + x_{\frac{q}{2}-1}(n)g^{q-1}, \tag{3.27}$$

we have the following lemma.

Lemma 13 *Let W be a primitive Nth root and let the "partial spectra" be already computed. Then, the computation of the transform*

$$\widehat{x}(m) = \sum_{n=0}^{N-1} x(n)W^{mn} = \sum_{\alpha=o}^{2^{s-1}-1} W^{\alpha m}\widehat{b}_{2\alpha+1}(m), \quad (m = 0, 1, \ldots, N-1)$$

requires no more than $O(2^s N)$ real multiplications.

From the above, we see that the calculation of the overlapped DFT consists of the following steps.

Step 1. The input sequence is broken down into subsequences that form the auxiliary sequence (3.27).

Step 2. The **A**-valued auxiliary transform of the T-periodic sequence is calculated $(T = q^{-1}N)$.

Step 3. The projections of the partial spectra of the auxiliary transform (i.e. of the "complex partial spectra") are found from the system of equalities (3.26) generated by the automorphisms of algebra **A**.

Step 4. The array $\widehat{x}(m)$ is reconstructed according to Lemma 10.

As a consequence, we have the following

Theorem 3 *For any positive integer s, $2 \leq s < r$, there exist constants λ_1, λ_2 and an algorithm for computing the transform*

$$\widehat{x}(m) = \sum_{n=0}^{N-1} x(n)W^{mn}, \quad (m = 0, 1, \ldots, N-1, \quad N = 2^r) \tag{3.28}$$

that requires

$$M(N) \leq \lambda_1 s^{-1} N \log_2 N + \lambda_2 2^s N \tag{3.29}$$

real multiplications.

For $s = s(N)$, $2^s \sim \log_2^\theta N$, $0 < \theta < 1$, $N \to \infty$, we have that (3.28) is dominated by the first term. As a consequence, we have

Corollary 2 *For any θ $(0 < \theta < 1)$, there exist constants δ_1, δ_2 depending only on θ, and an algorithm for computing (3.28), whose multiplicative complexity $M(N)$, with $N = 2^s > N_0(\theta)$, satisfies the asymptotic relationship*

$$M(N) \leq \delta_1 \frac{N \log_2 N}{\log_2 \log_2 N} + \delta_2 N \log_2^\theta N. \tag{3.30}$$

The algorithms described in this subsection are applicable not only for the algebra **A** over ℝ but also for the algebras over finite fields $\mathbf{F}_p$, see the number-theoretical transforms (NTT) [24]. We have the following interesting example.

In [28] the NTT was first applied to the fast multiplication of large integers. The most popular "school-method" of multiplication uses $O(N^2)$ operations to multiply $N-$digit integer, while the method introduced in Ref.[28] uses $O(N \log_2 N)$ operations.

The algorithm introduced in this subsection uses

$$O\left(\frac{N \log_2 N}{\log_2 \log_2 N}\right)$$

operations to multiply "very large" integers.

22.4   Conclusion

An analysis of the structure of the described fast algorithms for the discrete orthogonal transforms allows us to isolate the following three "hidden" steps.

Step 1. The values of the input array $x(n)$ are mapped into a set of elements of a group algebra $\mathbf{A}(\mathbf{F}, \mathbf{G})$, together with the basic functions $h_m(n)$ into a set of **A**-valued functions.

Step 2. The auxiliary transform is computed in $\mathbf{A}(\mathbf{F}, \mathbf{G})$ using to great advantage the algebraic properties of the group **G**, of the field **F** and the algebra $\mathbf{A}(\mathbf{F}, \mathbf{G})$).

Step 3. The projection of $\mathbf{A}(\mathbf{F}, \mathbf{G})$ into a Clifford algebra reduces the calculations to operations in the Clifford algebra.

I believe that the capabilities of the approach described in the article are not limited to the applications considered here. We have only

considered the case of group algebra projections to the algebra **C**. This method can be for $C(p, q)-$valued DOTs (in particular, **H**−valued DOTs). Significant emphasis should be placed on the development of multiplication rules like those of Lemmas 1 and 6 and the development of an appropriate mathematical formalism. One of the main goals of this article has been to attract the reader to the development and the possibilities of such a formalism.

Chapter 23

Counterexamples for Validation and Discovering of New Theorems

Pertti Lounesto

23.1 Introduction

This chapter describes an experiment which took place over the two year period 1997-1999. The chapter consists of an adaptation of excerpts from the two Web Pages

http://www.hit.fi/~lounesto/counterexamples.htm

http://www.hit.fi/~lounesto/sci.math.htm.

The first Web Page is itself a part of a previously published article in which published theorems were falsified by counterexamples. The posters of an Internet discussion group, *sci.math*, were challenged to validate, or invalidate, my counterexamples. The purpose of the challenge was to test the suitability of this new electronic media as a forum for scientific dialogues. Particular attention was paid to exploring whether this new media could replace scientific journals and conferences as a forum for settling disagreements between experts on a special topic, in this case Clifford algebras.

The second Web Page presents the conclusions of this study. It turned out that an Internet discussion group cannot come up with substantial or competent feedback for reasons which we will discuss below.

23.2 The Role of Counterexamples in Mathematics

MOTTO: In research, counterexamples show us that we are going the wrong way. They tell us where not to go in exploring a new domain.

The purpose of the author of this chapter is to demonstrate that many statements, published in mathematical literature as theorems, are false, by providing *counterexamples*. In mathematics, a theorem is either

1. a true statement, or

2. a statement with proof.

By either definition, a theorem cannot be falsified. In the following, I show that many statements published as theorems are not theorems, but rather just false statements.

In practice, a mathematician finds a proof for a statement. He evaluates his proof with a few colleagues and publishes his theorem in a refereed journal. The purpose of publication is to expose the theorem to public scrutiny. There follows a critical debate, which might result in a revision of the theorem in the literature. Mathematics is universal and effectively applied to the real world. This often leads mathematicians to a cognitive illusion: When several members of a research group have accepted a new statement as a theorem, the statement becomes unfalsifiable (in the minds of the members of the research group).

I informed almost all of the mathematicians about their errors, prior to exhibiting on my Web Page (www.hit.fi/ lounesto/counterexamples.htm). The mathematicians have mostly admitted their mistakes, after some reasoned dialogue, lasting for a few months or sometimes years. The course of events was usually as follows: I find that a theorem does not hold and work out the simplest non-trivial counterexample. I pay special attention to interpreting the text in the way the author has intended so as to make sure that my counterexample reveals an inner inconsistency. Then I send a letter to the author enclosing a detailed description of my counterexample. I also enclose a photocopy of the theorem in the envelope, underlining the false parts in red and marking in the margin the word WRONG. At first, the authors usually defend their theorems. After a few letters have been exchanged, most of the authors accept the validity of my counterexamples and admit their mistakes. When the mistake has been understood, the author usually explains the error away as casual and insignificant.

I argue that some of my findings are significant, according to the following criteria:

1. How important did the authors regard their "theorems" prior to realizing their mistakes?

2. How long did it take for the authors to understand and admit their mistakes?

The mistakes occurred at the frontiers of joint explorations of mathematicians, who still had inaccurate cognitive charts of the new domains they were exploring. Some mistakes were confined to individuals, who could be easily convinced about their mistakes. Some mistakes were common to groups of mathematicians, who had a collective cognitive illusion of "mathematical reality". Such groups often used a poor language to describe "mathematical reality". To break the cognitive illusions, I often had to learn the poor language and culture of the groups. Sometimes such groups defended their cognitive bugs vigorously.

The falsified "theorems" were seldom used in subsequent deductions and did not have an impact on the works of other researchers. Nevertheless, other researchers often repeated the same mistakes. From this observation, I come to the following main result of my findings:

Creative research mathematicians, exploring the frontiers of our common body of knowledge, tend to make similar mathematical mistakes.

This leads to collective cognitive bugs. In the realm of such bugs, statements collectively held true, although later proven to be false, cannot be distinguished from the correct ones. In other words, there is no practical possibility to make a distinction between

1. a theorem, and

2. a statement labelled as a theorem by all experts.

If there were, experts could just agree on classifying all statements labelled as theorems, in their speciality, into the above two classes.

Some of the counterexamples stem from the failure of the authors to check their statements for small numbers of indices or in low dimensions (typically 2). Quite a few of the counterexamples consist of exceptional cases in lower dimensions (typically 4,7,8). Counterexamples are also given in the cases where the author failed to notice a general pattern after some dimension (typically at and above 4,6).

Informing colleagues about their errors is more subtle. It offers them an opportunity to learn more mathematics. It might also result in a feeling of insufficiency, a cognitive conflict, and instigate a learning process, and thus indirectly lead to cognitive growth. See Ginsburg and Opper 1988 [18].

In order to benefit from the mathematical arguments presented, the reader should have the given references at hand, and follow the reasoning of the counterexamples line by line.

23.3   Clifford Algebras: An Outline

We first will give some preliminary notation, since some readers might be unfamiliar with Clifford algebras.

23.3.1   The Clifford algebra of the Euclidean plane

Consider the Euclidean plane $\mathbb{R}^2$ with a quadratic form sending a vector $x\mathbf{e}_1 + y\mathbf{e}_2$ to the scalar $x^2 + y^2$. The Clifford algebra Cl_2 of $\mathbb{R}^2$ is a real associative algebra of dimension 4 with the unit element 1. It contains copies of $\mathbb{R}$ and $\mathbb{R}^2$ in such a way that the square of the vector $x\mathbf{e}_1 + y\mathbf{e}_2$ equals the scalar $x^2 + y^2$. As an equation,

$$(x\mathbf{e}_1 + y\mathbf{e}_2)^2 = x^2 + y^2.$$

It follows that Cl_2 has the basis $\{1, \mathbf{e}_1, \mathbf{e}_2, \mathbf{e}_{12}\}$. The orthonormal unit vectors $\mathbf{e}_1, \mathbf{e}_2$ in $\mathbb{R}^2$ satisfy

$$\mathbf{e}_1^2 = \mathbf{e}_2^2 = 1,$$

and are anticommutative

$$\mathbf{e}_1\mathbf{e}_2 = -\mathbf{e}_2\mathbf{e}_1.$$

We write $\mathbf{e}_{12} = \mathbf{e}_1\mathbf{e}_2$. Computing the square of $\mathbf{e}_{12}$, we find

$$\mathbf{e}_{12}^2 = \mathbf{e}_1\mathbf{e}_2\mathbf{e}_1\mathbf{e}_2 = -\mathbf{e}_1^2\mathbf{e}_2^2 = -1.$$

The basis element $\mathbf{e}_{12}$ cannot be a scalar (in I R) nor a vector (in $\mathbb{R}^2$), nor even a linear combination of a scalar and a vector. It is a new kind of object, called a *bivector*. The basis elements

1

$\mathbf{e}_1, \mathbf{e}_2$

$\mathbf{e}_{12}$

span, respectively, the subspaces of scalars, vectors and bivectors (in Cl_2).

The Clifford algebra Cl_2 has a faithful matrix image, the matrix algebra $Mat(2, \mathbb{R})$ of 2×2-matrices with entries in I R. The basis elements $1, \mathbf{e}_1, \mathbf{e}_2, \mathbf{e}_{12}$ of Cl_2 can be represented by the matrices

$$E_0 = \begin{pmatrix} 1 & 0 \\ 0 & 1 \end{pmatrix}, E_1 = \begin{pmatrix} 1 & 0 \\ 0 & -1 \end{pmatrix}, E_2 = \begin{pmatrix} 0 & 1 \\ 1 & 0 \end{pmatrix}, E_{12} = \begin{pmatrix} 0 & 1 \\ -1 & 0 \end{pmatrix}.$$

In comparing Cl_2 and $Mat(2, \mathbb{R})$, which are isomorphic as associative algebras, it should be noted that Cl_2 has more structure: in the Clifford algebra Cl_2 there is a distinguished subspace, isometric to the Euclidean plane R^2. No such privileged subspace exist in $Mat(2, \mathbb{R})$.

23.3.2   The Clifford algebra of the Minkowski space-time

The Clifford algebra $Cl_{3,1}$ of the Minkowski space-time $\mathbb{R}^{3,1}$, with the quadratic form $x^2 + y^2 + z^2 - c^2t^2$, is a real associative algebra of dimension 16 with unit element 1. It has the basis elements

1

$\mathbf{e}_1, \mathbf{e}_2, \mathbf{e}_3, \mathbf{e}_4$

$\mathbf{e}_{12}, \mathbf{e}_{13}, \mathbf{e}_{14}, \mathbf{e}_{23}, \mathbf{e}_{24}, \mathbf{e}_{34}$

$\mathbf{e}_{123}, \mathbf{e}_{124}, \mathbf{e}_{134}, \mathbf{e}_{234}$

$\mathbf{e}_{1234}$

which span, respectively, the subspaces of scalars, vectors, bivectors, 3-vectors and 4-vectors. The vectors e_1, e_2, e_3, e_4 satisfy the following multiplication rules: they are unit vectors with squares

$e_1^2 = e_2^2 = e_3^2 = 1$ and $e_4^2 = -1$ and they anticommute $e_i e_j = -e_j e_i$ for $i \neq j$. We denote $e_{ij} = e_i e_j$ for $i \neq j$, and $e_{1234} = e_1 e_2 e_3 e_4$. These rules and conventions fix the computation rules of $Cl_{3,1}$,
for example

$e_1 e_2 e_1 e_3 = -e_1^2 e_2 e_3 = -e_{23}.$

The Clifford algebra $Cl_{3,1}$ of the Minkowski space-time $\mathbb{R}^{3,1}$ is isomorphic, as an associative algebra, to the real 4x4-matrix algebra $Mat(4, \mathbb{R})$. This isomorphism allows us to view $Cl_{3,1}$ through its faithful matrix image $Mat(4, \mathbb{R})$.

For the convenience of readers unfamiliar with Clifford algebras, I shall present the first counterexamples by means of a matrix algebra, namely $Mat(4, \mathbb{R})$, and then translate the presentation into the corresponding Clifford algebra $Cl_{3,1}$.

23.3.3 Clifford algebra viewed by means of the matrix algebra

The orthonormal basis e_1, e_2, e_3, e_4 of $\mathbb{R}^{3,1}$ can be represented by the matrices

$$E_1 = \begin{pmatrix} 1 & 0 & 0 & 0 \\ 0 & -1 & 0 & 0 \\ 0 & 0 & -1 & 0 \\ 0 & 0 & 0 & 1 \end{pmatrix}, \quad E_2 = \begin{pmatrix} 0 & 1 & 0 & 0 \\ 1 & 0 & 0 & 0 \\ 0 & 0 & 0 & 1 \\ 0 & 0 & 1 & 0 \end{pmatrix},$$

$$E_3 = \begin{pmatrix} 0 & 0 & 1 & 0 \\ 0 & 0 & 0 & -1 \\ 1 & 0 & 0 & 0 \\ 0 & -1 & 0 & 0 \end{pmatrix}, \quad E_4 = \begin{pmatrix} 0 & 0 & -1 & 0 \\ 0 & 0 & 0 & 1 \\ 1 & 0 & 0 & 0 \\ 0 & -1 & 0 & 0 \end{pmatrix}$$

satisfying the multiplication rules
$E_1^2 = E_2^2 = E_3^2 = I, E_4^2 = -I$ and $E_i E_j = -E_j E_i$ for $i \neq j$.
Take an element $a = (1+e_1)(1+e_{234})$ in $Cl_{3,1}$, represented by the matrix

$$A = (I + E_1)(I + E_2 E_3 E_4) = \begin{pmatrix} 2 & 2 & 0 & 0 \\ 0 & 0 & 0 & 0 \\ 0 & 0 & 0 & 0 \\ 0 & 0 & -2 & 2 \end{pmatrix}.$$

The so-called Clifford conjugation sending a in $Cl_{3,1}$ to a^- corresponds in

$Mat(4, \mathbb{R})$ to the anti-automorphism sending A to

$$A^- = E_4 A^T E_4^{-1} = \begin{pmatrix} 0 & 2 & 0 & 0 \\ 0 & 2 & 0 & 0 \\ 0 & 0 & 2 & 0 \\ 0 & 0 & -2 & 0 \end{pmatrix}.$$

Computing the products of A and A^- in different order we find:

$A^- A = 0$ but $AA^- = \begin{pmatrix} 0 & 8 & 0 & 0 \\ 0 & 0 & 0 & 0 \\ 0 & 0 & 0 & 0 \\ 0 & 0 & -8 & 0 \end{pmatrix} \neq 0$. In fact, AA^- is not even

diagonal, that is, it is not a scalar multiple of I. After this excursion into matrix algebra, the reader is hopefully prepared for Clifford algebra. Next, I will present some preliminary counterexamples by rewriting the above observation in terms of the Clifford algebra $Cl_{3,1}$.

23.4   Preliminary Counterexamples in Clifford Algebras

Consider the Clifford algebra $Cl_{3,1} = Mat(4, \mathbb{R})$ of the Minkowski space-time $\mathbb{R}^{3,1}$. Take an element

$$a = (1 + \mathbf{e}_1)(1 + \mathbf{e}_{234}) = 1 + \mathbf{e}_1 + \mathbf{e}_{234} + \mathbf{e}_{1234},$$

and apply Clifford-conjugation (the anti-automorphism of $Cl_{3,1}$ extending the map $\mathbf{x} \to -\mathbf{x}$ in $\mathbb{R}^{3,1}$)

$$a^- = (1 + \mathbf{e}_{234})(1 - \mathbf{e}_1) = 1 - \mathbf{e}_1 + \mathbf{e}_{234} + \mathbf{e}_{1234}.$$

Computing the products of a and a^- in different orders gives: $a^- a = 0$ although $aa^- = 4(\mathbf{e}_{234} + \mathbf{e}_{1234})$ is not zero, nor even a scalar in I R.

Harvey 1990 [20] claims on p. 202, in Lemma 10.45, that the following statements are equivalent: (c) $aa^- \in \mathbb{R}$, (d) $a^- a \in \mathbb{R}$. Compare the above result to the Lemma, claimed to have been proven by Harvey, and you have a counterexample to Harvey's lemma. In other words, my counterexample falsifies a result of Harvey 1990, Lemma 10.45, (c,d), p. 202, since $a^- a = 0$ is in I Rbut $aa^- = 4(\mathbf{e}_{234} + \mathbf{e}_{1234})$ is nonzero and not in I R. (Harvey introduces the Clifford-conjugation a^- on p. 183; he calls it a hat involution which he denotes by $\hat{a}$.)

Gilbert and Murray 1991 [17] denote $\delta(x) = x^- x$ and prove in Theorem 5.16 that for all x such that $\delta(x)$ is in I R, it necessarily follows that $\delta(x^-) = \delta(x)$ [p. 41, 1. 19] and in particular that $\delta(x) = 0$ forces $\delta(x^-) = 0$ [p.42, II. 2-3]. Choosing $x = a$ to find $\delta(a) = 0$ in I R, although $\delta(a^-) = (a^-)^- a^- = 4(\mathbf{e}_{234} + \mathbf{e}_{1234})$ is not 0, and therefore not in

I R. Compare this result to Theorem 5.16, claimed to have been proven by Gilbert and Murray [17], and you have a counterexample to Gilbert and Murray's theorem. In other words, Gilbert and Murray's Theorem 5.16, stating that $\delta(x^-) = \delta(x)$, has been falsified by my counterexample. (Gilbert and Murray's conjugation is the Clifford-conjugation, see p. 17.)

The element a^- also serves as a counterexample to Knus 1991 [24], page 228, line 13, since $(a^-)^- a^- = 4(\mathbf{e}_{234} + \mathbf{e}_{1234})$ is not in $Cl_{3,1}^+$, the even subalgebra of $Cl_{3,1}$. A simpler counterexample is $\mathbf{x} = \mathbf{e}_1 + \mathbf{e}_{23}$ in $Cl_3 = Mat(2, \mathbb{C})$, the real algebra 2x2-matrices with complex numbers as entries, for which $\tilde{\mathbf{x}}\mathbf{x} = -2\mathbf{e}_{123}$ is not in Cl_3^+. (Knus introduces the Clifford-conjugation $\bar{x}$ on p. 195; he calls it standard involution $\sigma(x)$.)

In particular, $\tilde{\mathbf{x}}\mathbf{x} = -2\mathbf{e}_{123}$ is not in I R for $\mathbf{x} = \mathbf{e}_1 + \mathbf{e}_{23}$ in Cl_3, and we have a counterexample to Dabrowski 1988 [8], page 7, line 12. In the Clifford algebra Cl_3 of the Euclidean space $\mathbb{R}^3$ there are elements whose exponentials are vectors, like $\mathbf{e}_3 = \exp[\pi/2(\mathbf{e}_{12} - \mathbf{e}_{123})]$. Therefore, the multivalued inverse of the exponential satisfies

$$\log \mathbf{e}_3 = \pi/2(\mathbf{e}_{12} - \mathbf{e}_{123}).$$

This shows that vectors can have logarithms in a Clifford algebra, and serves as a counterexample to Hestenes 1986 [22], p. 75 (the error is corrected in Hestenes 1987).

All the above counterexamples are trivial, in the sense that an expert will recognize the mistakes right away, except for maybe the last one. The detection of the last mistake, concerning functions in Clifford algebras, requires a knowledge of idempotents, nilpotents and minimal polynomials. A good place to start studying them is Sobczyk 1997 [40].

23.5 Counterexamples About Spin Groups

The Lipschitz group $\mathbf{L}_{p,q}$, also called the Clifford group although invented by Lipschitz 1880/86, can be defined as the subgroup in $Cl_{p,q}$ generated by invertible vectors $\mathbf{x} \in \mathbb{R}^{p,q}$, or equivalently by either of the following ways

$$\mathbf{L}_{p,q} = \{s \in Cl_{p,q}; \; for \; all \; \mathbf{x} \in \mathbb{R}^{p,q}, \; s\mathbf{x}\hat{s}^{-1} \in \mathbb{R}^{p,q}\},$$

$$\mathbf{L}_{p,q} = \{s \in Cl_{p,q}^+ \cup Cl_{p,q}^-; \; for \; all \; \mathbf{x} \in \mathbb{R}^{p,q}, \; s\mathbf{x}s^{-1} \in \mathbb{R}^{p,q}\}.$$

Note the presence of the grade involution: $s \to \hat{s}$ (the automorphism of $Cl_{p,q}$ extending the map $\mathbf{x} \to -\mathbf{x} \in \mathbb{R}_{p,q}$), and/or restriction to the even/odd parts $Cl_{p,q}^\pm$. The Lipschitz group $\mathbf{L}_{p,q}$ has a subgroup, normalized by the reversion: $s \to \check{s}$ (the anti-automorphism of $Cl_{p,q}$ extending the identity map $\mathbf{x} \to \mathbf{x} \in \mathbb{R}^{p,q}$),

$$\mathbf{Pin}(p, q) = \{s \in \mathbf{L}_{p,q}; s\check{s} = \pm 1\},$$

with an even subgroup

$$\mathbf{Spin}(p,q) = \mathbf{Pin}(p,q) \cap Cl_{p,q}^+,$$

which contains as a subgroup the two-fold cover

$$\mathbf{Spin}_+(p,q) = \{s \in \mathbf{Spin}(p,q); s\check{s} = 1\}$$

of the connected component $SO_+(p,q)$ of $SO(p,q) \subset O(p,q)$.

Although $SO_+(p,q)$ is connected, its two-fold cover $\mathbf{Spin}_+(p,q)$ need not be connected. In particular,

$$\mathbf{Spin}_+(1,1) = \{x + y\mathbf{e}_{12}; x, y \in \mathbb{R}, x^2 - y^2 = 1\}$$

has two components, two branches of a hyperbola (and so the group

$$\mathbf{Spin}(1,1) = \{x + y\mathbf{e}_{12}; x, y \in \mathbb{R}, x^2 - y^2 = \pm 1\}$$

has four components). This serves as a counterexample to Choquet-Bruhat et al. 1989 [6], p. 37, ll. 2-3, p. 38, ll. 22-23 (see also p. 27, ll. 4-5). Although the two-fold covers $\mathbf{Spin}(n) = \mathbf{Spin}(n,0) \simeq \mathbf{Spin}(0,n), n > 2$, and $\mathbf{Spin}_+(n-1,1) \simeq \mathbf{Spin}_+(1,n-1), n > 3$, are simply connected, $\mathbf{Spin}_+(3,3)$ is not simply connected, and therefore not a universal cover of $SO_+(3,3)$, since the maximal compact subgroup $SO(3) \times SO(3)$ of $SO_+(3,3)$ has a four-fold universal cover $\mathbf{Spin}(3) \times \mathbf{Spin}(3)$. The two-fold cover $\mathbf{Spin}_+(3,3)$ of $SO_+(3,3)$ is doubly connected, contrary to the claims of Lawson and Michelsohn 1989 [26], p. 57, l. 22, and Göckeler and Schücker 1987 [19], p.190, l. 17. Lawson and Michelsohn 1989 [26] give also correct information about the connectivity properties of the rotation groups $SO_+(p,q)$, see p. 20, ll. 6-8.

As a consequence, $\mathbf{Spin}_+(3,3) \simeq SL(4,\mathbb{R})$, and so $\mathbf{Spin}_+(3,3)/\{\pm 1\} \not\simeq SL(4,\mathbb{R})$ contrary to the claims of Harvey 1990 [20], p. 272, l. 24, and Lawson and Michelsohn 1989 [26], p. 56, l.21. Moreover, the element $\mathbf{e}_1\mathbf{e}_2 \cdots \mathbf{e}_6 \in \mathbf{Spin}(3,3) \setminus \mathbf{Spin}_+(3,3)$ is not in $\mathbf{Spin}_+(3,3)$, since it is a preimage of $-I \in SO(3,3) \setminus SO_+(3,3)$, contrary to the claims of Lawson and Michelsohn 1989 [26], p. 57, ll. 29-30.

23.5.1 Comment on Bourbaki 1959

The groups $\mathbf{Pin}(p,q)$ and $\mathbf{Spin}(p,q)$, obtained by normalizing the Lipschitz group $\mathbf{L}_{p,q}$, are two-fold coverings of the orthogonal and special orthogonal groups, $O(p,q)$ and $SO(p,q)$, respectively. If one defines, instead of the Lipschitz group, a slightly different group

$$\mathbf{G}_{p,q} = \{s \in Cl_{p,q}; \forall x \in \mathbb{R}^{p,q}, sxs^{-1} \in \mathbb{R}^{p,q}\},$$

one obtains, only in even dimensions, a cover of $O(p,q)$. Furthermore, for odd $n = p + q$, an element of $\mathbf{G}_{p,q}$ need not be even or odd, but might have an inhomogeneous central factor $x + y\mathbf{e}_{12\cdots n} \in \mathbb{R}^+ \wedge^n \mathbb{R}^{p,q}$. Thus Bourbaki 1959 [2], p. 151, Lemme 5, does not hold, as has been observed by Deheuvels 1981 [9], p. 355, Moresi 1988 [33], p. 621, and by Bourbaki himself (see Feuille d'Errata No. 10 distributed with Chapters 3,4 of Algèbre Commutative 1961). The confusion about a proper covering of $O(p,q)$ in $Cl_{p,q}$ pops up frequently.

In the Lipschitz group, every element $s \in \mathbf{L}_{p,q}$ is of the form $s = \rho g$, where $\rho \in \mathbb{R} \setminus \{0\}, g \in \mathbf{Pin}(p,q)$. The group $\mathbf{G}_{p,q}$ does not have this property in odd dimensions. For instance, the central element $z = x + y\mathbf{e}_{123} \in Cl_3$, with non-zero $x, y \in \mathbb{R}$, satisfies $z \in \mathbf{G}_3$, but $z \neq \rho g, g \in \mathbf{Pin}(3)$. This serves as a counterexample to Baum 1981 [1], p. 57, l. -1. (Baum's $C_{n,k}$ means $Cl_{k,n-k}$, see p. 51, and her $Pin(n,k)$ means $\mathbf{Pin}(k, n - k)$, see p. 53. Note that the two-fold cover of $O(3)$,

$$\mathbf{Pin}(3) = \mathbf{Spin}(3) \cup \mathbf{e}_{123}\mathbf{Spin}(3) \simeq SU(2) \cup iSU(2),$$

is a subgroup of $\mathbf{G}_3$, but since the actions are defined differently, $\mathbf{G}_3$ does not cover $O(3)$.

For all $s \in \mathbf{G}_3, s\check{s} > 0$. Therefore, if we normalize $\mathbf{G}_3$ by the reversion, the central factor is not eliminated, but instead we get the group $\{s \in \mathbf{G}_3; s\check{s} = 1\} \simeq U(2)$, which does not cover $O(3)$ but covers $SO(3)$ with kernel $\{x + y\mathbf{e}_{123}; x, y \in \mathbb{R}, x^2 + y^2 = 1\} \simeq U(1) \not\simeq \pm 1$. Compare this to Figueiredo 1994 [13], p. 230, ll. -4.

23.5.2   *Exponentials of bivectors*

There are two possibilities to exponentiate a bivector $B \in \wedge^2 \mathbb{R}^{p,q}$: the ordinary/Clifford exponential $\exp(B)$, and the exterior exponential $\exp^{\hat{B}}$, where the product is the exterior product. If the exterior exponential $\exp^{\hat{B}}$ is invertible with respect to the Clifford product, then it is in the Lipschitz group $\mathbf{L}_{p,q}$. For the ordinary exponential we always have $\exp(B) \in \mathbf{Spin}_+(p,q)$.

All the elements of the compact spin groups $\mathbf{Spin}(n,0) \simeq \mathbf{Spin}(0,n)$ are exponentials of bivectors (when $n > 1$). Among the other spin groups the same holds only for $\mathbf{Spin}_+(n-1,1) \simeq \mathbf{Spin}_+(1,n-1), n > 4$, see M. Riesz 1958/1993 [37] pp. 160, 172. In particular, the two-fold cover $\mathbf{Spin}_+(1,3) \simeq SL(2,C)$ of the Lorentz group $SO_+(1,3)$ contains elements which are not exponentials of bivectors: take $(\gamma_0 + \gamma_1)\gamma_2 \in \wedge^2\mathbb{R}^{1,3}, [(\gamma_0 + \gamma_1)\gamma_2]^2 = 0$, then

$$-\exp[(\gamma_0 + \gamma_1)\gamma_2] = -1 - (\gamma_0 + \gamma_1)\gamma_2 \neq \exp(B)$$

for any $B \in \wedge^2\mathbb{R}^{1,3}$.

Note, that in $\mathbf{Spin}_+(4,1) \simeq Sp(2,2)$ we have

$$-\exp((\mathbf{e}_1 + \mathbf{e}_5)\mathbf{e}_2) = -1 - (\mathbf{e}_1 + \mathbf{e}_5)\mathbf{e}_2 = \exp((\mathbf{e}_1 + \mathbf{e}_5)\mathbf{e}_2 + \pi\mathbf{e}_{34}).$$

However, all the elements of $\mathbf{Spin}_+(1,3)$ are of the form $\pm\exp(B), B \in \wedge^2\mathbb{R}^{1,3}$. Therefore, the exponentials of bivectors do not form a group, contrary to a statement of Dixon 1994 [10],p. 13, ll. 8-10.

Every element L of the Lorentz group $SO_+(1,3)$ is an exponential of an antisymmetric matrix, $L = \exp(A), gA^Tg^{-1} = -A$; a similar property is not shared by $SO_+(2,2)$, see M. Riesz 1958/1993 [37], pp. 150-152, 170-171. There are elements in $\mathbf{Spin}_+(2,2)$ which cannot be written in the form $\pm\exp(B), B \in \wedge^2\mathbb{R}^{2,2}$; for instance $\pm\mathbf{e}_{1234}\exp(\beta B), B = \mathbf{e}_{12} + 2\mathbf{e}_{14} + \mathbf{e}_{34}, \beta > 0$. This serves as a counterexample to Doran 1994 [11], p. 41, l. 26, formula (3.16).

Riesz also showed, by the same construction on pp. 170-171, that there are bivectors which cannot be written as sums of simple and completely orthogonal bivectors; for instance $B = \mathbf{e}_{12} + 2\mathbf{e}_{14} + \mathbf{e}_{34} \in \wedge^2\mathbb{R}^{2,2}$.

The above mistakes are not serious, in the sense that they could be rectified by stipulating the assertions, although such a correction is not obvious in the last examples. The above counterexamples should be easy to understand also for a non-expert, except maybe the last one by M. Riesz, which does require some knowledge of minimal polynomials of linear transformations. Good places to start studying minimal polynomials are Sobczyk 1997 [40, 41], and M. Riesz 1958/1993 [37], pp. 150-152, 170-171.

23.5.3   Internet as a scientific forum

Scientific knowledge is a result of public discussion/debate and scrutiny. In order to guarantee that serious scientists will participate in the discussion, it is essential that the author cannot alter or withdraw his writings after publication. The requirement of immutable dialogue, enabling later inspection, is not fulfilled in the media of Web Pages, which the author can modify, or Usenet newsgroups, where articles expire and can be cancelled.

23.5.4   How did I locate the errors and construct my counterexamples?

First, in trying to get a picture of what is new in a published work, I find something fishy. Then, I try to make sure that I have interpreted the text the way the author has intended. Then I make sure that there is an inner inconsistency and that the author has contradicted himself. Often I then checked formulas with CLICAL, a computer program designed for Clifford algebra calculations. Evaluating the left hand side with arbitrary arguments satisfying all the assumptions, and comparing the result to the right hand side, sometimes reveals a discrepancy. The next step is to find the simplest non-trivial counterexample in the lowest dimension and degree and with the smallest number of components. In discussions with authors about the fine points of their works, CLICAL has helped me to follow, verify or disqualify, the arguments presented, and to penetrate more quickly into the topic.

23.5.5   Progress in science via counterexamples

Ideally, scientists publish papers for the purpose of testing and evaluating their ideas by public scrutiny. This ideal has been somewhat obscured by the peer review /refereeing system, which pretends to guarantee correctness of ideas – prior to a public scrutiny, and the tendency to publish in order to get a position in academia. Traditionally, science has progressed through public debates about new ideas: statements, counterexamples, refined statements and new counterexamples, etc.

In mathematics, proving theorems, finding gaps and errors in the proofs, correcting the theorems, detecting errors in the corrected theorems, etc. is a normal activity. This is even more so in advanced mathematics because our cognitive charts are less accurate in the new frontiers of knowledge. See Lakatos 1976 [25].

In evaluating the validity of a mathematical theorem, one should either check every detail of its proof or point out a flaw in the chain of deductions or line of thoughts. After a counterexample has been presented, it is often easier to settle whether it fulfills all the assumptions than to check all the details of the proof. Just as in science, in mathematics we are faced with the fact that a single counterexample can falsify a theorem or a whole theory. See Popper 1972 [34].

The role of counterexamples in mathematics has been discussed by Lakatos 1976 [25], Dubnov 1963 [12] and Hauchecorne 1988 [21]. Lakatos focuses on the historical development of mathematics and Dubnov on various levels of abstraction. Both restrict themselves to a specific topic within mathematics. Hauchecorne gives counterexamples in almost all branches of mathematics. He also elaborates on virtues of counterexamples in teaching and in research: A theorem often necessitates several hypotheses – to chart out its domain of applications it is important to become convinced about the relevancy of each hypothesis. This can be done by dropping one assumption at a time, and giving a counterexample to each new 'theorem'. Counterexamples cannot be ignored on the basis that 'they do not treat the general case'. Counterexamples are not 'exceptions that confirm the rule'. In mathematical research, the negation of a theorem, the affirmation that it is false, is demonstrated by the existence of a case, where all the hypotheses are verified without the conclusion being valid. The mathematical justification for the falsity of a theorem is completed by presenting a counterexample; after verification of such a presentation further study in the same line at whatever generality is a useless and erroneous activity.

There are several books listing counterexamples in various branches of mathematics: Capobianco and Molluzo 1978 (graph theory)[3], Gelbaum and Olmsted 1964 (analysis) [15], Fornæss and Stensønes 1987 (several complex variables) [14], Khaleelulla 1982 (vector spaces) [23], Romano and Siegel 1986 (statistics) [38], Steen and Seebach 1970 (topology) [42], Stoyanov 1987 (probability) [43] and Wise and Hall 1993 (real analysis) [44] . Similarly as Lakatos, Dubnov and Hauchecorne these authors do not point out errors of contemporary mathematicians. My Web Pages differ from those studies in the respect that counterexamples are given to the works of living mathematicians, who can then participate themselves in a public debate about the correctness of my counterexamples.

23.6   Counterexamples on the Internet

Internet discussion groups are dominated by non-experts, sci.math by graduate students and young mathematicians who do not yet have their own speciality. Fallacies are common in sci.math. But, since many competent mathematicians regularly post in sci.math, it was reasonable to test its suitability for settling scientific debates. A test with seemingly controversial material, such as my counterexamples to proven theorems, was proposed by a friend. Thus came my challenge. The posters were given adequate time, two years, to digest the mathematical material outside of their own domains of expertise. The posters were given adequate guidance for relevant literature, such as my own book on Clifford algebras and spinors, and possibility to download a computer program, CLICAL, designed to check identities of Clifford numbers.

It was anticipated that the controversial issue would get attention and spark discussion. After all, my work seems eccentric: I claim that there are more mathematical mistakes in the literature than generally admitted and that the mistakes are more serious and significant. In addition, it is unusual to discuss openly about mistakes of living mathematicians. Although mathematicians do admit general fallibility, they are reluctant to admit specific mistakes and even to participate in discussions about mistakes of living mathematicians. Thus, I anticipated a lot of opposition and reprehension, in spite of the fact that open critics is the very nature of a public and self-correcting science.

However, in revealing that in my own research field that there were lots of mathematical mistakes, somebody might claim that my field consists of poor mathematicians. If such a claim was presented, I had to be prepared to challenge the poster in a competition over knowledge about Clifford algebras, and then nullify the claim by either winning the challenge or enticing the claimer into making a mistake about Clifford algebras. This actually happened.

During the two years of my challenge, sci.math was able to clear out only

1 of my 30 counterexamples, by Keith Ramsay on Dec 27, 1997. Ramsay scrutinized and verified one counterexample and mentioned verifying a few others. However, one swallow does not make a summer: although the overall performance of sci.math might have been more promising, if there had been more posters like Ramsay, the fact remains that only one counterexample was scrutinized by the users of sci.math during the two years of my challenge.

Instead of checking my counterexamples any further, the posters focused on explaining why the users of sci.math had not proceeded further in checking my counterexamples. The newsgroup sci.math ventilated its failure and frustration by the following arguments:

- Attack the person, instead of the argument, called an ad hominem attack. A typical ad hominem attack was to claim that by bringing forth math-mistakes I try to show off my greatness and superiority over mistake-makers. Those who appealed for ad hominem often compensated their failure to present relevant arguments.

- One cannot falsify theorems. But those who escaped by this diversion, failed to pick out the correct theorems from the whole of the published mathematical literature.

- My counterexamples are not valid. But those who took this position, failed to point out a single invalid counterexample in my list.

- I deal only with trivial mistakes/counterexamples. But those who choose this argument, could not scrutinize the presumably trivial counterexamples (with the exception of Keith Ramsay, who verified one trivial counterexample. Admittedly my list contains also trivial counterexamples, which were included to make the list also accessible to beginners).

- The errors I have detected are not significant. But those who claimed this, were not qualified on Clifford algebras and thus could not give an expert opinion about the significance or impact of my counter-examples.

- A specialized discussion should be carried out in a newsgroup geared toward research (on Clifford algebras). Debating in a newsgroup specializing on Clifford algebras would not have tested the possibility of neutral outsiders (= non-experts) being able to help to settle debates between disagreeing experts.

- The purpose of mathematics is to prove theorems, not to falsify theorems (= police out mistakes).

- The ultimate put-down:

- Clifford algebraists are not good mathematicians, because they make so many mistakes. This argument was presented repeatedly. I waited until it was used by one of the best mathematicians in the forum. Then, during a competition over Clifford algebras, I caught the poster of a math mistake which he could have avoided had he known Clifford algebras. Thus, the poster nullified his own agenda.

- Posters are not interested in examining my counterexamples or proofs. But, several posters had already presented their opinions about the math-content of my counterexamples prior to announcing their disinterest, and I did not impose but was rather challenged to present the proofs.

Soon the discussion in sci.math deteriorated to a repetition and a reeling out of the above responses. In sci.math strangers, who do not read Journal articles of the other posters, often participate in the discussions. Posters seldom engage in literature searches or detailed computation before posting, rather they just key in something which shows off their present state of "knowledge". Personal insults are common, because posters believe that they can attack others without being identified or harmed in return.

As a final remark, we believe that the newsgroup sci.math could not come up with substantial and competent critics in a matter requiring specialized knowledge of mathematics. This shows that the newsgroup sci.math cannot be used to settle scientific debates between disagreeing mathematicians. Thus, in our search for the scientific truth, scientific conferences and journals cannot be replaced by Usenet newsgroups.

Web-pages presenting articles of dialogues in sci.math:

    http://www.hit.fi/~lounesto/David.Ullrich
    http://www.hit.fi/~lounesto/Robin.Chapman
    http://www.hit.fi/~lounesto/Zdislav.Kovarik
    http://www.hit.fi/~lounesto/Lynn.Killingbeck.

Acknowledgments

I would like to thank Ron Bloom, Jeremy Boden, Gary Pratt and Keith Ramsay for their comments presented in the Internet discussion group sci.math. Their feedback has helped me in presenting this material as a www-page to serve a wider mathematical audience surfing in the Internet. As for the mathematical content, I am indebted to Johannes Maks, who came up with some of the counterexamples, and to Jacques Helmstetter, whose help I have benefited in finding the proofs. Finally thanks to E. Bayro–Corrochano and G. Sobczyk for their help to transform the Web Page articles into a chapter article.

Chapter 24

The Making of GABLE: A Geometric Algebra Learning Environment in Matlab

Stephen Mann, Leo Dorst, and Tim Bouma

24.1 Introduction

Geometric algebra extends Clifford algebra with geometrically meaningful operators with the purpose of facilitating geometrical computations. Present textbooks and implementation do not always convey this geometrical flavor or the computational and representational convenience of geometric algebra, so we felt a need for a computer tutorial in which representation, computation and visualization are combined to exhibit the intuition and the techniques of geometric algebra. Current software packages are either Clifford algebra only (CLICAL [11] and CLIFFORD [3]) or do not include graphics [2], so we decided to build our own. The result is GABLE (Geometric AlgeBra Learning Environment), a hands-on tutorial on geometric algebra that is suited for undergraduate students [7].

The GABLE tutorial explains the basics of geometric algebra. It starts with the outer product (as a constructor of subspaces), then treats the inner product (for perpendicularity), and moves via the geometric product (for invertibility) to the more geometrical operators such as projection, rotors, meet and join, and ends with the homogeneous model of Euclidean space. When the student is done he/she should be able to do simple Euclidean geometry of flat subspaces using the geometric algebra of homogeneous blades. For instance, the intersection between lines can be easily expressed in the basic operators:

```
e = e3;              % the homogeneous embedding
P = e+ e1/3+e2;      % point P
Q = e+ e1+e2/2;      % point Q
R = e+ e1/2-e2/4;    % point R
PQ = join(P,Q);      % line PQ
QR = join(Q,R);      % line QR
```

name	user group	platform	type	$\mathcal{C}\ell_{p,q,r}$	nature	graphics
CLICAL	mathematics	–	CA	$\mathcal{C}\ell_{p,q,r}$	numeric	no
CLIFFORD	mathematics	Maple	CA	$\mathcal{C}\ell_{p,q,r}$	symbolic	no
Cambridge	math. physics	Maple V	GA	mostly $\mathcal{C}\ell_{3q,q}$	symbolic	no
Clados	math. physics	Java	CA	$\mathcal{C}\ell_{p,q}$	numeric	no
GABLE	CS students	Matlab	GA	$\mathcal{C}\ell_{p,q,3-p-q}$	numeric	yes

TABLE 24.1. Four packages in Clifford algebra/geometric algebra compared. Note that although Maple can do graphics, these two Maple GA/CA packages provide no direct display of GA/CA objects. GA = geometric algebra; CA = Clifford algebra.

```
meet(PQ,QR)          % intersection of those lines
```

Drawing the objects of geometric algebra is an important part of GABLE. Using them appropriately in this example produces Figure 24.1, clarifying the relationship between the 3-D blade-based computations in the homogeneous model and their geometric semantics in 2-D Euclidean space.

It is mainly with such illustrations in mind that we designed GABLE, and limited it to the 3-dimensional algebras $\mathcal{C}\ell_{p,q,3-p-q}$ (although $\mathcal{C}\ell_{3,0,0}$ is somewhat better supported than the other 3-dimensional algebras). Since this software is meant for a tutorial, we did not have great efficiency concerns (though we did have some), and were most interested in ease of implementation and the creation of a software package that could be made widely available for teaching purposes. We therefore chose to implement GABLE in the *student version of Matlab*. We now find that GABLE's use extends beyond the purely tutorial, since the easy access to visualization helps build up intuition into Euclidean geometry (see also Section 24.6) and differential geometry. Thus the tutorial package may serve as a springboard towards a more professional 'GA toolkit' in the future, with a different focus than existing packages, see Table 24.1. But for now, we do not intend GABLE to be more than an accessible, visual implementation of 3-dimensional geometric algebra.

This chapter describes our experiences in developing the GABLE package. We obviously needed to decide on basic implementation issues such as representation, computational efficiency and stability of inverses; but we also ran into issues that stemmed from the insufficiently resolved structure of geometric algebra itself. Among these are the various inner products, and the precise definition and semantics of the `meet` and ⋈ operators in Euclidean geometry (an issue also treated in Chapter 2). This chapter motivates our decisions in these matters.

Our representation of geometric objects is a refinement of the 8×8 representation that has been presented by others, along the lines suggested by Lounesto and Ablamowicz [8](page 72), [3]. We compare this representation of geometric algebra to matrix representations of Clifford Algebras

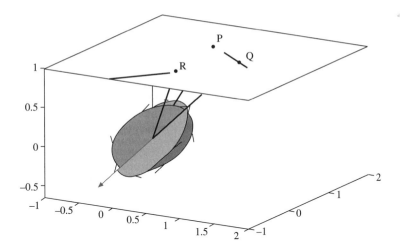

FIGURE 24.1. Intersection of lines in 2D as the meet of blades in a 3D homogeneous model of Euclidean geometry.

in Section 24.2. For the important but potentially expensive operation of inversion (or geometric division), we settled on a variation of a method proposed by Lounesto for $Cl_{3,0}$, which we extend in Section 24.3 to work for arbitrary signature (in 3 dimensions). At the higher level, Section 24.4 gives some detail on our implementation of the meet and ⋈ operations, extending them to the non-trivial cases of partially overlapping subspaces. Section 24.5 briefly discusses the graphics aspects of GABLE. The tutorial itself [7] is available on the World Wide Web[1]. The webpages contain both the Matlab package GABLE and the tutorial textbook.

24.2   Representation of Geometric Algebra

We first needed to decide on a representation for geometric algebra. In Matlab, a matrix representation would be most natural. Matrix representations for the geometric product in the Clifford algebras of various signatures are well studied [9]; for each signature a different matrix algebra results. That is slightly unsatisfactory. Moreover, our desire to make our algebra a proper *geometric* algebra implies that we should not only represent the geometric product, but also the outer and inner products, and preferably on a par with each other. These issues are discussed in more detail in Section 24.2.4; in brief, we ended up using a modified form of the

---

[1]Available at http://www.wins.uva.nl/~leo/clifford/gable.html
and http://www.cgl.uwaterloo.ca/~smann/GABLE/.

8×8 matrix representation.

24.2.1   The matrix representation of GABLE

In GABLE, we represent a multivector $\mathbf{A}$ as an 8×1 column matrix giving its coefficients relative to a basis for the Clifford algebra:

$$\mathbf{A} = [1, \; \mathbf{e}_1, \; \mathbf{e}_2, \; \mathbf{e}_3, \; \mathbf{e}_1 \wedge \mathbf{e}_2, \; \mathbf{e}_2 \wedge \mathbf{e}_3, \; \mathbf{e}_3 \wedge \mathbf{e}_1, \; \mathbf{e}_1 \wedge \mathbf{e}_2 \wedge \mathbf{e}_3] \begin{bmatrix} A_0 \\ A_1 \\ A_2 \\ A_3 \\ A_{12} \\ A_{23} \\ A_{31} \\ A_{123} \end{bmatrix},$$

where $\{\mathbf{e}_1, \mathbf{e}_2, \mathbf{e}_3\}$ form an orthogonal basis for the vector space of our algebra. We will use **bold** font for the multivector, and *math* font for its scalar-valued coefficients. The multivector $\mathbf{A}$ is thus represented by an 8×1 column matrix $[\mathbf{A}]$, which we will denote in shorthand as

$$\mathbf{A} \; \rightleftharpoons \; [\mathbf{A}].$$

Now if we need to compute the geometric product $\mathbf{AB}$, we view this as a linear function of $\mathbf{B}$ determined by $\mathbf{A}$, i.e., as the linear transformation $\underline{\mathbf{A}}^G(\mathbf{B})$, the '$G$' denoting the geometric product. Such a linear function can be represented by an 8×8 matrix, determined by $\mathbf{A}$ (and the fact that we are doing a geometric product) acting on the 8×1 matrix of $[\mathbf{B}]$. We thus expand the representation $[\mathbf{A}]$ of $\mathbf{A}$ to the 8×8 geometric product matrix $[\underline{\mathbf{A}}^G]$, and apply this to the 8×1 representation $[\mathbf{B}]$ of $\mathbf{B}$:

$$\mathbf{AB} \; \rightleftharpoons \; [\underline{\mathbf{A}}^G] [\mathbf{B}].$$

The result of the matrix product $[\underline{\mathbf{A}}^G][\mathbf{B}]$ is the 8×1 matrix representing the element $\mathbf{AB}$. The matrix entry $[\underline{\mathbf{A}}^G]_{\alpha,\beta}$ (so in column α and row β, with α and β running through the indices $\{0, 1, 2, 3, 12, 23, 31, 123\}$) can be computed in a straightforward manner from the multiplication table of the geometric product. This table depends on the signature. We denote the coefficients occurring in it by $c_{\alpha,\beta;\gamma}$ (they are known as the *structure coefficients* defining the algebra). The matrix $[\underline{\mathbf{A}}^G]$ is then computable through:

$$\mathbf{e}_\alpha \mathbf{e}_\beta \equiv c_{\alpha\beta;\gamma} \mathbf{e}_\gamma \quad \Longleftrightarrow \quad [\underline{\mathbf{A}}^G]_{\gamma,\beta} = c_{\alpha,\beta;\gamma} A_\alpha. \tag{2.1}$$

This 8×8 matrix $[\underline{\mathbf{A}}^G]$ can then be used to evaluate the (bilinear) product $\mathbf{AB}$ by applying it to the 8×1 column matrix $[\mathbf{B}]$ using the usual matrix multiplication:

$$[\mathbf{AB}]_\gamma = \left([\underline{\mathbf{A}}^G][\mathbf{B}]\right)_\gamma = \sum_\beta [\underline{\mathbf{A}}^G]_{\gamma,\beta} [\mathbf{B}]_\beta.$$

So for example the identity $\mathbf{e}_1\mathbf{e}_2 = \mathbf{e}_{12}$ leads to the matrix entry $[\underline{\mathbf{A}}^G]_{12,2} = A_1$; this is the only non-zero entry in column $\beta = 2$. In matrix multiplication between $\mathbf{A} = A_1\mathbf{e}_1$ and $\mathbf{B} = B_2\mathbf{e}_2$ this yields

$$[\mathbf{AB}]_{12} = [\underline{\mathbf{A}}^G]_{12,2}[\mathbf{B}]_2 = A_1B_2,$$

which is the correct contribution to the $\mathbf{e}_{12}$ component of the result.

Both the outer and the inner product can be implemented as matrix multiplications, since $\mathbf{A} \wedge \mathbf{B}$ and $\mathbf{A}\rfloor\mathbf{B}$ are also linear functions of $\mathbf{B}$, determined by $\mathbf{A}$. So we implement

$$\mathbf{A} \wedge \mathbf{B} \;\rightleftharpoons\; [\underline{\mathbf{A}}^O]\,[\mathbf{B}] \;\text{ and }\; \mathbf{A}\rfloor\mathbf{B} \;\rightleftharpoons\; [\underline{\mathbf{A}}^I]\,[\mathbf{B}].$$

The 8×8 matrices $[\underline{\mathbf{A}}^O]$ and $[\underline{\mathbf{A}}^I]$ are given below, and they are constructed according to Equation 2.1 for the outer product and inner product, respectively.

24.2.2   The representation matrices

Using the recipe of Equation 2.1, we can compute the actual matrices. The structure coefficients $c_{\alpha,\beta;\gamma}$ of Equation 2.1 contain signed products of σ_is. These represent the signature (and metric) through their definition as $\sigma_i \equiv \mathbf{e}_i\mathbf{e}_i$. We abbreviate $\sigma_{ij} \equiv \sigma_i\sigma_j$ and $\sigma_{ijk} \equiv \sigma_i\sigma_j\sigma_k$.

Geometric product matrix:

$$[\underline{\mathbf{A}}^G] =$$

$$
\begin{bmatrix}
A_0 & \sigma_1 A_1 & \sigma_2 A_2 & \sigma_3 A_3 & -\sigma_{12} A_{12} & -\sigma_{23} A_{23} & -\sigma_{13} A_{31} & -\sigma_{123} A_{123} \\
A_1 & A_0 & \sigma_2 A_{12} & -\sigma_3 A_{31} & -\sigma_2 A_2 & -\sigma_{23} A_{123} & \sigma_3 A_3 & -\sigma_{23} A_{23} \\
A_2 & -\sigma_1 A_{12} & A_0 & \sigma_3 A_{23} & \sigma_1 A_1 & -\sigma_3 A_3 & -\sigma_{31} A_{123} & -\sigma_{13} A_{31} \\
A_3 & \sigma_1 A_{31} & -\sigma_2 A_{23} & A_0 & -\sigma_{12} A_{123} & \sigma_2 A_2 & -\sigma_1 A_1 & -\sigma_{12} A_{12} \\
A_{12} & -A_2 & A_1 & \sigma_3 A_{123} & A_0 & \sigma_3 A_{31} & -\sigma_3 A_{23} & \sigma_3 A_3 \\
A_{23} & \sigma_1 A_{123} & -A_3 & A_2 & -\sigma_1 A_{31} & A_0 & \sigma_1 A_{12} & \sigma_1 A_1 \\
A_{31} & A_3 & \sigma_2 A_{123} & -A_1 & \sigma_2 A_{23} & -\sigma_2 A_{12} & A_0 & \sigma_2 A_2 \\
A_{123} & A_{23} & A_{31} & A_{12} & A_3 & A_1 & A_2 & A_0
\end{bmatrix}
$$

$$(2.2)$$

Outer product matrix:

$$[\underline{\mathbf{A}}^O] = \begin{bmatrix} A_0 & 0 & 0 & 0 & 0 & 0 & 0 & 0 \\ A_1 & A_0 & 0 & 0 & 0 & 0 & 0 & 0 \\ A_2 & 0 & A_0 & 0 & 0 & 0 & 0 & 0 \\ A_3 & 0 & 0 & A_0 & 0 & 0 & 0 & 0 \\ A_{12} & -A_2 & A_1 & 0 & A_0 & 0 & 0 & 0 \\ A_{23} & 0 & -A_3 & A_2 & 0 & A_0 & 0 & 0 \\ A_{31} & A_3 & 0 & -A_1 & 0 & 0 & A_0 & 0 \\ A_{123} & A_{23} & A_{31} & A_{12} & A_3 & A_1 & A_2 & A_0 \end{bmatrix} \tag{2.3}$$

Inner product matrix:

$$[\underline{\mathbf{A}}^I] = \begin{bmatrix} A_0 & \sigma_1 A_1 & \sigma_2 A_2 & \sigma_3 A_3 & -\sigma_{12} A_{12} & -\sigma_{23} A_{23} & -\sigma_{31} A_{31} & -\sigma_{123} A_{123} \\ 0 & A_0 & 0 & 0 & -\sigma_2 A_2 & 0 & \sigma_3 A_3 & -\sigma_{23} A_{23} \\ 0 & 0 & A_0 & 0 & \sigma_1 A_1 & -\sigma_3 A_3 & 0 & -\sigma_{13} A_{31} \\ 0 & 0 & 0 & A_0 & 0 & \sigma_2 A_2 & -\sigma_1 A_1 & -\sigma_{12} A_{12} \\ 0 & 0 & 0 & 0 & A_0 & 0 & 0 & \sigma_3 A_3 \\ 0 & 0 & 0 & 0 & 0 & A_0 & 0 & \sigma_1 A_1 \\ 0 & 0 & 0 & 0 & 0 & 0 & A_0 & \sigma_2 A_2 \\ 0 & 0 & 0 & 0 & 0 & 0 & 0 & A_0 \end{bmatrix}$$
$$\tag{2.4}$$

Note the relation between these matrices: the inner product matrix and the outer product matrix both have all non-zero elements taken from the geometric product matrix. Note also the lack of signature in the outer product matrix; this is in agreement with the fact that it forms a (non-metric) Grassmann algebra that may be viewed as a geometric algebra of null vectors, for which all σ_i equal 0.

The reader may realize from this matrix description that we have implemented an inner product that differs from Hestenes' inner product [6]: we prefer the *contraction* defined in [1], since we found that its geometric semantics is more straightforward (see [7, 10]). We have implemented other inner products as well (see the next section), but the contraction is the default.

24.2.3   The derived products

It is common to take the geometric product as basic, and define the other products using it by selecting appropriate grades. This can be the basis for an implementation; the Maple package at Cambridge [2] has been so constructed. For our comparative discussion below, we state the definitions; these can be used in a straightforward manner to derive the matrix representations.

- **Outer product**

$$\mathbf{A} \wedge \mathbf{B} = \sum_{r,s} \langle \langle \mathbf{A} \rangle_r \langle \mathbf{B} \rangle_s \rangle_{s+r}. \tag{2.5}$$

  where $\langle \cdot \rangle_r$ is the grade operator taking the part of grade r of a multivector.

- **Contraction inner product**

$$\mathbf{A} \rfloor \mathbf{B} = \sum_{r,s} \langle \langle \mathbf{A} \rangle_r \langle \mathbf{B} \rangle_s \rangle_{s-r}, \tag{2.6}$$

  where the grade operator for negative grades is zero (we thank Svenson [12] for this way of writing the contraction). Note that this implies that 'something of higher grade cannot be contracted onto something of lower grade'. For scalar α and a general non-scalar multivector $\mathbf{A}$ we get
  $\alpha \rfloor \mathbf{A} = \alpha \mathbf{A}$ and $\mathbf{A} \rfloor \alpha = 0$.

- **Modified Hestenes inner product**
  This is a variation of the Hestenes inner product, which fixes its odd behavior for scalars:

$$\mathbf{A} \cdot_M \mathbf{B} = \sum_{r,s} \langle \langle \mathbf{A} \rangle_r \langle \mathbf{B} \rangle_s \rangle_{|s-r|}. \tag{2.7}$$

  For scalar α and a general non-scalar multivector $\mathbf{A}$ we get
  $\alpha \cdot_M \mathbf{A} = \alpha \mathbf{A} = \mathbf{A} \cdot_M \alpha$.

- **Hestenes inner product**
  The original Hestenes product in [5] differs from Equation 2.7 in that the contributions of the scalar parts of $\mathbf{A}$ and $\mathbf{B}$ are explicitly set to zero.

Note that mixed-grade multivectors require expansion of a double sum in all these products.

24.2.4 *Representational issues in geometric algebra*

For non-degenerate Clifford algebras of arbitrary signature (p, q) (which means $p + q$ spatial dimensions, of which p basis vectors have a positive

	$q = 0$	$q = 1$	$q = 2$	$q = 3$
$p = 0$	$\mathbb{R}(1)$	$\mathbb{C}(1)$	$\mathbb{H}(1)$	$^2\mathbb{H}(1)$
$p = 1$	$^2\mathbb{R}(1)$	$\mathbb{R}(2)$	$\mathbb{C}(2)$	$\mathbb{H}(2)$
$p = 2$	$\mathbb{R}(2)$	$^2\mathbb{R}(2)$	$\mathbb{R}(4)$	$\mathbb{C}(4)$
$p = 3$	$\mathbb{C}(2)$	$\mathbb{R}(4)$	$^2\mathbb{R}(4)$	$\mathbb{R}(8)$
$p = 4$	$\mathbb{H}(2)$	$\mathbb{C}(4)$	$\mathbb{R}(8)$	$\mathbb{C}(8)$

TABLE 24.2. Matrix representations of Clifford algebras of signatures (p, q).

square, and q have a negative square) linear matrix representations have long been known.[2]

Those relevant to our discussion are repeated in Table 24.2, from [9]. In this table, $\mathbb{R}(n)$ are $n \times n$ real matrices, $\mathbb{C}(n)$ are $n \times n$ complex-valued matrices, $\mathbb{H}(n)$ are $n \times n$ quaternion-valued matrices. The 'dual real numbers' $^2\mathbb{R}(n)$ are ordered pairs of $n \times n$ real matrices, and similarly for the other number systems.

The various representations are non-equivalent, so the table can be used for arguments on unique representations. Note that the Clifford algebras for a 3-dimensional space can have many different representations depending on the signature. Though this is not a problem for implementations, it makes it harder to obtain a parametric overview on the metric properties of the various spaces, and a representation that contains the signature as parameters σ_i has our slight preference.

The outer product is linear and associative, and in each case isomorphic to a rather involved subalgebra of a matrix algebra, which we will not specify further. The inner product is not associative, and is therefore not isomorphic to a matrix (sub)algebra.

Our initial exclusive interest in $\mathcal{C}\ell_{3,0}$ suggested the representation $\mathbb{C}(2)$, with elements represented as

$$\left[\begin{array}{cc} (A_0 + A_3) + i(A_{12} + A_{123}) & (A_1 + A_{31}) + i(-A_2 + A_{23}) \\ (A_1 - A_{31}) + i(A_2 + A_{23}) & (A_0 - A_3) + i(-A_{12} + A_{123}) \end{array} \right],$$

but this works only for the geometric product; the other products would then have to be implemented using the grade operator. We prefer a representation in which all three products are representable on a par, and in which signatures are parameterized. This desire to represent arbitrary signatures parametrically necessitates viewing $\mathcal{C}\ell_{3,0}$ as a subalgebra of $\mathcal{C}\ell_{3,3}$,

---

[2]Less is known about *degenerate* Clifford algebras and their representations. In our preferred choice they are handled as easily as the non-degenerate algebras, so we will not discuss them in detail.

and therefore to choose a representation in $\mathbb{R}(8)$.

This algebra $\mathbb{R}(8)$ also contains a representation of the outer product as a certain kind of lower-triangular matrices (in fact, Equation 2.2 works nicely: the matrix product of two such matrices faithfully represents the outer product). For arbitrary signatures, there cannot exist a change of representation in which both the outer product matrices and the geometric product matrices could be reduced to a smaller size (i.e., brought onto a block-diagonal representation of the same kind), since we need the full $\mathbb{R}(8)$ to handle those signatures anyway.

Now the need to represent the inner product as well indicates that we can not represent the elements of the algebra by matrices in their function as both *operator* (i.e., first factor) and *operand* (i.e., second factor). We therefore switch to the view where each product is seen as a linear function of the operand, parameterized by the operator, as detailed in Section 24.2.1. We maintain the $\mathbb{R}(8)$-representation of these linear functions, but they now operate on 8-dimensional vectors representing the operand (rather than forming an algebra of operators). Thus we arrive at the representation we have chosen (also for the geometric product), with the operator matrices naturally defined as in Equation 2.1.

It should be clear that the same reasoning suggests an $\mathbb{R}(2^n)$ representation of the geometric algebra of n-dimensional space of arbitrary signatures, with matrices defined for the three products in the same way.

24.2.5 Computational efficiency

If we would represent our objects as 8×8 matrices of reals, the resulting matrix multiply to implement the geometric product would cost 512 multiplications and 448 additions. Further, using the 8×8 matrix representation, to compute the outer product and/or inner product, we would have to use the grade operator (or, for the outer product, pay the expansion cost to convert to the outer product 8×8 matrix representation). Addition and scalar multiplication of elements in this form require 64 additions and 64 multiplications respectively. This method is extremely inefficient and we will not discuss it further.

The computational efficiency of the 8×1 format is better and is summarized in Table 24.3, where the notation (a, m) denotes the number of additions and multiplications required. With the 8×1 format, the cost of computing a product is the computational cost of having to expand one of the one of the 8×1 matrices to an 8×8 matrix and then multiply it by an 8×1 matrix at a cost of 64 multiplications, 56 additions, and the cost of expansion. When we include the cost of signatures in the expansion cost, then the total cost is increased by 48 multiplications.

It is of course possible to use the table of Clifford algebra isomorphisms as a literal guide to the implementation. Let us consider the costs of implementing the special case of the 3-dimensional Clifford algebras; Table 24.2 shows that this involves implementation of $C(2)$, $^2\mathbb{R}(2)$ and $^2\mathbb{H}(1)$. In all representations the operations of addition and scalar multiplication have take 8 floating point additions and 8 floating point multiplications, respectively (in the $\mathbb{R}(8)$ representation, these operations are performed on the 8×1 matrices representing the objects).

To compute the geometric product we need to multiply elements. The cost of the basic multiplications is: one complex multiply is $(4, 2)$; one double real multiply takes $(2, 0)$; one quaternion multiply takes $(16, 12)$. For a full matrix implementation to produce the geometric product this yields for $C(2)$ a cost of $(32, 16)$; for $^2\mathbb{R}(2)$ a cost of $(16, 8)$; for $^2\mathbb{H}(1)$ a cost of $(32, 24)$. Depending on the structure of the algebra, one may thus be fortunate by a factor of two. These should be compared to our $\mathbb{R}(8)$ implementation acting on 8×1 matrices, which has a cost of $(110, 56)$, for general signatures. This is a factor of 3 worse than $^2\mathbb{H}(1)$, the most expensive of the other three representations. If we consider only $\mathcal{Cl}_{3,0}$, then we have no signature cost, and $\mathbb{R}(8)$ costs $(64, 56)$ compared to $(32, 16)$ for $C(2)$. Table 24.3 compares the cost of several operations of our $\mathbb{R}(8)$ implementation (without signature cost) to a $C(2)$-based implementation.

To implement a full geometric algebra, these specific geometric product implementations need to be augmented with a grade operation to extract elements of the grade desired, according to Equations 2.6 and 2.5. For $C(2)$ it takes $(0, 8)$ to extract the eight separate elements, and presumably the same for $^2\mathbb{H}(1)$ and $^2\mathbb{R}(2)$. For simplicity of discussion, when extracting a single grade, we will assume that it costs 3 additions (although for scalars and trivectors, the cost is only 1 addition).

This process of a geometric product followed by grade extraction is simple if the objects to be combined are *blades* (rather than general multivectors). Such an operation requires a geometric product followed by grade extraction, which for $C(2)$ has a total worst case cost of $(32, 19)$, although there may be some additional cost to test for the grade of the blade, etc., which would add $(0, 16)$ to the cost $((32, 35)$ total) if we need to perform a full grade extraction of each operand.

When taking the outer or inner product of multivectors that are *not* blades, the use of the geometric product and grade extraction becomes quite expensive, since we must implement a double sum (see Equations 2.5, 2.6, and 2.7). A naive implementation of this formula would require 16 geometric products and grade extractions, an additional 12 additions to combine the results for each grade, and 8 additions to reconstruct the result, for a total cost of $(512, 324)$. However, looking at Table 24.4, we see that that six of these geometric products will always be 0, and we can easily rewrite our code to take advantage of this. This modification to the code reduces the cost to $(320, 210)$.

By unbundling the loop and simplifying the scalar cases (i.e., multiplying B by the scalar portion of A reduces 4 geometric products to one floating

Operation	8×1	$C(2)$
addition	$(0, 8)$	$(0, 8)$
scalar multiplication	$(8, 0)$	$(8, 0)$
grade extraction	$(0, 0)$	$(0, 8)$
geometric product	$(64, 56)$	$(32, 16)$
other products of blades	$(64, 56)$	$(32, 19)$
other products of multivectors	$(64, 56)$	$(111, 79)$

TABLE 24.3. Comparison of costs for 8×1 and $C(2)$. Notation: (a, m) denotes number of additions a and multiplications m.

$a \wedge b$	0	1	2	3
0	0	1	2	3
1	1	2	3	(4)
2	2	3	(4)	(5)
3	3	(4)	(5)	(6)

TABLE 24.4. Grade of the outer product of blades. If the grade is greater than 3, then the result will be 0.

point addition (to extract the scalar) and 8 floating point multiplies, and multiplying A by the scalar portion of B reduces 3 more geometric products to one addition and 7 floating point multiplies) we can get the cost down to $(32 * 3 + 15, 19 * 3 + 2 + 12 + 8) = (111, 79)$. Further special casing of the vector and bivector terms can reduce this cost to $(33, 45)$ (details of the analysis can be found in [10]), but note that in doing this (a) we have left the complex representation for computing these products and (b) each product will need its own special case code.

The above discussion is on the cost of writing special case code for the outer product only. If we choose this route, we would also need to write special case code for each of the inner products and possibly for each dimensional space in which we wish to work. A reasonable compromise of special cases versus general code for the complex representation would be to handle the scalars as special cases and write the loops to avoid the combinations that will always give zero.

24.2.6 Asymptotic costs

If we are interested in arbitrary dimensional spaces, then we need to look at the asymptotic costs. Table 24.5 summarizes the costs of the complex and of the $2^n \times 1$ representation (where $n = p + q + r$ is the dimension of the

	Geometric Product	Other products on blades	Other products on multivectors
Complex	$2^{(3n+1)/2}$	$2^{(3n+1)/2}$	$n^2 2^{(3n+1)/2}$
$2^n \times 1$	2^{2n}	2^{2n}	2^{2n}

TABLE 24.5. Comparison of costs of various methods, with n being the dimension of the underlying vector space.

underlying vector space) for the geometric product and for the other products (e.g, inner and outer products) on blades and for the other products on general multivectors. In this table, we only give the top term in the cost expression, ignoring grade extraction, etc., for the complex representation of other products. Use of only this top order term also ignores the savings achieved for the complex representation by not computing the products whose grade is higher than n and special casing the scalar products; such optimizations roughly equate to savings of a factor of two. Note that we use the complex representation as a coarse representative of the other representations; in the other cases we would use the quaternion or double-real representation, which cost roughly a factor of two less than the complex representation.

From the table, we see that asymptotically the complex representation is always best. However, substituting numbers in these equations shows that for small n, the $2^n \times 1$ representation is best when performing inner or outer products of general multivectors, with the cross-over point being around $n = 14$. But when n equals 14, the cost of even the geometric product in the complex representation is extremely large, requiring roughly 3×10^6 multiplications, so this is unlikely to be used at all.

For smaller n, the complex representation is better than the $2^n \times 1$ representation for the geometric product and the products of blades, while the $2^n \times 1$ representation is computationally less expensive than the complex representation for the other products of general multivectors. However the other products of general multivectors are rarely (if ever) performed in our present understanding of what constitute geometrically significant combinations. Thus, in general the complex/quaternion/double-real representation will be more efficient than the $2^n \times 1$ representation by a factor of $2^{n/2}$. The conclusion must be that once one has decided on a particular geometry for one's application, reflected in a particular signature, it makes sense to implement it literally using the isomorphism of Table 24.2.

For our tutorial in 3 dimensional spaces, the cost of the 8×1 representation is only a factor of three more expensive than the complex representation. Since we were writing tutorial code, we felt this cost was more than offset by the explicitness of the signature and ease of implementation.

24.3 Inverses

In Matlab, the obvious way to compute the inverse of a geometric object $\mathbf{M}$ is to express it in the 8×8 geometric product matrix representation, $[\mathbf{M}]$. Then inversion of $[\mathbf{M}]$ may be done using the Matlab matrix inverse routine, and the first column of $[\mathbf{M}]^{-1}$ will be the representation of the inverse of $\mathbf{M}$. However, when we implemented this method for computing the inverse, we found that it introduced small numerical errors on even rather simple data, and thus was less stable than we would like. We investigated a method by Lounesto [8] to compute inverses in 3-dimensional Clifford algebras. This method proved more stable in our testing, and it is computationally considerably more efficient than a matrix inverse. We discuss it now, and extend it slightly.

Lounesto's trick is based on the observation that in three dimensions (and that is essential!) the product of a multivector $\mathbf{M}$ and its Clifford conjugate $\overline{\mathbf{M}}$ only has two grades, a scalar and a pseudoscalar (the Clifford conjugate is the grade involution of the reverse of a multivector). Let $\mathbf{M}_i$ denote the part of $\mathbf{M}$ of grade i, though we will write M_0 for the scalar part. Then we compute

$$
\begin{aligned}
\mathbf{M}\overline{\mathbf{M}} &= (M_0 + \mathbf{M}_1 + \mathbf{M}_2 + \mathbf{M}_3)(M_0 - \mathbf{M}_1 - \mathbf{M}_2 + \mathbf{M}_3) \\
&= (M_0^2 - \mathbf{M}_1^2 - \mathbf{M}_2^2 + \mathbf{M}_3^2) + 2(M_0\mathbf{M}_3 - \mathbf{M}_1 \wedge \mathbf{M}_2),
\end{aligned}
$$

and the first bracketed term is a scalar, the second a trivector.

Further, at least in Euclidean 3-space, if such an object of the form 'scalar plus trivector' $N_0 + \mathbf{N}_3$ is non-zero, then it has an inverse that is easily computed:

$$
(N_0 + \mathbf{N}_3)^{-1} = \frac{N_0 - \mathbf{N}_3}{N_0^2 - \mathbf{N}_3^2}.
$$

Please note that not all multivectors have an inverse, not even in a Euclidean space: for instance $\mathbf{M} = 1 + \mathbf{e}_1$ leads to $\mathbf{M}\overline{\mathbf{M}} = 0$, so this $\mathbf{M}$ is non-invertible. In a non-Euclidean space, the denominator may become zero even when N_0 and $\mathbf{N}_3$ are not, and we need to demand at least that $N_0^2 \neq \mathbf{N}_3^2$. When it exists, the inverse is unique. This follows using the associativity of the geometric product: if $\mathbf{A}$ and $\mathbf{A}'$ are left and right inverses of $\mathbf{B}$, respectively, then $\mathbf{A} = \mathbf{A}(\mathbf{B}\mathbf{A}') = (\mathbf{A}\mathbf{B})\mathbf{A}' = \mathbf{A}'$. Therefore any left inverse is a right inverse, and both are identical to *the* inverse.

These two facts can be combined to construct an inverse for an arbitrary multivector $\mathbf{M}$ (still in Euclidean 3-space) as follows:

$$
\mathbf{M}^{-1} = \overline{\mathbf{M}}\,\overline{\mathbf{M}}^{-1}\mathbf{M}^{-1} = \overline{\mathbf{M}}(\mathbf{M}\overline{\mathbf{M}})^{-1} = \frac{\overline{\mathbf{M}}\left(\langle\mathbf{M}\overline{\mathbf{M}}\rangle_0 - \langle\mathbf{M}\overline{\mathbf{M}}\rangle_3\right)}{\langle\mathbf{M}\overline{\mathbf{M}}\rangle_0^2 - \langle\mathbf{M}\overline{\mathbf{M}}\rangle_3^2}
$$

The following two lemmas and their proofs demonstrate the correctness of Lounesto's method in 3-dimensional spaces of arbitrary signature.

Lemma 24.1. $\mathbf{M}^{-1}$ *exists if and only if* $(\mathbf{M}\overline{\mathbf{M}})^{-1}$ *exists.*

Proof: First, assume that $\mathbf{M}^{-1}$ exists. Then $1 = \overline{\mathbf{M}^{-1}}\,\overline{\mathbf{M}} = (\overline{\mathbf{M}^{-1}\mathbf{M}^{-1}})(\mathbf{M}\overline{\mathbf{M}})$, so that $(\mathbf{M}\overline{\mathbf{M}})^{-1} = \overline{\mathbf{M}^{-1}}\,\mathbf{M}^{-1}$, which exists.

Secondly, assume that $(\mathbf{M}\overline{\mathbf{M}})^{-1}$ exists. Then we have $1 = (\mathbf{M}\overline{\mathbf{M}})(\mathbf{M}\overline{\mathbf{M}})^{-1} = \mathbf{M}(\overline{\mathbf{M}}(\mathbf{M}\overline{\mathbf{M}})^{-1})$, so that $\mathbf{M}^{-1} = \overline{\mathbf{M}}(\mathbf{M}\overline{\mathbf{M}})^{-1}$, which exists.   ☐

Lemma 24.2.   *Let* $\mathbf{N} = N_0 + \mathbf{N}_3$. *Then iff* $N_0^2 \neq \mathbf{N}_3^2$, $\mathbf{N}^{-1}$ *exists and equals*

$$(N_0 + \mathbf{N}_3)^{-1} = \frac{N_0 - \mathbf{N}_3}{N_0^2 - \mathbf{N}_3^2}$$

Proof: Assume $N_0^2 \neq \mathbf{N}_3^2$, then

$$
\begin{aligned}
(N_0 + \mathbf{N}_3)\,(N_0 - \mathbf{N}_3)/(N_0^2 - \mathbf{N}_3^2) &= (N_0^2 + \mathbf{N}_3 N_0 - N_0 \mathbf{N}_3 - \mathbf{N}_3^2)/(N_0^2 - \mathbf{N}_3^2)^2 \\
&= (N_0^2 - \mathbf{N}_3^2)/(N_0^2 - \mathbf{N}_3^2) = 1,
\end{aligned}
$$

so $\mathbf{N}^{-1}$ is as stated. Now assume that $\mathbf{N}^{-1}$ exists. Then if $\mathbf{N}_3 = 0$ the result is trivial. If $\mathbf{N}_3 \neq 0$ and $N_0 = 0$ the result is trivial. So take $\mathbf{N}_3 \neq 0$ and $N_0 \neq 0$. Let $\mathbf{K}$ be the inverse of $\mathbf{N} = N_0 + \mathbf{N}_3$. Then it needs to satisfy

$$(N_0 + \mathbf{N}_3)(K_0 + \mathbf{K}_1 + \mathbf{K}_2 + \mathbf{K}_3) = 1,$$

so, written out in the different grades

$$(N_0 K_0 + \mathbf{N}_3 \mathbf{K}_3) + (N_0 \mathbf{K}_1 + \mathbf{N}_3 \mathbf{K}_2) + (N_0 \mathbf{K}_2 + \mathbf{N}_3 \mathbf{K}_1) + (N_0 \mathbf{K}_3 + \mathbf{N}_3 K_0) = 1$$

Straightforward algebra on the terms of grade 0 and 3 yields

$$(N_0^2 - \mathbf{N}_3^2)\mathbf{K}_3 + \mathbf{N}_3 = 0,$$

and since $\mathbf{N}_3 \neq 0$ this gives $N_0^2 \neq \mathbf{N}_3^2$. Then the case above shows that the inverse is $\mathbf{N}^{-1} = (N_0 - \mathbf{N}_3)/(N_0^2 - \mathbf{N}_3^2)$.   ☐

Table 24.6 summarizes the costs to compute the inverse for both the 8×1 representation and for the $\mathtt{C}(2)$ representation. In this table, we give three algorithms for each representation: a naive algorithm that does not try to exploit any extra knowledge we have about the terms we are manipulating; a good algorithm that exploits the structure of $(\mathbf{M}\overline{\mathbf{M}})^{-1}$, which is a scalar plus a pseudo-scalar, and thus does not require a full product when multiplied by $\mathbf{M}$; and a scalar version that can be used when $(\mathbf{M}\overline{\mathbf{M}})^{-1}$ is

Term	8 × 1			C(2)		
	Naive	Good	Scalar	Naive	Good	Scalar
$\overline{\mathbf{M}}$	(0,0)	(0,0)	(0,0)	(0,8)	(0,8)	(0,8)
$\mathbf{M}\overline{\mathbf{M}}$	(64,56)	(64,56)	(64,56)	(32,16)	(32,16)	(32,16)
$\mathbf{M}\overline{\mathbf{M}}^{-1}$	(4,1)	(4,1)	(0,0)	(4,1)	(4,1)	(0,0)
$\mathbf{M}(\mathbf{M}\overline{\mathbf{M}})^{-1}$	(64,56)	(16,8)	(8,0)	(32,16)	(16,8)	(8,0)
Total	(132,113)	(84,65)	(72,56)	(64,41)	(52,33)	(40,24)

TABLE 24.6. Cost of Lounesto's inverse.

a scalar. This last case occurs when **M** is a blade, a scalar plus a bivector, or a vector plus the pseudo-scalar, which covers most of the geometrically significant objects we manipulate.

Note that in this table we have omitted the cost of the six negations needed to compute the Clifford conjugate. Also note that the complex representation requires 8 additions when computing the Clifford conjugate because it has to separate and recombine the scalar and pseudo-scalar part of the geometric object.

Lounesto's method is computationally much cheaper than the matrix inverse method, with a good implementation of Lounesto's method requiring 149 Matlab floating point operations for the 8 × 1 representation, while the Matlab matrix inverse routine on 8 × 8 matrices requires 1440 Matlab floating point operations. Lounesto's method really makes convincing use of the special structure of our matrices. While a faster matrix inversion routine may be available, it is unlikely that there will be a general routine capable of inverting our special 8 × 8 matrix in fewer than 149 floating point operations (which is after all little more than twice the number of matrix elements!). Further, in practice we found our modified Lounesto inverse to compute a more numerically stable inverse than the matrix inverse routine provided by Matlab (perhaps not surprisingly, since it involves fewer operations).

Had we used the C(2) representation of elements in our geometric algebra, the cost of matrix inversion would have dropped dramatically, with Matlab requiring only 260 floating point operations to invert a 2 × 2 complex matrix. However, Lounesto's method using the complex representation only requires 85 floating point operations. Thus Lounesto's inversion method is also less expensive in the C(2) representation.

24.4   Meet and Join

The geometric intersection and union of subspaces is done by the meet and ⋈ operations. These have mostly been used by others in the context of *projective geometry*, which has led to the neglect of some scalar factors and signs (since they do not matter in that application). This issue was partly treated in [4], but the development of the tutorial required more investigation of those scalar factors. This section reports on that.

24.4.1 Definition

The meet and ⋈ operations are geometrical 'products' of a higher order than the elementary products treated before. They are intended as geometrical intersection and union operators on (sub)spaces of the algebra. Since subspaces are represented by pure blades, *these operations should only be applied to blades.*

Let blades $\mathbf{A}$ and $\mathbf{B}$ contain as a common factor a blade C of maximum grade (this is like a 'largest common divisor' in the sense of the geometric product), so that we can write

$$\mathbf{A} = \mathbf{A}' \wedge C \text{ and } \mathbf{B} = C \wedge \mathbf{B}'$$

(note the order!). We chose $\mathbf{A}'$ and $\mathbf{B}'$ to be perpendicular to C, so that we could write the factorization in terms of the geometric product:
$\mathbf{A} = \mathbf{A}'C$ and $\mathbf{B} = C\mathbf{B}'$ (but note that $\mathbf{A}'$ and $\mathbf{B}'$ are in general *not* mutually perpendicular!). If $\mathbf{A}$ and $\mathbf{B}$ are disjoint, then C is a scalar (a 0-blade). We now define meet and ⋈ as

$$\bowtie (\mathbf{A}, \mathbf{B}) = \mathbf{A}' \wedge C \wedge \mathbf{B}' \text{ and } \text{meet}(\mathbf{A}, \mathbf{B}) = C.$$

Note that the factorization is *not* unique: we may multiply C by a scalar γ. This affects the ⋈ result by $1/\gamma$ and the meet by γ, so meet and ⋈ are *not well-defined blades.* (Since γ may be negative, not even the orientation of the results is defined unambiguously.) So these operations are hard to define in a Clifford algebra; but in an algebra intended for practical geometry, they are definitely desired. Fortunately, many geometric constructions are insensitive to the magnitude and/or sign of the blade representing the subspace. (A prime example is the projection $(\mathbf{x} \rfloor \mathbf{A})/\mathbf{A}$ onto the subspace represented by $\mathbf{A}$ – there is not problem using for $\mathbf{A}$ the outcome of a meet or ⋈.) All we need to guarantee is that meet and ⋈ of the same subspaces can be used consistently. Therefore we need to base both operations on the same factorization.

It is straightforward to make the computational relationships between meet and ⋈ explicit. The definition gives for the ⋈, given the meet, is

$$\bowtie (\mathbf{A}, \mathbf{B}) = \frac{\mathbf{A}}{\text{meet}(\mathbf{A}, \mathbf{B})} \wedge \mathbf{B}, \tag{4.8}$$

where the fraction denotes right-division.[3]

By duality relative to $\bowtie (\mathbf{A}, \mathbf{B})$ and symmetry of a scalar-valued contraction (or inner product) it follows from Equation 4.8 that

$$1 = \frac{\mathbf{A}}{\mathtt{meet}(\mathbf{A}, \mathbf{B})} \rfloor \frac{\mathbf{B}}{\bowtie (\mathbf{A}, \mathbf{B})} = \frac{\mathbf{B}}{\bowtie (\mathbf{A}, \mathbf{B})} \rfloor \frac{\mathbf{A}}{\mathtt{meet}(\mathbf{A}, \mathbf{B})}.$$

The division by $\mathtt{meet}(\mathbf{A}, \mathbf{B})$ can be factored out (this is due to the containment relationship of the factors of the contraction and easy to prove using the techniques in [4]) and we obtain

$$\mathtt{meet}(\mathbf{A}, \mathbf{B}) = \frac{\mathbf{B}}{\bowtie (\mathbf{A}, \mathbf{B})} \rfloor \mathbf{A}. \tag{4.9}$$

Thus we can start from either $\mathtt{meet}$ or $\bowtie$ and compute the other in a consistent manner. The symmetry of the equations means that either way is equally feasible.[4] Algorithms for $\mathtt{meet}$ and $\bowtie$ are discussed in [7]. For a more detailed discussion on the algebra of incidence see Chapter 13 and Chapter 7.

24.5   Graphics

Since we wanted a visual tutorial, we created graphical representations for all blades, and used Matlab rendering commands to draw them. The following table summarizes our representations:

Type	Representation	Orientation
scalar	Text above window	Sign
vector	Line from origin	Arrow head
bivector	Disk centered at origin	Arrows along edge
trivector	Line drawn sphere	Line segments going out or in

Figure 24.2 illustrates the vector, bivector, and trivector; the axes are put in automatically by Matlab.

---

[3] If $\boldsymbol{C}$ is a null blade (i.e., a blade with norm 0, non-invertible) then we cannot compute $\mathbf{A}'$ in terms of $\mathbf{A}$ from the factorization equation $\mathbf{A} = \mathbf{A}'\boldsymbol{C}$, and therefore not compute $\bowtie (\mathbf{A}, \mathbf{B}) = \mathbf{A}' \wedge \mathbf{B}$ from the $\mathtt{meet}$ (or vice versa, by a similar argument). We thus have to limit $\bowtie$ and $\mathtt{meet}$ to non-null blades; which means that we restrict their use to (anti-)Euclidean spaces only.

[4]Equation 4.9 is frequently extended to provide a 3-argument $\mathtt{meet}$ function relative to a general blade $\mathbf{I}$: $\mathtt{meet}(\mathbf{A}, \mathbf{B}, \mathbf{I}) \equiv (\mathbf{B}/\mathbf{I}) \cdot \mathbf{A}$. However, since the geometric significance of using anything but $\bowtie (\mathbf{A}, \mathbf{B})$ as third argument is unclear, we will not use it. Also, beware that some writers may switch the order of the arguments in this formula!

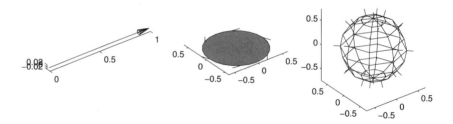

FIGURE 24.2. Graphical representation of vector, bivector, and trivector.

We chose the disk as our representation for bivectors since with our matrix representation of the geometric objects, we do not necessarily have the defining vectors for the bivector (which may not even exist, as is the case if the bivector was created as the dual of a vector). Without such vectors, we can not use the standard parallelogram representation of the bivector. There is a similar issue with the trivector (i.e., we were unable to use a parallelepiped as its representation) and thus we used the sphere. However, we also provide demonstration routines to illustrate the more standard representations of bivectors and trivectors; the user must then provide the basis on which to decompose them.

Objects of mixed grade presented a more difficult problem. While it is easy to draw the scalar, vector, bivector, and trivector components independently, this is not particularly illustrative. In particular, we needed to find a way to illustrate the operations of the inner, outer, and geometric products. The first two are fairly easy to demonstrate: we have two subwindows, in the former we draw the operands and in the latter we draw the result. The geometric product is more difficult to illustrate. So in addition to providing a routine to show the operands and result of the geometric product, we presented examples of using the geometric product as an *operator* to perform rotations and interpolation between orientations, rather than as a (composite) *object* by itself.

24.6   Example: Pappus's Theorem

As a more complete example of GABLE, we present an illustration of Pappus's theorem, which says take any two lines and three points on each line (P1 P2 P3 and Q1 Q2 Q3), cross-join the point (i.e., build the line segments P1Q2, P1Q3, P2Q1, P2Q3, P3Q1, and P3Q2) and compute the intersection of the three cross-joined pairs of segments (i.e., intersect P1Q2 with P2Q1, P2Q3 with P3Q2, and P3Q1 with P1Q3), and then these three points of intersection will be collinear.

To illustrate this theorem, we first need to construct six points and draw the relevant line segments. We choose to e3 as the homogeneous coordinate:

```
>> P1 = e3+e1; P2 = e3+2*e1; P3 = e3+4*e1;
>> Q1 = e3+e2; Q2 = e3+e1+2*e2; Q3 = e3+2*e1+3*e2;
>> DrawPolyline({P1,P3},'r'); DrawPolyline({Q1,Q3},'r');
>> DrawPolyline({P1,Q2},'k'); DrawPolyline({P1,Q3},'k');
>> DrawPolyline({P2,Q1},'k'); DrawPolyline({P2,Q3},'k');
>> DrawPolyline({P3,Q1},'k'); DrawPolyline({P3,Q2},'k');
```

The DrawPolyline calls draw the line segments of Pappus's theorem. Next we want to compute the intersection of corresponding line segments. As a first step, we need to compute each line segment (the ⋈ of two points on the segment) and as a second step we need to intersect pairs of line segment (the meet of the two segments). Note that the meet will give us a homogeneous point, and we need to normalize its coordinates to put the point back in the homogeneous plane:

```
>> %...
>> H3 = meet(join(P1,Q2),join(P2,Q1)); A3 = H3/inner(H3,e3);
>> H2 = meet(join(P1,Q3),join(P3,Q1)); A2 = H2/inner(H2,e3);
>> H1 = meet(join(P2,Q3),join(P3,Q2)); A1 = H1/inner(H1,e3);
>> DrawHomogeneous(e3,H1,'n','g');
>> DrawHomogeneous(e3,H2,'n','g');
>> DrawHomogeneous(e3,H3,'n','g');
>> DrawPolyline({A1,A3},'b')
```

The resulting GABLE drawing appears in Figure 24.3. Two other examples that appear in our tutorial are an illustration of Napoleon's theorem and an illustration of Morley's triangle.

24.7   Conclusions

In GABLE, our Matlab package for the geometric algebra tutorial, we have chosen an 8×1 representation of multivectors, to be expanded to an 8×8 matrix representation when they are used as operands in the elementary products (geometric product, inner product, outer product). In our detailed comparison of the complexity of this representation with representations based on the isomorphisms of Clifford algebras with matrix algebras, this choice appeared not always the most efficient for software used in an actual application (rather than a tutorial), especially if the

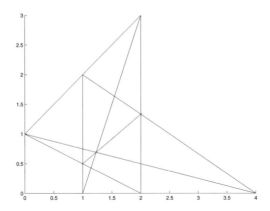

FIGURE 24.3. Illustration of Pappus's theorem created by GABLE.

signature of the space required could be known beforehand, and if one would deal mostly with pure blades. Further developments in the practical use of geometric algebra should show whether blades are indeed sufficient for our needs. If applications would require many inner and outer products of multivectors of mixed grade, then our explicit representation of these products by matrices should be considered.

GABLE is an implementation of 3-dimensional Clifford algebras with arbitrary signature. The generalization to arbitrary dimensions is readily obtained from Equation 2.1. However, as noted earlier in this chapter, high dimensional Clifford algebras are computationally expensive, and rather than use our $2^n \times 1$ representation, a specific complex, etc., representation may be preferred.

For the geometric division, we have extended Lounesto's method to compute inverses in 3-dimensional spaces of arbitrary signatures; but it should be emphasized that the method does *not* work in spaces of higher dimensions since it is based on properties of the Clifford conjugation that do not generalize to such spaces. In those spaces, an inversion of the geometric product matrix might be required.

The need to make geometrical macros for intersection and connection of geometrical objects necessitated a detailed study of the ⋈ and **meet** operations and their relationship. We have now embedded them properly into the geometric algebra of blades, even though each is only determined up to a scalar factor; the key is to realize that both are based on the same factorization of blades. The tutorial shows that despite this unknown scalar, geometrically significant quantities based on them are unambiguously determined. We noted that the GABLE **meet** and ⋈ operators only work for Euclidean signature; further research is needed to extend these operations to arbitrary signature.

At the start of this project, we thought it would be straight-forward to implement this software using results in the literature. However, we found the literature lacking in several areas, which we have partly addressed in this chapter. As a result of our work, we now have GABLE, a Matlab package and tutorial that should facilitate learning geometric algebra by people new to the subject. Moreover, we have found the package useful for testing out ideas and verifying results in our own research.

Acknowledgments

We gratefully acknowledge the funding of Stephen Mann's sabbatical by the Dutch Organization for Scientific Research (NWO), and the funding of Tim Bouma's summer stay by the University of Amsterdam.

Chapter 25

Helmstetter Formula and Rigid Motions with CLIFFORD

Rafal Ablamowicz

25.1 Introduction

CLIFFORD is a Maple package for symbolic computations in Clifford algebras $C\ell(B)$ of an arbitrary symbolic or numeric bilinear form B. The purpose of this paper is to show usability and power of CLIFFORD when performing computer-based proofs and explorations of mathematical aspects of Clifford algebras and their applications. It is intended as an invitation to engineers, computer scientists, and robotics to use Clifford algebra methods as opposed to coordinate/matrix methods. CLIFFORD has been designed as a tool to promote and facilitate explorative mathematics among non Clifford-algebra specialists. As an example of the power of CLIFFORD, we restate a formula due to Helmstetter which relates the product in $C\ell(g)$, the Clifford algebra of the symmetric part of B, to the product in $C\ell(B)$. Then, with CLIFFORD, we prove it in dimension 3. Clifford algebras of a degenerate quadratic form provide a convenient tool with which to study groups of rigid motions in $\mathbb{R}^3$. Using CLIFFORD we will actually explicitly describe all elements of $\mathbf{Pin}(3)$ and $\mathbf{Spin}(3)$. Rotations in $\mathbb{R}^3$ can then be generated by unit quaternions realized as even elements in $C\ell_{0,3}^+$. Simple computations using quaternions are then performed with CLIFFORD. Throughout this paper we illustrate actual CLIFFORD commands and steps undertaken to solve the problems.

A first working version of a Maple package CLIFFORD was presented in Banff in 1995 [1]. From a modest program capable of symbolic computations in Clifford algebras of an arbitrary bilinear form, CLIFFORD has grown to include 96 main procedures, 21 new Maple types, over $4,000$ lines of code written in the Maple programming language, and an extensive online documentation [4]. There is a number of special-purpose extensions available to CLIFFORD such as GTP for certain computations in graded tensor products of Clifford algebras, OCTONION for computations with octonions considered as para-vectors in $C\ell_{0,7}$, CLI4PLUS for computations in and conversions to a Clifford basis rather than a Grassmann basis. In fact, anyone who uses Maple can easily write additional procedures to tackle specific problems.

There are major advantages in using CLIFFORD with its symbolic capabilities over, for example, CLICAL, a semi-symbolic calculator for computations in Clifford algebras of a quadratic form [14]. One is its ability to solve algebraic equations in any Clifford algebra. This feature is useful when looking for the most general element satisfying certain conditions. We will use it in Section 25.3 where we will systematically search for all types of elements in **Pin**(3). Five general types, not entirely exclusive, are eventually found through a systematic search and analysis. Then, the elements of **Spin**(3) are computed and related to unit quaternions. The package was also extensively used in [5] where Young operators in the Hecke algebra $H_{\mathbb{F}}(3, q)$ were found by solving systematically equations which define them. Rotations in coordinate planes and in a plane orthogonal to an arbitrary non-zero axis vector are described using quaternions realized as elements of $C\ell_{0,3}^+$. A symbolic formula describing the most general rotation is derived. Finally, using the ability of CLIFFORD to compute in Clifford algebras of a degenerate quadratic form, the semi-direct product **Spin**(3) $\rtimes \mathbb{R}^3$ is shown to generate rigid motions on a suitable subspace of the Clifford algebra $C\ell_{1,0,3}$ [1].

The second advantage of using symbolic program like CLIFFORD is its ability to compute with expressions containing totally undefined symbolic coefficients. It is possible, like in Section 25.3, to impose additional conditions on these coefficients when needed (by defining aliases for roots of polynomial equations). In Section 25.2 we verify one of Helmstetter's formulas [12] that relates Clifford product in $C\ell(B)$, the Clifford algebra of an arbitrary bilinear form B, to the Clifford product in $C\ell(g)$ where $g = g^T$ is the symmetric part of B. We re-word the Helmstetter formula and explicitly show the form of a bivector element F needed to relate the two products. In fact, while this problem turned up to be a challenge for CLIFFORD in view of its complexity, it also has helped to fine-tune the program to make such computations feasible. We will only prove our result in dimension 3; however, similar proofs in dimension up to 9 have been successfully completed with CLIFFORD and are left for the Reader.

25.2 Verification of the Helmstetter Formula

In his paper [12] Helmstetter studies canonical isomorphisms between Clifford algebras $C\ell(Q)$ and $C\ell(Q')$ of two quadratic forms Q and Q' defined on the same (real or complex) vector space V. The forms are related via

---

[1] $C\ell_{d,p,q}$ denotes the Clifford algebra of the quadratic space $(V, Q) = V' \perp V^\perp$ where V' is endowed with the non-degenerate part of Q of signature (p, q) and $\dim V^\perp = d$. When $d = 0$ we will write $C\ell_{p,q}$ instead of $C\ell_{0,p,q}$.

the identity $Q'(\mathbf{x}) = Q(\mathbf{x}) + B(\mathbf{x}, \mathbf{x})$ for every $\mathbf{x} \in V$ and some bilinear form B on V. Helmstetter constructs a *deformed* Clifford product $*$ on $C\ell(Q)$ by extending the Clifford product $\mathbf{x}\mathbf{y}$ of two elements $\mathbf{x}$ and $\mathbf{y}$ in $V \hookrightarrow C\ell(Q)$

$$\mathbf{x} * \mathbf{y} = \mathbf{x}\mathbf{y} + B(\mathbf{x}, \mathbf{y})$$

to all elements in $C\ell(Q)$. Together with the new product $*$, the Clifford algebra $C\ell(Q)$ becomes a deformed Clifford algebra $C\ell(Q, B)$. Given now two different bilinear forms B and B' on the quadratic space (V, Q) such that $B(\mathbf{x}, \mathbf{x}) = B'(\mathbf{x}, \mathbf{x})$ for every $\mathbf{x} \in V$, Helmstetter proves only that there *exists* $F \in \bigwedge^2 V$ such that

$$B'(\mathbf{x}, \mathbf{y}) - B(\mathbf{x}, \mathbf{y}) = < F, \mathbf{x} \wedge \mathbf{y} >$$

and that the mapping

$$\phi : C\ell(Q, B) \to C\ell(Q, B'), \quad u \mapsto e^{\wedge F} \lrcorner u \tag{2.1}$$

gives an isomorphism from $C\ell(Q, B)$ to $C\ell(Q, B')$ which acts as an identity on V. In the above, $e^{\wedge F} \lrcorner u$ denotes the *left contraction* of u by the exterior exponential of F (see [7], [15]). A special case of (2.1) occurs when B is symmetric, that is, $B = g = g^T$ and $B' = g + A$ for some antisymmetric form A.

Thus, with a slight change of notation, let

$$B = g + A, \; g^T = g, \; A^T = -A$$

and let us consider two Clifford algebras $C\ell(g)$ and $C\ell(B)$ on the same vector space V. We have therefore three left contractions:
$\mathbf{x} \underset{B}{\lrcorner} \mathbf{y} = B(\mathbf{x}, \mathbf{y})$, $\mathbf{x} \underset{A}{\lrcorner} \mathbf{y} = A(\mathbf{x}, \mathbf{y})$, and $\mathbf{x} \underset{g}{\lrcorner} \mathbf{y} = g(\mathbf{x}, \mathbf{y})$ with respect to B, A, and g respectively. Then, the B-dependent Clifford product $\underset{B}{uv}$ of any two elements u and v in $C\ell(B)$ can be written [16] solely in terms of the operations in $C\ell(g)$ as

$$\underset{B}{uv} = ((u \underset{g}{\llcorner} e^{\wedge F})(v \underset{g}{\llcorner} e^{\wedge F})) \underset{g}{\llcorner} e^{\wedge(-F)} \tag{2.2}$$

where $u \underset{g}{\llcorner} e^{\wedge F}$ denotes the *right contraction* of u by $e^{\wedge F}$ with respect to g. The product of $u \underset{g}{\llcorner} e^{\wedge F}$ and $v \underset{g}{\llcorner} e^{\wedge F}$ in (2.2) is taken in $C\ell(g)$. Let $\dim_{\mathbb{R}}(V) = n$. We claim that the element $F \in \bigwedge^2 V$ may be chosen as

$$F = \sum_{K, L} (-1)^{|\pi(K, L)|} A_K \mathbf{e}_L \mathbf{j}^{-1} \tag{2.3}$$

where $\mathbf{j}^{-1}$ denotes the inverse of the unit pseudoscalar $\mathbf{j} = \mathbf{e}_1 \wedge \mathbf{e}_2 \wedge \cdots \wedge \mathbf{e}_n$ in $C\ell(g)$, the product $\mathbf{e}_L \mathbf{j}^{-1}$ is taken in $C\ell(g)$, and the summation is taken over all multi indices $K = [k_1, k_2]$ and $L = [l_1, l_2, \ldots, l_s]$ satisfying the following relations:

$$K \cap L = \emptyset, \ K \cup L = \{1, 2, \ldots, n\}, \ n = 2 + s, \ K \text{ and } L \text{ ordered by } <.$$

$\pi(K, L)$ denotes a permutation which puts the list $[k_1, k_2, l_1, l_2, \ldots, l_s]$ in the standard order $[1, 2, \ldots, n]$ and $|\pi(K, L)|$ equals 0 or 1 depending whether $\pi(K, L)$ is an even or odd element of S_n. In (2.3) we have also adopted notation $\mathbf{e}_L = \mathbf{e}_{l_1 l_2 \ldots l_s} = \mathbf{e}_{l_1} \wedge \mathbf{e}_{l_2} \wedge \cdots \wedge \mathbf{e}_{l_s}$.

Before we proceed to verify formulas (2.2) and (2.3) with CLIFFORD [4], let's observe the following properties of the left and right contraction:

$$u \underset{g}{\llcorner} v = \mathbf{j}^{-1}((\mathbf{j}\,u) \wedge v), \ u \underset{g}{\lrcorner} v = (u \wedge (v\mathbf{j}))\mathbf{j}^{-1}, \ \mathbf{j}^{-1} = \frac{\tilde{\mathbf{j}}}{\det(g)}, \qquad (2.4)$$

where tilde $\tilde{\ }$ denotes the g-dependent *reversion* in $C\ell(g)$ [2]. Observe also that since $F \in \bigwedge^2 V$, we have

$$e^{\wedge F} = \sum_{k=0}^{N} \frac{F^{\wedge k}}{k!}, \quad N = \lfloor n/2 \rfloor \qquad (2.5)$$

where $F^{\wedge k} = F \wedge F \wedge \cdots \wedge F$ is the exterior product of F computed k-times and $\lfloor \cdot \rfloor$ denotes the floor function. For example, for different values of n, F has the following form:

$$n = 2, \ F = -A_{12}\,\mathbf{e}_{12}^{-1},$$

$$n = 3, \ F = -(A_{12}\,\mathbf{e}_3 - A_{13}\,\mathbf{e}_2 + A_{23}\,\mathbf{e}_1)\,\mathbf{e}_{123}^{-1}, \qquad (2.6)$$

$$n = 4,$$

$$F = -(A_{12}\,\mathbf{e}_{34} - A_{13}\,\mathbf{e}_{24} + A_{14}\,\mathbf{e}_{23} + A_{23}\,\mathbf{e}_{14} + A_{34}\,\mathbf{e}_{12} - A_{24}\,\mathbf{e}_{13})\,\mathbf{e}_{1234}^{-1},$$

and so on. We will verify the validity of (2.2) and the choice for F in a numeric and a symbolic case. In the Maple symbolic language, formula (2.2) becomes: [3]

```
cmul(u, v) =
    RCg(cmulg(RCg(u, wexp(F, N)), RCg(v, wexp(F, N)), wexp(-F, N)))
```

with the CLIFFORD procedures cmulg and RCg representing the Clifford product and the right contraction in $C\ell(g)$ and wexp giving the exterior exponential in $\bigwedge V$. We limit our two examples to $n = 3$. Computations presented in the following two sections have been extended by the author to higher dimensions but will not be shown.

---

[2] From now on in this section we assume that $\det(g) \neq 0$.

[3] In CLIFFORD, the Clifford product uv of two elements u and v can be entered as $u \,\&c\, v$ (the infix form) or as cmul(u, v).

25.2.1 Numeric example when $n = 3$

Let's first assign an arbitrary matrix to B, split B into its symmetric and antisymmetric parts g and A, and compute the bivector F with a procedure makeF :

> dim:=3:eval(makealiases(dim,'ordered')):

    B:=matrix(dim,dim,[4,8,3,0,9,5,-2,1,7]);

$$B := \begin{bmatrix} 4 & 8 & 3 \\ 0 & 9 & 5 \\ -2 & 1 & 7 \end{bmatrix}$$

> g,A:=splitB(B);

$$g, A := \begin{bmatrix} 4 & 4 & \dfrac{1}{2} \\ 4 & 9 & 3 \\ \dfrac{1}{2} & 3 & 7 \end{bmatrix}, \begin{bmatrix} 0 & 4 & \dfrac{5}{2} \\ -4 & 0 & 2 \\ \dfrac{-5}{2} & -2 & 0 \end{bmatrix}$$

> F:=makeF(dim);

$$F := \frac{2}{91}\, e13 + \frac{86}{455}\, e12$$

Notice that F displayed above has the form (2.6). We find next the exterior exponentials of F and $-F$ which we assign to the Maple variables F1 and F2 respectively.

> N:=floor(dim/2):F1:=wexp(F,N);F2:=wexp(-F,N);

$$F1 := 1 + \frac{2}{91}\, e13 + \frac{86}{455}\, e12, \quad F2 := 1 - \frac{2}{91}\, e13 - \frac{86}{455}\, e12.$$

Let u and v be two arbitrary elements in $C\ell(B)$:

> u:=2+e1-e23+e123;v:=3-e3+e12+e23;

$$u := 2 + e1 - e23 + e123, \quad v := 3 - e3 + e12 + e23.$$

The Clifford product $\underset{B}{uv}$ of u and v in $C\ell(B)$ displayed in the left hand side of (2.2) can be computed with a procedure cmul :[4]

> cmul(u,v);

$$-48\, e3 + 79\, Id + 13\, e13 - 6\, e12 + 81\, e2 - 8\, e123 - 81\, e1$$

while the right hand side of (2.2), as expected, gives the same result:

> RCg(cmulg(RCg(u,F1),RCg(v,F1)),F2);

$$-48\, e3 + 79\, Id + 13\, e13 - 6\, e12 + 81\, e2 - 8\, e123 - 81\, e1$$

25.2.2 Symbolic computations when $n = 3$

A purely symbolic verification of (2.2) when $n = 3$ will look as follows. Matrix B will be now defined as an arbitrary symbolic 3×3 matrix with a

---

[4] In the following, Id denotes the unit element in $C\ell(B)$.

symmetric part g and an antisymmetric part A, and F is again computed using the procedure makeF. The exterior exponentials of F and $-F$ are again denoted in Maple by F1 and F2 respectively. All symbolic parameters in B are assumed to be real or complex.

```
> dim:=3:eval(makealiases(dim)):
```

```
  B:=matrix(dim,dim,[g11,g12+A12,g13+A13,
                     g12-A12,g22,g23+A23,
                     g13-A13,g23-A23,g33]):
```

```
> g,A:=splitB(B);
```

$$g,\, A := \begin{bmatrix} g11 & g12 & g13 \\ g12 & g22 & g23 \\ g13 & g23 & g33 \end{bmatrix}, \begin{bmatrix} 0 & A12 & A13 \\ -A12 & 0 & A23 \\ -A13 & -A23 & 0 \end{bmatrix}$$

```
> F:=map(normal,clicollect(makeF(dim)));
```

$$F := -\frac{(A23\, g11\, A12\, g13 - A13\, g12)\, e23}{\%1}$$
$$+ \frac{(-A23\, g13 + A13\, g23 - A12\, g33)\, e12}{\%1}$$
$$+ \frac{(A23\, g12 - A13\, g22 + A12\, g23)\, e13}{\%1}$$

$$\%1 := -g33\, g22\, g11 + g22\, g13^2 - 2\, g13\, g23\, g12 + g33\, g12^2 + g23^2\, g11$$

```
> N:=floor(dim/2):F1:=wexp(F,N):F2:=wexp(-F,N):
```

We will now define two general elements u and v in $C\ell(B)$ by decomposing them over the Grassmann basis of $C\ell(B)$ provided by a procedure cbasis. Coefficients uu_i and vv_i, $1,\ldots,8$, in these two expansions are assumed to be real or complex.

```
> cbasis(dim); #Grassmann basis for Cl(B)
```

$$[Id,\, e1,\, e2,\, e3,\, e12,\, e13,\, e23,\, e123]$$

```
> u:=add(uu[k]*cbasis(dim)[k],k=1..2^dim);
```

$$u := uu_1\, Id + uu_2\, e1 + uu_3\, e2 + uu_4\, e3 + uu_5\, e12 + uu_6\, e13 + uu_7\, e23$$
$$+ uu_8\, e123$$

```
> v:=add(vv[k]*cbasis(dim)[k],k=1..2^dim);
```

$$v := vv_1\, Id + vv_2\, e1 + vv_3\, e2 + vv_4\, e3 + vv_5\, e12 + vv_6\, e13 + vv_7\, e23$$
$$+ vv_8\, e123$$

The Clifford product of u and v in $C\ell(B)$ is then collected and assigned to a constant res1 which we won't display due to its length.

```
> res1:=clicollect(cmul(u,v)):
```

As before, we finish by computing the right hand side of (2.2). By assigning it to res2, we can then easily find that res1 − res2 = 0 as expected.

```
> res2:=clicollect(RCg(cmulg(RCg(u,F1),RCg(v,F1)),F2)):
  res1-res2;
```

$$0$$

Therefore we have proven that formula (2.2) is valid in the case when $n = 3$. Similarly, using CLIFFORD, one can prove its validity for $n \leq 9$.

25.3   Rigid Motions with Clifford Algebras

Clifford algebras $C\ell(V, Q)$ on a quadratic space (V, Q) endowed with a *degenerate* quadratic form Q and associated groups **Spin**, **Pin**, Clifford, etc., were studied in [3, 6] and [2, 8, 9]. In contrast to the Clifford algebras of a non-degenerate quadratic form, these algebras possess a non-trivial two-sided nilpotent ideal called *Jacobson radical*. The Jacobson radical J is generated by the null-vectors in V which are orthogonal to the entire space V (that is, J is generated by the orthogonal complement $V^{\perp}$ of V). It is known [11, 13] that J contains every nilpotent left and right ideal in $C\ell(V, Q)$. From the point of view of the spinorial representation theory of Clifford algebras, an important difference is that $C\ell(V, Q)$ does not possess faithful matrix representation when Q is degenerate.

Let $V = V' \perp V^{\perp}$ where V' is endowed with a non-degenerate part Q' of Q of signature (p, q). Let $\dim V^{\perp} = d$, hence $p + q + d = \dim V$. Let's denote $C\ell(Q)$ as $C\ell_{d,p,q}$. Then we have a direct sum decomposition $C\ell_{d,p,q} = C\ell_{p,q} \oplus J$ into two $C\ell_{p,q}$-modules. It was shown in [2] that when $d = 1$ this decomposition is responsible for a semi-direct product structure of the group of units $C\ell^*_{1,p,q}$ of $C\ell_{1,p,q}$ and of all of its subgroups such as the *Clifford group* $\Gamma(1, p, q)$ and the *special Clifford groups* $\Gamma^{\pm}(1, p, q) = \Gamma(1, p, q) \cap C\ell^{\pm}_{1,p,q}$, where $C\ell^+_{1,p,q}$ (resp. $C\ell^-_{1,p,q}$) denotes the even (resp. odd) part of $C\ell_{1,p,q}$. The Clifford group is defined as $\Gamma(1, p, q) = \{g \in C\ell^*_{1,p,q} \,|\, gvg^{-1} \in V, v \in V\}$, that is, without a twist.[5] Let $N : \Gamma(1, p, q) \to C\ell^*_{1,p,q}$ be defined as $N(g) = \tilde{g}g$. Then we define the *reduced Clifford groups* as $\Gamma^{\pm}_0(1, p, q) = \ker N \cap \Gamma^{\pm}(1, p, q)$. The **Pin**$(1, p, q)$ and **Spin**$(1, p, q)$ groups are then:

$$\begin{aligned}
\mathbf{Pin}(1, p, q) &= \{g \in \Gamma(1, p, q) \,|\, N(g) = \pm 1\}, \\
\mathbf{Spin}(1, p, q) &= \{g \in \Gamma^+(1, p, q) \,|\, N(g) = \pm 1\}.
\end{aligned} \tag{3.7}$$

In preparation for our computations below, from now on we assume that $p + q$ is an odd positive integer. Let $G = \{1 + v\mathbf{e}_1 | v \in V', \mathbf{e}_1^2 = 0\}$ be a subgroup of $C\ell^*_{1,p,q}$. Then it was proven in [2] that

$$\mathbf{Pin}(1, p, q) = G \rtimes \Gamma^{\pm}_0(p, q), \quad \mathbf{Spin}(1, p, q) = G \rtimes \mathbf{Spin}(p, q). \tag{3.8}$$

---

[5] See Crumeyrolle [10] for a definition of the Clifford group with the twist given by α : in Crumeyrolle's notation α denotes the *principal automorphism* or the *grade involution* in $C\ell(Q)$. Then the *twisted Clifford group* is defined as $\Gamma_{\alpha}(1, p, q) = \{g \in C\ell^*_{1,p,q} \,|\, \alpha(g)vg^{-1} \in V, v \in V\}$.

In the above, symbol $\rtimes$ denotes a semi-direct product with the group on the right acting on the group on the left. For example, it will be of interest to us to note that the homogeneous Galileian group of rigid motions $G = \mathbb{R}^3 \rtimes SO(3)$ in $\mathbb{R}^3$ is isomorphic to $SO^+(1,0,3)$ and it is doubly covered by $\mathbf{Spin}^+(1,0,3)$, the identity component of $\mathbf{Spin}(1,0,3)$. For a similar result to (3.8) when one considers the twisted Clifford group $\Gamma_\alpha(1,p,q)$ and a twisted map $N_\alpha : \Gamma_\alpha(1,p,q) \to C\ell^*_{1,p,q}$ defined as $N_\alpha(g) = \bar{g}g$ where $\bar{}$ denotes the *conjugation* in $C\ell_{1,p,q}$, see [8, 17]. In the following two sections we will use approach and notation from Selig [18] where the author denotes the degenerate Clifford algebra $C\ell_{d,p,q}$ as $C(p,q,d)$. Furthermore Selig uses twisted groups and defines the **Pin** and **Spin** groups as follows:

$$\mathbf{Pin}(n) = \{\mathbf{g} \in C(0,n,0) : \mathbf{g}\mathbf{g}^* = \mathbf{1} \text{ and } \alpha(\mathbf{g})\mathbf{x}\mathbf{g}^* \in V, \mathbf{x} \in V\}, \qquad (3.9)$$

$$\mathbf{Spin}(n) = \{\mathbf{g} \in C^+(0,n,0) : \mathbf{g}\mathbf{g}^* = \mathbf{1} \text{ and } \mathbf{g}\mathbf{x}\mathbf{g}^* \in V, \mathbf{x} \in V\}, \qquad (3.10)$$

with * denoting the conjugation $\bar{}$ in the Clifford algebra $C(p,q,d)$. It is implicit in the definitions above that the actions of **Pin** and **Spin** on V are $\mathbf{x} \mapsto \alpha(\mathbf{g})\mathbf{x}\mathbf{g}^*$ and $\mathbf{x} \mapsto \mathbf{g}\mathbf{x}\mathbf{g}^*$ respectively.

25.3.1   Group **Pin**(3)

In this section we will perform some computations with **Pin**(3). It is known [17] that elements of the Clifford group $\Gamma(p,q)$ of a non-degenerate finite dimensional real quadratic space V are representable as the product of a finite number of non-isotropic vectors in V. In particular, the same is true of the elements of **Pin**(3) while in the case of the spin groups the number of elements in the product is obviously even. Using our computer-based approach, we will find explicitly all forms of the elements in **Pin**(3) and later in **Spin**(3). This will illustrate how one can find the most general elements in any Clifford algebra satisfying certain conditions. The same approach was used in [5] where Young idempotents of prescribed symmetries in a deformed Clifford algebra were found.

We begin by assigning a diagonal matrix to the bilinear form B. Grassmann basis for $C\ell_{0,3}$ will be stored in the variable `clibas`. Following Selig we re-name Clifford conjugation as a procedure `star` and define a Euclidean norm on $V = \mathbb{R}^3$ as a procedure `Enorm`. We will also define some additional Maple procedures that will be useful below.

```
> B:=diag(-1$3);eval(makealiases(3)):clibas:=cbasis(3);
```

$$B := \begin{bmatrix} -1 & 0 & 0 \\ 0 & -1 & 0 \\ 0 & 0 & -1 \end{bmatrix}$$

$$clibas := [Id, \, e1, \, e2, \, e3, \, e12, \, e13, \, e23, \, e123]$$

```
>    star:=proc(x) conjugation(x) end: #conjugation
     Enorm:=v->simplify(scalarpart(v &c star(v))): #norm in V
     alpha:=proc(x) gradeinv(x) end: #grade involution
     scalprod:=(x,y)->
             scalarpart(1/2*(x &c star(y) + star(y) &c x)):
     Pin_action:=(x,g)->
             clicollect(simplify(alpha(g) &c x &c star(g)));
```

$$Pin\_action := (x, g) \rightarrow \text{clicollect}(\text{simplify}((\alpha(g) \text{ '\&c' } x) \text{ '\&c' } star(g)))$$

```
>    Spin_action:=(x,g)->
             clicollect(simplify(g &c x &c star(g)));
```

$$Spin\_action := (x, g) \rightarrow \text{simplify}((g \text{ '\&c' } x) \text{ '\&c' } star(g))$$

Let $\mathbf{v}, \mathbf{v}_1, \mathbf{v}_2$ be three arbitrary vectors in $\mathbb{R}^3$ with some undetermined coefficients expressed in a pseudo-orthonormal basis $\{\mathbf{e}_1, \mathbf{e}_2, \mathbf{e}_3\}$:

```
>    v:=c1*e1+c2*e2+c3*e3:
     v1:=c11*e1+c12*e2+c13*e3:v2:=c21*e1+c22*e2+c23*e3:
```

Then the Euclidean norm in $\mathbb{R}^3$ is:

```
>    Enorm(v);
```

$$c1^2 + c2^2 + c3^2$$

The action of **Pin** on $C\ell_{0,3}$ is realized as the procedure Pin_action defined above. Let's verify Selig's claim ([18], page 153) that when $\mathbf{x}, \mathbf{g}$ are both in V, then the product of three elements $\mathbf{gxg}^*$ automatically belongs to V :

```
>    Pin_action(v,v1);
```

$$(c11^2 \, c3 - c13^2 \, c3 - 2 \, c13 \, c12 \, c2 + c12^2 \, c3 - 2 \, c13 \, c11 \, c1) \, e3$$
$$+ (-c11^2 \, c1 - 2 \, c11 \, c13 \, c3 + c12^2 \, c1 + c13^2 \, c1 - 2 \, c11 \, c12 \, c2) \, e1$$
$$+ (c11^2 \, c2 - 2 \, c12 \, c11 \, c1 - 2 \, c12 \, c13 \, c3 + c13^2 \, c2 - c12^2 \, c2) \, e2$$

As we can see from the above, the output of Pin_action(v, v1) belongs to V. In order to check that indeed the action of **Pin** in V preserves the scalar product in V, we will first find all possible forms of $\mathbf{g} \in \mathbf{Pin}(3)$. Recall that according to (3.9) every element $\mathbf{g} \in \mathbf{Pin}(n)$ must satisfy two conditions:

$$(1) \ \mathbf{gg}^* = \mathbf{1}, \quad \text{and} \quad (2) \ \alpha(\mathbf{g})\mathbf{vg}^* \in V \ \text{for any} \ \mathbf{v} \in V.$$

Let $\mathbf{g}$ be any element in $C\ell_{0,3}$ expressed in the Grassmann basis. [6]

```
>    g:=add(x.i * clibas[i],i=1..nops(clibas));
```

$$g := x1 \, Id + x2 \, e1 + x3 \, e2 + x4 \, e3 + x5 \, e12 + x6 \, e13 + x7 \, e23 + x8 \, e123$$

---

[6] Recall that $\mathbf{e}_{12}, \mathbf{e}_{13}, \mathbf{e}_{23}, \mathbf{e}_{123}$ are aliases for the wedge products $\mathbf{e}_1 \wedge \mathbf{e}_2, \mathbf{e}_1 \wedge \mathbf{e}_3, \mathbf{e}_2 \wedge \mathbf{e}_3, \mathbf{e}_1 \wedge \mathbf{e}_2 \wedge \mathbf{e}_3$. They are defined in CLIFFORD with the command makealiases.

We will now attempt to find conditions that the coefficients $x_i, i = 1, \ldots, 8$, must satisfy so that $\mathbf{gg}^* = \mathbf{1}$. We will again use the command `clisolve2`. In order to shorten its output, it is convenient to define additional aliases $\kappa_j, j = 1, \ldots, 5$, which have been collected in Appendix 25.4. The first condition (1) gives:

```
>  sol:=clisolve2(cmul(g,star(g))-Id,[x.(1..8)]);
```

$$sol := [\{x1 = -\frac{-x7\,\kappa 5 - x5\,x4 + x6\,x3}{x8}, \ x8 = x8, \ x4 = x4, \ x7 = x7,$$

$$x6 = x6, x5 = x5, x2 = \kappa 5, x3 = x3\}, \{x8 = 0, \ x1 = \frac{\kappa 4}{x7}, \ x4 = x4,$$

$$x7 = x7, \ x6 = x6, x5 = x5, \ x3 = x3, \ x2 = \frac{-x5\,x4 + x6\,x3}{x7}\}, \{x7 = 0,$$

$$x8 = 0, \ x1 = \frac{\kappa 3}{x6}, x3 = \frac{x4\,x5}{x6}, \ x2 = x2, \ x4 = x4, \ x6 = x6, \ x5 = x5\},$$

$$\{x4 = 0, \ x7 = 0, \ x6 = 0, \ x8 = 0, \ x2 = x2, \ x5 = x5, \ x3 = x3, \ x1 = \kappa 2\},$$

$$\{x7 = 0, \ x6 = 0, \ x5 = 0, \ x8 = 0, \ x2 = x2, \ x4 = x4, \ x3 = x3, \ x1 = \kappa 1\}]$$

Thus, there are five different solutions to $\mathbf{gg}^* = \mathbf{1}$, three of which requiring respectively that x_6, x_7 and x_8 be non-zero. Let's substitute these solutions into $\mathbf{g}$. This way we will have five different expressions for $\mathbf{g}$ satisfying condition (1).

```
>  for i to nops(sol) do g.i:=subs(sol[i],g) od;
```

$$g1 := -\frac{(-x7\,\kappa 5 - x5\,x4 + x6\,x3)\,Id}{x8} + \kappa 5\,e1 + x3\,e2 + x4\,e3 + x5\,e12$$
$$+ \ x6\,e13 + x7\,e23 + x8\,e123$$

$$g2 := \frac{\kappa 4\,Id}{x7} + \frac{(-x5\,x4 + x6\,x3)\,e1}{x7} + x3\,e2 + x4\,e3 + x5\,e12 + x6\,e13$$
$$+ \ x7\,e23$$

$$g3 := \frac{\kappa 3\,Id}{x6} + x2\,e1 + \frac{x4\,x5\,e2}{x6} + x4\,e3 + x5\,e12 + x6\,e13$$

$$g4 := \kappa 2\,Id + x2\,e1 + x3\,e2 + x5\,e12$$

$$g5 := \kappa 1\,Id + x2\,e1 + x3\,e2 + x4\,e3$$

```
>  for i to nops(sol) do simplify(cmul(g.i,star(g.i))) od;
```

$$Id, \ Id, \ Id, \ Id, \ Id$$

We need to make sure now that each $\mathbf{g1}, \mathbf{g2}, \mathbf{g3}, \mathbf{g4}, \mathbf{g5}$ above satisfies also the second condition (2). We begin with the simplest element $\mathbf{g5}$. Its action on an arbitrary vector $\mathbf{v} \in V$ gives:

```
>  Pin_action(v,g5);
```

$$(2\,\kappa 1\,c3\,x4 + 2\,\kappa 1\,c1\,x2 + 2\,\kappa 1\,c2\,x3)\,Id$$
$$+ (-2\,x3\,x4\,c2 - 2\,x4\,x2\,c1 - 2\,x4^2\,c3 + c3)\,e3$$
$$+ (-2\,x4\,x2\,c3 - 2\,x3\,x2\,c2 - 2\,x2^2\,c1 + c1)\,e1$$
$$+ (-2\,x3^2\,c2 - 2\,x4\,x3\,c3 - 2\,x2\,x3\,c1 + c2)\,e2$$

It should be clear from the above that in order for $\alpha(\mathbf{g5})\mathbf{vg5}^*$ to be in V, the coefficient of Id must be zero for any c_1, c_2, c_3. Thus, either $x_2 = x_3 = x_4 = 0$, which would imply $\mathbf{v} = 0$, or $\kappa_1 = 0$. Let $\varepsilon = \pm 1$ which will appear in the following Maple outputs as eps.[7] Then, $\mathbf{g} = \kappa_1 Id = \pm Id = \pm 1$ when $x_2 = x_3 = x_4 = 0$,

```
>   g51:=subs({x2=0,x3=0,x4=0},g5);
```

$$g51 := eps\ Id$$

or, $\mathbf{g} = x_2\mathbf{e}_1 + x_3\mathbf{e}_2 + \lambda_1\mathbf{e}_3$

```
>   g52:=subs({kappa1=0,x4=lambda1},g5);
```

$$g52 := x2\ e1 + x3\ e2 + \lambda 1\ e3$$

where $\lambda_1 = \pm\sqrt{1 - x_2^2 - x_3^2}$.[8] We will collect all **Pin** group elements in a set Pin_group beginning with g51 and g52.

```
>   Pin_group:={g51,g52}:
```

Similarly, we consider g4. We assign the identity coefficient of the **Pin** action $\alpha(\mathbf{g4})\mathbf{vg4}^*$ to a variable eq and find a solution to the resulting two equations that will be parameterized by c_1, c_2 :

```
>   a:=Pin_action(v,g4);
```

$$a := \left(2\,x3\,\kappa2\,c2 - 2\,x5\,x2\,c2 + 2\,x5\,x3\,c1 + 2\,x2\,\kappa2\,c1\right)Id + c3\ e3$$
$$+ \left(-2\,x2^2\,c1 - 2\,x5\,\kappa2\,c2 - 2\,x3\,x2\,c2 + c1 - 2\,x5^2\,c1\right)e1$$
$$+ \left(-2\,x5^2\,c2 - 2\,x2\,x3\,c1 + 2\,\kappa2\,x5\,c1 - 2\,x3^2\,c2 + c2\right)e2$$

```
>   eq:=collect(coeff(a,Id),{c1,c2}):
    eq1:=coeff(eq,c1):eq2:=coeff(eq,c2):
    sol:=[solve({eq1,eq2},{x2,x5,x3})];
```

$$sol := [\{x5 = 0,\ x2 = \lambda2,\ x3 = x3\},\ \{x3 = 0,\ x2 = 0,\ x5 = x5\}]$$

In the above, $\lambda_2 = \pm\sqrt{1 - x_3^2}$. Likewise, we set $\lambda_3 = \pm\sqrt{1 - x_5^2}$.[9] The two new elements g41, g42 of **Pin** obtained this way we add to Pin_group.

```
>   for i to nops(sol) do g.4.i:=simplify(subs(sol[i],g4)) od:
```

```
    Pin_group:=Pin_group union {g41,g42}:
```

In order to continue with g3 displayed above, we must make the assumption $x_6 \neq 0$ known to Maple. Then, the action of g3 on a vector can be again computed with the procedure Pin_action as before (we won't display it due to its length).

```
>   assume(x6>0,x6<0):a:=Pin_action(v,g3):
```

As before, the quantity a is expressed in terms of $\{Id, e1, e2, e3\}$. We will isolate the coefficient of the identity element in a, assign it to a variable eq, and then determine for which values of x_2, x_4, x_5, x_6 it will be automatically zero for every choice of c_1, c_2, c_3. This will require that the coefficients of

---

[7] In Maple one way to make $\varepsilon = \pm 1$ is to define alias(eps=RootOf(_Z^2-1)):.

[8] In Maple we define alias(lambda1=RootOf(_Z^2+x2^2+x3^2-1)):.

[9] alias(lambda2=RootOf(_Z^2-1+x3^2)):, alias(lambda3=RootOf(_Z^2-1+x5^2)):.

c_1, c_2, c_3 in `eq` all be zero. We assign these three coefficients to three Maple constants `eq1, eq2, eq3`, set them all equal to zero, and solve the resulting equations for x_2, x_4, x_5, x_6 [10].

```
>   cliterms(a);
```
$$\{Id, \ e1, \ e2, \ e3\}$$
```
>   eq:=collect(coeff(a,Id),{c.(1..3)});
```
$$eq := -2\,\frac{\left(x6^{\sim 3}\,x2 - \lambda 4\,x4\,x6^{\sim}\right)c3}{x6^{\sim 2}} - 2\,\frac{\left(-\lambda 4\,x5\,x4 + x2\,x5\,x6^{\sim 2}\right)c2}{x6^{\sim 2}}$$
$$-\,2\,\frac{\left(-\lambda 4\,x2\,x6^{\sim} - x5^2\,x4\,x6^{\sim} - x4\,x6^{\sim 3}\right)c1}{x6^{\sim 2}}$$
```
>   for i to 3 do eq.i:=coeff(eq,c.i) od:
    sol:=solve({eq.(1..3)},{x6,x4,x5,x2});
```
$$sol := \{x2 = 0,\ x6^{\sim} = x6^{\sim},\ x5 = x5,\ x4 = 0\}$$
This time we only have one solution which we call `g31` :
```
>   g31:=subs(sol,g3);
```
$$g31 := \frac{\lambda 5\,Id}{x6^{\sim}} + x5\,e12 + x6^{\sim}\,e13$$
where λ_5 is another alias displayed in Appendix 25.4.
```
>   Pin_group:=Pin_group union {g31}:
```
By continuing in the similar fashion with `g2` and `g1`, one can find all five explicit forms of all elements in **Pin**(3). These forms are:
```
>   'Pin_group'=Pin_group;
```
$$Pin\_group := \{\lambda 3\,Id + x5\,e12,\ \frac{\lambda 5\,Id}{x6^{\sim}} + x5\,e12 + x6^{\sim}\,e13,$$
$$\frac{\lambda 7\,Id}{x7^{\sim}} + x5\,e12 + x6\,e13 + x7^{\sim}\,e23,\ \lambda 9\,e1 + x3\,e2 + x4\,e3 + x8^{\sim}\,e123,$$
$$\lambda 1\,e3 + x2\,e1 + x3\,e2,\ \lambda 2\,e1 + x3\,e2,\ eps\,Id\}$$
where, to shorten the output, we have defined λ_7 and λ_9 as some additional aliases (see Appendix 25.4). It is a simple matter now to verify that all elements of **Pin** displayed in *Pin_group* satisfy both conditions (1) and (2) from the definition (3.9).
```
>   for g in Pin_group do evalb(cmul(g,star(g)=Id)) od;
```
$$true,\ true,\ true,\ true,\ true,\ true$$
```
>   for g in Pin_group do
    evalb(Pin_action(v,g)=vectorpart(Pin_action(v,g),1)) od;
```
$$true,\ true,\ true,\ true,\ true,\ true$$
We are now in position to verify Selig's claim [18], page 153, that the

---

[10] In the following outputs Maple reminds us that the assumption $x_6 \neq 0$ has been made by displaying x_6 as $x6^{\sim}$ rather than as $x6$.

scalar product in $V = \mathbb{R}^3$ defined in $C\ell_{d,p,q}$ as

$$\mathbf{v}_1 \cdot \mathbf{v}_2 = \frac{1}{2}(\mathbf{v}_1\mathbf{v}_2^* + \mathbf{v}_2\mathbf{v}_1^*), \quad \text{for any } \mathbf{v}_1, \mathbf{v}_2 \in \mathbb{R}^3,$$

is preserved under the action of the **Pin** group. Let $\mathbf{v}_1, \mathbf{v}_2$ be the two arbitrary 1-vectors defined earlier and let `scalprod` be a simple Maple procedure that gives the scalar product in V :

```
>    scalprod:=(x,y)->
           scalarpart(1/2*(x &c star(y) + star(y) &c x)):
     for g in Pin_group do
         simplify(scalprod(Pin_action(v1,g),Pin_action(v2,g))
             -scalprod(v1,v2))
         od;
```
$$0,\ 0,\ 0,\ 0,\ 0,\ 0$$

Thus, **Pin**(3) preserves the scalar product in $\mathbb{R}^3$ and, therefore, we have a homomorphism from **Pin**(3) to O(3) which is known to be a double-covering map. In the process, we have found all types of elements in **Pin**(3).

25.3.2   Group **Spin**(3)

In this section we will perform a few computations with **Spin**(3). After finding all general elements in **Pin**(3) it is much easier to find elements in **Spin**(3). Recall from (3.10) that

$$\mathbf{Spin}(3) = \{\mathbf{g} \in C\ell_{0,3}^+ : \mathbf{gg}^* = \mathbf{1} \text{ and } \mathbf{gxg}^* \in \mathbb{R}^3 \text{ for all } \mathbf{x} \in \mathbb{R}^3\}.$$

Let's find $\mathbf{g}$, a general element in **Spin**(3). Since $\mathbf{Spin}(3) \subset C\ell_{0,3}^+$, we will begin by decomposing $\mathbf{g}$ over even basis elements in $C\ell_{0,3}$:

```
>    clibaseven:=cbasis(3,'even');
```
$$clibaseven := [Id,\ e12,\ e13,\ e23]$$
```
>    g:=c0*Id+c3*e12+c2*e13+c1*e23;
```
$$g := c0\ Id + c3\ e12 + c2\ e13 + c1\ e23$$

Notice that under the action of the **Spin** group defined as a procedure `Spin_action`

```
>    Spin_action:=(x,g)-> simplify(g &c x &c star(g));
```
$$Spin\_action := (x, g) \rightarrow simplify((g\ `\&c`\ x)\ `\&c`\ star(g))$$

1-vectors are automatically mapped into 1-vectors since the action of $\mathbf{g}$ on a vector $\mathbf{v} \in V$ contains only the 1-vector part, or, in another words, it belongs back to V :

```
>    Spin_action(v,g)-vectorpart(Spin_action(v,g),1);
```
$$0$$

We just need to make sure that $\mathbf{gg^*} = \mathbf{1}$ for each $\mathbf{g} \in \mathbf{Spin}(3)$. To simplify
Maple output, we define $\kappa = \sqrt{1 - c_1^2 - c_2^2 - c_3^2}$ as a Maple alias.

```
>    alias(kappa=sqrt(-c1^2-c2^2-c3^2+1)):

     sol:=clisolve2(cmul(g,star(g))-Id,[c.(0..3)]);
```

$$sol := [\{c1 = c1,\ c2 = c2,\ c3 = c3,\ c0 = \kappa\},$$
$$\{c1 = c1,\ c2 = c2,\ c3 = c3,\ c0 = -\kappa\}]$$

```
>    g:=eps*kappa*Id+c3*e12+c2*e13+c1*e23;
```

$$g := eps\ \kappa\ Id + c3\ e12 + c2\ e13 + c1\ e23$$

where $eps = \varepsilon = \pm 1$. Thus, the most general element in $\mathbf{Spin}(3)$ is just

$$\mathbf{g} = \varepsilon\kappa\mathbf{1} + c_3\mathbf{e}_{12} + c_2\mathbf{e}_{13} + c_1\mathbf{e}_{23}.$$

Notice, that the defining properties of $\mathbf{g}$ are easily checked:

```
>    simplify(cmul(g,star(g)));
```

$$Id$$

```
>    evalb(Spin_action(v,g)=vectorpart(Spin_action(v,g),1));
```

$$true$$

In fact, element $\mathbf{g} \in \mathbf{Spin}(3)$ could be identified with a unit quaternion
spanned over the basis $\{\mathbf{1}, \mathbf{e}_{12}, \mathbf{e}_{13}, \mathbf{e}_{23}\}$. Then, the * conjugation becomes
the quaternionic conjugation. It can be easily checked by hand or with
CLIFFORD that the basis bivectors anticommute and square to $-\mathbf{1}$.

```
>    quatbasis:=[e12,e13,e23];
```

$$quatbasis := [e12,\ e13,\ e23]$$

```
>    M:=matrix(3,3,(i,j)->cmul(quatbasis[i],quatbasis[j]));
```

$$M := \begin{bmatrix} -Id & e23 & -e13 \\ -e23 & -Id & e12 \\ e13 & -e12 & -Id \end{bmatrix}$$

We have, therefore, that unit quaternions on a unit sphere in $\mathbb{R}^4$ are isomorphic to $\mathbf{Spin}(3)$ while the even part of $C\ell_{0,3}$ is isomorphic with the
quaternionic division ring $\mathbb{H}$. $\mathbf{Spin}(3)$ acts on $\mathbb{R}^3$ through the rotations.
In Appendix 25.4 one can find a procedure $\mathtt{rot}$ which takes as its first
argument an arbitrary vector $\mathbf{v}$ in $\mathbb{R}^3$ while as its second argument it
takes a quaternion. For example, a counter-clockwise rotation in the plane
spanned by $\{\mathbf{e}_1, \mathbf{e}_2\}$ is accomplished with a help of a unit quaternion
$\cos(\frac{\theta}{2}) + \sin(\frac{\theta}{2})\mathbf{e}_{12}$:

```
>    rot(e1,cos(theta/2)+sin(theta/2)*e12);

     rot(e2,cos(theta/2)+sin(theta/2)*e12);

     rot(e3,cos(theta/2)+sin(theta/2)*e12);
```

$$\cos(\theta)\ e1 + e2\sin(\theta),\ -e1\sin(\theta) + \cos(\theta)\ e2,\ e3$$

Let's now take a general element from $\mathbf{Spin}(3)$ and act on all three unit
vectors $\mathbf{e}_1, \mathbf{e}_2, \mathbf{e}_3$. We can easily verify that the new elements $\mathbf{e}_{11}, \mathbf{e}_{22}, \mathbf{e}_{33}$
provide another orthonormal basis with the same orientation:

```
>   e11:=rot(e1,g):e22:=rot(e2,g):e33:=rot(e3,g):
>   e11 &c e22 + e22 &c e11, e11 &c e33 + e33 &c e11,
    e22 &c e33 + e33 &c e22;
```
$$0, \, 0, \, 0$$

```
>   e1 &w e2 &w e3,e11 &w e22 &w e33;
```
$$e123, \; e123$$

Length of any vector **v** under the action of **Spin**(3) is of course preserved:

```
>   Enorm(v) = Enorm(Spin_action(v,g));
```
$$c1^2 + c2^2 + c3^2 \; = \; c1^2 + c2^2 + c3^2$$

Example 1: Rotations in coordinate planes.

Let's define unit quaternions responsible for the rotations in the coordinate planes. These are counter-clockwise rotations when looking down the rotation axis. In CLIFFORD, a pure-quaternion basis will be denoted by $\{qi = e_2 \wedge e_3, \; qj = e_1 \wedge e_3, \; qk = e_1 \wedge e_2\}$ in place of traditionally used $\{\mathbf{i}, \mathbf{j}, \mathbf{k}\}$.

```
>   qi:=e23:qj:=e13:qk:=e12:
    q12:=cos(alpha/2)*Id+sin(alpha/2)*'qk';#xy-plane
```
$$q12 := \cos(\frac{1}{2}\alpha)\,Id + \sin(\frac{1}{2}\alpha)\,qk$$

```
>   q13:=cos(beta/2)*Id+sin(beta/2)*'qj';#xz-plane
```
$$q13 := \cos(\frac{1}{2}\beta)\,Id + \sin(\frac{1}{2}\beta)\,qj$$

```
>   q23:=cos(gamma/2)*Id+sin(gamma/2)*'qi';#yz-plane
```
$$q23 := \cos(\frac{1}{2}\gamma)\,Id + \sin(\frac{1}{2}\gamma)\,qi$$

Notice that to rotate by an angle $n\alpha$ it is enough to find the n-th Clifford power of the appropriate quaternion and then apply it to the given vector.

```
>   q12 &c q12; #rotation by the angle 2*alpha
```
$$\cos(\alpha)\,Id + e12\sin(\alpha)$$

```
>   q12 &c q12 &c q12; #rotation by the angle 3*alpha
```
$$\cos(\frac{3}{2}\alpha)\,Id + \sin(\frac{3}{2}\alpha)\,e12$$

Let's see now how these basis rotations in the coordinate planes act on an arbitrary vector $\mathbf{v} = a\mathbf{e}_1 + b\mathbf{e}_2 + c\mathbf{e}_3$:

```
>   v:=a*e1+b*e2+c*e3;
```
$$v := a\,e1 + b\,e2 + c\,e3$$

The norm of **v** is $\|\mathbf{v}\| = \sqrt{\mathbf{v}\mathbf{v}^*}$ and it can be defined in CLIFFORD as follows:

```
>   vlength:=sqrt(scalarpart(v &c star(v)));
```
$$vlength := \sqrt{c^2 + a^2 + b^2}$$

Certainly, rotations do not change length. For example, let's rotate **v** with the product of two quaternions q13 and q12 :

```
> v123:=rot(v,q13 &c q12); #rotation q12 followed by q13
```
$$v123 := (-b\sin(\beta)\sin(\alpha) + a\sin(\beta)\cos(\alpha) + c\cos(\beta))\, e3$$
$$+ (-b\sin(\alpha)\cos(\beta) - c\sin(\beta) + a\cos(\beta)\cos(\alpha))\, e1$$
$$+ (b\cos(\alpha) + a\sin(\alpha))\, e2$$

```
> vlength:=sqrt(scalarpart(v123 &c star(v123)));
```
$$vlength := \sqrt{c^2 + a^2 + b^2}$$

Thus, the length of $\mathbf{v}_{123}$ is the same as the length of **v**. However, rotations do not commute. We will show that by applying quaternions q12 and q13 to **v** in the reverse order and by comparing the result with q123 :

```
> v132:=rot(v,q12 &c q13); #rotation q13 followed by q12
```
$$v132 := (a\sin(\beta) + c\cos(\beta))\, e3 + (a\cos(\beta)\cos(\alpha) - c\sin(\beta)\cos(\alpha)$$
$$- b\sin(\alpha))\, e1 + (b\cos(\alpha) - c\sin(\beta)\sin(\alpha) + a\sin(\alpha)\cos(\beta))\, e2$$

```
> clicollect(v123-v132);
```
$$(-b\sin(\beta)\sin(\alpha) + a\sin(\beta)\cos(\alpha) - a\sin(\beta))\, e3$$
$$+ (-b\sin(\alpha)\cos(\beta) - c\sin(\beta) + c\sin(\beta)\cos(\alpha) + b\sin(\alpha))\, e1$$
$$+ (a\sin(\alpha) + c\sin(\beta)\sin(\alpha) - a\sin(\alpha)\cos(\beta))\, e2$$

As it can be seen, $\mathbf{v}_{123} \neq \mathbf{v}_{132}$.

Example 2: Counter-clockwise rotation by an angle α around the given axis.

In this example we will find a way to rotate a given vector $\mathbf{v} \in \mathbb{R}^3$ by an angle α in a plane orthogonal to the given axis vector $\mathbf{axis} = a_1\mathbf{e}_1 + a_2\mathbf{e}_2 + a_3\mathbf{e}_3$. This rotation will be counter-clockwise when looking down the axis towards the origin $(0,0,0)$ of the coordinate system. In order to derive symbolic formulas, we will assume that the symbolic vector **axis** has been normalized by defining $\lambda = \pm\sqrt{1 - a_1^2 - a_2^2}$ and $\mathbf{axis} = a_1\mathbf{e}_1 + a_2\mathbf{e}_2 + \lambda\mathbf{e}_3$.

```
> alias(lambda=RootOf(-a1^2-a2^2-_Z^2+1)):
  axis:=a1*e1+a2*e2+lambda*e3;
```
$$axis := a1\, e1 + a2\, e2 + \lambda\, e3$$

```
> simplify(axis &c star(axis));
```
$$Id$$

Thus, in the symbolic case, we always have $\|\mathbf{axis}\| = 1$. Notice that in order to represent a rotation around **axis** we need to find a *dual unit quaternion* which we call **qaxis**. It will need to be defined in such a way as to give the desired orientation for the rotation. Since we have opted for counter-clockwise rotations, we define $\mathbf{qaxis} = -\mathbf{axis}\,\mathbf{e}_{123}$ where $\mathbf{e}_{123}$ is a unit pseudoscalar in $C\ell_{0,3}$. In the following we will refer to the $\mathbf{axis} = a_1\mathbf{e}_1 + a_2\mathbf{e}_2 + a_3\mathbf{e}_3$ as the triple (a_1, a_2, a_3).

```
> qaxis:=axis &c (-e123);
```
$$qaxis := \lambda\, e12 + a1\, e23 - a2\, e13$$

In Appendix 25.4 Reader can find procedure `qrot` which finds the dual quaternion **qaxis**. The first three arguments to `qrot` are the (numeric or symbolic) components of the **axis** vector in the basis $\{e_1, e_2, e_3\}$ while the fifth argument is the angle of rotation. For example, we can define various rotation quaternions:

```
>   q100:=qrot(1,0,0,theta);#rotation about the axis (1,0,0)
```

$$q100 := \cos(\tfrac{1}{2}\,\theta)\,Id + \sin(\tfrac{1}{2}\,\theta)\,e23$$

```
>   q010:=qrot(0,1,0,theta);#rotation about the axis (0,1,0)
```

$$q010 := \cos(\tfrac{1}{2}\,\theta)\,Id - \sin(\tfrac{1}{2}\,\theta)\,e13$$

```
>   q101:=qrot(1,0,1,theta);#rotation about the axis (1,0,1)
```

$$q101 := \cos(\tfrac{1}{2}\,\theta)\,Id + \sin(\tfrac{1}{2}\,\theta)\,(\tfrac{1}{2}\,e12\,\sqrt{2} + \tfrac{1}{2}\,\sqrt{2}\,e23)$$

```
>   q111:=qrot(1,1,1,theta);#rotation about the axis (1,1,1)
```

$$q111 := \cos(\tfrac{1}{2}\,\theta)\,Id + \sin(\tfrac{1}{2}\,\theta)\,(\tfrac{1}{3}\,e12\,\sqrt{3} - \tfrac{1}{3}\,\sqrt{3}\,e13 + \tfrac{1}{3}\,\sqrt{3}\,e23)$$

For example, let's rotate the first basis vector $\mathbf{e}_1$ around various axes listed above by some angle α. In the next example we will find a general formula for the components of the rotated arbitrary vector **v**.

```
>   v:=e1:
    vnew:=rot(v,q100); #rotation around the axis (1,0,0)
```

$$vnew := e1$$

```
>   vnew:=rot(v,q010); #rotation around the axis (0,1,0)
```

$$vnew := -\sin(\theta)\,e3 + \cos(\theta)\,e1$$

```
>   eval(subs(theta=Pi/2,vnew));
```

$$-e3$$

```
>   vnew:=rot(v,q101); #rotation around the axis (1,0,1)
```

$$vnew := -\frac{1}{2}\,(-1 + \cos(\theta))\,e3 + \frac{1}{2}\,(\cos(\theta) + 1)\,e1 + \frac{1}{2}\,\sqrt{2}\,e2\,\sin(\theta)$$

```
>   eval(subs(theta=Pi/2,vnew));
```

$$\frac{1}{2}\,e3 + \frac{1}{2}\,e1 + \frac{1}{2}\,e2\,\sqrt{2}$$

```
>   vnew:=rot(v,q111); #rotation around the axis (1,1,1)
```

$$vnew := -\frac{1}{3}\,(\sqrt{3}\sin(\theta) - 1 + \cos(\theta))\,e3 + \frac{1}{3}\,(2\cos(\theta) + 1)\,e1$$
$$+ \frac{1}{3}\,(\sqrt{3}\sin(\theta) + 1 - \cos(\theta))\,e2$$

```
>   eval(subs(theta=Pi/2,vnew));
```

$$-\frac{1}{3}\,(\sqrt{3} - 1)\,e3 + \frac{1}{3}\,e1 + \frac{1}{3}\,(\sqrt{3} + 1)\,e2$$

```
>   eval(subs(theta=Pi,vnew));
```

$$\frac{2}{3} \, e3 - \frac{1}{3} \, e1 + \frac{2}{3} \, e2$$

Example 3: General rotations.

Finally, we derive a general formula for a rotation of an arbitrary vector $\mathbf{v} = v_1\mathbf{e}_1 + v_2\mathbf{e}_2 + v_3\mathbf{e}_3$ around an arbitrary $\mathbf{axis} = a_1\mathbf{e}_1 + a_2\mathbf{e}_2 + a_3\mathbf{e}_3$ and by an arbitrary angle α. In the purely symbolic case we assume that $\mathbf{axis}$ is of unit length, that is, $a_3 = \lambda = \pm\sqrt{1 - a_1^2 - a_2^2}$.

```
>   v:=v1*e1+v2*e2+v3*e3; #an arbitrary vector
```

$$v := v1 \; e1 + v2 \; e2 + v3 \; e3$$

The rotated vector will have the following components:

```
>   qnew:=clicollect(rot(v,qrot(a1,a2,lambda,theta)));
```

$$
\begin{aligned}
qnew := \;& (-\lambda \, v1 \, a1 \cos(\theta) - \lambda \, v2 \, a2 \cos(\theta) + a1^2 \, v3 \cos(\theta) - v1 \, a2 \sin(\theta) \\
& + \lambda \, v1 \, a1 + \lambda \, v2 \, a2 + a2^2 \, v3 \cos(\theta) + v2 \, a1 \sin(\theta) + v3 - a2^2 \, v3 \\
& - a1^2 \, v3)e3 + (v1 \cos(\theta) - a1 \, v2 \, a2 \cos(\theta) - \lambda \, v3 \, a1 \cos(\theta) + \lambda \, v3 \, a1 \\
& + a1^2 \, v1 - a1^2 \, v1 \cos(\theta) - v2 \, \lambda \sin(\theta) + v3 \, a2 \sin(\theta) + a1 \, v2 \, a2)e1 \\
& + (a2^2 \, v2 + a1 \, v1 \, a2 - v3 \, a1 \sin(\theta) - a2^2 \, v2 \cos(\theta) + \lambda \, v3 \, a2 \\
& + v1 \, \lambda \sin(\theta) - \lambda \, v3 \, a2 \cos(\theta) + v2 \cos(\theta) - a1 \, v1 \, a2 \cos(\theta))e2
\end{aligned}
$$

For example, let's rotate $\mathbf{v} = \mathbf{e}_1 - 2\mathbf{e}_2 + 4\mathbf{e}_3$ around the axis $(2, -3, 4)$ by an angle $\alpha = \pi/4$.

```
>   clicollect(rot(e1-2*e2+4*e3,qrot(2,-3,4,Pi/4)));
```

$$
\begin{aligned}
& (-\frac{1}{58} \sqrt{29} \sqrt{2} + \frac{10}{29} \sqrt{2} + \frac{96}{29}) \, e3 + (-\frac{2}{29} \sqrt{29} \sqrt{2} + \frac{48}{29} - \frac{19}{58} \sqrt{2}) \, e1 \\
& + (\frac{7}{29} \sqrt{2} - \frac{2}{29} \sqrt{29} \sqrt{2} - \frac{72}{29}) \, e2
\end{aligned}
$$

Thus, in this section we have shown how easy it is to derive vector rotation formulas from vector analysis using elements of $\mathbf{Spin}(3)$ considered as unit quaternions. It has been very helpful to be able to embed $\mathbf{Spin}(3)$ in $C\ell_{0,3}$ and consider quaternions $\mathbb{H}$ as isomorphic to the even subalgebra of $C\ell_{0,3}$.

25.3.3 Degenerate Clifford algebra and the proper rigid motions

In this final section we will use the ability of CLIFFORD to perform computations in Clifford algebras of an arbitrary quadratic form including, of course, degenerate forms. We will consider the semi-direct product $\mathbf{Spin}(3) \ltimes \mathbb{R}^3$ that double covers the group of proper rigid motions SE(3). We will follow notation used in [18], page 156, except that our basis vector that squares to 0 will be $\mathbf{e}_4$ and not $\mathbf{e}$. We begin by defining B as a degenerate diagonal form $\mathrm{diag}(-1,-1,-1,0)$ of signature $(1,0,3)$. Recall from the previous section that the procedure $\mathtt{star}$ gives conjugation in $C\ell_{1,0,3}$. In [18], the Clifford algebra $C\ell_{1,0,3}$ is denoted as $C(0,3,1)$.

```
>   dim:=4:n:=dim-1:eval(makealiases(dim)):B:=diag(-1$n,0);
```

$$B := \begin{bmatrix} -1 & 0 & 0 & 0 \\ 0 & -1 & 0 & 0 \\ 0 & 0 & -1 & 0 \\ 0 & 0 & 0 & 0 \end{bmatrix}$$

Let the vector basis in $\mathbb{R}^3$ be stored in $\mathtt{vbasis}$ and let $\mathbf{t}$ an arbitrary vector in $\mathbb{R}^3$.

```
>   vbasis:=cbasis(n,1):t:=add(t.i*vbasis[i],i=1..n);
```

$$t := t1 \ e1 + t2 \ e2 + t3 \ e3$$

Elements of the form $\mathbf{1}+\mathbf{t}\,\mathbf{e}_4$ are invertible in $C\ell_{1,0,3}$ since $u\,\mathbf{e}_4$ and $\mathbf{e}_4\,u$ are nilpotent for any u in $C\ell_{1,0,3}$. That is, the Jacobson radical J in $C\ell_{1,0,3}$ is generated by $\mathbf{e}_4$. The symbolic inverse of $\mathbf{1}+\mathbf{t}\,\mathbf{e}_4$ can be computed with the procedure $\mathtt{cinv}$.

```
>   cinv(1+t &c e4);#symbolic inverse of 1+te4
```

$$Id - t1 \ e14 - t2 \ e24 - t3 \ e34$$

Let $\kappa = \sqrt{-c_1^2 - c_2^2 - c_3^2 + 1}$ and $\varepsilon = \pm 1$ as before. We will verify now statements made on page 156 in [18]. We know from the previous section that the most general element $\mathbf{g}$ in $\mathbf{Spin}(3)$ has the form:

```
>   alias(kappa=sqrt(-c1^2-c2^2-c3^2+1)):
    alias(eps=RootOf(_Z^2-1)):
    g:=eps*kappa*Id+c3*e12+c2*e13+c1*e23;
```

$$g := eps \ \kappa \ Id + c3 \ e12 + c2 \ e13 + c1 \ e23$$

We consider a subgroup G of the group of units of $C\ell_{1,0,3}$ of the form $\mathbf{g}+\frac{1}{2}\mathbf{t}\mathbf{g}\mathbf{e}_4$ where $\mathbf{g}$ belongs to $\mathbf{Spin}(3)$ and $\mathbf{t}$ is a 1-vector in $\mathbb{R}^3$. Elements in G will be given by the procedure $\mathtt{ge}$.

```
>   ge:=proc(g,t) clicollect(simplify(g+1/2*t &c g &c e4)) end:
```

For the most general $\mathbf{g} \in \mathbf{Spin}(3)$ and $\mathbf{t} \in \mathbb{R}^3$, procedure $\mathtt{ge}$ gives:

```
>   'ge(g,t)'=ge(g,t);
```

$$ge(g, t) = (\frac{1}{2}\, t2\, \kappa\, eps - \frac{1}{2}\, t1\, c3 + \frac{1}{2}\, t3\, c1)\, e24$$

$$+ (\frac{1}{2}\, t3\, \kappa\, eps - \frac{1}{2}\, t2\, c1 - \frac{1}{2}\, t1\, c2)\, e34 + (\frac{1}{2}\, t2\, c3 + \frac{1}{2}\, t3\, c2 + \frac{1}{2}\, t1\, \kappa\, eps)\, e14$$

$$+ (-\frac{1}{2}\, t2\, c2 + \frac{1}{2}\, t3\, c3 + \frac{1}{2}\, t1\, c1)\, e1234 + eps\, \kappa\, Id + c3\, e12$$

$$+ c2\, e13 + c1\, e23$$

First, let's verify that

$$(\mathbf{g} - \frac{1}{2}\mathbf{t}\,\mathbf{g}\,\mathbf{e}_4)^* = \mathbf{g}^* + \frac{1}{2}\mathbf{g}^*\,\mathbf{t}\,\mathbf{e}_4. \tag{3.11}$$

Notice that the left-hand-side in (3.11) is just the conjugation of $ge(\mathbf{g}, -\mathbf{t})$ while the right-hand-side is equal to $ge(\mathbf{g}^*, -\mathbf{t})$ where $\mathbf{g}^*$ denotes the conjugate of $\mathbf{g}$.

```
>   L:=clicollect(star(ge(g,-t))):
    R:=simplify(star(g)+1/2*star(g) &c t &c e4):simplify(L-R);
```
$$0$$

Thus, (3.11) has been verified.

Next, we define the action of the group G on the subspace of $C\ell_{1,0,3}$ consisting of the elements of the form $1 + \mathbf{x}\,\mathbf{e}_4$ as follows:

$$(\mathbf{g} + \frac{1}{2}\mathbf{t}\,\mathbf{g}\,\mathbf{e}_4)(1 + \mathbf{x}\,\mathbf{e}_4)(\mathbf{g} - \frac{1}{2}\mathbf{t}\,\mathbf{g}\mathbf{e}_4)^* = 1 + (\mathbf{g}\,\mathbf{x}\,\mathbf{g}^* + \mathbf{t})\,\mathbf{e}_4 \tag{3.12}$$

where $\mathbf{x}, \mathbf{t} \in \mathbb{R}^3$ and $\mathbf{g} \in \mathbf{Spin}(3)$. The identity (3.12) can be shown as follows. We define a procedure $\texttt{rigid}$ which will give this action on $\mathbb{R}^3$:
$\mathbf{x} \mapsto \mathbf{g}\,\mathbf{x}\,\mathbf{g}^* + \mathbf{t}$.

```
>   rigid:=proc(x,g,t) local p;
    if not evalb(x=vectorpart(x,1)) or
            not evalb(t=vectorpart(t,1)) then
            ERROR('x and t must be vectors')
    fi:
    if not type(g,evenelement) then ERROR('g must be even') fi;
    RETURN(clicollect(simplify(cmul(g,x,star(g))+t)))
    end:
```

This action will be the *rigid motion* on $\mathbb{R}^3$. We can compute the right-hand-side of (3.12) by using $\texttt{rigid}$ while the left-hand-side will be computed directly.

```
>   x:=add(x.i*vbasis[i],i=1..n);
```
$$x := x1\ e1 + x2\ e2 + x3\ e3$$

```
>  LHS:=simplify(ge(g,t) &c (1 + x &c e4) &c star(ge(g,-t)))):
   RHS:=simplify(Id+rigid(x,g,t) &c e4):
   simplify(LHS-RHS);
```
$$0$$

Finally, we will verify directly that the action $\mathbf{x} \mapsto \mathbf{g}\,\mathbf{x}\,\mathbf{g}^* + \mathbf{t}$ is a rigid motion. If we denote by $\mathbf{x}_p, \mathbf{y}_p$ the images of $\mathbf{x}, \mathbf{y}$ under this action, we will need to show that $\|\mathbf{x}_p - \mathbf{y}_p\| = \|\mathbf{x} - \mathbf{y}\|$ in the Euclidean norm. We can compute the norm $\|\mathbf{x} - \mathbf{y}\|$ by taking $\sqrt{(\mathbf{x} - \mathbf{y})(\mathbf{x} - \mathbf{y})^*}$ in $C\ell_{1,0,3}$, or by using a procedure `distance`.

```
>  y:=add(y.i*vbasis[i],i=1..n):
   distance:=proc(x,y)
       sqrt(simplify(scalarpart(cmul(x-y,star(x-y))))) end:
   xp:=rigid(x,g,t):yp:=rigid(y,g,t):
   evalb(distance(x,y)=distance(xp,yp));
```
$$true$$

Thus the action of G defined on $\mathbb{R}^3$ as $\mathbf{x} \mapsto \mathbf{g}\,\mathbf{x}\,\mathbf{g}^* + \mathbf{t}$ is a rigid motion. In view of the presence of the non-trivial radical in $C\ell_{1,0,3}$, this group G is in fact a semi-direct product of $\mathbf{Spin}(3)$ and $\mathbb{R}^3$ that are responsible for rotations and translations respectively. It is well known of course that $\mathbf{Spin}(3) \rtimes \mathbb{R}^3$ doubly covers the group of proper rigid motions SE(3). We leave it as an exercise for the Reader to check in CLIFFORD that the composition of two rigid motions is a rigid motion.

25.4 Summary

The main purpose of this paper has been to show a variety of computational problems that can be approached with the symbolic package CLIFFORD. Relation between the Clifford products in $C\ell(g)$ and $C\ell(B)$ through the Helmstetter's formula appears more clear once we proved with CLIFFORD our choice for the bivector F. Applications presented in Section 25.3 show that CLIFFORD is a convenient tool to carry out practical computations in the low dimensional algebras such as $C\ell_{0,3}$ and $C\ell_{1,0,3}$ needed in vector rotations and rigid motions in $\mathbb{R}^3$. In addition, we found the explicit form of the most general elements in $\mathbf{Pin}(3)$ and $\mathbf{Spin}(3)$.

Appendix A

Aliases needed in Section (25.3.1):

```
>   alias(kappa1=RootOf(_Z^2-1+x3^2+x2^2+x4^2)):
    alias(kappa2=RootOf(_Z^2-1+x5^2+x3^2+x2^2)):
    alias(kappa3=RootOf(-x6^2+x5^2*x6^2+x4^2*x5^2+x6^4+x2^2*x6^2
    +x4^2*x6^2+_Z^2)):

>   alias(kappa4=RootOf(-x7^2+x5^2*x7^2+x3^2*x7^2+x6^2*x7^2+x7^4
    +x4^2*x5^2-2*x4*x5*x6*x3+x6^2*x3^2+x4^2*x7^2+_Z^2)):
    alias(kappa5=RootOf((x7^2+x8^2)*_Z^2+(2*x5*x4*x7-2*x3*x6*x7)*_Z
    -x8^2+x6^2*x3^2+x8^4+x4^2*x5^2-2*x4*x5*x6*x3+x6^2*x8^2
    +x7^2*x8^2+x5^2*x8^2+x3^2*x8^2+x4^2*x8^2)):
    alias(eps=RootOf(_Z^2-1)):
    alias(lambda1=RootOf(_Z^2+x2^2+x3^2-1)):
    alias(lambda2=RootOf(_Z^2-1+x3^2)):
    alias(lambda3=RootOf(_Z^2-1+x5^2)):
    alias(lambda4=RootOf(-x6^2+x5^2*x6^2+x4^2*x5^2+x6^4+x2^2*x6^2
    +x4^2*x6^2+_Z^2)):
    alias(lambda5=RootOf(_Z^2-x6^2+x5^2*x6^2+x6^4)):
    alias(lambda6=RootOf(-x7^2+x5^2*x7^2+x3^2*x7^2+x6^2*x7^2+x7^4
    +x4^2*x5^2-2*x4*x5*x6*x3+x6^2*x3^2+x4^2*x7^2+_Z^2)):
    alias(lambda7=RootOf(_Z^2-x7^2+x5^2*x7^2+x6^2*x7^2+x7^4)):
    alias(lambda8=RootOf((x7^2+x8^2)*_Z^2+(2*x5*x4*x7-2*x3*x6*x7)*_Z
    -x8^2+x6^2*x3^2+x8^4+x4^2*x5^2-2*x4*x5*x6*x3+x6^2*x8^2
    +x7^2*x8^2+x5^2*x8^2+x3^2*x8^2+x4^2*x8^2)):
    alias(lambda9=RootOf(_Z^2*x8^2-x8^2+x8^4+x3^2*x8^2+x4^2*x8^2)):
```

Appendix B

Procedure `clisolve2` was used extensively throughout this paper to solve linear equations in the Clifford algebra $C\ell(B)$.

```
>   clisolve2:=proc(eq,indet) local i,T,vars,sol,sys;
    if type(indet,list) then
            vars:=convert(indet,set) else
            vars:=select(type,indets(indet),indexed)
    fi;
    T:=cliterms(eq);
    sys:={coeffs(clicollect(simplify(eq)),T)};
    sol:=[solve(sys,vars)];
    if type(indet,list) then RETURN(sol) else
        RETURN([seq(subs(sol[i],indet),i=1..nops(sol))]);
    fi;
    end:
```

We display the code of two procedures `rot` and `qrot` that were needed in Section 25.3.2. Procedure `rot` performs a rotation of a vector by a quaternion through a certain angle.

```
>  rot:=proc(v,q) local qs;
   qs:=star(q):
   RETURN(map(factor,clicollect(simplify(cmul(q,v,qs)))))
   end:
```

Procedure qrot finds a unit quaternion that is dual to the rotation axis vector **axis**.

```
>  qrot:=proc(p1,p2,p3,theta) local bas,c,e,k,l,i,q,n;
   global qaxis;
   if type(p1,name) or type(p2,name) then
           RETURN(cos(theta/2)*Id+sin(theta/2)*qaxis) fi;
   if evalb(simplify(p1^2+p2^2+p3^2)=1) then
           q:=simplify(subs({a1=p1,a2=p2},qaxis)) fi;
   n:=sqrt(p1^2+p2^2+p3^2):
   if n=0 then ERROR('axis vector must be a non-zero vector')
           elif has(n,RootOf) then n:=max(allvalues(n)) fi;
   q:=simplify(subs({a1=p1/n,a2=p2/n},qaxis));
   bas:=[Id,e12,e13,e23]:c:=[]:k:=0:
   for i to 4 do l:=coeff(q,bas[i]);
                   if has(l,RootOf) then k:=i fi:
                   c:=[op(c),l]
               od:
   if k<>0 then e:=allvalues(c[k]);
      if p3>0 then c:=subsop(k=max(e),c) elif
          p3<0 then c:=subsop(k=min(e),c) else ERROR('p3=0') fi;
   fi;
   q:=add(c[i]*bas[i],i=1..4);
   RETURN(cos(theta/2)*Id+sin(theta/2)*q)
   end:
```

References

Chapter 1

[1] H. Li, D. Hestenes, A. Rockwood, Generalized Homogeneous Coordinates for Computational Geometry. In: G. Sommer (Ed.), *Geometric Computing with Clifford Algebra* (Springer-Verlag, Heidelberg, 2000), p. 25–58.

[2] D. Hestenes, The design of linear algebra and geometry, *Acta Appl. Math.* **23**: 65–93, 1991.

[3] H.Grassmann, Linear Extension Theory (Die Lineale Ausdehnungslehre), translated by L. C. Kannenberg. In: *The Ausdehnungslehre of 1844 and Other Works*, Chicago, La Salle: Open Court Publ. 1995.

[4] D. Hestenes and R. Ziegler, Projective Geometry with Clifford Algebra, *Acta Appl. Math.* **23**: 25–63, 1991.

[5] K. Menger, New foundation of Euclidean geometry, *Am. J. Math.* **53**: 721–745, 1931

[6] A. Dress , T. Havel, Distance Geometry and Geometric Algebra, *Foundations of Physics* **23**: 1357–1374, 1993.

[7] C. Doran, D. Hestenes, F. Sommen, Van Acker, N.: Lie Groups as Spin Groups, *Journal of Mathematical Physics*, 1993.

[8] D. Hestenes, *New Foundations for Classical Mechanics*, D. Reidel, Dordrecht/Boston, 2nd edition, 1999.

[9] D. Hestenes, Invariant body kinematics I: Saccadic and compensatory eye movements, *Neural Networks* **7**: 65–77, 1994.

[10] D. Hestenes, Invariant body kinematics II: Reaching and neurogeometry, *Neural Networks* **7**: 79–88, 1994.

[11] J. M. McCarthy, *An introduction to Theoretical Mechanics*, MIT Press, Cambridge, 1990.

[12] J. M. Selig, *Geometrical Methods in Robotics*, Springer, New York, 1996.

Chapter 2

[1] M. Barnabei, A. Brini and Rota G-C. On the exterior calculus of invariant theory. Journal of Algebra, 96, p. 120–160.

[2] W. E. Baylis, 1996. Clifford (Geometric) Algebras With Applications in Physics, Mathematics, and Engineering. Birkhäuser, Boston.

[3] J. Cnops, 1996. Vahlen Matrices for Non-definite Metrics in Clifford Algebras with Numeric and Symbolic Computations Editors: R. Ablamowicz, P. Lounesto, J.M. Parra Birkhauser, Boston.

[4] A. W. M. Dress, T. F. Havel, 1993. Distance Geometry and Geometric Algebra Foundations of Physics, Vol. 23, No. 10, 1993.

[5] W. Fulton and J. Harris, 1991. Representation Theory: A First Course. Springer–Verlag, New York.

[6] C. Doran, D. Hestenes, F. Sommen and N. Van Acker. Lie groups and spin groups. J. Math. Phys. 34 (8), August 1993.

[7] D. Hestenes, and G. Sobczyk, 1984. Clifford Algebra to Geometric Calculus: A Unified Language for Mathematics and Physics. *D. Reidel*, Dordrecht.

[8] D. Hestenes, 1991. The Design of Linear Algebra and Geometry. Acta Applicandae Mathematicae 23:65-93, Kluwer Academic Publishers.

[9] D. Hestenes and R. Ziegler, 1991. Projective geometry with Clifford algebra. *Acta Applicandae Mathematicae*, 23: 25–63.

[10] P. Lounesto, 1997. Clifford Algebras and Spinors. Cambridge University Press, Cambridge.

[11] P. Lounesto, 1987.   CLICAL software paquet and user manual. Helsinki University of Technology of Mathematics, Research Report A248.

[12] P. Lounesto and A. Springer, 1989. Mobius Transformations and Clifford Algebras of Euclidean and Anti-Euclidean Spaces J. Lawrynowicz (ed.), Deformations of Mathematical Structures, 79-90, Kluwer Academic Publishers.

[13] J. M. Selig, Clifford algebra of points, lines and planes. South Bank University, School of Computing, Information Technology and Maths, Technical Report SBU–CISM–99–06, 1999.

[14] I. R. Porteous, 1995. Clifford Algebras and the Classical Groups Cambridge University Press

[15] M. Riesz 1993. Clifford Numbers and Spinors. Lecture Series No. 38, University of Maryland, 1958. Reprinted as facsimile (eds. : E.F. Bolinder and P. Lounesto) by Kluwer, Dordrecht, The Netherlands.

[16] G. Sobczyk, 1995. A universal geometric algebra in *Advanced Mathematics: Computations and Applications* A. S. Alekseev and N. S. Bakhvalov (Editors), pp. 447-455, 1995 NCC Publisher.

[17] G. Sobczyk, 1997. Spectral integral domains in the classroom Aportaciones Matemáticas, Serie Comunicaciones 20 (1997), p.169-188.

[18] G. Sobczyk, 1997. The Generalized Spectral Decomposition of a Linear Operator The College Mathematics Journal, 28:1 (1997) 27-38.

Chapter 3

[1] W. E. Baylis, 1996. Clifford (Geometric) Algebras With Applications in Physics, Mathematics, and Engineering. Birkhauser, Boston.

[2] J. Cnops, 1996. Vahlen Matrices for Non-definite Metrics in Clifford Algebras with Numeric and Symbolic Computations Editors: R. Ablamowicz, P. Lounesto, J.M. Parra Birkhauser, Boston.

[3] P.J. Davis, 1974. The Schwarz Function and its Applications. The Mathematical Association of America

[4] A. W. M. Dress, T. F. Havel, 1993. Distance Geometry and Geometric Algebra Foundations of Physics, Vol. 23, No. 10, 1993.

[5] J. Haantjes, 1937. Conformal representations of an n-dimensional euclidean space with a non-definite fundamental form on itself *Proc. Ned. Akad. Wet (Math)*, 40, 700–705, 1937.

[6] T. Havel, 1995. Geometric Algebra and Mobius Sphere Geometry as a Basis for Euclidean Invariant Theory Editor: N.L. White, Invariant Methods in Discrete and Computational Geometry, p. 245-256, Kluwer.

[7] D. Hestenes and G. Sobczyk, 1984. Clifford Algebra to Geometric Calculus: A Unified Language for Mathematics and Physics. *D. Reidel*, Dordrecht.

[8] D. Hestenes, 1991. The Design of Linear Algebra and Geometry. Acta Applicandae Mathematicae 23:65-93, Kluwer Academic Publishers.

[9] D. Hestenes and R. Ziegler, 1991. Projective geometry with Clifford algebra. *Acta Applicandae Mathematicae*, 23: 25–63.

[10] O. Kobayashi, M. Wada, 1999. Circular Geometry and The Schwarzian 5th International Clifford Algebra Conference, Ixtapa

[11] P. Lounesto, 1997. Clifford Algebras and Spinors. Cambridge University Press, Cambridge.

[12] P. Lounesto, 1987. CLICAL software paquet and user manual. Helsinki University of Technology of Mathematics, Research Report A248.

[13] P. Lounesto and A. Springer, 1989. Mobius Transformations and Clifford Algebras of Euclidean and Anti-Euclidean Spaces J. Lawrynowicz (ed.), Deformations of Mathematical Structures, 79-90, Kluwer Academic Publishers.

[14] J. G. Maks, 1989. Modulo (1,1) Periodicity of Clifford Algebras and Generalized Mobius Transformations Technical University of Delft (Dissertation)

[15] I. R. Porteous, 1995. Clifford Algebras and the Classical Groups Cambridge University Press

[16] J. W. Young, 1930. Projective Geometry The Open Court Publishing Company, Chicago, ILL.

Chapter 4

[1] J. W. Cannon, W. J. Floyd, R. Kenyon and W. R. Parry (1997): Hyperbolic geometry, in *Flavors of Geometry*, S. Levy (ed.), Cambridge University Press.

[2] T. E. Cecil (1992). *Lie Sphere Geometry*. Springer, New York.

[3] A. Crumeyrolle (1990): *Orthogonal and Symplectic Clifford Algebras*, D. Reidel, Dordrecht, Boston.

[4] R. Delanghe, F. Sommen and V. Soucek (1992): *Clifford Algebra and Spinor-Valued Functions*, D. Reidel, Dordrecht, Boston.

[5] W. Fenchel (1989): *Elementary Geometry in Hyperbolic Space*, Walter de Gruyter and Company.

[6] M. J. Greenberg (1980): *Euclidean and Non-Euclidean Geometries*, W. H. Freeman and Company, 2nd edition.

[7] T. Havel (1991): Some examples of the use of distances as coordinates for Euclidean geometry, *J. Symbolic Computat.* **11**: 579–593.

[8] T. Havel (1995): Geometric algebra and Möbius sphere geometry as a basis for Euclidean invariant theory, in *Invariant Methods in Discrete and Computational Geometry*, N. L. White (ed.), pp. 245–256, D. Reidel, Dordrecht.

[9] D. Hestenes (1966): *Space-Time Algebra*, Gordon and Breach, New York.

[10] D. Hestenes and G. Sobczyk (1984): *Clifford Algebra to Geometric Calculus*, D. Reidel, Dordrecht, Boston.

[11] D. Hestenes (1987): *New Foundations for Classical Mechanics*, D. Reidel, Dordrecht, Boston.

[12] D. Hestenes and R. Ziegler (1991): Projective geometry with Clifford algebra, *Acta Appl. Math.* **23**: 25–63.

[13] D. Hestenes (1991): The design of linear algebra and geometry, *Acta Appl. Math.* **23**: 65–93.

[14] D. Hestenes, H. Li and A. Rockwood (1999): New algebraic tools for classical geometry, in *Geometric Computing with Clifford Algebra*, G. Sommer (ed.), Springer.

[15] B. Iversen (1992): *Hyperbolic Geometry*, Cambridge University Press.

[16] H. Li (1996b): Clifford algebra and Lobachevski geometry, in *Clifford Algebras and Their Applications in Mathematical Physics*, pp. 239–245, V. Dietrich et al. (ed.), D. Reidel, Dordrecht.

[17] H. Li (1997): Hyperbolic geometry with Clifford algebra, *Acta Appl. Math.*, **48**(3): 317–358.

[18] H. Li (1998): Some applications of Clifford algebra to geometries. *Automated Deduction in Geometries*, LNAI **1669**, X.-S. Gao, D. Wang, L. Yang (eds.), pp. 156-179.

[19] H. Li, D. Hestenes and A. Rockwood (1999a): Generalized homogeneous coordinates for computational geometry, in *Geometric Computing with Clifford Algebra*, G. Sommer (ed.), Springer.

[20] H. Li, D. Hestenes and A. Rockwood (1999b): A universal model for conformal geometries of Euclidean, spherical and double-hyperbolic spaces, in *Geometric Computing with Clifford Algebra*, G. Sommer (ed.), Springer.

[21] H. Li, D. Hestenes and A. Rockwood (1999c): Spherical conformal geometry with geometric algebra, in *Geometric Computing with Clifford Algebra*, G. Sommer (ed.), Springer.

[22] S. Lie (1872): Über Komplexe, inbesondere Linien- und Kugelkomplexe, mit Anwendung auf der Theorie der partieller *Differentialgleichungen. Math. Ann.* **5**: 145–208, 209–256. (Ges. Abh. **2**: 1–121)

[23] B. Mourrain and N. Stolfi (1995a): Computational symbolic geometry, in *Invariant Methods in Discrete and Computational Geometry*, N. L. White (ed.), pp. 107–139, D. Reidel, Dordrecht, Boston.

[24] B. Mourrain and N. Stolfi (1995b): Applications of Clifford algebras in robotics, in *Computational Kinematics '95*, J.-P. Merlet and B. Ravani (eds.), pp. 141–150, D. Reidel, Dordrecht, Boston.

[25] U. Pinkall (1981): Dupin'sche Hyperflächen. Dissertation, Univ. Freiburg.

[26] B. Rosenfeld (1988): *A History of Non-Euclidean Geometry*, Springer, New York.

[27] J. Seidel (1952): Distance-geometric development of two-dimensional Euclidean, hyperbolic and spherical geometry I, II, Simon Stevin **29**: 32–50, 65–76.

[28] W. P. Thurston (1997): *Three-Dimensional Geometry and Topology*, Princeton.

[29] L. Yang, X.-S. Gao, S.-C. Chou and J.-Z. Zhang (1996): Automated production of readable proofs for theorems in non-Euclidean geometries, in *ADG'96*, pp. 171–188, LNAI **1360**.

Chapter 5

[1] T. Boy de la Tour, S. Fèvre, D. Wang, (1999): Clifford term rewriting for geometric reasoning in 3D. In: *Automated Deduction in Geometry* (Gao, X.-S., Wang, D., Yang, L., eds.), LNAI **1669**, Springer, Berlin Heidelberg, pp. 130–155.

[2] R. S. Boyer, J. S. Moore, (1988): *A Computational Logic Handbook*. Academic Press, Boston San Diego (2nd edition, 1998).

[3] S. -C. Chou, X. -S. Gao, J. -Z. Zhang, (1993): Automated geometry theorem proving by vector calculation. In: *Proc. ISSAC '93* (Kiev, Ukraine, July 6–8, 1993), ACM Press, New York, pp. 284–291.

[4] H. Crapo, J. Richter-Gebert, (1995): Automatic proving of geometric theorems. In: *Invariant Methods in Discrete and Computational Geometry* (White, N. L., ed.), Kluwer, Dordrecht, pp. 167–196.

[5] D. Fearnley-Sander, (1999): Plane Euclidean reasoning. In: *Automated Deduction in Geometry* (Gao, X.-S., Wang, D., Yang, L., eds.), LNAI **1669**, Springer, Berlin Heidelberg, pp. 86–110.

[6] D. Fearnley-Sander, T. Stokes, (1997): Area in Grassmann geometry. In: *Automated Deduction in Geometry* (Wang, D., ed.), LNAI **1360**, Springer, Berlin Heidelberg, pp. 141–170.

[7] S. Fèvre, D. Wang, (1998): Proving geometric theorems using Clifford algebra and rewrite rules. In: *Proc. CADE-15* (Lindau, Germany, July 5–10, 1998), LNAI **1421**, Springer, Berlin Heidelberg, pp. 17–32.

[8] S. Fèvre, D. Wang, (1999): Combining Clifford algebraic computing and term-rewriting for geometric theorem proving. *Fundamenta Informaticae* **39**: 85–104.

[9] T. F. Havel, (1991): Some examples of the use of distances as coordinates for Euclidean geometry. *J. Symb. Comput.* **11**: 579–593.

[10] T. F. Havel, (1997): Computational synthetic geometry with Clifford algebra. In: *Automated Deduction in Geometry* (Wang, D., ed.), LNAI **1360**, Springer, Berlin Heidelberg, pp. 102–114.

[11] D. Hestenes, G. Sobczyk, (1984): *Clifford Algebra to Geometric Calculus.* D. Reidel, Dordrecht Boston.

[12] H. Li, (2000): Vectorial equation-solving for mechanical geometry theorem proving. *J. Automat. Reason.* (to appear)

[13] H. Li, (2000): Clifford algebra approaches to mechanical geometry theorem proving. Chapter 8 in: *Mathematics Mechanization and Applications* (Gao, X.-S., Wang, D., eds.), Academic Press, London.

[14] H. Li, Cheng, M.-t. (1997): Proving theorems in elementary geometry with Clifford algebraic method. *Adv. Math.* (Beijing) **26**: 357–371.

[15] B. Mourrain, (1999): New aspects of geometrical calculus with invariants. *Adv. Math.* (to appear).

[16] B. Mourrain, N. Stolfi, (1995): Computational symbolic geometry. In: *Invariant Methods in Discrete and Computational Geometry* (White, N. L., ed.), Kluwer, Dordrecht, pp. 107–139.

[17] J. Richter-Gebert, (1995): Mechanical theorem proving in projective geometry. *Ann. Math. Artif. Intell.* **13**: 139–172.

[18] W. Shen, B. Wall, D. Wang, (1992): Manipulating uncertain mathematical objects: The case of indefinite sums and products. *RISC-Linz Report Series* No. 92-23, Johannes Kepler University, Austria.

[19] S. Stifter, (1993): Geometry theorem proving in vector spaces by means of Gröbner bases. In: *Proc. ISSAC '93* (Kiev, Ukraine, July 6–8, 1993), ACM Press, New York, pp. 301–310.

[20] D. Wang, (1997): Clifford algebraic calculus for geometric reasoning with application to computer vision. In: *Automated Deduction in Geometry* (Wang, D., ed.), LNAI **1360**, Springer, Berlin Heidelberg, pp. 115–140.

[21] D. Wang, (1998): Clifford algebraic computing and term-rewriting for geometric theorem proving. In: *Proc. 2nd Int. Theorema Workshop* (RISC-Linz, Austria, June 29–30, 1998), RISC-Linz Report no. 98-10 (Buchberger, B., Jebelean, T., eds.), Johannes Kepler University, Austria.

[22] N. L. White, (1991): Multilinear Cayley factorization. *J. Symb. Comput.* **11**: 421–548.

[23] W. -t. Wu, (1994): *Mechanical Theorem Proving in Geometries: Basic Principle* (translated from the Chinese edition of 1984), Springer, Wien New York.

[24] H. Yang, S. Zhang, G. Feng, (1999): A Clifford algebraic method for geometric reasoning. In: *Automated Deduction in Geometry* (Gao, X.-S., Wang, D., Yang, L., eds.), LNAI **1669**, Springer, Berlin Heidelberg, pp. 111–129.

Chapter 6

[1] J. W. Cannon, W. J. Floyd, R. Kenyon and W. R. Parry (1997): Hyperbolic geometry, in *Flavors of Geometry*, S. Levy (ed.), Cambridge University Press.

[2] T. E. Cecil (1992): *Lie Sphere Geometry.* Springer, New York.

[3] A. Crumeyrolle (1990): *Orthogonal and Symplectic Clifford Algebras*, D. Reidel, Dordrecht, Boston.

[4] R. Delanghe, F. Sommen and V. Soucek (1992): *Clifford Algebra and Spinor-Valued Functions*, D. Reidel, Dordrecht, Boston.

[5] W. Fenchel (1989): *Elementary Geometry in Hyperbolic Space*, Walter de Gruyter and Company.

[6] M. J. Greenberg (1980): *Euclidean and Non-Euclidean Geometries*, W. H. Freeman and Company, 2nd edition.

[7] T. Havel (1991): Some examples of the use of distances as coordinates for Euclidean geometry, *J. Symbolic Computat.* **11**: 579–593.

[8] T. Havel (1995): Geometric algebra and Möbius sphere geometry as a basis for Euclidean invariant theory, in *Invariant Methods in Discrete and Computational Geometry*, N. L. White (ed.), pp. 245–256, D. Reidel, Dordrecht.

[9] D. Hestenes (1966): *Space-Time Algebra*, Gordon and Breach, New York.

[10] D. Hestenes and G. Sobczyk (1984): *Clifford Algebra to Geometric Calculus*, D. Reidel, Dordrecht, Boston.

[11] D. Hestenes (1987): *New Foundations for Classical Mechanics*, D. Reidel, Dordrecht, Boston.

[12] D. Hestenes and R. Ziegler (1991): Projective geometry with Clifford algebra, *Acta Appl. Math.* **23**: 25–63.

[13] D. Hestenes (1991): The design of linear algebra and geometry, *Acta Appl. Math.* **23**: 65–93.

[14] D. Hestenes, H. Li and A. Rockwood (1999): New algebraic tools for classical geometry, in *Geometric Computing with Clifford Algebra*, G. Sommer (ed.), Springer.

[15] B. Iversen (1992): *Hyperbolic Geometry*, Cambridge University Press.

[16] H. Li (1996b): Clifford algebra and Lobachevski geometry, in *Clifford Algebras and Their Applications in Mathematical Physics*, pp. 239–245, V. Dietrich et al. (ed.), D. Reidel, Dordrecht.

[17] H. Li (1997): Hyperbolic geometry with Clifford algebra, *Acta Appl. Math.*, **48**(3): 317–358.

[18] H. Li (1998): Some applications of Clifford algebra to geometries. *Automated Deduction in Geometries*, LNAI **1669**, X.-S. Gao, D. Wang, L. Yang (eds.), pp. 156-179.

[19] H. Li, D. Hestenes and A. Rockwood (1999a): Generalized homogeneous coordinates for computational geometry, in *Geometric Computing with Clifford Algebra*, G. Sommer (ed.), Springer.

[20] H. Li, D. Hestenes and A. Rockwood (1999b): A universal model for conformal geometries of Euclidean, spherical and double-hyperbolic spaces, in *Geometric Computing with Clifford Algebra*, G. Sommer (ed.), Springer.

[21] H. Li, D. Hestenes and A. Rockwood (1999c): Spherical conformal geometry with geometric algebra, in *Geometric Computing with Clifford Algebra*, G. Sommer (ed.), Springer.

[22] S. Lie (1872). Über Komplexe, inbesondere Linien- und Kugelkomplexe, mit Anwendung auf der Theorie der partieller *Differentialgleichungen*. *Math. Ann.* **5**: 145–208, 209–256. (Ges. Abh. **2**: 1–121).

[23] B. Mourrain and N. Stolfi (1995a): Computational symbolic geometry, in *Invariant Methods in Discrete and Computational Geometry*, N. L. White (ed.), pp. 107–139, D. Reidel, Dordrecht, Boston.

[24] B. Mourrain and N. Stolfi (1995b): Applications of Clifford algebras in robotics, in *Computational Kinematics '95*, J.-P. Merlet and B. Ravani (eds.), pp. 141–150, D. Reidel, Dordrecht, Boston.

[25] U. Pinkall (1981): Dupin'sche Hyperflächen. Dissertation, Univ. Freiburg.

[26] B. Rosenfeld (1988): *A History of Non-Euclidean Geometry*, Springer, New York.

[27] J. Seidel (1952): Distance-geometric development of two-dimensional Euclidean, hyperbolic and spherical geometry I, II, Simon Stevin **29**: 32–50, 65–76.

[28] W. P. Thurston (1997): *Three-Dimensional Geometry and Topology*, Princeton.

[29] L. Yang, X.-S. Gao, S.-C. Chou and J.-Z. Zhang (1996): Automated production of readable proofs for theorems in non-Euclidean geometries, in *ADG'96*, pp. 171–188, LNAI **1360**.

Chapter 7

[1] M.A.J. Ashdown 1998. Maple code for geometric algebra Available from http://www.mrao.cam.ac.uk/~maja.

[2] E. Bayro–Corrochano and J. and Lasenby, 1995. Object modelling and motion analysis using Clifford algebra. In *Proceedings of Europe-China Workshop on Geometric Modeling and Invariants for Computer Vision*, Ed. Roger Mohr and Wu Chengke, Xi'an, China, April, pp. 143–149.

[3] E. Bayro–Corrochano, J. Lasenby and G. Sommer G. 1996. Geometric Algebra: A framework for computing point and line correspondences and projective structure using n uncalibrated cameras. *IEEE Proceedings of ICPR'96 Viena, Austria*, Vol. I, August, pp. 334-338.

[4] E. Bayro–Corrochano and J. Lasenby, 1998. Geometric techniques for the computation of projective invariants using n uncalibrated cameras. In *Proceedings of the Indian Conference on Computer Vision and Image Processing*, New delhi, India, 21–23 December, pp. 95–100.

[5] J. Lasenby and E. Bayro–Corrochano, 1999. Analysis and Computation of Projective Invariants fromMultiple Views in the Geometric Algebra Framework. Special Issue on Invariants for Pattern Recognition and Classification, ed. M.A. Rodrigues. *Int. Journal of Pattern Recognition and Artificial Intelligence*, Vol 13, No 8, December 1999, pp. 1105–1121.

[6] S. Carlsson, 1994. The Double Algebra: and effective tool for computing invariants in computer vision. *Applications of Invariance in Computer Vision*, Lecture Notes in Computer Science 825; Proceedings of 2nd-joint Europe-US workshop, Azores, October 1993. Eds. Mundy, Zisserman and Forsyth. Springer-Verlag.

[7] W.K. Clifford, 1878. Applications of Grassmann's extensive algebra. *Am. J. Math.* 1: 350–358.

[8] L. Dorst, S. Mann and T. Bouma. 1999. GABLE: A Matlab tutorial for geometric algebra. available from

http://www.carol.wins.uva.nl/~gable.

[9] O. Faugeras, 1995. Stratification of three-dimensional vision: projective, affine and metric representations. *J.Opt.Soc.Am.A*, 465-484.

[10] H. Grassmann, 1877. Der Ort der Hamilton'schen Quaternionen in der Ausdehnungslehre. *Math. Ann.*, 12: 375.

[11] R. Hartley, 1994. Lines and Points in three views – a unified approach. In *ARPA Image Understanding Workshop*, Monterey, California.

[12] R. Hartley, 1998. The Quadrifocal tensor In *ECCV98*, LNCS, Springer-Verlag.

[13] D. Hestenes, 1986. A unified language for mathematics and physics. *Clifford algebras and their applications in mathematical physics.* Eds. J.S.R. Chisholm and A.K. Common, D.Reidel, Dordrecht, p1.

[14] D. Hestenes, 1991. The Design of Linear Algebra and Geometry. *Acta Applicandae Mathematicae*, 23: 65–93.

[15] D. Hestenes and G. Sobczyk, 1984. Clifford Algebra to Geometric Calculus: A unified language for mathematics and physics. *D. Reidel*, Dordrecht.

[16] D. Hestenes and R. Ziegler, 1991. Projective Geometry with Clifford Algebra. *Acta Applicandae Mathematicae*, 23: 25–63.

[17] A.N. Lasenby, 1994. A 4D Maple package for geometric algebra manipulations in spacetime. Available from http://www.mrao.cam.ac.uk/~clifford.

[18] J. Lasenby, A.N. Lasenby, C.J.L Doran and W.J Fitzgerald. New geometric methods for computer vision – an application to structure and motion estimation. *International Journal of Computer Vision*, 26(3), 191-213. 1998.

[19] Q. T. Luong and O.D. Faugeras, 1995. The fundamental matrix: theory, algorithms, and stability analysis. *Int. J. Comput. Vision* 17(1), 43–76.

[20] C.B.U. Perwass, 2000. *Applications of geometric algebra in computer vision*. Ph.D. Thesis, University of Cambridge.

[21] A. Shashua and M. Werman, 1995. Trilinearity of three perspective views and its associated tensor. In proceedings of *ICCV'95*, MIT.

[22] W. Triggs, 1995. Matching Constraints and the Joint Image. In proceedings of *ICCV'95*, MIT.

[23] Z. Zhang Z. and O.D. Faugeras 1992. *3–D Dynamic Scene Analysis*. Springer Verlag.

Chapter 8

[1] E. Bayro-Corrochano and J. Lasenby. Geometric techniques for the computation of projective invariants using n uncalibrated cameras. Proceedings of *Indian Conference on Computer Vision, Graphics and*

Image Processing (ICVGIP), New Delhi, India, pp. 95-100. 21,22 Dec. 1998.

[2] P.E. Debevec, C.J. Taylor, J. and Malik. Facade: Modeling and Rendering Architecture from Photographs. *SIGGRAPH96*, 1996. Also, see: http://www.cs.berkeley.edu/~debevec/Research

[3] C.J.L. Doran and A.N. Lasenby. Lecture Notes to accompany 4th year undergraduate course on *Physical Applications of Geometric Algebra*. 2000. Available at http://www.mrao.cam.ac.uk/~clifford/ptIIIcourse/.

[4] L. Dorst, S. Mann and T. Bouma. GABLE: A Matlab tutorial for Geometric Algebra. 2000. Available at http://www.carol.wins.uva.nl/~leo/clifford/gablebeta.html.

[5] D. Hestenes. New Foundations for Classical Mechanics – second edition. *D. Reidel*, Dordrecht. 1999.

[6] D. Hestenes and G. Sobczyk, Clifford Algebra to Geometric Calculus: A unified language for mathematics and physics. *D. Reidel*, Dordrecht. 1984.

[7] J. Lasenby, W.J. Fitzgerald, A.N. Lasenby and C.J.L. Doran. New geometric methods for computer vision – an application to structure and motion estimation. *International Journal of Computer Vision*, 26(3), 191-213. 1998.

[8] J. Lasenby and A.N. Lasenby. Estimating tensors for matching over multiple views. *Phil.Trans.Roy.Soc.-A*, 356, 1267-1282. Ed. J. Lasenby, A. Zisserman, H.C. Longuet-Higgins and R. Cipolla. 1998.

[9] J. Lasenby and E. Bayro–Corrochano. Analysis and Computation of Projective Invariants from Multiple Views in the Geometric Algebra Framework. *International Journal of Pattern Recognition and Artificial Intelligence – Special Issue on Invariants for Pattern Recognition and Classification*, Vol.13, No.8, 1105-1121. Ed. M.A. Rodrigues. 1999.

[10] H.C. Longuet-Higgins. A computer algorithm for reconstructing a scene from two projections. *Nature*, 293: 133–138. 1981.

[11] D.P. Robertson and R. Cipolla. Photobuilder. 2000. available at `http://svr-www.eng.cam.ac.uk/photobuilder/`.

[12] J. Weng, T.S. Huang and N. Ahuja. Motion and Structure from Two Perspective Views: Algorithms, Error Analysis and Error Estimation. *IEEE Trans.Pattern Anal.Mach.Intelligence*, 11(5): 451–476. 1989.

Chapter 9

[1] K. Arun, T.S. Huang, and S.D. Blostein. Least squares fitting of two 3-D point sets. *IEEE Trans. PAMI*, 9:698–700, 1987.

[2] J.F. Cornwell. *Group Theory in Phsics II*. Academic Press Ltd., London, 1984.

[3] C. J. L. Doran. *Geometric Algebra and its Application to Mathematical Physics*. PhD thesis, Cambridge University, 1994.

[4] C.J.L. Doran, A.N. Lasenby, S.F. Gull, and J. Lasenby. Lectures in geometric algebra. In W.E. Baylis, editor, *Clifford (Geometric) Algebras*, pages 65–236. Birkhauser, Boston, 1996.

[5] D. Hestenes and G. Sobczyk. *Clifford Algebra to Geometric Calculus*. Reidel, Dordrecht, 1984.

[6] B.K.P. Horn. Closed-form solution of absolute orientation using unit quaternions. *J. Opt. Soc. Am.*, 4:629–642, 1987.

[7] T.S. Huang and A.N. Netravali. Motion and structure from feature correspondences: A review. *Proc. of the IEEE,*, 82(2):252–268, 1994.

[8] J. Lasenby, W.J. Fitzgerald, A.L. Lasenby, and C.J.L. Doran. New geometric methods for computer vision: An application to structure and motion estimation. *Int. J. Comp. Vision*, 26(3):191, 1998.

[9] J. Lasenby and A. Stevenson. Using geometric algebra in optical motion capture. In E. Bayro and G. Sobczyk, editors, *Geometric algebra: A geometric approach to computer vision, neural and quantum computing, robotics and engineering*. Birkhauser, 2000.

[10] J. Ponce and Y. Genc. Epipolar geometry and linear subspace methods: A new approach to weak calibration. *Int. J. of Comp. Vision*, 28(3):223–243, 1998.

[11] B. Sabata and J.K. Aggarwal. Estimation of motion from a pair of range images: A review. *CVGIP: Image Understanding*, 54(3):309–324, 1991.

[12] D.S. Sivia. *Data Analysis, A Bayesian tutorial*. Oxford University Press, 1996.

[13] J. Weng, T.S. Huang, and N. Ahuja. Motion and structure from two perspective views: Algorithms, error analysis and error estimation. *IEEE Trans. Pattern Anal. Mach. Intelligence*, 11(5):451–476, 1989.

[14] Z. Zhang. Determining the epipolar geometry and its uncertainty: A review. *Int. J. of Comp. Vision*, 27(2):161–195, 1998.

Chapter 10

[1] E. Bayro–Corrochano, and J. Lasenby, 1998. Geometric techniques for the computation of projective invariants using n uncalibrated cameras. In *Proceedings of the Indian Conference on Computer Vision and Image Processing*, New delhi, India, 21–23 December, pp. 95–100.

[2] S. Carlsson, 1994. The Double Algebra: and effective tool for computing invariants in computer vision. *Applications of Invariance in Computer Vision*, Lecture Notes in Computer Science 825; Proceedings of 2nd-joint Europe-US workshop, Azores, October 1993. Eds. Mundy, Zisserman and Forsyth. Springer-Verlag.

[3] S. Carlsson, 1998. Symmetry in Perspective. In *Proceedings of the European Conference on Computer Vision*, Freiburg, Germany, pp. 249–263.

[4] G. Csurka and O. Faugeras 1998. Computing three dimensional project invariants from a pair of images using the Grassmann–Cayley algebra *Journal of Image and Vision Computing*, 16, pp. 3-12.

[5] R. I. Hartley 1994. Projective Reconstruction and Invariants from Multiple Images *IEEE Trans.PAMI*, Vol 16, No.10, 1036-1041.

[6] Hestenes, D. and Sobczyk, G. 1984. Clifford Algebra to Geometric Calculus: A unified language for mathematics and physics. *D. Reidel*, Dordrecht.

[7] D. Hestenes, and R. Ziegler, 1991. Projective Geometry with Clifford Algebra. *Acta Applicandae Mathematicae*, 23: 25–63.

[8] J. Lasenby, E. J. Bayro–Corrochano, A. Lasenby and G. Sommer 1996. A new methodology for computing invariants in computer vision. IEEE Proceedings of ICPR'96, Viena, Austria, Vol. I, pages 393-397, August, 1996.

[9] J. Laseby and E. Bayro–Corrochano 1999. Analysis and Computation of Projective Invariants from Multiple Views in the Geometric Algebra Framework. In Special Issue on Invariants for Pattern Recognition and Classification, ed. M.A. Rodrigues. *Int. Journal of Pattern Recognition and Artificial Intelligence*, Vol 13, No 8, December 1999, pp. 1105-1121.

[10] J. Mundy and A. Zisserman, (Eds.) 1992 Geometric Invariance in Computer Vision. MIT Press.

[11] C. J. Poelman and T. Kanade 1994. A paraperspective factorization method for shape and motion recovery. In J–O. Eklundh, editor, *European Conference on Computer Vision*, Stockholm, pp. 97–108.

[12] L. Quan 1994. Invariants of 6 points from 3 uncalibrated images. In *Proc. of the European Conference of Computer Vision*, Vol. II, pp. 459–470.

[13] A. Shashua, 1994. Projective structure from uncalibrated images: structure from motion and recognition PAMI, 16(8), 778:790, August.

[14] G. Sparr, 1994. Kinetic depth. In *Proc. of the European Conference of Computer Vision*, Vol. II, pp. 471–482.

[15] C. Tomasi and T. Kanade 1992. Shape and motion from image streams under orthography: a factorization method. *Int. J. Computer Vision*, 9(2), pp. 137–154.

[16] W. Triggs, 1995. Matching Constraints and the Joint Image. In proceedings of *ICCV'95*, MIT.

Chapter 11

[1] W. Blaschke. *Kinimatik und Quaternionen*. Deutscher Verlag der Wissenschaften, Berlin 1960.

[2] W.K. Clifford. Preliminary sketch of biquaternions. *Proceedings of the London Mathematical Society*, iv(64/65):381–395, 1873.

[3] J. Duffy. *Analysis of Mechanisms and Robot Manipulators*. Edward Arnold, London, 1980.

[4] C.G. Gibson. *Elementary geometry of algebraic curves*. Cambridge University Press, Cambridge, 1998.

[5] H.W. Guggenheimer. *Differential Geometry*. McGraw-Hill, New York, 1963.

[6] K.H. Hunt and I.A. Parkin. Finite displacements of points, planes, and lines via screw theory. *Mechanism and Machine Theory*, 30(2):177–192, 1995.

[7] D.L. Pieper. *The Kinematics of Manipulators under Computer Control*. PhD thesis, Mech. Eng., Stanford University, 1968.

[8] I.R. Porteous. *Topological Geometry*. Cambridge University Press, Cambridge, second edition, 1981.

[9] J.M. Selig. *Geometrical Methods in Robotics*. Springer Verlag, New York, 1996.

[10] J.M. Selig. Clifford algebra of points, lines and planes. *Robotica*, to appear.

[11] D.R. Smith. *Design of Solvable 6R Manipulators*. PhD thesis, Georgia Institute of Technology, 1990.

[12] E. Study. von den Bewegungen und Umlegungen. *Math. Ann.*, 39:441–566, 1891.

Chapter 12

[1] W. K. Clifford, *Preliminary Sketch of Biquaternions* (1873), In Mathematical Papers, edited by R. Tucker, Macmillan, London, 1882, pp. 658.

[2] P. DeCasteljau, *Outillages methodes calcul*, technical report, A. Citroen, Paris, 1959.

[3] P. DeCasteljau, *Courbes et surfaces a poles*, technical report, A. Citroen, Paris, 1963.

[4] K. R. Etzel, and J. M. McCarthy, Interpolation of Spatial Displacements Using the Clifford Algebra of E4, *ASME Journal of Mechanical Design*, Vol. 121, 1999.

[5] G. Farin, Curves and Surfaces for CAGD, Academic Press, Inc., Chestnut Hill, MA pp. 33-43, 122-125, 1997.

[6] Q. J. Ge, On the Matrix Algebra Realization of the Theory of Biquaternions, *Mechanism Synthesis and Analysis*, volume ASME Publication DE-Vol. 70, 425-432, 1994.

[7] Q. J. Ge, and Kang, D., Motion Interpolation With G^2 Composite Bezier Motions, *ASME Journal of Mechanical Design*, 117:520-525, 1995.

[8] Q. J. Ge, and Ravani, B., 1994 Geometric Construction of Bézier Motions, *ASME Journal of Mechanical Design*, 116:749-755, 1994.

[9] W. R. Hamilton, *Elements of Quaternions* (1860), Reprinted by Chelsea Pub., 1969, New York.

[10] C. Innocenti, Polynomial Solution of the Spatial Burmester Problem. DE-Vol. 70, *Mechanism synthesis and Analysis,* Proceedings of the ASME Design Engineering Technical Conferences, Minneapolis, MN, Sept. 1994.

[11] J. M. McCarthy, and Q. Liao, On the Seven Position Synthesis of a 5-SS Platform Linkage, ASME *Journal of Mechanical Design* (in press).

[12] J. M. McCarthy, *An Introduction to Theoretical Kinematics,* MIT Press, Cambridge, MA, 1990.

[13] K. Shoemake, Animating Rotation with Quaternion Curves, *ACM Siggraph,* 19(3):245-254, 1985.

[14] C. H. Suh, andC. W. Radcliffe, *Kinematics and Mechanism Design.* John Wiley and Sons, New York, 1978.

Chapter 13

[1] E. Bayro–Corrochano, 2000. Geometric algebras of 2D and 3D kinematics. To appear in Geometric Computing with Clifford Algebras, Gerald Sommer (ed.), chap. 18. Springer Verlag, Germany.

[2] W.E. Baylis, 1996. Clifford (Geometric) Algebras With Applications in Physics, Mathematics, and Engineering. Birkhauser, Boston.

[3] W. Fulton and J. Harris. 1991. Representation Theory: A First Course. Springer–Verlag, New York.

[4] C. Doran, D. Hestenes, F. Sommen and N. Van Acker. Lie groups and spin groups. J. Math. Phys. 34 (8), August 1993, pp. 3642–3669.

[5] D. Hestenes and G. Sobczyk. 1984. Clifford Algebra to Geometric Calculus: A Unified Language for Mathematics and Physics. *D. Reidel,* Dordrecht.

[6] D. Hestenes. 1991 The Design of Linear Algebra and Geometry. Acta Applicandae Mathematicae 23:65-93, Kluwer Academic Publishers.

[7] D. Hestenes and R. Ziegler. 1991. Projective geometry with Clifford algebra. *Acta Applicandae Mathematicae,* 23: 25–63.

[8] P. Lounesto. 1987. Pertti Lounesto. *CLICAL User Manual.* Helsinki University of Technology, Institute of Mathematics, Research Report A428.

[9] P. Lounesto. 1997. Clifford Algebras and Spinors. Cambridge University Press, Cambridge.

[10] J. Belinfante and B. Kolman, 1972. Lie Groups and Lie Algebras: With Applications and Computational Methods. *SIAM*, Philadelphia.

[11] W. Blaschke. Kinematik und Quaternionen. *VEB Deutscher Verlag der Wissenschaften*, Berlin, 1960.

[12] B.C. Hoffman. The Lie Algebra of Visual Perception. *Journal of Mathematical Psychology*, 3, 1966.

[13] H. Li, D. Hestenes, and A. Rockwood. 2000. Generalized Homogeneous Coordinates for Computational Geometry. In: G. Sommer (Ed.), *Geometric Computing with Clifford Algebra* (Springer-Verlag, Heidelberg, 2000), pp. 25–58.

[14] P. Lounesto. 1987. CLICAL software paquet and user manual. Helsinki University of Technology of Mathematics, Research Report A248.

[15] P. Lounesto and A. Springer, 1989. Mobius Transformations and Clifford Algebras of Euclidean and Anti-Euclidean Spaces J. Lawrynowicz (ed.), Deformations of Mathematical Structures, 79-90, Kluwer Academic Publishers.

[16] T. McMillian and N. White. The dotted straightening algorithm. *J. Symbolic Comput.*, 11:471-482, 1991.

[17] J.M. Selig. Clifford algebra of points, lines and planes. South Bank University, School of Computing, Information Technology and Maths, Technical Report SBU–CISM–99–06, 1999.

[18] W. Miller, 1968. Lie Theory and Special Functions. Academic Press, New York.

[19] M. Riesz, 1993. Clifford Numbers and Spinors. Lecture Series No. 38, University of Maryland, 1958. Reprinted as facsimile (eds. : E.F. Bolinder and P. Lounesto) by Kluwer, Dordrecht, The Netherlands.

[20] G. Sobczyk. 1997. The Generalized Spectral Decomposion of a Linear Operator, *The College Mathematics Journal*, 28:1, pp. 27-38.

Chapter 14

[1] *Physics world (special issue on quantum information)*, March, 1998.

[2] A. Barenco, C. H. Bennett, R. Cleve, D. P. DiVincenzo, N. Margolus, P. Shor, T. Sleator, J. A. Smolin and H. Weinfurter, Elementary gates for quantum computation, *Phys. Rev. A*, 52 (1995), pp. 3457–3467.

[3] P. Benioff, The computer as a physical system: A microscopic quantum mechanical model of computers as represented by turing machines, *J. Stat. Phys.*, 22 (1980), pp. 563–591.

[4] C. H. Bennett, E. Bernstein, G. Brassard and U. Vazirani, Strengths and weaknesses of quantum computing, SIAM *J. Comput.*, 26 (1997), pp. 1510–1523.

[5] C. H. Bennett and P. W. Shor, Quantum information theory, *IEEE Trans. Info. Th.*, 44 (1998), pp. 2724–2742.

[6] K. Blum, *Density Matrix Theory and Applications*, Plenum Pub. Corp. (2nd ed.), 1996.

[7] B. Blümich and H. W. Spiess, Quaternions as a practical tool for the evaluation of composite rotations, *J. Maget. Reson.*, 61 (1985), pp. 356–362.

[8] B. Boulat and M. Rance, Algebraic formulation of the product operator formalism in the numerical simulation of the dynamic behavior of multispin systems, *Mol. Phys.*, 83 (1994), pp. 1021–1039.

[9] H. E. Brandt, Qubit devices and the issue of decoherence, *Prog. Quantum Electronics*, 22 (1998), pp. 257–370.

[10] S. L. Braunstein, C. M. Caves, R. Jozsa, N. Linden, S. Popescu and R. Schack, Separability of very noisy mixed states and implications for NMR quantum computing, *Phys. Rev. Lett.*, 83 (1999), pp. 1054–1057.

[11] M. Brooks, ed., *Quantum Computing and Communications*, Springer Verlag, London, U.K., 1999.

[12] I. L. Chuang, N. Gershenfeld, M. G. Kubinec and D. W. Leung, Bulk quantum computation with nuclear magnetic resonance: Theory and experiment, *Proc. R. Soc. Lond. A*, 454 (1998), pp. 447–467.

[13] D. G. Cory, A. F. Fahmy and T. F. Havel, Ensemble quantum computing by nuclear magnetic resonance spectroscopy, *Proc. Natl. Acad. Sci.*, 94 (1997), pp. 1634–1639.

[14] D. G. Cory, W. Maas, M. Price, E. Knill, R. Laflamme, W. H. Zurek, T. F. Havel and S. S. Somaroo, Experimental quantum error correction, *Phys. Rev. Lett.*, 81 (1998), pp. 2152–2155.

[15] D. G. Cory, M. D. Price and T. F. Havel, Nuclear magnetic resonance spectroscopy: An experimentally accessible paradigm for quantum computing, *Physica D*, 120 (1998), pp. 82–101.

[16] C. Counsell, M. H. Levitt and R. R. Ernst, Analytical theory of composite pulses, *J. Magn. Reson.*, 63 (1985), pp. 133–141.

[17] D. Deutsch, Quantum theory, the Church-Turing principle and the universal quantum computer, *Proc. R. Soc. Lond. A*, 400 (1985), pp. 97–117.

[18] D. Deutsch and R. Jozsa, Rapid solution of problems by quantum computation, *Proc. R. Soc. Lond. A*, 439 (1992), pp. 553–558.

[19] D. Divincenzo, Quantum computation, *Science*, 270 (1995), pp. 255–261.

[20] C. J. L. Doran, A. N. Lasenby and S. F. Gull, States and operators in the spacetime algebra, *Found. Phys.*, 23 (1993), pp. 1239–1264.

[21] C. J. L. Doran, A. N. Lasenby, S. F. Gull, S. S. Somaroo and A. D. Challinor, Spacetime algebra and electron physic, in *Advances in Imaging and Electron Physics*, P. Hawkes, ed., Academic Press, Englewood Cliffs, NJ, 1996, pp. 271–386.

[22] A. Einstein, B. Podolsky and N. Rosen, Can quantum-mechanical description of physical reality be considered complete?, *Phys. Rev.*, 47 (1935), pp. 777–780.

[23] A. Ekert and R. Jozsa, Quantum computation and Shor's factorizing algorithm, *Rev. Mod. Phys.*, 68 (1996), pp. 733–753.

[24] R. R. Ernst, G. Bodenhausen and A. Wokaun, *Principles of Nuclear Magnetic Resonance in One and Two Dimensions*, Oxford Univ. Press, U.K., 1987.

[25] R. P. Feynman, Simulating physics with computers, *Int. J. Theor. Phys.*, 21 (1982), pp. 467–488.

[26] R. Freeman, *Spin Choreography*, Oxford University Press, 1998.

[27] M. R. Garey and D. S. Johnson, *Computers and Intractability: A Guide to the Theory of NP-Completeness*, W. H. Freeman, San Francisco, 1979.

[28] N. A. Gershenfeld and I. L. Chuang, Bulk spin-resonance quantum computation, *Science*, 275 (1997), pp. 350–356.

[29] D. Giulini, E. Joos, C. Kiefer, J. Kupsch, I. Stamatescu and H. D. Zeh, *Decoherence and the Appearance of a Classical World in Quantum Theory*, Springer-Verlag, 1996.

[30] L. K. Grover, Quantum mechanics helps in searching for a needle in a haystack, *Phys. Rev. Lett.*, 79 (1997), pp. 325–328.

[31] U. Haeberlen and J. S. Waugh, Coherent averaging effects in magnetic resonance, *Phys. Rev.*, 175 (1968), pp. 453–467.

[32] T. F. Havel, S. S. Somaroo, C. Tseng and D. G. Cory, *Principles and demonstrations of quantum information processing by NMR spectroscopy*, 1999. In press (see `quant/ph-9812086`).

[33] D. Hestenes, *Space-Time Algebra*, Gordon and Breach, New York, NY, 1966.

[34] D. Hestenes, *New Foundations for Classical Mechanics*, D. Reidel Pub. Co., Dordrecht, NL, 1986.

[35] J. A. Jones and E. Knill, Efficient refocussing of one spin and two spin interactions for nmr quantum computation, *J. Magn. Reson.*, 141 (1999), pp. 322–325.

[36] E. Knill, I. Chuang and R. Laflamme, Effective pure states for bulk quantum computation, *Phys. Rev. A*, 57 (1998), pp. 3348–3363.

[37] S. Lloyd, Quantum-mechanical computers, *Sci. Am.*, 273 (Oct. 1995), pp. 140–145.

[38] ——, Universal quantum simulator, *Science*, 273 (1996), pp. 1073–1078.

[39] A. Peres, *Quantum Theory: Concepts and Methods*, Kluwer Academic Pub., Amsterdam, NL, 1993.

[40] J. Preskill, Quantum computing: Pro and con, *Proc. R. Soc. Lond. A*, 454 (1998), pp. 469–486.

[41] M. D. Price, S. S. Somaroo, A. E. Dunlop, T. F. Havel and D. G. Cory, Generalized (controlled) ^{n}-NOT quantum logic gates, *Phys. Rev. A*, 60 (1999), pp. 2777–2780.

[42] M. D. Price, S. S. Somaroo, C. Tseng, J. C. Gore, A. F. Fahmy, T. F. Havel and D. G. Cory, Construction and implementation of NMR quantum logic gates for two-spin systems, *J. Magn. Reson.*, 140 (1999), pp. 371–378.

[43] R. Schack and C. M. Caves, Classical Model for Bulk-Ensemble NMR Quantum Computation, *Phys. Rev. A*, 60 (1999), pp. 4354–4362.

[44] P. W. Shor, Scheme for reducing decoherence in quantum computer memory, *Phys. Rev. A*, 52 (1995), pp. R2493–R2496.

[45] ——, Polynomial-time algorithms for prime factorization and discrete logarithms on a quantum computer, SIAM *J. Comput.*, 26 (1997), pp. 1484–1509.

[46] D. J. Siminovitch, Rotations in NMR: Part I. Euler – Rodrigues parameters and quaternions, *Concepts Magn. Reson.*, 9 (1997), pp. 149–171.

[47] ——, Rotations in NMR: Part II. Applications of the Euler – Rodrigues parameters, *Concepts Magn. Reson.*, 9 (1997), pp. 211–225.

[48] S. Somaroo, A. Lasenby and C. Doran, Geometric algebra and the causal approach to multiparticle quantum mechanics, *J. Math. Phys.*, 40 (1999), pp. 3327–3340.

[49] S. S. Somaroo, D. G. Cory and T. F. Havel, Expressing the operations of quantum computing in multiparticle geometric algebra, *Phys. Lett. A*, 240 (1998), pp. 1–7.

[50] S. S. Somaroo, C. Tseng, T. F. Havel, R. Laflamme and D. G. Cory, Quantum simulations on a quantum computer, *Phys. Rev. Lett.*, 82 (1999), pp. 5381–5384.

[51] O. W. Sörensen, G. W. Eich, M. H. Levitt, G. Bodenhausen and R. R. Ernst, Product operator formalism for the description of NMR pulse experiments, *Prog. NMR Spect.*, 16 (1983), pp. 163–192.

[52] A. M. Steane, Error correcting codes in quantum theory, *Phys. Rev. Lett.*, 77 (1996), pp. 793–797.

[53] ——, Quantum computing, *Rep. Prog. Theor. Phys.*, 61 (1998), pp. 117–173.

[54] M. E. Stoff, A. J. Vega and R. W. Vaughan, Explicit demonstration of spinor character for a spin 1/2 nucleus via NMR interferometry, *Phys. Rev. A*, 16 (1977), pp. 1521–1524.

[55] T. Toffoli, *Reversible computing*, in Automata, Languages and Programming, J. W. de Bakker and J. van Leeuwen, eds., Springer-Verlag, 1980, pp. 632–644.

[56] C. H. Tseng, S. Somaroo, Y. Sharf, E. Knill, R. Laflamme, T. F. Havel and D. G. Cory, Quantum simulation of a three-body interaction Hamiltonian on an NMR quantum computer, *Phys. Rev. A*, 61 (2000), article no. 012302.

[57] F. J. M. van de Ven and C. W. Hilbers, A simple formalism for the description of multiple-pulse experiments. Application to a weakly coupled two-spin ($I = 1/2$) system, *J. Magn. Reson.*, 54 (1983), pp. 512–520.

[58] L. M. K. Vandersypen, C. S. Yannoni, M. H. Sherwood and I. L. Chuang, Realization of logically labeled effective pure states for bulk quantum computation, *Phys. Rev. Lett.*, 83 (1999), pp. 3085–3088.

[59] W. S. Warren, The usefulness of NMR quantum computing, *Science*, 277 (1997), pp. 1688–1689 See also response by N. Gershenfeld and I. Chuang, pp. 1689–1690.

[60] C. P. Williams and S. H. Clearwater, *Explorations in Quantum Computing*, Springer-Verlag, New York, NY, 1998.

[61] W. Zhang and D. G. Cory, First direct measurement of the spin diffusion rate in a homogenous solid, *Phys. Rev. Lett.*, 80 (1998), pp. 1324–1327.

[62] W. H. Zurek, Decoherence, einselection and the existential interpretation (the rough guide), *Phil. Trans. R. Soc. Lond. A*, (1998), pp. 1793–1821.

Chapter 15

[1] P. Arena, R. Caponetto, L. Fortuna, G. Muscato and M.G. Xibilia 1996. Quaternionic multilayer perceptrons for chaotic time series prediction. IEICE Trans. Fundamentals. Vol. E79-A. No. 10 October, pp. 1-6.

[2] E. Bayro–Corrochano. 1996. Clifford selforganizing Neural Network, Clifford Wavelet Network. *Proc. 14th IASTED Int. Conf. Applied Informatics*, Feb. 20-22, Innsbruck, Austria, p. 271-274.

[3] E. Bayro–Corrochano E., G. Sommer and K. Daniilidis. 2000. Motor algebra for 3D kinematics. The case of the hand–eye calibration To appear in *Int. Journal of Mathematical Imaging and Vision*.

[4] E. Bayro–Corrochano, S. Buchholz and G. Sommer. 1996. Selforganizing Clifford neural network *IEEE ICNN'96 Washington, DC*, June, pp. 120-125.

[5] G. Cybenko, 1989. Approximation by superposition of a sigmoidal function. *Mathematics of control, signals and systems*, Vol. 2, 303:314.

[6] C.J.L. Doran. 1994. Geometric algebra and its applications to mathematical physics. *Ph.D. Thesis*, University of Cambridge.

[7] D. Hestenes. 1966. Space-Time Algebra. *Gordon and Breach.*

[8] D. Hestenes. 1993. Invariant body kinematics I: Saccadic and compensatory eye movements. Neural Networks, Vol. 7, 65-77.

[9] D. Hestenes. 1993. Invariant body kinematics II: Reaching and neurogeometry. Neural Networks, Vol. 7, 79-88.

[10] K. Hornik. 1989. Multilayer feedforward networks are universal approximators. Neural Networks, Vol. 2, pp. 359-366.

[11] D. Hestenes and G. Sobczyk, 1984. Clifford Algebra to Geometric Calculus: A Unified Language for Mathematics and Physics.

[12] G. M. Georgiou and C. Koutsougeras, 1992. Complex domain backpropagation. *IEEE Trans. on Circuits and Systems*, 330:334.

[13] J.J. Koenderink. 1990. The brain a geometry engine. Psychological Research, Vol. 52, pp. 122-127.

[14] J.K. Pearson and D.L. Bisset. 1992. Back Propagation in a Clifford Algebra. *Artificial Neural Networks, 2, I. Aleksander and J. Taylor (Ed.)*, 413:416.

[15] A. Pellionisz and R. Llinàs. 1980. Tensorial approach to the geometry of brain function: cerebellar coordination via a metric tensor. Neuroscience Vol. 5, pp. 1125-1136

[16] A. Pellionisz and R. Llinàs. 1985. Tensor network theory of the metaorganization of functional geometries in th central nervous system. Neuroscience Vol. 16, No. 2, pp. 245-273.

[17] S.J. Perantonis and P.J.G. Lisboa. 1992. Translation, rotation, and scale invariant pattern recognition by high-order neural networks and moment classifiers. IEEE Trans. on Neural Networks, Vol. 3, No. 2, March, pp 241-251.

[18] T. Poggio and F. Girosi, 1990. Networks for approximation and learning. *IEEE Proc.*, Vol. 78, No. 9, 1481:1497, Sept.

[19] I.R. Porteous. 1995. Clifford Algebras and the Classical Groups. *Cambridge University Press*, Cambridge.

[20] D.E. Rumelhart and J.L. McClelland. 1986. Parallel Distributed Processing: Explorations in the Microstructure of Cognition. 2 Vols. Cambridge: MIT Press.

[21] V. Vapnik, 1998. Statistical Learning Theory. Wiley, New York.

[22] W.R. Hamilton. 1853. Lectures on Quaternions. Hodges and Smith, Dublin.

Chapter 16

[1] Mitrea Marius Clifford Wavelets, *Singular Integrals and Hardy Spaces.* Lecture Notes in Mathematics 1575 Springer Verlag 1994.

[2] Chui Charles, *An Introduction to wavelets*, Academic press, 1992.

[3] L. Traversoni, Quaternion Wavelet Problems. *Proceedings of the VIII Texas International Symposium on Approximation Theory.*

[4] L. Traversoni and A. Gonzalez, Object compression. Extended abstract for the conference "Complex annalysis and related topics" organized by *CINVESTAV I.P.N.Cuernavaca, México, November 1996.*

[5] K. Guerlebeck and W. Sprössig, *Quaternion and Clifford Calculus for Phycisists and Engineers.* J Wiley 1998.

[6] Sukumar Narrasimham, *PhD Thesis, Northwestern University*, 1998.

[7] L. Traversoni, An algorithm for natural spline interpolation. *Numerical Algorithms V5 num 1-4 November 1993.* J.C. Baltzer.

[8] R. Delanghe, F. Brackx , F. Sommen. Clifford Analysis. *Pitman advanced publishing program.* 1982

[9] L. Schumaker, *Reconstructing 3D objects from cross-sections. Computation of curves and surfaces.* Kugwer 1990.

[10] E. Bayro–Corrochano, K. Daniilidis and G. Sommer, [1997]. Hand-eye calibration in terms of motion of lines using geometric algebra. *In the 10th Scandinavian Conference on Image Analysis.* June 9-11, Lapperanta, Finland, pp. 397-404.

[11] B. Jüttler, M. G. Wagner, Computer Aided Design with spatial rational B-Spline motions. *Journal of Mexhanical Design,* June 1966, vol 118.

Chapter 17

[1] L. Dorst and R. van den Boomgaard, The support cone: a representational tool for the analysis of boundaries and their interactions. *IEEE-PAMI,* vol.22, no.2, 2000.

[2] A.W.M. Dress and T.F. Havel, *Distance geometry and geometric algebra,* Found. Phys., 23:1357-1374, 1993.

[3] D. Hestenes and G. Sobczyk, *Clifford algebra to geometric calculus,* D. Reidel, Dordrecht, 1984.

[4] H. Li, D. Hestenes and A. Rockwood, Generalized homogeneous coordinates for computational geometry in: *"Geometric Computing with Clifford Algebra",* ed. G. Sommer, Springer Series in Information Science, to be published 2000.

[5] V.S. Nalwa, Representing oriented piecewise C^2 surfaces. *Int. J. Computer Vision,* vol.3, pp. 131–153, 1989.

[6] B. Widrow and S. D. Stearns, *Adaptive Signal Processing* . Prentice Hall, 1985.

Chapter 18

[1] P. Engel, *Geometric Crystallography: an axiomatic introduction to crystallography,* Kluwer Academic Publishers, Dordretch, 1986.

[2] M. Kléman, Curved Crystals, *Adv. Phys.* 38, 605, 1989.

[3] D. Shechtman, I. Blech, D. Gratias and J.W. Cahn, Metallic Phase with Long-Range Orientational Order and No Translational Symmetry, *Phys. Rev. Lett.* 53, 1951, 1984.

[4] A. L. Mackay, The Generalization of Orthodox Crystallography, *Mater. Sci. Forum* 150-151, 1, 1994.

[5] H. Brown, R. Bülow, J. Neubüser, H. Wondratschek and H. Zassenhaus, *Crystallographic Groups of 4-Dimensional Space*, Wiley, New York, 1978.

[6] C. Janot, *Quasicrystals. A Primer.*, Clarendon Press, Oxford, 1992.

[7] W. Steurer, The Structure of Quasicrystals, *Z. Krist.* 190, 179, 1990.

[8] L. X. He, X. Z. Li, Z. Zhang and K. H. Kuo, one-Dimensional Quasicrystal in Rapidly Solidified Alloys, *Phys. Rev. Lett.* 61, 1116, 1988.

[9] R. Penrose, The Role of Aesthetics in Pure and Applied Mathematical Research, *Bull. Inst. Math. Appl.* 10, 266, 1974.

[10] A. L. Mackay, Crystallography and the Penrose Pattern, *Physica A* 114, 609, 1982.

[11] P. A. Kalugin, A. Y. Kitaev and L. S. Levitov, $Al_{0.86}Mn_{0.14}$: a six-Dimensional Crystal, *JETP Lett.* 41, 145, 1985.

[12] A. Katz and M. Duneau, Quasiperiodic Patterns and Icosahedral Symmetry, *J. Physique* 47, 181, 1986.

[13] D. Martinais, Classification des Groupes Cristallographiques de Type Icosaedrique en 6D, *C.R. Acad. Sci. Paris* 304, 789, 1987.

[14] J.H. Conway and N.A. Sloane, 1988, *Sphere Packings, Lattices and Groups*, Springer-Verlag, New York, 1988.

[15] A. P. Tsai, A. Inoue and T. Masumoto, A Stable Quasicrystal in Al-CuFe System, *Jap. J. Appl. Phys.* 26, L1505, 1987.

[16] A. P. Tsai, A. Inoue and T. Masumoto, New Stable Icosahedral Al-CuRu and AlCuOs Alloys, *Jap. J. Appl. Phys.* 27, L1587, 1988.

[17] A. P. Tsai, A. Inoue and T. Masumoto, Stable Icosahedral AlPdMn and AlPdRe Alloys, *Mat. Trans. JIM* 31, 98, 1990.

[18] B.K. Vainshtein, *Modern Crystallography I. Symmetry of Crystals. Methods of structural crystallography*, Springer Series in Solid-State Sciences, Springer-Verlag, New York, 1981.

[19] B. Dubost, J-M. Lang, M. Tanaka, P. Sainfort and M. Audier, Large AlCuLi Single Quasicrystal with Triacontahedral Solidification Morphology, *Nature* 324, 48, 1986.

[20] W. Ohashi, F. Spaepen, Stable Ga-Mn-Zn, Quasiperiodic Crystals with Pentagonal Dodecahedral Solidification Morphology, *Nature* 330, 555, 1987.

[21] S. Ranganathan, On the Geometry of Coincidence-Site Lattices, *Acta Cryst.* 21, 197, 1966.

[22] M. A. Fortes, N-Dimensional Coincidence-Site-Lattice Theory, *Acta Cryst. A* 21, 351, 1983.

[23] M. Duneau, C. Oguey and A. Thalal, Coincidence Lattices and Associated Shear Transformations, *Acta Cryst. A* 48, 772, 1992.

[24] M. Baake, Solutions of the Coincidence Problem in Dimensions $d \leq 4$, in *The Mathematics of Aperiodic Order*, ed. R.V. Moody, Kluwer Academic Publishers, Dordretch, 1997.

[25] J. L. Aragón, D. Romeu, L. Beltrán and A. Gómez, Grain Boundaries as Projections from Higher-Dimensional Lattices, *Acta Cryst. A* 53, 772, 1997.

[26] I. Porteous, *Clifford Algebras and the Classical Groups*, Cambridge University Press, Cambridge, 1995.

Chapter 19

[1] S.L. Altman. *Rotations, Quaternions, and Double Groups.* Clarendon Press, Oxford, 1986.

568   References

[2] L.I. Aramanovitch. Application of quaternions to relative orientation of a stereo-pair. *Izvestia Vuzov, Geodesy and Aerophotography*, 4:99–109,1990.

[3] L.I. Aramanovitch. A method of optimal filtering for quaternion systems. *Izvestia Vuzov, Geodesy and Aerophotography* , 5:101–108,1991.

[4] L.I. Aramanovitch. Solution of the problem of photo orientation in aerial surveying by means of quaternion algebra. *Manuscripta Geodaetica*, 17:334–341, 1992.

[5] L.I. Aramanovitch. Quaternion non-linear filter for estimation of rotating body attitude. *Mathematical Methods in the Applied Sciences*, 18:1239–1255, 1995.

[6] L.I. Aramanovitch. Spacecraft orientation based on space object observations by means of quaternion algebra. *Journal of Guidance, Control and Dynamics*, 18(4):859–866, 1995.

[7] V.N. Branetz and I.P. Shmiglevsky. *Introduction to the Theory of Strap-down Inertial Navigation*. Nauka, Moscow, 1992.

[8] A.E. Bryson and Y.C. Ho. *Applied Optimal Control*, Hemisphere, New York, 1975.

[9] J. Dambeck. *Diagnose und Therapie geodätischer Trägheitsnavigationssysteme* . Dissertation, Universität Stuttgart, 1998.

[10] G.H. Golub and U. von Matt. Quadratically constrained least squares and quadratic problems. *Numerishce Mathematik*, 59:561–580, 1991.

[11] E.W. Grafarend. *Reference Frame Rotation - Regularized Theory by Quaternions and Spinors*. Geodesy in Transition, a volume dedicated to H.Moritz on the occasion of his 50th birthday,ed.by K.P.Schwarz and G. Lachapelle, 60002, The University of Calgary, 1983.

[12] D. Hestenes. *New Foundations for Classical Mechanics* . D.Reidel, Dordrecht, 1986.

[13] D. Hestenes and G. Sobczyk. *Clifford Algebra to Geometric Calculus* . D.Reidel, Dordrecht, 1984.

[14] F.E. Hohn. *Elementary Matrix Algebra*. MacMillan Comp., New York, 1973.

[15] R.E. Kalman. A new approach to linear filtering and prediction problems. *Trans.ASME*, 82 D, 1960.

[16] J.B. Kuipers. *Quaternions and Rotations Sequences*. Princeton Univ.Press., Princeton, 1998.

[17] J. Lasenby. *Geometric Algebra: Application in Engineering*, In W.Baylis, ed. *Clifford (Geometric) Algebras*, Birkhäuser, 1996.

[18] L.A. Ljusternik and V.I. Sobolev. *Elements of Functional Analysis*. Nauka, Moscow, 1965.

[19] L. Meister. *Quaternions and Their Application in Photogrammetry and Navigation*. Habilitationsschrift, TU Bergakademie, Freiberg, 1998.

[20] F. Sanso. An exact solution of the roto-translation problem. *Photogrammetria*, 29:203–216, 1973.

[21] J.M. Selig. *Geometrical Methods in Robotics*. Springer, 1996.

[22] M.D. Shuster. A survey of attitude representation. *The Journal of the Astronautical Sciences*, 41(4):439–517, 1993.

[23] V.N. Tutubalin. Correlative analysis of random quaternions. *Vestnik MGU, Ser.I Mathematics.Mechanics.*, 2:15–23, 1992.

[24] M.S. Urmaev. Applying of quaternion algebra in photogrammetry. *Izv. Vuzov, Geodesy and Aerophotography*, 2:81–90,1986.

Chapter 20

[1] S. L. Altmann. *Rotations, Quaternions, and Double Groups*, Oxford University Press, New York, 1986.

[2] A. O. Barut. *Electrodynamics and Classical Theory of Fields and Particles*, Dover, New York, 1980.

[3] W. E. Baylis, editor. *Clifford (Geometric) Algebra with Applications to Physics, Mathematics, and Engineering*, Boston, 1996. Birkhäuser.

[4] W. E. Baylis. "Special relativity with 2×2 matrices." *Am. J. Phys.* **48**, 918–925 (1980).

[5] W. E. Baylis. *Electrodynamics: A Modern Geometric Approach.* Birkhäuser, Boston, MA, 1999.

[6] W. E. Baylis and Y. Yao. "Relativistic dynamics of charges in electromagnetic fields: An eigenspinor approach." *Phys. Rev. A* **60**, 785–795 (1999).

[7] D. J. Griffiths and M. A. Heald. "Time-dependent generalizations of the Coulomb-Faraday and Biot-Savart laws." *Am. J. Phys.* **59**, 111–117 (1991) and **60**, 393–394 (1992).

[8] D. Hestenes. *Spacetime Algebra*, Gordon and Breach, New York, 1966.

[9] D. Hestenes. Proper dynamics of a rigid point particle. *J. Math. Phys.* **15**, 1778–1786 (1974).

[10] P. Lounesto. *Clifford Algebras and Spinors.* Cambridge University Press, Cambridge, 1997.

[11] J. G. Maks. PhD thesis, Technische Universiteit Delft, the Netherlands, 1989.

[12] C. W. Misner, K. S. Thorne, and J. A. Wheeler. *Gravitation.* W. H. Freeman and Co., San Francisco, 1970.

[13] R. Penrose and W. Rindler. *Spinors and Spacetime*, Vol. I. Cambridge University: Cambridge, 1984.

Chapter 21

[1] E. Bayro–Corrochano, G. Sommer and K. Daniilidis. 2000. Motor algebra for 3D kinematics. The case of the hand–eye calibration. To appear in *Int. Journal of Mathematical Imaging and Vision.*

[2] D. Hestenes. 2000. Old wine in new bottles: a new algebraic framework for computational geometry. In *This volume*. Birkhauser.

[3] L. Dorst. 2000. Objects in contact: boundary collisions as geometric wave propagation. In *This volume*. Birkhauser.

[4] K.R. Etzel and M. McCarthy. 1999. Interpolation of spatial displacements using the Clifford algebra of E^4. *J.Mech.Design.* pp.121–39.

[5] G.F.R. Ellis and M. Bruni.1989. *Phys.Rev.D.* 40, p.1804.

[6] J. Lasenby and E.D. Bayro–Corrochano. 1999. Analysis and computation of projective invariants from multiple views in the geometric algebra framework. *International Journal of Pattern Recognition and Artificial Intelligence*, 13(8) p.1105. (Special Issue on Invariants for Pattern Recognition and Classification).

[7] L. Dorst. 2000. Honing geometric algebra for its use in the computer sciences. In G. Sommer, editor, *Geometric Computing with Clifford Algebras*. Springer. (In Press).

[8] D. Hestenes. 1966. *Space-Time Algebra*. Gordon and Breach, New York.

[9] S.F. Gull, A.N. Lasenby, and C.J.L. Doran. 1993. Imaginary numbers are not real — the geometric algebra of spacetime. *Found. Phys.*, 23(9) p.1175.

[10] A.N. Lasenby, C.J.L. Doran, and S.F. Gull. 1993. A multivector derivative approach to Lagrangian field theory. *Found. Phys.*, 23(10) p.1295.

[11] S.F. Gull, A.N. Lasenby, and C.J.L. Doran. 1993. Electron paths, tunnelling and diffraction in the spacetime algebra. *Found. Phys.*, 23(10) p.1329.

[12] C.J.L. Doran, A.N. Lasenby, S.F. Gull, and J. Lasenby. 1996. Lectures in geometric algebra. In W.E. Baylis, editor, *Clifford (Geometric) Algebras*, pp. 65–236. Birkhauser, Boston.

[13] A.D. Challinor, A.N. Lasenby, S.S. Somaroo, C.J.L. Doran, and S.F. Gull. 1997. Tunnelling times of electrons. *Phys. Lett. A*, pp. 227–143.

[14] A.M. Lewis, A.N. Lasenby, and C.J.L. Doran. 2000. Electron scattering in the spacetime algebra. In R. Ablamowicz and B. Fauser, editors, *Proceedings of the 5th International Conference on Applications of Clifford Algebras, Mexico, 1999.* (In press.).

[15] C.J.L Doran, A.N. Lasenby, S.F. Gull, S.S. Somaroo, and A.D. Challinor. 1996. Spacetime algebra and electron physics. *Adv. Imag. & Elect. Phys.*, pp. 95–271.

[16] A.N. Lasenby, C.J.L. Doran, and S.F. Gull. 1998. Gravity, gauge theories and geometric algebra. *Phil. Trans. R. Soc. Lond. A*, 356, pp.487–582.

[17] A.D. Challinor and A.N. Lasenby. 1999. Cosmic microwave background anisotropies in the cold dark matter model: a covariant and gauge-invariant approach. *Astrophys.J.*, 513, p.1.

[18] C.J.L. Doran, A.N. Lasenby, and S.F. Gull. 1996. Physics of rotating cylindrical strings. *Phys. Rev. D*, 54(10), p.6021.

[19] A.N. Lasenby, C.J.L. Doran, Y. Dabrowski, and A.D. Challinor. 1997. Rotating astrophysical systems and a gauge theory approach to gravity. In N. Sánchez and A. Zichichi, editors, *Current Topics in Astrofundamental Physics, Erice 1996*, p.380. World Scientific, Singapore.

[20] Y. Dabrowski, A. Fabian, K. Iwasawa, A. Lasenby, and C. Reynolds. 1997. The profile and equivalent width of the x-ray iron emission line from a disk around a kerr black hole. *Mon. Not. R. Astron. Soc.*, 288, L11.

[21] C. Doran. 2000. New form of the Kerr solution. To appear in *Phys. Rev. D*, D 61(6).

[22] J. Clements. 1999, Beam buckling using geometric algebra. M.Eng. project report, Cambridge University Engineering Department.

[23] R. Penrose and W. Rindler. 1986. *Spinors and space-time, Volume II: spinor and twistor methods in space-time geometry.* Cambridge University Press.

[24] A.N. Lasenby, C.J.L. Doran, and S.F. Gull. 1993. 2-spinors, twistors and supersymmetry in the spacetime algebra. In Z. Oziewicz, B. Jancewicz, and A. Borowiec, editors, *Spinors, Twistors, Clifford Algebras and Quantum Deformations*, p.233. Kluwer Academic, Dordrecht.

[25] D. Hestenes and G. Sobczyk. 1984. *Clifford Algebra to Geometric Calculus*. Reidel, Dordrecht.

Chapter 22

[1] E. Artin. *Geometric Algebra*. Interscience Publ, Inc., New York, London, 1957.

[2] E. Bayro–Corrochano and G. Sommer. Object Modelling and Collision Avoidance Using Clifford Algebra. In: V. Hlavac, R. Sara (Eds.). *Computer Analysis of Image and Pattern*, Springer Verlag (Lecture Note Computer Science), 970: 699-704, 1995.

[3] E. Bayro–Corrochano and S. Buchholz. Geometric Neural Networks. In: G. Sommer and J.J. Koenderink (Eds.). *Algebraic Frames for Perception-Action Cycle*, Springer Verlag (Lecture Note Computer Science), 1395: 379-394, 1997.

[4] R.E. Blahut. *Fast Algorithms for Digital Signal Processing*. Addison-Wesley, 1985.

[5] T. Buelow and G. Sommer. Multi-Dimensional Signal Processing Using an Algebraically Extended Signal Representation. In: G. Sommer and J.J. Koenderink (Eds.). *Algebraic Frames for Perception-Action Cycle*, Springer Verlag (Lecture Note Computer Science), 1395: 148-163, 1997.

[6] V.M. Chernov. Fast Algorithms of Discrete Orthogonal Transforms for Data Representation in Cyclotomic Fields. *Pattern Recognition and Image Analysis*, 3(4): 455-458, 1993.

[7] V.M. Chernov. Arithmetic Methods in the Theory of Discrete Orthogonal Transforms. *Proceedings SPIE*, 2363: 134-141, 1994.

[8] V.M. Chernov. Discrete Orthogonal Transforms with Data Representation in Composition Algebras. In: *Proceedings of The 9th Scandinavian Conference on Image Analysis*. Uppsala, Sweden, 1: 357-364, 1995.

[9] V.M. Chernov. On the Group Algebras' Hierarchy Pertaining to the Parametrization of Fast Algorithms of Discrete Orthogonal Transforms. In: V. Hlavac, R. Sara (Eds.). *Computer Analysis of Image and Pattern*, Springer Verlag (Lecture Note Computer Science), 970: 655-660.

[10] V.M. Chernov. Parametrization of Some Classes of Fast Algorithms for Discrete Orthogonal Transforms (1). *Pattern Recognition and Image Analysis*, 2(5): 238-245, 1995.

[11] V.M. Chernov. A Metric Unified Treatment of Two-Dimensional FFT. In: *13th Int. Conf. on Pattern Recognition*, 2: 662-669, Vienna, ICPR'96, 1996.

[12] V.M. Chernov, T. Buelow and M. Felsberg. Synthesis of Fast Algorithms for Discrete Fourier-Clifford Transform. *Pattern Recognition and Image Analysis*, 8(2): 274-275, 1998.

[13] M.A. Chichyeva and M.V. Pershina. On Various Schemes of 2D-DFT Decomposition with Data Representation in the Quaternion Algebra. *Image Processing and Communications*, 2(1): 13-20, 1996.

[14] W.K. Clifford. Applications of Grassmann's Extensive Algebra. *Amer. J. Math.* 1: 350-358, 1878.

[15] K. Daniilidis and E. Bayro–Corrochano. The Dual Quaternion Approach to Hand-Eye Calibration. In: *13th Int. Conf. on Pattern Recognition*, 1: 318-322, Vienna, ICPR'96, 1996.

[16] C.J.L. Doran. *Geometric Algebra and its Application to Mathematical Physics*. PhD thesis, University of Cambridge, 1994.

[17] L. Duhamel and H. Hollman. Split-radix FFT Algorithm. *Electronics Letters*. 20 (1): 14-16, 1984.

[18] D.F. Elliot and K.R. Rao. *Fast Transforms*. Academic, New York, 1982.

[19] M. Hall. *The Theory of Groups*. The Mcmillan Company. New York, 1959.

[20] D. Hestenes. *Space-Time Algebra*. Gordon and Breach, New York, 1966.

[21] D. Hestenes and G. Sobczyk. *Clifford Algebra to Geometric Calculus*. D.Redel Publ. Comp., Dordrecht, Boston, Lancaster, Tokyo, 1984.

[22] D. Hestenes. *New Foundation for Classical Mechanics*. D.Redel Publ. Comp., Dordrecht, Boston, Lancaster, Tokyo, 1986.

[23] J. Lasenby, E. Bayro–Corrochano, A.N.Lasenby and G. Sommer. A New Methodology for Computing Invariants in Computer Vision. In: *13th Int. Conf. on Pattern Recognition*, 1: 393-397, Vienna, ICPR'96, 1996.

[24] H.J. Nussbaumer. *Fast Fourier Transform and Convolution Algorithms*. Springer-Verlag, Berlin, 1982.

[25] R. Plymen and P. Robinson. *Spinors in Hilbert Space*. Cambridge Univ. Press, Cambridge, 1994.

[26] I.R. Porteous. *Clifford Algebras and the Classical Groups.* Cambridge University Press, Cambridge, 1995.

[27] S.J. Sangwine. Fourier Transforms of Color Images Using Quaternion, or Hypercomplex, Numbers. *Electronics Letters,* 32(21): 1979-1980, 1996.

[28] A. Schoenhage and V. Strassen. Schnelle Multiplikation grossen Zahlen. *Computing,* 7(3/4): 281-292, 1966.

Chapter 23

[1] H. Baum: *Spin-Strukturen und Dirac-Operatoren über pseudorie-mannschen Mannigfaltigkeiten*. Teubner, Leipzig, 1981.

[2] N. Bourbaki: *Algèbre, Chapitre 9, Formes sesquilinéaires et formes quadratiques.* Hermann, Paris, 1959.

[3] M. Capobianco, J. Molluzo: *Examples and Counterexamples in Graph Theory.* North Holland, Amsterdam, 1978.

[4] C. Chevalley: *Theory of Lie Groups.* Princeton University Press, Princeton, 1946.

[5] C. Chevalley: *The Algebraic Theory of Spinors.* Columbia University Press, New York, 1954.

[6] Y. Choquet-Bruhat, C. DeWitt-Morette: *Analysis, Manifolds and Physics, Part II.* North Holland, Amsterdam, 1989.

[7] A. Crumeyrolle: *Orthogonal and Symplectic Clifford Algebras, Spinor Structures.* Kluwer Academic Publishers, Dordrecht, 1990.

[8] L. Dabrowski: *Group Actions on Spinors.* Bibliopolis, Napoli, 1988.

[9] R. Deheuvels: *Formes quadratiques et groupes classiques.* Presses Universitaires de France, Paris, 1981.

[10] G.M. Dixon: Division Algebras:*Octonions, Quaternions, Complex Numbers and the Algebraic Design of Physics.* Kluwer Academic Publishers, Dordrecht, 1994.

[11] C. Doran: *Geometric Algebra and its Applications to Mathematical Physics.* Thesis, Univ. Cambridge, 1994.

[12] Ya.S. Dubnov: *Mistakes in Geometric Proofs.* Heath, Boston, 1963.

[13] V. Figueiredo: Clifford algebra approach to Cayley-Klein matrices; pp. 230-236 in P.S. Letelier, W.A. Rodrigues (eds.): Gravitation: The Space-Time Structure, SILARG VIII. *Proc. 8th Latin American Symposium on Relativity and Gravitation (Brazil, 1993).* World Scientific, Singapore, 1994.

[14] J.E. Fornæss, B. Stensønes: *Lectures on Counterexamples in Several Complex Variables.* Princeton University Press, Princeton, NJ, 1987.

[15] B.R. Gelbaum, J.H.M. Olmsted: *Counterexamples in Analysis.* Holden-Day, San Francisco, 1964.

[16] B.R. Gelbaum, J.H.M. Olmsted: *Theorems and Counterexamples in Mathematics.* Springer, New York, 1990.

[17] J. Gilbert, M. Murray: *Clifford Algebras and Dirac Operators in Harmonic Analysis.* Cambridge Studies in Advanced Mathematics, Cambridge University Press, Cambridge, 1991.

[18] H.P. Ginsburg, S. Opper: *Piaget's Theory of Intellectual Development.* Prentice Hall, Englewood Cliffs, NJ, 1988.

[19] M. Göckeler, Th. Schücker: *Differential Geometry, Gauge Theories, and Gravity.* Cambridge University Press, Cambridge, 1987.

[20] F.R. Harvey: *Spinors and Calibrations.* Academic Press, San Diego, 1990.

[21] B. Hauchecorne: *Les contre-exemples en mathématiques.* Ellipses, Paris, 1988.

[22] D. Hestenes: *New Foundations for Classical Mechanics.* Reidel, Dordrecht, 1986, 1987, 1990.

[23] S.M. Khaleelulla: *Counterexamples in Topological Vector Spaces.* Springer, Berlin, 1982.

[24] M.-A. Knus: *Quadratic and Hermitian Forms over Rings.* Springer, Berlin, 1991.

[25] I. Lakatos: *Proofs and Refutations. The Logic of Mathematical Discovery.* Cambridge University Press, Cambridge, 1976.

[26] H.B. Lawson, M.-L. Michelsohn: *Spin Geometry.* Universidade Federal do Ceará, Brazil, 1983. Princeton University Press, Princeton, NJ, 1989.

[27] R. Lipschitz: Principes d'un calcul algébrique qui contient comme espèces particulières le calcul des quantités imaginaires et des quaternions. C.R. *Acad. Sci. Paris* 91 (1880), 619-621, 660-664. Reprinted in *Bull. Soc. Math.* (2) 11 (1887), 115-120.

[28] R. Lipschitz: Untersuchungen über die Summen von Quadraten. Max Cohen und Sohn, Bonn, 1886, pp. 1-147. The first chapter of pp. 5-57 translated into French by J. Molk: Recherches sur la transformation, par des substitutions réelles, d'une somme de deux ou troix carrés en elle-même. *J. Math. Pures Appl.* (4) 2 (1886), 373-439. French résumé of all three chapters in *Bull. Sci. Math.* (2) 10 (1886), 163-183.

[29] R. Lipschitz (purpoted author): Correspondence. *Ann. of Math.* 69 (1959), 247-251.

[30] P. Lounesto: *Counterexamples in Clifford algebras with CLICAL*, pp. 3-30 in R. Ablamowicz et al. (eds.): *Clifford Algebras with Numeric and Symbolic Computations*. Birkhäuser, Boston, 1996.

[31] P. Lounesto: Counterexamples in Clifford algebras. *Advances in Applied Clifford Algebras* 6 (1996), 69-104.

[32] P. Lounesto: *Clifford Algebras and Spinors*. Cambridge University Press, Cambridge, 1997.

[33] R. Moresi: A remark on the Clifford group of a quadratic form; pp. 621-626 in *Stochastic Processes, Physics and Geometry*. Ascona/Locarno, 1988.

[34] K. Popper: *Objective Knowledge*. Oxford University Press, Oxford, 1972.

[35] I.R. Porteous: *Topological Geometry*. Van Nostrand Reinhold, London, 1969. Cambridge University Press, Cambridge, 1981.

[36] I.R. Porteous: *Clifford Algebras and the Classical Groups*. Cambridge University Press, Cambridge, 1995.

[37] M. Riesz: Clifford Numbers and Spinors. The Institute for Fluid Dynamics and Applied Mathematics, *Lecture Series* No. 38, University of Maryland, 1958. Reprinted as facsimile (eds.: E.F. Bolinder, P. Lounesto) by Kluwer Academic Publishers, 1993.

[38] J.P. Romano, A.F. Siegel: *Counterexamples in Probability and Statistics*. Wadsworth & Brooks, Monterey, CA, 1986.

[39] E. Schechter: *The most common mathematical errors of undergraduate students.*

[40] G. Sobczyk: Spectral integral domains in the classroom, *Aportaciones Matematicas, Serie Comunicaciones* 20 (1997), 169-188.

[41] G. Sobczyk: The generalized spectral decomposition of a linear operator, *The College Mathematics Journal,* 28:1 (1997), 27-38.

[42] L.A. Steen, J.A. Seebach: *Counterexamples in Topology.* Holt, Rinehart and Winston, New York, 1970. Springer, New York, 1978.

[43] J. Stoyanov: *Counterexamples in Probability.* John Wiley, Chichester, 1987, 1997.

[44] G.L. Wise, E.B. Hall: *Counterexamples in Probability and Real Analysis.* Oxford University Press, New York, 1993.

Chapter 24

[1] Pertti Lounesto, Marcel Riesz's work on Clifford algebras, *Clifford numbers and spinors*, Kluwer Academic, 1993, pp 119–241.

[2] Anthony Lasenby and M. Ashdown et al., *GA package for Maple V*, Available at http://www.mrao.cam.ac.uk/~clifford/software/GA/, 1999.

[3] R. Ablamowicz, Clifford Algebra Computations with Maple, *Clifford (Geometric) Algebras with Applications to Physics, Mathematics and Engineering*, W.E. Baylis, Birkhäuser, 1996.

[4] Leo Dorst, Honing geometric algebra for its use in the computer sciences, *Geometric Computing with Clifford Algebra*, G. Sommer, Springer, expected 2000, Preprint available at http://www.wins.uva.nl/~leo/clifford/.

[5] David Hestenes and Garrett Sobczyk, *Clifford Algebra to Geometric Calculus*, Reidel, 1984.

[6] David Hestenes, *New Foundations for Classical Mechanics*, Reidel, 1986, ISBN 90-277-2090-8.

[7] Leo Dorst and Stephen Mann and Tim Bouma, GABLE: *A Matlab Tutorial for Geometric Algebra*, 1999. Available at http://www.wins.uva.nl/~leo/clifford/gable.html.

[8] Pertti Lounesto, *Clifford Algebras and Spinors*, Cambridge University Press, London Mathematical Society Lecture Note Series 239, 1997.

[9] Ian R Porteous, *Topological Geometry*, Cambridge university Press, Cambridge, 1981.

[10] Stephen Mann, Leo Dorst and Tim Bouma, *The Making of a Matlab Geometric Algebra*, University of Waterloo, CS-99-27, December, 1999. Available at ftp://cs-archive.uwaterloo.ca/cs-archive/CS-99-27/.

[11] Pertti Lounesto and Risto Mikkola and Vesa Vierros, *CLICAL User Manual: Complex Number, Vector Space and Clifford Algebra Calculator for MS-DOS Personal Computers*, Institute of Mathematics, Helsinki University of Technology, 1987, 80 pages, A248, ISBN 951-754-198-8.

[12] Lars Svenson, *Personal communication at ACACSE'99, Ixtapa, Mexico*, 1999.

Chapter 25

[1] R. Ablamowicz. Clifford algebra computations with Maple, *Proc. Clifford (Geometric) Algebras*, Banff, Alberta Canada, 1995. Ed. W. E. Baylis, Birkhäuser, Boston, 1996, pp. 463–501.

[2] R. Ablamowicz. Structure of spin groups associated with degenerate Clifford algebras, *Journal of Mathematical Physics*, Vol. **27**, No. **1**, January 1986, pp. 1–6.

[3] R. Ablamowicz. Deformation and contraction in Clifford algebras, *Journal of Mathematical Physics*, Vol. **27**, No. **2**, January 1986, pp. 1–6.

[4] R. Ablamowicz. CLIFFORD - Maple V package for Clifford algebra computations, ver. 4 (Copyright 1995-1999) and two supplementary packages SUPPL and ASVD are available from http://math.tntech.edu/rafal/cliff4/.

[5] R. Ablamowicz and B. Fauser. Hecke algebra representations in ideals generated by q-Young Clifford idempotents, *Proceedings of the 5th International Conference on Clifford Algebras*, Ixtapa, Mexico, 1999, Vol. 1 (submitted) and math.QA/9908062.

[6] R. Ablamowicz and P. Lounesto. Primitive idempotents and indecomposable left ideals in degenerate Clifford algebras, *Proc. of "Clifford Algebras and Their Applications in Mathematical Physics* (Canterbury, 1985), pp. 61–65, NATO Adv. Sci. Inst. Ser. C: Math. Phys. Sci., 183, Reidel, Dordrecht-Boston, Mass.

[7] R. Ablamowicz and P. Lounesto. *On Clifford algebras of a bilinear form with an antisymmetric part*, in: R. Ablamowicz, P. Lounesto, and J. M. Parra, eds., *On Clifford Algebras with Numeric and Symbolic Computations* (Birkhäuser, Boston, 1996), pp. 167–188.

[8] J.A. Brooke. A Galileian formulation of spin. I. Clifford algebras and Spin groups, *J. Math. Phys.* **19**, no. **5**, 952–959 (1978); A Galileian formulation of spin. I. Explicit realizations, *J. Math. Phys.* **21**, no. **4**, pp. 617–621 (1980).

[9] J.A. Brooke. Spin groups associated with degenerate orthogonal spaces, *Proc. of "Clifford Algebras and Their Applications in Mathematical Physics* (Canterbury, 1985), pp. 93–102, NATO Adv. Sci. Inst. Ser. C: Math. Phys. Sci., 183, Reidel, Dordrecht-Boston, Mass.

[10] A. Crumeyrolle. *Orthogonal and Symplectic Clifford Algebras: Spinor Structures*, Kluwer, Dordrecht, 1990.

[11] C.W. Curtis and I. Reiner. *Methods of Representation Theory with Applications to Finite Groups and Orders*, Vol. 1, Wiley-Interscience, Now York, 1981.

[12] J. Helmstetter. Mono"ides de Clifford et déformations d'algèbres de Clifford, *Journal of Algebra*, Vol. **111** (1987), pp. 14–48.

[13] I.N. Herstein. Noncommutative Rings, *The Carus Mathematical Monographs*, Number 15, Mathematical Association of America, Chicago, 1968.

[14] P. Lounesto, R. Mikkola, and V. Vierros. 1987, 'CLICAL User Manual', Helsinki University of Technology, Institute of Mathematics, Research Reports **A248**, Helsinki.

[15] P. Lounesto. *Clifford Algebras and Spinors*, Cambridge University Press, Cambridge, 1997.

[16] P. Lounesto. Private communication, 1997.

[17] I. R. Porteous. *Clifford Algebras and the Classical Groups*, Cambridge University Press, Cambridge, 1995.

[18] J.M. Selig. *Geometrical Methods in Robotics*, Monographs in Computer Science, Springer-Verlag, New York, 1996.

Index